Introduction to Kinesiology

Studying Physical Activity

FIFTH EDITION

Shirl J. Hoffman, EdD
UNIVERSITY OF NORTH CAROLINA AT GREENSBORO,
PROFESSOR EMERITUS

Duane V. Knudson, PhD
TEXAS STATE UNIVERSITY

EDITORS

HUMAN
KINETICS

Library of Congress Cataloging-in-Publication Data

Names: Hoffman, Shirl J., 1939- editor. | Knudson, Duane V., 1961- editor.
Title: Introduction to kinesiology / Shirl J. Hoffman, Duane V. Knudson, editors.
Description: Fifth edition. | Champaign, IL : Human Kinetics, [2018] | Includes bibliographical references and index.
Identifiers: LCCN 2017009248 (print) | LCCN 2017007457 (ebook) | ISBN 9781492549925 (print) | ISBN 9781492549932 (ebook)
Subjects: | MESH: Kinesiology, Applied | Motor Activity | Physical Education and Training
Classification: LCC QP303 (ebook) | LCC QP303 (print) | NLM WE 103 | DDC 612.7/6--dc23
LC record available at https://lccn.loc.gov/2017009248

ISBN: 978-1-4925-4992-5 (print)

The web addresses cited in this text were current as of June 2017, unless otherwise noted.

Acquisitions Editors: Myles Schrag and Bridget Melton; **Developmental Editor:** Ragen E. Sanner; **Managing Editor:** Amanda S. Ewing; **Copyeditor:** Tom Tiller; **Indexer:** Andrea Hepner; **Permissions Manager:** Dalene Reeder; **Senior Graphic Designer:** Nancy Rasmus; **Graphic Designer:** Sean Roosevelt; **Cover Designer:** Keri Evans; **Photograph (cover):** Anion/fotolia.com; **Photo Asset Manager:** Laura Fitch; **Photo Production Manager:** Jason Allen; **Senior Art Manager:** Kelly Hendren; **Illustrations:** © Human Kinetics, unless otherwise noted; **Printer:** Walsworth

Printed in the United States of America 10 9 8 7 6 5 4

The paper in this book was manufactured using responsible forestry methods.

Human Kinetics
1607 N. Market Street
Champaign, IL 61820
USA

United States and International
Website: **US.HumanKinetics.com**
Email: info@hkusa.com
Phone: 1-800-747-4457

Canada
Website: **Canada.HumanKinetics.com**
Email: info@hkcanada.com

E7052

CONTENTS

PREFACE

Welcome to the fifth edition of *Introduction to Kinesiology*. Let us introduce you to the goals, updates, and features of this new edition of an outstanding text. Along with the usual updating of content, this revision includes a blend of previous and new contributors, editors, and features. The result is a streamlined yet comprehensive introduction to the exciting and diverse discipline of kinesiology.

Goals of the Book

The primary goals of *Introduction to Kinesiology* are to give students a comprehensive overview of the discipline and to inspire them to pursue a major professional career related to it. In order to accurately summarize the diverse and growing discipline of kinesiology, we use a unique model that addresses both the sources of knowledge and the primary subdisciplinary fields of this area of study. The last six chapters highlight characteristics of professionals and several professional careers open to kinesiology graduates. We hope this book inspires you to continuously study physical activity and join us in promoting physical activity for the benefit of all people.

Organization

This fifth edition retains the three-part structure addressing the sources of kinesiology knowledge, the primary subdisciplines of kinesiology, and the major career areas for kinesiology graduates. Chapter 1 provides an overview of the book, offers key definitions, and presents the model on which the text is organized. Part I has two chapters that introduce seven spheres of physical activity experiences and explore the importance of such experience for participation in and knowledge of physical activity.

Part II reviews seven of the most common subdisciplines of kinesiology: philosophy, history, sociology, motor behavior, psychology, biomechanics, and physiology. Each chapter gives a brief overview of major historical events in the development of the subdiscipline; the research methods used in the subdiscipline; what professionals such as biomechanists, exercise physiologists, and others do in the course of their professional work; and how students' current knowledge can form a foundation for more advanced study. Each of these chapters also presents practical, real-world applications from the relevant subdiscipline and is organized to help readers understand why the subdiscipline is important and how it may relate to a variety of professional endeavors.

Part III presents characteristics of professionals, followed by a look at professional opportunities in five major career areas or clusters: health and fitness, therapeutic exercise, physical education, coaching and sport instruction, and sport management.

Updates in the Fifth Edition

Textbooks do not get to fifth editions unless they contribute meaningfully to the field and are well received by both students and faculty. The editors and contributors of this fifth edition have built on the strong foundation of the previous edition by listening to helpful feedback from instructors while determining how to update the content and adapt the features of the book and its ancillaries. As a result, this new edition is shorter even as it includes new content.

Each chapter includes the latest research and updated data (including more than 190 new or updated references) from the relevant subdisciplines or professions. For example, this edition addresses both recent expansion of kinesiology opportunities in public health and likely future changes in other allied health careers. In addition, the multiple sources of kinesiology knowledge emphasized in previous editions of the book have been supplemented in this revision to illustrate evidence-based practice and different styles of scholarship and reporting across subdisciplines and professions. New graphics have also been developed to illustrate the multidimensional model of kinesiology knowledge and the subdisciplinary lenses used to study physical activity.

Moreover, this new edition includes two new features designed to help students see consistent themes across the subdisciplines and professions of kinesiology and to picture themselves in future careers. The Professional Issues in Kinesiology feature highlights how knowledge from a given chapter contributes to the research and major problems addressed by kinesiology. The Kinesiology Colleagues feature provides engaging stories about real-world professionals and how their careers have been influenced by their mastery of kinesiology.

The textbook isn't the only item to receive updates. The delivery of the instructor ancillaries is more robust, and the web study guide offers students more opportunities to engage with the content and assess understanding of the material.

Features of the Book

The fifth edition of *Introduction to Kinesiology* provides several features to help students organize and understand essential knowledge about kinesiology and its professions. Readers will find key points, engaging sidebars, figures, and photos that contribute to an enjoyable reading experience. Each chapter begins with objectives and an opening scenario, which, taken together, provide a road map for the content. Take time to read the many sidebars sprinkled throughout the text. These features will stimulate your thinking and reinforce what you have learned by reading the previous paragraphs. Don't miss the unique opportunity they offer for learning!

In addition, focus on the key points in every chapter. These points briefly summarize the most important points made in the preceding pages. Read each key point carefully and think about what you have read in the chapter. Similarly, when you reach the end of a chapter, a set of review questions will direct you to the most important concepts.

Another powerful resource is the web study guide. For students, this guide is essential to developing full understanding of course content and contributing to online or in person class discussions. For instructors, it allows you to grade and assess students' progress and their comprehension of the material so that they are better prepared for exams and for further study. The web study guide provides a variety of interactive, multimedia experiences to help students learn, understand, and apply the information presented in the text.

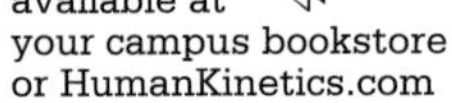

- Many activities use a variety of approaches—audio, video, drag-and-drop activities, self-ratings, interviews, and more—to demonstrate various aspects of kinesiology and bring the content alive.
- Sometimes you'll be asked to search the web for specific information and then complete an associated activity that connects your online experience with what you've learned in the book.
- The web study guide includes a key points activity in each chapter that tests how well you understand the material. You can also print out the key points as a study aid.
- After reading the Kinesiology Colleagues profile in the chapter, you can complete an associated activity in the web study guide to reflect on your own professional development with a short written response or contribution to a discussion forum.
- Understanding how to read and evaluate research will be critical as you continue your studies in kinesiology. Working in tandem with the Professional Issues in Kinesiology sidebars in the book, the web study guide activities expose you to actual research studies that contribute to the body of knowledge in kinesiology and guide professionals in their everyday practice.
- Practical Plug-Ins present you with a problem or challenge faced by a practicing professional. You'll consider how the professional might handle the situation, then examine how he or she actually resolved it. Allow time to use these lessons as you go through the course and when you face similar situations or challenges in the field.

The web study guide is available at www.HumanKinetics.com/IntroductionToKinesiology.

Instructor Resources

In this fifth edition, *Introduction to Kinesiology* is again supported by a complete set of ancillaries: presentation package, image bank, instructor guide, test package, and chapter quizzes.

- The presentation package provides more than 340 PowerPoint slides with selected illustrations and tables from the text.
- The image bank contains all of the figures, tables, and content photos. You can use these images to supplement lecture slides, create handouts, or develop your own presentations and teaching materials.

- The instructor guide contains information about how to use the various ancillaries; a syllabus; and chapter-by-chapter files that include chapter outlines with key points, additional student assignments, ideas for additional teaching topics and guest speakers, resources, and answers to the in-text review questions.
- The test package contains more than 330 questions in a mix of true-or-false, multiple-choice, fill-in-the-blank, and essay formats.
- The ready-made chapter quizzes allow you to check students' understanding of the most important chapter concepts.

All of these instructor ancillaries can be accessed at www.HumanKinetics.com/IntroductionToKinesiology.

ACKNOWLEDGMENTS

A project of this sort involves the work of many hands, and we have been fortunate to play a central role in updating it. Our knowledgeable contributors, along with a team of highly competent professionals at Human Kinetics, have made this fifth edition into a streamlined, updated, and improved introduction to the diverse discipline of kinesiology.

Although this book features the work of many contributors, it is not an anthology or a compilation. To the contrary, it is a textbook in every sense of the word. It is organized around a central theme; structured to accomplish specific purposes; and integrated in its terms, concepts, objectives, and graphics. At the same time, each contributor brings to the project a unique professional and academic background and a track record of achievement in his or her respective specialization, thus creating the richness and breadth that are largely responsible for the book's success. Achieving consistency and flow—and explaining what are sometimes difficult concepts from diverse perspectives—presented substantial challenges for the contributors, the editors, and the editorial staff at Human Kinetics.

At the risk of inadvertently omitting a name or two, we would like to give special acknowledgment to those who made key contributions in the production of *Introduction to Kinesiology.* Our contributors caught the spirit of the project, sensed its importance, and produced highly readable and informative chapters. Special credit goes to Janet Harris, who shared editorial duties for the first edition and whose initial influence continues throughout this edition. The time investment required of the editors in guiding this project would not have been possible without support from our wives, Lois Knudson and Claude Mourot. We appreciate their support, as well as the support of numerous colleagues who have enriched our professional and personal lives.

Only those who have been fortunate enough to author books published by Human Kinetics can appreciate the talents of its publication teams. Those specifically involved with this book are listed on the copyright page, and each did his or her work superbly.

1

Introduction to Kinesiology

Duane V. Knudson and Shirl J. Hoffman

The authors acknowledge the contributions of Janet C. Harris to this chapter.

CHAPTER OBJECTIVES

In this chapter we will

- help you appreciate the pervasiveness and diversity of physical activity in human life;
- introduce you to ways of defining and thinking about physical activity;
- discuss the discipline of kinesiology and its relationship to physical activity;
- familiarize you with the types of knowledge about physical activity that are acquired through physical activity experience, scholarly study, and professional experience; and
- help you gain a preliminary understanding of what a profession is and of the career possibilities centering on physical activity.

Professional computer gaming is a growing spectator sport; it is also big business. Although some people have difficulty seeing computer gamers as athletes pursuing a sport, their activities require considerable mental and physical skill, as well as strategy, and in some cases teamwork—for example, when a group competes with another team in, say, *League of Legends*. From this perspective, computer gaming constitutes a sport. Even though it does not involve much whole-body movement or energy expenditure, it does involve physical activity through the use of fast perceptual and fine motor skills. As a result, it may also apply to the use of other kinds of computer-controlled systems, such as those found in construction, materials handling, and national defense settings.

The wide variety of expressions of physical activity means that numerous opportunities exist for kinesiology professionals. The scholarly discipline of kinesiology focuses on creating a body of knowledge about physical activity—that is, about voluntary human movement performed intentionally in order to achieve a goal.

Darryl Dennis/Icon Sportswire

You may not have fully appreciated it until now, but performing physical activity consumes most of your daily life. Even if you don't go to the gym or athletic field or engage in hard labor on a given day, you will probably get out of bed, walk to the bathroom, brush your teeth, get dressed, eat breakfast, and make your way to class. After your morning classes, you will probably eat lunch, visit the library, go back to your room, and perhaps surf the web. All of these are forms of physical activity.

If you take a moment to reflect on how physically active you are, you will see that your life involves an endless variety of physical activity. You walk, reach, run, lift, leap, throw, grasp, wave, push, pull, move your fingers and toes, adjust your head for a better line of vision, adjust your posture, and perform thousands of other movements as part of living a normal human existence. As a result, physical activity is essential in your work, whether you perform hard physical labor or low-energy tasks. We also use physical activity to express ourselves in gesture, art, and dance.

Indeed, physical activity is part of human nature. It is an important means by which we explore and discover our world. Linking movement with complex cognitive plans helps us define ourselves as human beings. A significant part of our lifetime is spent in learning to master a broad range of physical activities, from the earliest skills of reaching, grasping, and walking to enormously complex skills such as hitting a baseball, performing a somersault, or playing the piano. Most of us master a broad range of physical activities at a moderate level of competence. Others concentrate on a limited number of skills, and this focus can lead to extraordinary performances. For example, NBA star

Steph Curry's ability to sink three-point shots consistently from 30 feet (9 meters) away is the result of intense practice and motivation, as is Yo Yo Ma's skill in positioning the bow on the cello's strings or a pilot's ability to land a fighter jet on the runway of an aircraft carrier that is being tossed by the sea.

In this chapter, we talk about physical activity in general terms and explain its relation to the field of **kinesiology**, which is a discipline or body of knowledge focused on physical activity (Newell, 1990a). Taking time to read the chapter carefully will help you appreciate the complexity and diversity of physical activity, as well as its importance to human life. The chapter will also help you understand how the discipline of kinesiology is organized. If you've been physically active throughout your life, you already have some knowledge of physical activity. This background will be of enormous benefit to you as you roll up your sleeves and begin to probe the depths of knowledge of kinesiology. However, prior experiences can also hinder your understanding, especially when you are required to think about those experiences in new ways. At times, therefore, you will have to set your assumptions aside so that you can examine physical activity from a fresh and exciting point of view. This endeavor may sometimes be more challenging than you might imagine.

Interest in the Discipline of Kinesiology

Because people are now more aware than ever of the importance of physical activity, enrollment in college and university curriculums devoted to the study of physical activity has been on the rise. According to one study, the number of undergraduate students majoring in kinesiology increased by 50 percent between 2003 and 2008 (Wojciechowska, 2010), which made it one of the fastest-growing majors in higher education; in fact, in some universities, the kinesiology department is one of the largest academic units on campus. This surge in interest has resulted from two major reasons. First, career opportunities have expanded greatly for college-trained professionals who possess in-depth knowledge of the scientific and humanistic bases of physical activity. Before the 1990s, most departments of kinesiology (then referred to as "physical education") were designed primarily for preparing physical education teachers and coaches. Now, however, kinesiology serves as the academic base for a diverse assortment of careers, such as physical education teaching and coaching, physical therapy, cardiac rehabilitation, sport management, athletic training, fitness leadership and management, public health, and more.

The growth in kinesiology also derives from increasing awareness of the importance of physical activity and the realization that it deserves to be studied just as seriously and systematically as do other disciplines in higher education, such as biology, psychology, and sociology. No doubt you've heard the word *discipline,* but you may not fully understand what it means in this context. A **discipline** is a body of knowledge organized around a certain theme or focus (see figure 1.1); it embodies knowledge that learned people consider worthy of study. The focus of a discipline identifies what is studied by those who work in the discipline. For example, biology focuses on life forms, psychology on the mind and mental and emotional processes, and anthropology on cultures. Although debates continue about the focus of kinesiology, it is now generally regarded as the discipline that focuses on human physical activity.

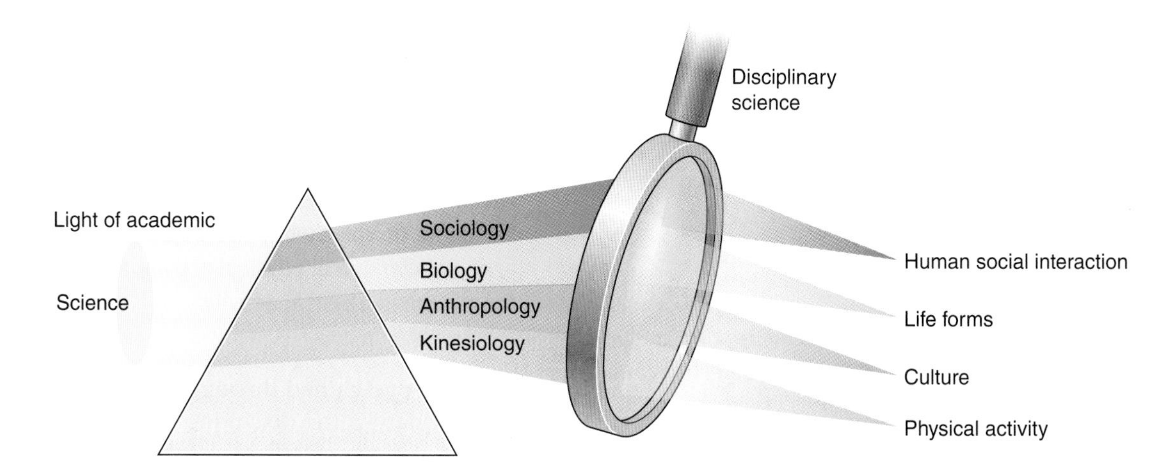

FIGURE 1.1 The disciplines of science each focus on knowledge of one topic, and kinesiology focuses on the topic of human physical activity.

Physical Activity: The Focus of Kinesiology

In your college courses, you may have noticed that disciplines are not all learned or studied in the same way. Art, for example, may be studied through reading, writing, and experimentation with studio projects. People learn history, literature, and philosophy largely through reading, writing, memorization, and discussion. The same activities are important when learning chemistry and biology, but these disciplines also involve active participation in laboratory exercises.

People learn kinesiology in three different but related ways (see figure 1.2), one of which is through **physical activity experience**. Just as students in art and music learn to appreciate their disciplines in part by watching, listening, and performing, you can develop your understanding of kinesiology in part through the direct personal experience of watching or performing physical activity. A second way of developing an understanding of kinesiology is through **scholarship of physical activity**. This way of learning involves researching, reading about, studying, and discussing with colleagues both theoretical and practical aspects of physical activity; it also involves laboratory experiences. These forms of study are necessary in order to master the various subjects included in the kinesiology curriculum. Where does the knowledge contained in such subjects come from? Mostly, it comes from work done by people in the field of kinesiology who have developed and added to the knowledge base through systematic research and scholarship. Scholars who conduct research in biomechanics, motor learning, and many other subdisciplines of kinesiology develop important foundational knowledge for the scholarly study of physical activity.

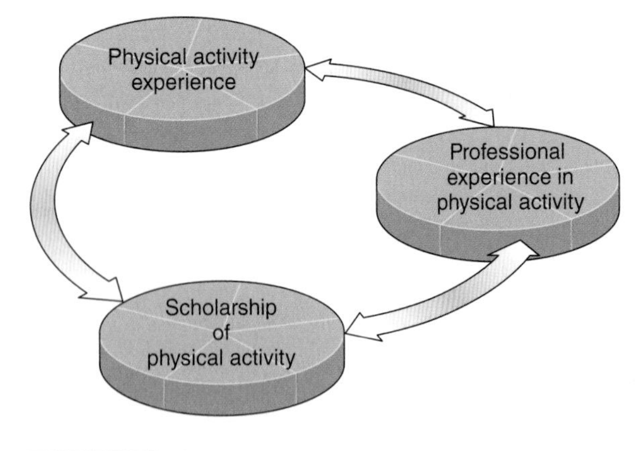

FIGURE 1.2 Three interrelated sources of knowledge in kinesiology.

A **subdiscipline** in kinesiology is often related to a broader, more established "parent" discipline, such as psychology, physiology, sociology, biology, history, or philosophy. For instance, exercise physiology draws on basic concepts and theories from physiology, the study of motor behavior draws on psychology, and the philosophy of physical activity draws on the general field of philosophy. These relationships mean that kinesiology students must develop a working knowledge of the language, theories, and conceptual frameworks of a number of major disciplines and learn to apply them to physical activity. Some subdisciplines focus on the effects of physical activity in particular populations, such as older adults, children, or persons affected by disease or disability.

A third way of learning about physical activity is through **professional experience in physical activity**. Here, the focus is placed not so much on learning to *perform* physical activity or on learning *about* it but on learning by designing and implementing physical activity programs for clients in one's professional practice. For example, professionals such as physical education teachers, personal trainers, and cardiac rehabilitation specialists systematically manipulate the physical activity experiences of students, clients, patients, and others whom they serve in order to help them achieve personal goals.

These three sources of knowledge about kinesiology provide the organizational structure for this book. Part I explores the knowledge gained through physical activity experience—for example, learning a human movement by participating in an organized sport, a physical activity class, or a recreational physical activity. Part II then focuses on seven subdisciplines of the scholarship of physical activity that are common in university curriculums, and part III describes the professional experience in physical activity that is involved in common career areas for kinesiology graduates.

KEY POINT

The three sources of knowledge that constitute the discipline of kinesiology are physical activity experience, scholarship of physical activity, and professional experience in physical activity.

Knowledge gained through these three sources becomes part of the discipline of kinesiology only when it is embedded in a college or university curriculum in kinesiology or is universally accepted and used by kinesiologists in their research. This caveat is given in order to clarify precisely what is considered part of the "official" discipline and what is not. Many people experience physical activity (e.g., mowing the lawn, playing golf), study it informally (e.g., read *Sports Illustrated* or popular trade books on fitness or sport), or engage in some form of physical activity leadership (e.g., volunteering as a Little League coach) outside the confines of the university curriculum. However, these activities do not constitute the practice of kinesiology, and the people engaging in them are not kinesiologists in the strict definition used in this text.

To be sure, these activities may be important and valuable in their own right, but they do not constitute practicing kinesiology any more than the use of elementary psychological principles by a businessperson to motivate her sales force constitutes practicing psychology. The discipline of psychology remains tied to the college and university curriculum and to the research conducted by psychologists. Similarly, people may use the principles of kinesiology outside of the discipline, but kinesiology per se remains a function of curriculums and research in colleges and universities.

KEY POINT

Only knowledge about physical activity that is included in a college or university curriculum or used in research is considered to be part of the body of knowledge of kinesiology.

Another reason for limiting our definition of kinesiology lies in the fact that the knowledge you acquire in your major curriculum is more highly organized and more scientifically verifiable than the knowledge of physical activity held by laypersons. Universities use rigorous methods to organize and monitor the authenticity of the knowledge included in their curriculums and in the research conducted by their faculty members. Think about it: Would you have more confidence in recommendations made by a university kinesiologist who specializes in fencing than in recommendations made by someone who fences as a hobby? Similarly, would you have more confidence in the scientific accuracy of recommendations for exercise programs offered by an exercise physiologist than in recommendations offered by a television exercise guru who lacks formal training in kinesiology? Would you be more likely to trust the recommendations of a university specialist in pedagogy when organizing physical activity instruction for a large group of young children than the recommendations of a volunteer coach who has no formal training in kinesiology? Given your decision to invest several years of hard work in preparing for a career in the specialized field of kinesiology, your answer to all three questions is likely to be yes.

What Is Physical Activity?

You may think it silly to ask what is meant by the term *physical activity*. Everybody knows what it is, so why waste time defining it? However, definitions are important, especially in scientific and professional fields where terms are often defined differently than they are in everyday language. These **technical definitions** ensure that people working in a particular profession or science share a common understanding of key terms. Even so, kinesiologists do not all agree on the technical definition of *physical activity*. Therefore, before we go any further, let us ensure that we have the same understanding of what physical activity is.

In everyday language, almost any muscular action is considered physical activity. Throwing a javelin, driving a car, walking, performing a cartwheel, swimming, digging a ditch, hammering a nail, scratching your head—all are examples of physical activity. So are the blinking of your eye and the kick you exhibit when a doctor taps your patellar tendon to test your reflexes; some people would even use the term *physical activity* to describe the contraction of your diaphragm when you sneeze or the peristaltic action of your small

Physical Activity and Doing

The distinguished neurophysiologist Sir Charles Sherrington wrote the following in his classic book *Man on His Nature:* "All [humans] can do is to move things, and [their] muscle contraction is [their] sole means thereto" (1940, p. 107). What do you think Sherrington meant when he said that all we can do is "move things"? Can you think of anything that you can do without moving? What does your answer imply about the importance of physical activity in our lives? (Hint: What is the dictionary definition of *doing*?)

intestine brought on by muscular contractions. Do all of these muscular actions warrant equal attention from kinesiologists? Not really. Although all are examples of human movement, they are too diverse in form and purpose to be studied by any single discipline. Indeed, if kinesiology focused on all forms of human movement, then kinesiologists would study everything that humans do, because living is moving (see the sidebar titled Physical Activity and Doing).

For this reason, kinesiologists use a narrower definition of physical activity than people typically use in everyday language. The discipline requires a definition that is neither too inclusive (e.g., all human movement) nor too exclusive (e.g., only human movement related to sport). The definition used in this text takes its cue from Karl Newell's formulation of **physical activity** as "intentional, voluntary movement directed toward achieving an identifiable goal" (1990b).

Notice three things about Newell's definition: First, it does not stipulate anything about the energy requirements of the movements used to produce the activity. The highest levels of energy are typically required for large-muscle activities, such as swimming, lifting weights, and running marathons. However, Newell's definition does not limit physical activity to such activities. Indeed, smaller-scale activities such as typing, handwriting, sewing, and surgery are every bit as much forms of physical activity as are large-muscle activities such as chopping wood.

Second, the setting in which physical activity takes place is irrelevant. Surely, shooting a basketball is a form of physical activity, and so is tossing a piece of paper into a wastebasket. Pole-vaulting is a physical activity, and so is jumping over a fence. Playing hopscotch is a physical activity, and so is swinging a sledgehammer. In other words, just as physical activity takes place in many settings, it also takes many forms. Wrestling and skiing differ greatly from typing and performing sign language, but all are forms of physical activity. In chapter 2, you will learn more about the wide range of physical activity that holds interest for people in the field of kinesiology.

Even so—and this is the third point to note about Newell's definition—simply moving your body does not constitute physical activity. Rather, the movements must be directed toward a purposeful end. This distinction can be confusing, especially when we consider that the term *kinesiology* is derived from the Greek words *kinesis* (movement) and *kinein* (to move). **Movement** consists of any change in the position of one's body parts relative to each other. Of course, physical activity requires that we move our bodies or body parts, but movement by itself does not constitute physical activity. One way to think about the relationship between movement and physical activity is this: Movement is a necessary but not sufficient condition for physical activity.

Only movement that is intentional and voluntary—purposefully directed toward an identifiable goal—meets the technical definition of physical activity that is used in this text. This formulation excludes all involuntary reflexes and all physiological movements controlled by involuntary muscles, such as peristalsis, swallowing, and reflexes such as eye blinking. It also excludes voluntary movements that people perform without having a goal in mind. For example, a thoughtless scratch of the head, absentminded twirling of one's hair, and the repetitive movements of a compulsive-obsessive psychiatric patient constitute human movement but fall outside of the technical definition of physical activity because they are not designed to achieve a goal.

Defining Physical Activity

Although we are all familiar with physical activity, we rarely think deeply about it. Before you continue reading this chapter, think carefully about the following questions. Take time to compare your answers with those of your classmates.

- How would you define physical activity?
- What do you consider to be the most important characteristics of physical activity?
- What methods can you think of that kinesiologists might use to study physical activity?
- Name five professions that use physical activity.

After you have thought about or discussed these questions, rewrite your definition of physical activity.

KEY POINT

Physical activity consists of human movement that is intentional, voluntary, and directed toward achieving an identifiable goal; it does not include movements performed aimlessly or without a specific purpose.

As you read the literature in kinesiology, you will soon discover that kinesiologists and scholars in related fields define physical activity in various ways. This variation is to be expected given the diversity of human movements and the range of disciplines that pursue interests related to physical activity. For example, the Centers for Disease Control and Prevention (CDC) has been very influential in keeping the U.S. public aware of the importance of physical activity for healthful living. The CDC (2015) defines physical activity as "any bodily movement produced by skeletal muscle that increases energy expenditure above a basal level. In these guidelines, [the term] physical activity generally refers to the subset of physical activity that enhances health."

Now, let us take the CDC's definition apart. First, it limits physical activity to voluntary movement (performed by skeletal muscles), which seems sensible because most kinesiologists are interested in voluntary rather than involuntary movements of the body. At the same time, however, it seems too narrow, because it limits physical activity to movements that increase energy expenditure above a base level. Obviously, kinesiologists are interested in high-energy forms of physical activity, such as running, lifting weights, and exercising. However, they do not limit their interest to these health- and longevity-related aspects of human movement; they also study skilled movement that may require little energy expenditure.

For example, motor control researchers may be interested in measuring fine motor skills or reaction times, such as the time required to move one's hand a short distance under different experimental conditions. Similarly, kinesiology professionals practice in broad areas of physical activity. Physical education teachers and coaches, for instance, often teach bowling and archery, whereas occupational and physical therapists may treat stroke patients by using gentle movements or teach elderly residents of a nursing home how to use a knife and fork, how to bathe, or how to comb their hair—none of which requires the expenditure of large amounts of energy. Thus, although the CDC's definition is appropriate for that organization's health-promotion mission, it is too restrictive for kinesiology scholars and professionals because it eliminates many types of physical activity.

Of course, we could limit our technical definition of physical activity to physical activities that relate directly to exercise and sport. Indeed, this definition might hold some appeal, both because kinesiology is rooted historically in exercise and sport (see chapter 5) and because many students enter the field with their sights set on a career in exercise or sport. Even so, kinesiology scholars, who have long argued these issues, have generally concluded that this limited definition would be incorrect because the field embraces a far wider range of physical activity than sport and exercise.

For example, kinesiologists study basic postural mechanisms, the physiology and body mechanics of work, the development of reaching and grasping behaviors in infants, and daily life-support activities (e.g., balance) in elderly people. Similarly, physical education teachers teach children how to perform both fundamental movement patterns (e.g., hopping, running, skipping) and expressive physical activities (e.g., dancing). Kinesiologists working in rehabilitation programs teach patients to recover lost capacities, such as walking, sitting, rising from a chair, and driving a car. These activities do not relate strictly to sport or exercise. Moreover, as the range of occupations pursued by kinesiology graduates continues to expand, it is clear that we require a technical definition of physical activity that reaches beyond exercise and sport.

Context-Based Definitions

Sometimes kinesiologists focus so closely on the specific area of physical activity in which they work that they define physical activity in terms that make sense only to them. For example, the definition promoted by the CDC was written by physiologists and public health experts who were concerned about the health ramifications of failing to engage in daily vigorous exercise and who therefore included the qualification about energy expenditure. (The health benefits of physical activity are roughly proportional to the level of energy expenditure.) How might physical activity be defined by coaches using only their focused experiences of physical activity? By physical therapists? By dance teachers?

What Is Kinesiology?

Departments of kinesiology are the only academic units at colleges and universities that identify the study of physical activity as their sole mission and offer an organized curriculum to teach it. Beyond the academic realm, you have surely noticed the many websites, blogs, books, and magazines that address sport and fitness. Many of these sources feature material provided by people who lack in-depth education in kinesiology; as a result, this material may have no basis in either science or systematic analysis. Thus, we cannot consider them as part of the discipline of kinesiology.

The American Kinesiology Association (AKA), an association of academic departments of kinesiology at North American universities, defines kinesiology as "an academic discipline which involves the study of physical activity and its impact on health, society, and quality of life" (2017). As a discipline, kinesiology draws on several sources of knowledge, including personal and corporate physical activity experiences, professional practices centered on physical activity, and scholarly study of and research about physical activity itself. Although the discipline is most often associated with the third source—scholarly study and research—AKA recognizes that the body of knowledge of kinesiology is informed by and defined by the other two sources as well.

Ultimately, the uniqueness of kinesiology as a discipline derives from its embrace and integration of multidimensional study and application of physical activity—including biological, medical, and health-related aspects but also psychological, social-humanistic, and various professional perspectives as well. Although individual departments may choose to shape their curriculums and research agendas on the basis of selected aspects of the discipline, such institutional preferences should not be interpreted as reflecting a comprehensive definition of the discipline.

Some Focuses of Physical Activity in Kinesiology

Kinesiology focuses on a variety of human physical activities, including dance, exercise, fundamental movements, sport, and therapy. You already know much about these forms or categories of physical activity. Indeed, positive experiences in one or more of these forms are often what bring students to kinesiology majors. Beyond your personal experiences, it is also important for you to appreciate the variety of other activities that hold interest for kinesiologists.

To that end, consider the following examples, while noting that they do not constitute an exhaustive classification of all physical activity and that the categories they include are not mutually exclusive. One example is dance, in which people may participate for artistic expression, exercise, competition, or social recreation. Similarly, individuals may participate in competitive judo partly because they enjoy it and also because of its health benefits. Other examples include prehabilitation training programs and conditioning programs aimed at reducing one's risk of injury due to imbalances in flexibility or strength. More specifically, several biomechanical and sports medicine studies have shown that specific, specialized training programs (usually implemented by athletic trainers or strength and conditioning professionals) can reduce one's risk of certain traumatic or overuse injuries (Elliott & Khangure, 2002; Herman et al., 2009; Hewett, Lindenfield, Riccobene, & Noyes, 1999; Noehren, Scholz, & Davis, 2011). Use these categories not as hard-and-fast distinctions but as guides for understanding and appreciating the types of physical activity that concern kinesiologists.

KEY POINT

Kinesiology is a discipline that focuses on a variety of physical activity forms, including dance, exercise, fundamental movements, sport, and therapy.

AKA Core Curriculum Elements

The American Kinesiology Association has established four major elements of a common core curriculum for an undergraduate degree in kinesiology. Notwithstanding the unique features of some majors in kinesiology departments, the following areas of study are common to kinesiology degree programs: (1) the importance of physical activity in health, wellness, and quality of life; (2) scientific foundations of physical activity; (3) cultural, historical, and philosophical dimensions of physical activity; and (4) the practice of physical activity. How do the courses in your major program align with these four core elements?

Dance

The expressive movement of **dance** is a compelling form of physical activity that is common to all cultures. Dance involves body movements, often timed to music, that express messages, emotions, or artistic values. It is a form of physical activity performed both for its personal intrinsic value and for its performance value to spectators. The art and science of dance have a long history with kinesiology, given that early physical education departments, where kinesiology originated, often included faculty who focused on the study and instruction of dance.

Exercise

Another important form of physical activity is exercise. People engage in **exercise** to improve their physical performance, improve their health, or regain performance ability that has been reduced as a result of injury, disease, or aging. Because exercise involves many types of physical activity focused on improving movement function, it is helpful to break the concept down into subcategories.

› **Training** consists of exercise performed for the express purpose of improving athletic, military, work-related, or recreation-related performance (see chapter 3). Physical activity as training is particularly important to kinesiology graduates who embark on careers as strength and conditioning specialists for university, Olympic, or professional sport teams. Kinesiology graduates may take the test for the Certified Strength and Conditioning Specialist (CSCS) and other advanced credentials affiliated with the National Strength and Conditioning Association (NSCA).

› **Health-related exercise** is undertaken specifically to develop or maintain a sound working body and reduce the risk of disease for the purpose of healthy longevity. Physical activity as health-related exercise serves as the primary focus for kinesiology graduates who work as fitness leaders and personal trainers.

› **Cosmetic exercise** consists of training intended to reshape a person's body for aesthetic reasons. This kind of systematic training is used by bodybuilders, models, and persons wanting to lose or gain weight for a summer trip. It is sometimes difficult to determine whether trying to achieve, say, a smaller waistline is more a cosmetic effort or a health-related endeavor. Even so, it is important for kinesiology professionals to understand the motivations and difficulties involved in exercising to change one's body shape or size because these goals are often difficult to achieve.

Fundamental Movements

Given the wide variety of physical activity options, kinesiology often focuses on the human-movement building blocks of activities. These building blocks, known as **fundamental movement patterns**, are the broad categories of skilled human movements that involve large muscles and are used for general purposes (Knudson, 2013). Examples include carrying, catching, jumping, kicking, running, throwing, and walking. Scholars study fundamental movements in order to understand the mechanisms and generate theories of human movement and skill development. Such knowledge is used by kinesiology professionals to help their clients. For example, elementary physical education teachers help 1st graders learn how to perform fundamental movement patterns such as skipping, throwing, and hopping. When children develop skill in these fundamental movements (e.g., overarm throw) at an early age, they may later be able to achieve higher levels of proficiency in sport (e.g., baseball pitching) than they would if they experienced more limited practice in fundamental movements.

Sport

Sport involves skilled movement organized in game contexts, where players try to achieve a goal in a manner specified by rules and usually involving competition. This definition includes four elements that bear close consideration. First, the physical activity found in sport is "skilled," which means that it is performed efficiently and effectively. Not all forms of physical activity require a great deal of skill, but in every type of sport the advantage belongs to competitors who have learned to move their bodies skillfully: the soccer player who passes the ball deftly to her teammate, the golfer who strikes the ball squarely, and the gymnast who completes a double rotation on his dismount.

Second, rules are essential in sport. Rules set by national or international governing bodies provide a level playing field for competitors. They are also essential for organizing games; without rules, players could do whatever they felt like doing, and the game would become chaotic. Third, the physical activities performed in sport tend to be framed by competition—that is, performed against other teams, other individuals, established records, or one's personal best. Fourth, this definition of sport does not require that the performer use a certain number of muscle groups or body parts or expend a certain amount of metabolic energy.

Kinesiology Colleagues

Kevin M. Guskiewicz, Kenan Distinguished Professor, Exercise and Sport Science

Kevin Guskiewicz

Have you ever imaged yourself working for the National Football League to help understand the causes of concussions and other brain injuries? How about waking up one day to find that you have been given several hundred thousand dollars to pursue that work? That's exactly what happened to Kevin Guskiewicz, dean of the College of Arts and Sciences and Kenan Distinguished Professor of Athletic Training in the Department of Exercise and Sport Science at the University of North Carolina at Chapel Hill. He is an athletic trainer and researcher who has been a longtime scientific advisor to the NFL thanks to his research on sport-related concussions. His work has been so influential that in 2011 he received a MacArthur Foundation Fellowship or "Genius Grant." These unannounced awards are given to scholars and artists whose work is deemed to show great promise for improving the world; the award's prestige and funding make it the academic equivalent of a Nobel Prize. Guskiewicz and other kinesiology scholars are changing the world for the better.

Therapy

Therapy or **therapeutic exercise** involves specialized and individualized exercise performed to restore or develop physical capacities that have been lost due to injury, disease, behavioral patterns, or aging. For example, individualized therapeutic physical activity programs are prescribed and implemented by kinesiology graduates who complete additional education as athletic trainers or occupational or physical therapists. Stroke patients, for instance, sometimes require many sessions of physical therapy to help them regain the ability to perform important movements of daily living.

Kinesiology and Your Career

Kinesiology opens the door to a wide range of careers. Because kinesiology programs expose you to broad knowledge about physical activity, they provide excellent preparation for careers in fitness leadership and consulting, teaching and coaching, cardiac and neuromuscular rehabilitation, sport management, strength and conditioning, and numerous other areas of physical activity. Undergraduate programs in kinesiology also provide critical preparation for students who wish to further their studies with graduate or professional training in such fields as athletic training, physical therapy, occupational therapy, podiatry, chiropractic care, medicine, public health, nutrition, and, of course, kinesiology itself.

Many college graduates pursue master's and doctoral degrees in kinesiology (see figure 1.3). Some, like those preparing to be physical education teachers, do so to become more knowledgeable about their profession and to meet certification requirements. Others pursue graduate work to meet the educational requirements of a profession; in fact, a master's degree is increasingly viewed as the minimal requirement for most allied health professions. For example, all nationally accredited programs in athletic training will be moving to master's entry level by 2023. Sometimes master's training enhances a graduate's profile to qualify for additional training in professional doctoral programs (MD or DPT). Sometimes students continue their studies beyond the master's level in order to obtain a research oriented doctoral degree (PhD) in kinesiology so that they can become college or university faculty members or researchers.

Professionals trained in kinesiology share a common interest in and curiosity about physical activity in its broadest dimensions. Physical education teachers may hold the broadest perspective because of the enormous breadth of their teaching assignments, which can range from fitness and sport skills to dance and developmental skills.

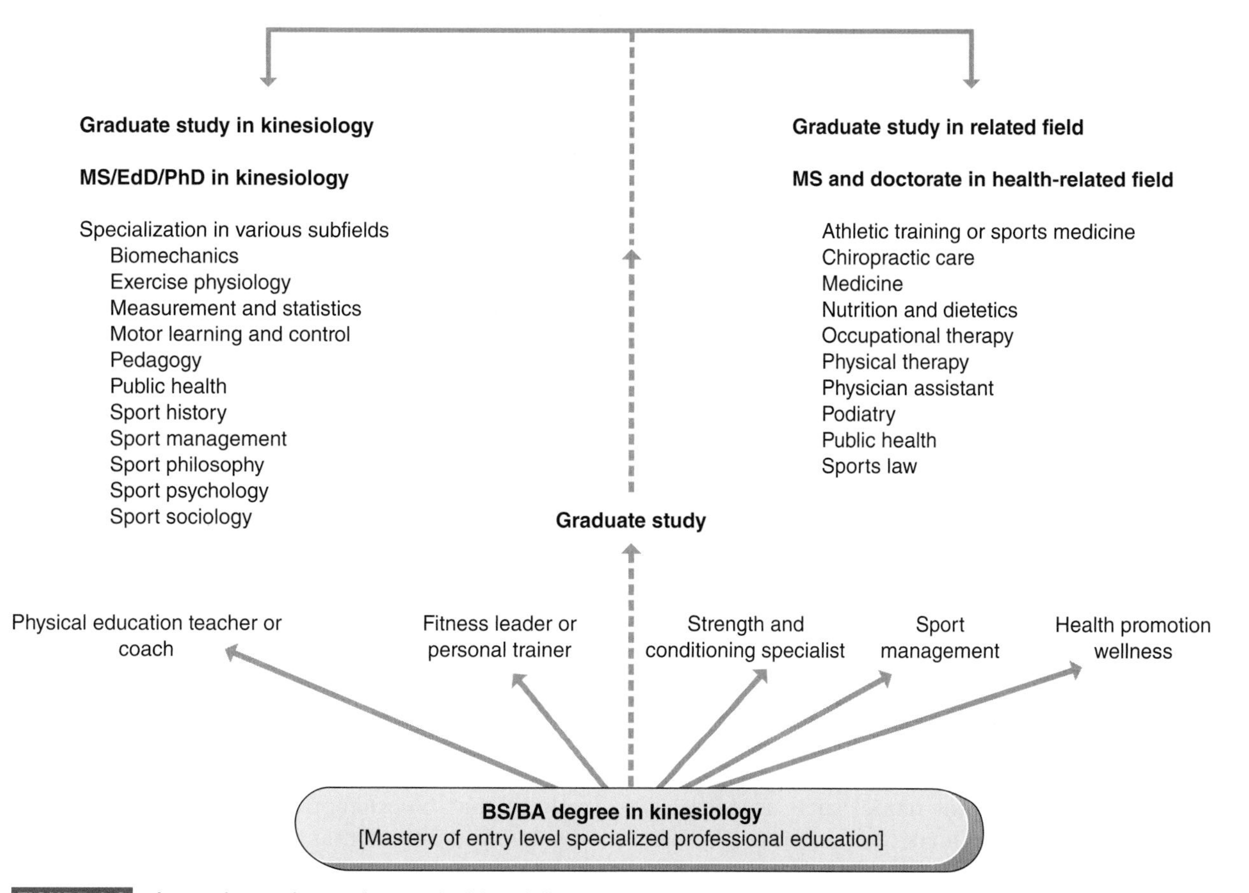

FIGURE 1.3 An undergraduate degree in kinesiology can serve as the starting point for many physical activity careers, some of which require additional training in graduate programs.

In addition, these practitioners may take special interest in using physical activity to develop social responsibility, academic achievement, and other desirable personal traits in children.

Other practitioners may pursue much narrower interests. For example, fitness leaders focus especially on cardiorespiratory fitness, strength, and exercise involving the body's large muscles. These practitioners are particularly interested in how such activities can alter physiological functioning. Similarly, professionals working in athletic training or rehabilitation exercise focus on physical activity as a medium for rehabilitation. In contrast, sport marketers have an altogether different interest in physical activity—namely, how physical activity performances by elite athletes can be made more appealing to paying audiences. Thus, in addition to developing your understanding of and appreciation for physical activity in general, you should seek to understand physical activity in the context of whatever specialized professional practice you choose to enter.

Why *Kinesiology?*

Scholars have debated at length which term best characterizes an academic discipline focused broadly on physical activity. For many years, the term *physical education* was considered appropriate for the holistic study of physical activity; in the first half of the 1900s, however, the use of the term became limited to the preparation of school physical education teachers. Today, physical education is generally understood to refer to broad-based instructional programs in sport and exercise offered by public schools, colleges, and universities. Most colleges and universities continue to sponsor an activity program (sometimes referred to as "the basic instruction program") for the general student body. In addition, the term *physical education teacher education (PETE)* is used to refer to the academic pedagogy specialization in many kinesiology departments that prepares teachers of these programs.

However, because most departments originally referred to as physical education departments

Surging Support for *Kinesiology*

The American Kinesiology Association currently identifies more than 170 departments of kinesiology at U.S. colleges and universities. Most of the departments at prestigious doctoral universities have adopted the term *kinesiology* because it best characterizes a discipline that deals with many forms of physical activity in diverse professional settings. This term has gained support from the National Academy of Kinesiology, an honorary society of some 140 elected scholars that was founded in 1930 as the American Academy of Physical Education but changed its name in 2010 to use the term *kinesiology*. A similar change was made in 2012 by the National Association for Kinesiology and Physical Education in Higher Education, which shortened its name to the National Association for Kinesiology in Higher Education. Both of these changes reflect the growing acceptance of *kinesiology* as a name for the discipline. Moreover, according to the last several AKA surveys, *kinesiology* is the most common department name for this discipline in universities in the United States (e.g., Mahar & Crenshaw, 2015). The Canadian Kinesiology Alliance also promotes kinesiology as a stand-alone allied health profession supported by the national health care program. Interest in kinesiology as a discipline and as its own profession in physical activity promotion is growing.

now prepare students for a panorama of careers in physical activity, the term *physical education* fails to capture the field's breadth. Many alternative names have been proposed for the discipline, and the emerging label—which is the one used in this text—is *kinesiology*. Still, your department might be called by a different name, such as exercise and sport science, human performance, health and human performance, health and kinesiology, human movement science, sport studies, exercise science, physical education, nutrition and kinesiology, or some combination of such terms. A given department's name is influenced most heavily by the mission, politics, and academic structure of its host college or university.

The diverse range of departmental titles can be confusing, which is a problem not faced by older fields, such as anthropology, sociology, psychology, and history. Because kinesiology is a young, evolving discipline—and because disciplines often need a long time to define themselves—it is not surprising that various names are still used to identify the field. Variation in departmental names also stems from the fact that in some departments—usually at smaller colleges—kinesiology is but one of several majors offered to students. Thus, a department that also prepares health educators or recreation specialists may call itself something like "kinesiology and leisure studies" or "health, exercise science, and recreation" to reflect the multiple degree programs that are offered. For example, at one large U.S. university, kinesiology degrees are offered along with two other degrees in the Department of Nutrition, Food, and Movement Sciences. Thus, we must be careful not to confuse the titles of departments at particular universities with the name of the field as a whole as it is coming to be known in the United States.

None of the various names used for our field are wrong, but most scholars emphasize the need for a single term that is broad enough to describe the discipline in full. Although not all scholars believe that *kinesiology* is the best name for the discipline, we think that it is. Therefore, regardless of the name of the department in which you are enrolled, we encourage you to refer to the discipline as *kinesiology*.

Potential Sources of Confusion About *Kinesiology*

One could argue that the best term for our field is *academic kinesiology,* which would distinguish it from what might be called *clinical kinesiology*. This latter term is still used sometimes to refer to functional anatomy courses taught in medical and physical therapy programs that help students learn how muscles and bones produce specific movements.

In addition, some forms of alternative medicine have recently begun to use the term *kinesiology* in their titles. For example, some practitioners of chiropractic use "applied kinesiology," which refers to an alternative medicine in which muscle testing determines the course of therapy. Similarly, "dental kinesiology" applies virtually the same techniques in analyzing the motions of the jaw and tongue, "behavioral kinesiologists" treat patients using biofeedback, and "spiritual kinesiologists" claim to promote healing by uniting body and soul. This use of the term kinesiology is rare and not the primary academic meaning that is older and more widely recognized by the world.

So far, these alternative meanings of kinesiology have not been problematic, given the lack of evidence for their claims as compared with the robust, international body of research in academic kinesiology. Still, the AKA (2017) has clarified the differences between *academic kinesiology* and other practices labeled as "kinesiology" in the following statement:

> The academic discipline of kinesiology is taught and researched in colleges and universities and is to be differentiated from Applied Kinesiology and other fields that use the term "kinesiology" (dental, spiritual, holistic, bio-spiritual), some of which lack grounding in the scientific study of physical activity. The AKA supports the academic discipline of kinesiology.

Allied Fields

Kinesiology has historical bonds with disciplines such as health education, dance, and recreation. Over time, however, these fields have begun to separate into their own specializations. In this context, in 2013, the original professional organization for all of these related fields in the United States—the American Alliance for Health, Physical Education, Recreation and Dance (AAHPERD)—changed its name to the Society of Health and Physical Educators (SHAPE America). The Canadian Association for Health, Physical Education, Recreation and Dance changed its name to Physical and Health Education Canada (PHE Canada) in 2008. These two organizations now focus more on professional health and physical education than on kinesiology or on older multidisciplinary units combining health, kinesiology, recreation, and dance.

Of course, health problems that arise from inactivity hold great interest both for physical activity professionals and for professionals in public health and health education. However, health professionals are also interested in an array of other problems (e.g., communicable diseases, smoking cessation, and drug abuse) that have little to do with physical activity. Similarly, recreation specialists are interested in physical activity as a leisure pursuit, but leisure pursuits stretch far beyond those in which physical activity is a primary concern. They also include, for example, travel, crafts, outdoor education, and nature study; organized recreation programs that go beyond physical activity are offered by parks, counties, cities, corporations, universities, and private companies. Finally, dance professionals often hold membership in artistic or other specialized organizations beyond kinesiology. In addition, although dancers are interested in physical activity, they focus on a much narrower slice of the physical activity pie—namely, the expressive and artistic forms of movement.

From the vantage point of today, it may seem strange that this assortment of fields was ever combined into a single professional organization, yet these fields are marked by a long historic relationship. Indeed, for most of the 20th century, they were represented in a single department at colleges and universities throughout most of the world; in fact, in some cases, they are still housed in that department. Today, however, each of these fields has become more specialized and isolated and has developed its own professional organizations and journals. As a result, in many cases, these fields are now taught in separate departments. Still, because of their shared history and the fact that they are often interested in the same problems, faculty from these areas frequently work together as colleagues.

Holistic Nature of Kinesiology

We hope that by the time you complete this introductory study of kinesiology, you will be convinced of the holistic nature of physical activity. **Holism** refers to the interdependence of mind, emotion, body, and spirit. Although some people think that kinesiology deals exclusively with the body and with bodily movement, in reality it spans a much broader range. Physical activity involves our minds, our emotions, and our souls as much as it does our bodies. It is convenient to speak of *physical* activity because the physical aspects are so easily observed, yet physical activity is also *cognitive* activity, *emotional* activity, and even *spiritual* activity for many people.

Thus, studying kinesiology will take you far beyond the study of the biological aspects of physical activity. Kinesiologists also study the psychological antecedents and outcomes of physical activity; the sociological, philosophical, and historical foundations of sport and physical activity; the dynamics of skill development, performance, and learning; and the human processes involved in teaching and learning physical activity.

KEY POINT

Although bodily aspects of kinesiology receive the most attention, a kinesiology program must account for the fact that human beings are holistic, multidimensional creatures characterized by the interrelated elements of cognition, emotion, body, and soul.

Think about the last time you went for a run or a brisk walk. Wherever you were—perhaps letting your mind wander as you enjoyed the scenery of a lake or the beauty of a mountain trail—this physical activity may have seemed so easy that you underestimated the complexity of its underlying physiological and psychological components. Physical activity is so much a part of our everyday lives that we often forget how wonderfully complex the human body is. Even so, most students who enroll in kinesiology programs are actively curious about the functioning of the human body. In your kinesiology program, you will study anatomy and physiology, as well as exercise physiology, biomechanics, sport and exercise psychology, and motor behavior. These classes will help you understand and appreciate the many mechanisms and systems that enable our bodies to perform physical activity.

You may also be required to take a course exploring the philosophy of physical activity. If so, you are fortunate because this course will help you understand how truly holistic physical activity is. By thinking philosophically, you will come to understand more fully the personal meaning that you discover in physical activity. This thinking will also sensitize you to myriad ethical issues associated with sport, exercise, and other types of physical activity.

In addition, your study of the psychology and sociology of physical activity will help you appreciate how attitudes and social settings influence our interpretations of our physical activity experiences. It is often easy to think about our bodies (or about specific body parts, such as the heart, muscles, and bones) simply as machines or instruments used by our minds or souls to achieve our purposes. In reality, however, your body is an indivisible part of your humanity. Indeed, it may not be an exaggeration to say that no other discipline is so diverse in its aims, so interdisciplinary in its subject matter, or so complex in its organization as is kinesiology.

The three-dimensional analysis of kinesiology offered in this book—in terms of physical activity experience, scholarship of physical activity, and professional experience in physical activity—is designed to help you organize your thinking about the discipline of kinesiology and the diversity of physical activity. This approach will not only help you develop a framework for understanding physical activity and the physical activity professions; it will also help you understand your course work in kinesiology more deeply and plan and pursue your career goals.

Wrap-Up

With this introductory knowledge under your belt, you are now ready to dig into the topics of physical activity and kinesiology more deeply. In this chapter, you have learned that kinesiology is the study of physical activity, and you now know that for kinesiologists, the term *physical activity* has

Experiencing the beauty of the world around us speaks to our holistic nature, which is an important focus of the kinesiology discipline.

Julien Leblay/fotolia.com

Professional Issues in Kinesiology

Equality, Evidence-Based Practice, Expertise, and Overload

Kinesiology professionals routinely grapple with the issues of equality, evidence-based practice, expertise, and overload. As a result, each chapter of this book illustrates how kinesiology knowledge can be applied to make professional decisions related to these issues. All three sources of kinesiology knowledge are illustrated, and key references for further study are cited in these features. The book also integrates a variety of areas of kinesiology in order to apply kinesiology knowledge to diverse areas of professional practice.

"Professional Issues in Kinesiology" sidebars also provide examples of different formats and styles for scholarly citations and references that are common in the field of kinesiology with its numerous subdisciplines and professional applications. Many behavioral sciences in kinesiology use the American Psychological Association (APA) format, whereas publications in sports medicine and athletic training often use the American Medical Association (AMA) format. At times, kinesiology professionals also need to write reports or read research from other areas with different reporting styles. For more information about the APA and AMA styles, please see the following sources:

› American Medical Association. (2007). *AMA manual of style* (10th ed.). Philadelphia: Williams & Wilkins.
› American Psychological Association. (2010). *Publication manual of the American Psychological Association* (6th ed.). Washington, DC: Author.

a technical definition that is quite different from our use of the term in everyday life. You also have learned that the expressions of physical activity studied by kinesiology vary widely, as do the professional careers related to the field.

In addition, you have learned something about how the field of kinesiology is organized. Specifically, the three main sources of knowledge in kinesiology are the experience that comes from participating in and watching physical activity, the scholarly study of physical activity, and the knowledge that comes from professional experience in physical activity.

REVIEW QUESTIONS

1. What is the difference between movement and physical activity? Give an example of an instance in which human movement does not meet the technical definition of physical activity.
2. What is meant when kinesiology is described as a holistic discipline?
3. What forms or categories of physical activity are studied in kinesiology? Which receive the most attention, and how do they relate to each other?
4. What are the three sources of knowledge of kinesiology?

Go online to complete the web study guide activities assigned by your instructor.

PART I
Experiencing Physical Activity

One of the everlasting mysteries of sport is why people watch and perform it. Several decades ago, in an effort to get to the bottom of his own attraction to sport, Joseph Epstein penned a popular essay titled "Obsessed With Sport," in which he formulated the question as follows: "What is the fascination? Why is it that, with the prospect of a game to watch in the evening or [on the] weekend, the day seems lighter and brighter?" (1976, p. 67). Despite decades of work by philosophers, journalists, and kinesiology scholars, there is still no agreed-upon explanation for why some of us sit for hours in front of the television, put on silly hats, and paint our faces to watch our favorite teams—all while putting off other things in life.

Nor, despite years of thinking and research, have we determined precisely why some of us become so hooked on exercising or participating in sport that we prioritize it in our lives. The avid golfer who sneaks off to the course for a "quick 18 holes"; the dedicated tennis player who fills up every weekend with matches and tournaments; and the runner who seems unable to stop herself from running in the rain, the snow, or the unbearable heat—all are propelled by some force that is not easy to explain. A former editor of *Runner's World*, trying to understand his attraction to running, put it this way: "I write because the thoughts inside have to be put in more visible form. I run because it's inside pushing to get out" (Sheehan 1978, p. 74).

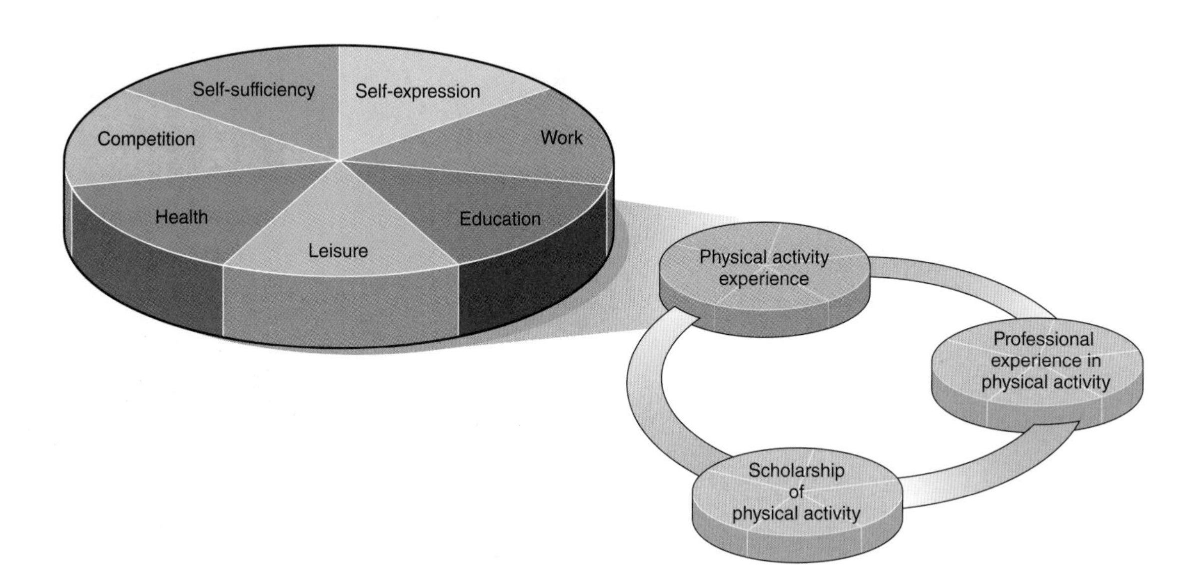

The chapters included in part I explore the critical importance, and the wide variety, of experience in physical activity and the subsequent knowledge acquired. Chapter 2 describes the variety of ways in which physical activity penetrates our daily lives. More specifically, it introduces you to the seven spheres of physical activity experience—distinct aspects of our lives in which physical activity plays an important role. Chapter 3 previews key kinesiology concepts that address how physical experience can be manipulated to bring about specific changes in health or performance capacity. It also examines the subjective dimension of physical activity experiences—that is, the interior experiences of physical activity—which are important because they affect our thoughts and feelings, as well as our future physical activity choices. Individuals who do not enjoy physical activity experiences are unlikely to participate voluntarily in the future.

Physical activity experience is so prevalent in our lives that we tend to take it for granted. Yet it provides the foundation for the study of kinesiology. In the following chapters, we invite you to view physical activity experience from an entirely new perspective, beginning with the seven spheres of physical activity.

Spheres of Physical Activity Experience

Shirl J. Hoffman

CHAPTER OBJECTIVES

In this chapter, we will

- broaden your understanding of the universe of physical activities,
- spark your thinking about the various types of physical activity undertaken for specific purposes in different social settings or "spheres," and
- introduce you to some of the potential benefits and limitations of physical activity.

Vashti Cunningham is a phenomenal 16-year-old athlete, winner of the IAAF World Indoor Championships with a jump of 6 feet 6 1/4 inches (2 meters), and qualifier for the U.S. Olympic team that competed in Rio de Janeiro in 2016. Although she struggled to clear 6 feet 4 inches (1.93 meters) at the Games and failed to earn a medal, she told reporters that the loss would simply motivate her to perform better at the next Olympics.

Which aspect of Cunningham's performance captures your attention? Is it the height of the jump, the way her body is draped over the bar, or perhaps the deep arch of her back? Perhaps your attention is drawn to her lead arm as it thrusts upward and forward to lead her body over the bar, or perhaps you noticed her lagging arm as it seems to simply hang on for the ride. Some observers marvel at the display of sheer athleticism, while others are drawn to the beauty of the performance.

Such questions provide a good introduction to kinesiology because they ask you to examine a form of physical activity more closely and systematically than you might be accustomed to doing. In one way or another, all kinesiologists—whatever their area of expertise—are intelligent evaluators of movement, whether they do so in a teaching, coaching, exercise, or clinical setting. By pursuing a four-year major in kinesiology, you too will become a skilled observer and evaluator of physical activity.

The photograph of Vashti Cunningham also helps illustrate the main point of this chapter: the study of kinesiology encompasses an enormous range of movements occurring in a wide variety of contexts. Sport and exercise are of special interest to kinesiologists, but the photo could just as easily have shown an elderly person walking with a cane, a woman mowing a lawn, a miner loading coal, or a young child taking his first steps. The domain of kinesiology encompasses any form of human movement that meets the definition of physical activity provided in chapter 1.

AP Photo/Charlie Riedel, File

If you're like most people, you probably haven't given much thought to how important physical activity is in your daily life; indeed, you may think seriously about the subject only when your capacity for moving is limited by disease or injury. For example, nothing helps us appreciate the importance of the ankle joint quite like spraining it. When we stop to think about it, we realize that physical activity is involved in our lives in thousands of ways. Thus, physical activity professionals are concerned not only with physical activity that occurs on sport fields or in fitness centers but also with physical activity that occurs in workplaces, rehabilitation centers, dance studios, nursing homes, and many other venues.

This chapter will stretch your conception of physical activity. In reading it, you embark on an expedition through a vast expanse of physical activities performed for various reasons and in various settings. The purpose is both to expand your frame of reference for thinking about physical activity and to increase your appreciation of how critically important kinesiology is to our daily lives. One way to begin to appreciate the enormous variety of physical activities—and the way they intersect with your daily life—is to ask yourself how often and in what ways you were physically active during the past week. In answering this question, you might first reflect on the amount of physical activity you performed. For example, you might think, "I waited on 50 customers at the delicatessen where I work," or "I ran 4 miles on a wooded trail."

But it isn't simply the volume of physical activity that matters; the variety of physical activity experiences is also important. For example, the types of physical activity and movement performed in working at a delicatessen—such as slicing ham, making sandwiches, and operating a cash register—are all quite different from those involved in running or swimming. These tasks at the deli require rather precise movements of your arms but rarely tax your cardiovascular system. So, as you relate this chapter to your everyday experiences, don't think only about how much physical activity you performed; think also about your various kinds of physical activity.

As noted in chapter 1, physical activity consists of human movement intended to accomplish a specific purpose. As you act out your daily life—whether doing something as simple as moving to the other side of a room or as complicated as juggling three balls—the purposes you wish to achieve usually determine both the type and the amount of physical activity. This may sound simple enough; however, in thinking about the enormous variety of goals and purposes of physical activity, and the various contexts in which that activity occurs, it is easy to feel overwhelmed. Therefore, in order to get a handle on the importance and pervasiveness of physical activity in our lives, we need a system for organizing our physical activity experiences—that is, a conceptual framework.

To that end, physical activities can be classified into seven spheres of physical activity experience (see the figure presented in the part I opener):

1. Self-sufficiency
2. Self-expression
3. Work
4. Education
5. Leisure
6. Health
7. Competition

These spheres represent dimensions of everyday life in which physical activity plays an important and distinctive role. Generally, the spheres identify various purposes of physical activity that usually relate to certain social contexts. For example, the work sphere includes all physical activity done in the workplace in order to meet the demands of employment, whereas the education sphere includes physical activity carried out in educational settings for the purpose of learning.

As you may notice, some spheres are characterized less definitively in terms of the social context. The self-sufficiency sphere, for example, includes physical activity carried out in order to survive and live an independent life. Although most of these activities take place in the home, they can also be performed at work, during leisure time, or in educational settings. Similarly, physical activity in the self-expression sphere may occur in a variety of social contexts but always serves the purpose of allowing us to express our emotions.

As you study the spheres, hold in mind the following considerations:

- The spheres are not intended to uniquely classify specific types of physical activity. They simply highlight the different compartments of life experience in which physical activity plays an important part.
- Some activities may be common in more than one sphere of experience. For example, if you run 3 miles (about 5 kilometers) per day, this activity might be assigned to the sphere of leisure but would also relate to your health and perhaps your self-expression.
- The purpose of this chapter is not to compartmentalize the activities themselves but to provide you with a general framework for thinking about the importance and pervasiveness of physical activity in all aspects of your life.

The following sections explore each of the seven spheres of physical activity experience by examining some activities that they encompass. We discuss both positive and negative issues related to the spheres and describe professionals who specialize in addressing some of those issues. Remember that the spheres serve merely to help you look at the many ways and contexts in which you experience physical activity. We hope this discussion challenges you to think more deeply about the many levels of this field.

Sphere of Self-Sufficiency

You spend a significant part of your life simply taking care of yourself, and you do so largely through the critical means of physical activity—for example, eating breakfast, walking to the

Cooking is an example of an instrumental activity of daily living.
Monkey Business/fotolia.com

bathroom, and brushing your teeth. Though often performed perfunctorily, such activities make it possible for you to live independently. If you were unable to do them, you would be forced to rely on help from others; in other words, you would lose your self-sufficiency. Self-sufficiency and personal comfort are also aided by other activities that may not be essential for survival or independent living, such as cleaning your home, washing the dishes, ironing clothes, and driving a car.

Physical and occupational therapists help people who, through injury or disease, have lost the ability to perform important tasks of daily life that enable self-sufficiency. Typically, therapists divide self-sufficient physical activity into two major categories: **activities of daily living (ADLs)** and **instrumental activities of daily living (IADLs)**. The category of ADLs includes personal behaviors related to grooming, using the toilet, dressing, eating, and walking. People who require daily assistance with such tasks may be disabled by injury or disease or frail because of age. The category of IADLs includes less personal activities, such as telephoning, shopping, cooking, and doing laundry (Katz, Ford, Moskowitz, Jackson, & Jaffe, 1963). Generally, IADLs are more physically demanding than ADLs. Health insurance companies often use these two classifications to gauge a patient's level of disability and determine the type of health support to cover. For example, an older person may not qualify to collect on long-term health care insurance coverage unless a therapist documents that he or she is unable to perform two or more ADLs.

Of course, self-sufficiency may also depend on performing some physical activities around one's home that are more complicated than typical ADLs and IADLs and may demand more energy. Examples include **home maintenance activities** such as shoveling snow, fixing an automobile, painting one's home, and repairing electrical fixtures. All of these activities require relatively high levels of energy or skill or both. Although many people hire others to perform home maintenance activities for them (e.g., gardeners, painters, automobile mechanics, plumbers), the explosion of online do-it-yourself videos and home improvement television programs suggests that many people want to perform such activities themselves.

KEY POINT

To live functional, independent lives, we must perform activities of daily living (ADLs) and instrumental activities of daily living (IADLs).

Limitations in Physical Activity for Self-Sufficiency

A variety of movements are required in order to carry out daily tasks such as cleaning the house, doing laundry, bathing, cooking, opening jars, writing checks, and shopping for groceries. For example, using a vacuum cleaner includes the fundamental movements of walking, standing, grasping, reaching, pushing, and maintaining an upright posture. Remaining self-sufficient also requires us to transport objects of different weights, ascend or descend stairways, and perform other fairly complex actions.

When our movement capabilities are compromised by disease or injury, we can work with occupational and physical therapists to learn or relearn how to perform self-maintenance activities.

Before devising treatment plans, or interventions, to rehabilitate a person affected by physical disability, a physical or occupational therapist must completely understand the physical activity requirements of each task. This understanding can be obtained by thoroughly analyzing the movements that a person must perform in order to carry out a given task. Only then can the therapist decide which muscles the patient must strengthen, or which movement patterns the patient must master, in order to regain self-sufficiency. Such activity analyses can also help us understand the amazing complexity of what may seem at first to be simple tasks.

KEY POINT

Injury or disease can hinder a person's ability to perform daily physical activities. To help the person recover as much functionality as possible, a physical or occupational therapist creates therapeutic strategies based on activity analyses.

For example, let us consider an activity analysis of the seemingly simple self-care task of standing up from a seated position, which may pose a major challenge to someone who has experienced a stroke. The activity analysis reveals at least four critical phases: (1) the feet must be placed on the floor in a good position to receive weight evenly divided between the two legs; (2) the trunk must be flexed forward at the waist while remaining extended; (3) the knees must move forward of the ankles; and (4) the hips and knees must extend for final alignment (Carr & Shepherd, 1987). By skillfully comparing the patient's performance of these movements with a model of healthy performance, the therapist can design a training program to speed the patient's recovery in this fundamental task. Such activity analyses require in-depth knowledge of anatomy and biomechanics.

Self-Sufficiency and Aging

You may have given little thought to the importance of ADLs and IADLs in your life because you may have experienced little trouble in performing them. However, a surprisingly large portion of the population requires assistance with these basic self-care tasks, a condition that deprives them of independence in their daily living. Roughly 9.4 million noninstitutionalized people have difficulty performing one ADL, and 5 million of those need assistance to do so. Similarly, more than 15 million noninstitutionalized adults find it difficult to perform one or more IADLs. In addition, according to the Centers for Disease Control and Prevention (CDC, 2015a), 13 percent of men and 21 percent of women aged 75 to 84 experience difficulty in doing errands on their own. Such limitations may result from accident, injury, congenital disorder, or aging.

In addition to having difficulty with ADLs and IADLs, older people are often injured during their attempts to perform such activities. Common accidents include falling on stairs or in bathtubs and suffering burns and scalds. Most of these accidents are caused by movement limitations, but they can also result from environmental factors such as poorly lit stairways, frayed rugs, and poorly maintained homes. Because older people often recognize the high risk of having an accident around the home, many simply stop performing certain ADLs and IADLs. This decision can result in a severely diminished quality of life (Czaja, 1997).

The prospect of a growing population of individuals who depend on others to carry out their daily tasks also looks undesirable from an economic standpoint. When major segments of the population lose their ability to perform ADLs and IADLs, the national health care system takes on a heavy burden. An elderly population that requires assistance to perform ADLs exerts a significant effect on the costs of social security programs. You have a real stake in those costs as a taxpayer.

KEY POINT

Limitations in performing ADLs and IADLs require many elderly people to depend on others, or on institutions, to perform their tasks of daily living. This problem holds great importance in both personal and economic terms.

These trends pose a challenge for people in the physical activity profession who work with older populations. Although the number of elderly people will inevitably increase over the next few decades, that trend need not by accompanied by an increase in the proportion of this population that is hampered by limitations in physical activity. Many of the people who suffer the most limitations in later life are those who failed to make physical activity a daily part of

Physical Activity as Frontline Defense Against Self-Sufficiency Limitations

Research on elderly populations is booming as scientists begin to focus on the many problems that accompany aging. Here, we consider three studies examining the benefits of exercise by older adults in preventing dementia.

Physical activity limitations among elderly people often go hand in hand with the development of Alzheimer's disease, which is a burgeoning health problem. In 2013, Alzheimer's was responsible for more than 84,000 U.S. deaths, and in 2016 an estimated 700,000 people died with the disease (Alzheimer's Association, 2016). Not only does it rob people of their ability to perform IADLs; it can also be devastating to family members, who must take on major health care responsibilities.

Among efforts to deal with this problem, one bright spot can be found in the promising results produced by research on exercise and Alzheimer's, particularly a study by J.C. Smith and colleagues (2014). Their results suggest that exercise may help ward off and possibly delay the neurodegeneration associated with genetic risk for Alzheimer's disease. The researchers tracked four groups of healthy adults between the ages of 65 and 89 with normal cognitive abilities over an 18-month period. Before and after the study period, they used MRI technology to measure the volume of the hippocampus, which is a region of the brain responsible for memory and spatial orientation. Groups were partitioned into high- and low-risk categories based on the presence or absence of a genetic marker thought to increase risk for the disease; they were also categorized according to high or low levels of physical activity as measured by a subjective, self-report survey. Low activity was defined as two or fewer days per week of low-intensity exercise, and high activity was defined as three or more days per week of moderate to vigorous activity.

The findings suggested an important role for exercise in preventing Alzheimer's disease. Participants who had a high genetic disposition for Alzheimer's and did not exercise evidenced a 3 percent decrease in the volume of the hippocampus. In contrast, no significant reduction in hippocampal volume was seen in the other three groups, including the high-exercise group with a genetic disposition for Alzheimer's. The researchers believe that their findings warrant recommending increased levels of physical activity as a way to maintain brain integrity by reducing atrophy in the region of the brain that is critical for the formation of episodic memory.

The value of exercise in slowing brain degeneration may show up much earlier than one would expect. In one study along these lines, a Swedish cohort of more than 1 million 18-year-old male conscripts underwent exams between 1968 and 2005 and were followed for up to 42 years (mean follow-up time was 28 years). The study, led by Dr. Jenny Nyberg, was conducted by a research team at the University of Gothenburg, who examined the relationship between objective data on cardiovascular and cognitive fitness collected at the beginning of the study and the risk of early-onset dementia and mild cognitive impairment later in life (Nyberg et al., 2014). The results were striking: Men who were originally identified with poor cardiovascular fitness were more than twice as likely to develop early-onset dementia later in life. The risk levels endured even after controlling for other risk factors such as heredity, medical history, and socioeconomic circumstances. Moreover, the combination of low cardiovascular fitness and poor cognitive performance in early adulthood was associated with a 14-fold increase in risk of early-onset mild cognitive impairment and an 8-fold increase in risk of early-onset dementia.

Beyond objective measures of fitness, it appears that even one's self-perception of physical health may be a reasonable predictor of developing dementia in later life. A team of Finnish researchers followed a group of more than 3,500 individuals over a 30-year period and found that people who rated themselves as having poor physical fitness at the mean age of 50 years were four times more likely to experience dementia during the next three decades than were people who rated themselves as being in good shape (Kulmala et al., 2014). The association was strongest among people who did not carry a protein that is a known genetic marker for dementia.

life over the years. To minimize this outcome, recreational activities and exercise programs designed and administered by physical activity professionals will play an increasing role in preventing and rehabilitating age-related disabilities.

Sphere of Self-Expression

The urge to express our feelings is one of the most basic human instincts. All of us would like to demonstrate, in one way or another, what is unique about us, what makes us special. In doing so, of course, we face limitations. People often hesitate to express themselves in speech, and few possess the talent of a poet, songwriter, or visual artist. We can, however, give outlet to our feelings by moving our bodies.

In fact, the way in which we move can speak volumes about how we feel. We walk rapidly and with a spring in our step when we feel happy but move in a slow, plodding manner when we feel sad. We jump up and down for joy, stretch our arms overhead in elation, hang our heads in sorrow, and clap our hands in appreciation. In these ways and many others, we use body movements to express our emotional states. Moreover, our body movements seem to be wired directly to our emotions so that changes in our movements may occur without voluntary effort on our part. Although such movements point to intriguing aspects of human identity, they do not meet our technical definition of physical activity because they do not involve deliberate or voluntary movement directed toward a goal.

Your body language may speak louder than your words.
Photodisc/Getty Images

Many times, however, we use movement deliberately in order to express a feeling or mood or to convey another type of message. Have you ever thought about how much you depend on physical activity to help you communicate when you are talking? By using gestures, hand signals, and changes in body posture, we can deliberately and intentionally emphasize such verbal messages as describing shapes and showing directions. We also use physical activity intentionally to express our feelings and emotions in dance, religious liturgy, and various other ceremonies. Even when standing, our posture may send signals about our feelings or attitudes. Let's now take a closer look at some ways in which we use physical activity to express ourselves.

Gestures

Gestures are movements of the hands, fingers, or other body parts in order to communicate our intentions to others. We can use gestures either in place of or in conjunction with talking. Examples include beckoning someone to approach by flexing and extending the index finger with the palm up, signaling someone to stop by holding up a hand with the palm facing away, and nodding instead of saying yes. Scientists who study nonverbal communication distinguish between three types of intentional gestures: emblems, illustrators, and regulators.

Emblems

Emblems are body movements, usually hand movements, that can be directly translated into words and are easily understood by people in the culture or subculture in which they are used (Morris, 1994). People may use emblems with accompanying words, but they can convey a great deal of meaning even when used without words. For example, they can be used to communicate at a distance or in environments in which verbal communication is difficult or impossible. For example, gestures can be used by crane operators in construction settings to communicate with workers on the ground, by coaches in noisy athletic arenas to send signals to players on the court, and by scuba divers to communicate with each other under water. Similarly, referees and umpires use emblems to indicate various information to players, coaches, and audiences—for instance, that a player has committed a foul, that a base runner is out or safe, that a tennis shot is in or out, or that the game clock has expired.

Illustrators

Illustrators are gestures that we use to illustrate or complement what we are saying. For example, if you are talking about yourself, you might point to yourself; if you are talking about someone else in the room, you might point to him or her. People also use illustrators to describe the motions of objects. At a baseball game, you might use a gesture to describe the path of a foul ball that narrowly missed your head as you sat along the third-base line. Illustrators can also be used to convey a particular tone in a verbal message. When a coach pounds the fist of one hand into the palm of the other while talking, the gesture adds a sense of determination and seriousness to the verbal message.

Regulators

Regulators are body movements used to guide the flow of conversation. Examples include hand and body movements used in greeting (e.g., handshake, wave, nod) or parting (e.g., handshake, wave, hug). A person might also use regulators to signal that he or she is finished with a conversation—for instance, shifting weight from one foot to the other or turning toward the door. In contrast, a person might use a regulator (e.g., hand gesture, shift in gaze, head movement) to signal that he or she has *not* yet finished talking or that it is the other person's turn to talk.

Cultural Differences in Gestures

The meaning of a gesture is often specific to a particular culture, and being unaware of that meaning may lead to embarrassing situations. For example, beckoning someone with the forefinger and an upraised palm is an obscene gesture in some cultures, as is making an O shape with your thumb and index finger (as if to signal "okay" in U.S. culture). Some of the most culturally specific gestures are regulators, especially those used for greetings. For example, if you live in the United States and extend a hand for a handshake with an acquaintance, you might think it rude if he responds by folding his arms across his chest; in Malaysia, however, this movement indicates a respectful greeting. Similarly, in Eskimo country, a hit on the shoulder doesn't mean that you are being challenged to a fight; rather, it means "hello." Whereas nodding your head up and down means yes in most countries, it means no in Greece and Bulgaria.

Even within a given culture, the meanings attached to a gesture can change. Fifty years ago, if you saw two Americans bumping the knuckles of their clenched fists together after one of them made a spectacular golf shot, you might have thought it was some form of secret greeting. Today, of course, the fist bump is a popular form of congratulation. Similarly, after scoring a touchdown, teammates in American football often perform what is known as the "back to back" or "shoulder tap" by running toward each other and then jumping and turning their backs.

KEY POINT

We use physical activity in the form of gestures as a means of communication and expression, either in combination with or in place of words.

Dance

We often express our feelings through the ways in which we execute physical activity. More specifically, we may combine expressive elements with the movements that we use for accomplishing specific tasks. Even though we may not run, swim, or lift weights specifically to express something about ourselves, we often cannot help doing so. For instance, the manner in which a basketball player bounces a basketball before shooting a free throw involves both instrumental and expressive movements. **Instrumental movements** are the critical movements required to attain the goal of an activity, whereas **expressive movements** are idiosyncratic movements that are not required for goal attainment but that express something about the individual. It may be impossible to separate instrumental movements from expressive movements in some athletic performances, such as a twisting layup or a run around the bases, because athletes tend to employ the movements that are most effective for them and those movements often include a sizeable expressive component.

The back to back is a modern equivalent of a high five to say "good job."
Doug Murray/Icon Sportswire

Sometimes, however, we use physical activity deliberately to express sentiment and emotion with no instrumental goal in mind. In dance and in ritual, for example, physical activity serves the express purpose of conveying feelings and symbolic meaning. In fact, people have used physical activity to express emotion and symbolic meaning from the earliest periods of recorded time. One reason is that human movement enables us to communicate complex thoughts and feelings that are difficult or impossible to convey verbally. Dancers use the mediums of force, time, and sequencing in much the same way that painters use color and form, sculptors use texture and shape, and musicians use tone and rhythm—all to express aesthetic messages. The aesthetic characteristics of a dance are also affected by the shape and dimensions of the dancer's body. For instance, a leap evokes much different emotions in observers when performed by a tall, slender dancer than it does when executed by a short, overweight dancer. In either case, however, the moving body can tell a story that may be difficult to express in words or song.

The physical requirements of dance can be as exacting as—or even more exacting than—those of the most strenuous sport or exercise routines. For instance, the energy requirements of ballet, disco, jazz, Latin, modern, and tap dancing are approximately the same as those for hunting, kayaking, gymnastics, climbing, and team sports (Ainsworth et al., 2000). Dancing calls for muscular and cardiorespiratory endurance, flexibility, strength, balance, agility, and coordination. In addition, ballet dancers must adhere to long, arduous practice and conditioning regimens in order to be able to achieve the standard body positions and movements. Intense training and conditioning are also required in order to perform the free-flowing, creative routines of

What Your Style of Walking Says About You

Research has shown that the way in which you walk conveys a great deal about you (May, 2012). In a series of experiments, Ambady and Rosenthal (1993) asked undergraduates to evaluate teachers on the basis of 30-second video clips of their performances in which the audio was muted so that the students' judgments would be made strictly on the basis of nonverbal cues. (The teachers shown in the video had not served as instructors for the participants.) The students rated the teachers in terms of variables such as honesty, warmth, competence, and optimism. The ratings not only were consistent across raters but also correlated significantly with ratings given by the teachers' own students who had sat in their classrooms for an entire semester. These correlations persisted even when the video clip was shortened to 10 seconds. In fact, watching teachers perform (without audio) for as little as two seconds led to approximately the same ratings given by students who had attended class for a whole semester!

Clearly, the ways in which the teachers moved sent strong messages to the students watching the video. However, could the students also have been reacting to other information, such as the teacher's clothes, gender, race, or level of attractiveness? To find out, Thoresen, Vuong, and Atkinson (2012) eliminated all extraneous information from video-recorded images of male and female volunteers who walked in their natural style for a distance of about 25 feet (7.5 meters). They did so by using a method known as "point light representation," in which a moving image is constructed by attaching reflective markers to body joints and using dark clothing and a dark background. This approach results in a video that depicts only the light dots moving in relation to each other; nothing else is visible. Based only on these representations, study participants rated each walking example on six factors: adventurousness, extroversion, neuroticism, trustworthiness, warmth, and approachability (Thoresen et al., 2012).

Amazingly, the ratings reflected general consensus. That is, if one rater judged a given point-light display to represent a person who was approachable, adventurous, or extroverted, other raters were likely to agree. (The ratings did not agree, however, with the self-ratings of the people depicted in the representations.) Follow-up studies asked observers to determine each walker's gender, estimate his or her age, and rate the walker on such variables as attractiveness and excitability. The raters were highly consistent and accurate in their determinations of gender but much less accurate in guessing the walkers' ages. These perceptions of light-point displays appeared to influence observers' judgments of the personalities of the walkers. For example, walkers whose locomotion patterns led to their being identified as masculine were also perceived as emotionally stable, those identified as attractive were described as approachable, and those judged to be calm were identified as warm.

These experiments underscore the fact that the ways in which we move send out signals that tend to be interpreted similarly, whether accurately or not, by those who watch us. Would similar findings be observed from light-point recordings of tennis serves, golf drives, and gymnastics performances?

modern dance, in which the dancer may have more freedom in interpreting the choreographer's notations. Regardless of the form of dance performed, the basic result is the same: "the presentation of a significant emotional concept through formal movement materials" (Phenix 1964, p. 197).

KEY POINT

The art form of dance uses physical activity to express attitudes and feelings that may be difficult to express in words. Rituals often use physical activity to express sacred values or beliefs in symbolic form.

Sometimes we dance not to communicate something to others but simply to give outlet to our feelings, even when we are alone. In these cases, we do not require an audience, a partner, or a band in order to enjoy dancing. Have you ever danced to music when no one else was in the room? At other times, we dance spontaneously to express an emotion—for example, when football players dance in the end zone after scoring a touchdown. You may also enjoy social dances, such as the swing, the tango, the electric slide, or country line dances, in which you can demonstrate your skill at moving your body in relation to a partner or the beat of the music. Perhaps dance allows you to communicate something about yourself that you can't communicate in any other way.

Sphere of Work

For most of us, work constitutes a considerable portion of our total life experience, and it always involves at least a small degree of physical activity. What different kinds of work have you done in your life? Have you worked as a construction laborer, a truck loader for a shipping company, a lawn maintenance technician, a pizza delivery person, a house painter, a server at a restaurant, a kitchen worker for a fast food chain, an office assistant, or perhaps a checkout attendant at a grocery store? What kinds of physical activity have your jobs required?

In thinking about the answer to this question, you probably recognized that the type and intensity of physical activity can vary enormously from job to job. For example, a checkout clerk and a construction laborer both engage in physical activity, but they are active in very different ways. Moreover, although you may easily recognize the importance of physical activity to a pizza delivery person or truck loader, it may be a little harder to appreciate the equal importance of physical activity (though of a different kind) to the work of a secretary, accountant, computer consultant, or graphic artist. Yet physical actions such as using a computer, though not strenuous, require precise positioning and movement of the limbs and fingers.

Throughout recorded history, societies have associated manual labor and its high demand for physical activity with lower socioeconomic status while according higher status to managerial and supervisory jobs that require little physical activity. Yet some low-status jobs that require high levels of physical activity may also offer a health bonus for workers. Specifically, increased levels of some forms of physical activity may ward off certain diseases associated with physical inactivity (sometimes called *hypokinetic* or *low-physical-activity diseases*). For example, when one list of the 10 worst jobs in 2010 included the position of mail carrier partly because of its physical demands, the list discounted the health bonus that carriers receive from their daily walking activity (Strieber, 2010). Similarly, a classic study of drivers and conductors of English double-decker buses revealed that the drivers, who were mostly sedentary (and held the easier, higher-status job), had a higher incidence of heart disease than did the conductors (the lower-status workers), who were constantly in motion as they walked up and down stairs (Morris, Heady, Raffle, Roberts, & Parks, 1953). These findings suggest that the types of work valued highly by a society—often those with a low amount and intensity of physical activity—may not be conducive to good health.

KEY POINT

As technology continues to shape the character of work, the amount of physical activity required on the job is likely to decrease, thus placing workers at higher risk for diseases brought on by physical inactivity.

The amount of physical activity required for performing a job will generally continue to decrease as workplaces increasingly adopt technological innovation. In addition, current workforce projections suggest that the jobs of the future are much more likely to be found in the service and managerial sectors, in which the physical activity requirements tend to be more modest than they are in manufacturing, construction, and agriculture. Furthermore, workers in the United States continue to log more working hours than do their counterparts in other countries; in fact, according to one report, Americans work about 1 hour more each day than do European workers (Bick, Brüggemann, & Fuchs-Schündeln, 2016). Add to this scenario the fact that the average commute to work has increased to 46 minutes per day in the United States, and it is clear that U.S. workers may not have the time required to exercise daily (Carroll, 2007). This reality has led some health scientists to recommend that people put more emphasis on integrating physical activity into their working lives—for example, by parking their cars farther from their offices or by walking up the stairs rather than taking the elevator (Dong, Block, & Mandel, 2004).

With such concerns in mind, one recent innovation involves the use of adjustable desks in the workplace to enable people to work in different positions. Sitting at a desk for extended periods not only increases the possibility of spine and hip problems but also has been related to such conditions as elevated blood pressure, high blood sugar, and the accumulation of excessive fat around the waist. Ergonomic engineers have responded to this crisis by designing desks and desk-mounted lifts that can be elevated to different positions that range from sitting to standing height. A substantial market has also emerged for treadmill desks, which

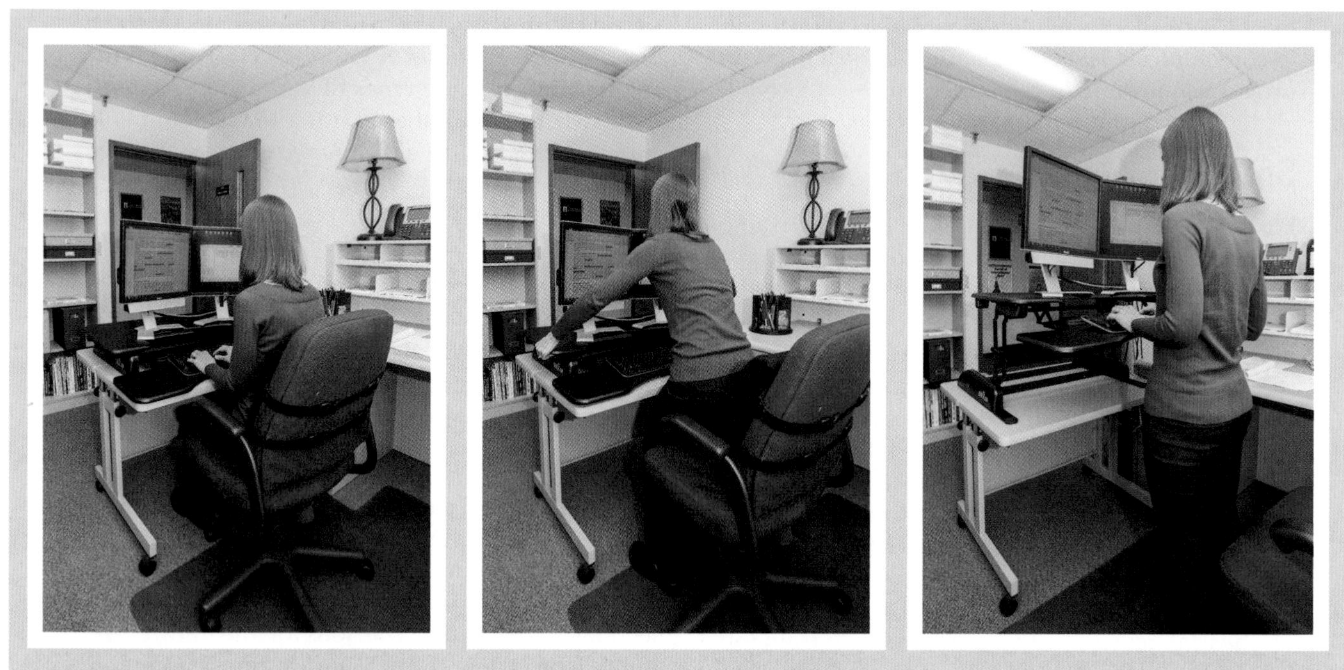

Adjustable desks allow employees to move easily between sitting and standing positions.
© Human Kinetics

spread the workspace across and above a treadmill so that the worker can exercise and "push papers" at the same time.

Physical Activity, Efficiency, and Injury in the Workplace

Work can be viewed as a process in which workers trade mental and physical activity for compensation. Employers, in turn, translate workers' activity into production of goods or services—and the more efficient this production is, the greater the profit enjoyed by the employer. Thus, workers' physical activity, whether vigorous and expansive or mild and limited in scope, serves as the foundation on which the commercial enterprise rests.

Businesses can increase production efficiency in two ways. The first way is to replace physical activity that is inefficient with technology that costs less. For example, machines and robots now allow car manufacturers to produce many more cars per day at lower cost than when people performed all of the assembly through manual labor. In the tech sector, Foxconn, a major supplier for Apple and Samsung, announced in 2016 that it was cutting its job force from 110,000 to 50,000 thanks to a new generation of factory robots (Wakefield, 2016). Of course, technology cannot replace all human physical activity; indeed, in many jobs, movements performed by workers remain critical to the efficient production of goods or services. In these cases, organizations seek to increase production by improving the efficiency and safety of workers' movements, tools, and working conditions. Such changes can provide benefits in diverse work situations, ranging from assembly line employees who package television sets, to agricultural laborers who pick fruit from trees or vegetables in fields, to machine operators who work in furniture and textile factories.

Professionals who specialize in improving efficiency, safety, and well-being among workers are called **ergonomists** or **human factors engineers**. The particular roles of these specialists continue to be discussed by experts; specifically, some claim that the study of human factors is more theoretical and relies more heavily on psychology, whereas ergonomics is more practical and relies more on anatomy and biomechanics. The trend, however, is for the two terms to be used synonymously (Kroemer, Kroemer, & Kroemer-Elbert, 1994).

Ergonomists, or human factors engineers, apply their engineering knowledge along with psychology and kinesiology (i.e., anatomy, biomechanics, and exercise physiology) to study the demands of various jobs and tools. These engineers then recommend certain changes to make in the movements used to perform the task, in the work environment, or in the training programs provided to workers. In many cases, workers' efficiency can be improved markedly through simple measures such as rearranging the location of objects to be assembled, adjusting the heights of workers' chairs or tables, redesigning the shapes of workstations, or modifying the order in which workers grasp components in the assembly process. Ergonomists are usually expert observers of the movements used in various kinds of work, and they base their recommendations for change on critical analysis of those movements. They often look for risks inherent in the work. In order to make such analyses, they require thorough knowledge of the structure of the body, as well as a complete understanding of how the body moves. A typical risk analysis of a work site is presented in the sidebar titled Typical Ergonomic Analysis of a Worksite.

Typical Ergonomic Analysis of a Worksite

Ergonomists often evaluate workplaces in order to improve safety conditions for workers. When doing a very general review of a worksite, an ergonomist might use a worksheet similar to the one presented here.

Job: General Maintenance Worker in a Manufacturing Company

Work includes cleaning and mopping (wet or dry) floors and stairways; wiping, vacuuming, scrubbing, and polishing surfaces; receiving and storing shipments; shoveling snow; making minor repairs of blinds; cleaning windows; and maintaining equipment.

Musculoskeletal Demands

Does the job require the following? (Indicate yes or no for each.)

- Awkward posture
- Bending or twisting at the waist to handle tools, loads, or parts
- Lifting or carrying loads beyond worker's capacity
- Ladder climbing
- Frequent, repetitive motions
- Flexing of the neck, shoulder, elbow, wrist, or finger joints
- Forceful or sudden movements
- Sustained contraction of muscles
- Excessive flexing or extending of joints

Working Environment

Can the workplace be described as follows? (Indicate yes or no for each.)

- Too hot or too cold
- Poorly or too brightly lit
- Noisy such that it distracts or produces hearing loss
- Crowded so as to prevent comfortable work positions or movements
- Dirty, slippery, cluttered, or obstructed

Manipulation of Tools

Can the tools required for the job be described as follows? (Indicate yes or no for each.)

- Unreasonably heavy
- Exposing worker to risk of cuts, punctures, or strains
- Difficult to operate
- Requiring severe flexing of the wrist, neck, or back
- Vibrating or jerking excessively

Productivity in the workplace suffers its greatest decline when workers are injured and miss work. This circumstance is especially troublesome when an injury results directly from performing the physical activity required to carry out the job. Many such injuries involve musculoskeletal disorders that arise from overexertion or repetitive motion. Women are particularly susceptible to such injuries and suffer an estimated 63 percent of all work-related repetitive motion injuries (National Institute for Occupational Safety and Health, 2000). Among all workers, the incidence rate increased 4 percent from 2009 to 2010 for musculoskeletal injuries due to overexertion that resulted in days away from work; the rate increased 10 percent, however, among nurses and orderlies, who often lift and transport patients (U.S. Bureau of Labor Statistics, 2011). Overexertion injuries result from application of maximal force by the body, usually when the worker is either physically incapable of such exertion (e.g., due to weak back muscles) or poorly prepared from a positional standpoint (e.g., lifts without bending the knees). Most of these injuries occur when a worker is lifting, pushing, pulling, holding, carrying, or turning an object (USDHHS, 1997). However, even sitting for long periods at a desk can lead to musculoskeletal injury.

Cumulative trauma disorders result from repeated physical stress in joints and muscles, which damages tendons, nerves, and skeletal structures requiring rehabilitative therapy. These disorders can afflict assembly line workers who package items and therefore perform the same arm, hand, finger, and trunk movements several hundred (or even several thousand) times per day. Even the small range of motion involved in typing can produce cumulative trauma disorder. Although our bodies are remarkably durable and adaptable, sometimes work asks too much of them, whereupon they suffer deterioration or collapse.

One of the most common types of cumulative trauma disorder is **carpal tunnel syndrome**. This painful wrist injury often occurs in workers who pack, type, drive, lift, hammer, manually assemble products, or simply sit at a desk for long periods of clerical work. One way in which ergonomic specialists try to prevent this disorder is to redesign tools and workspaces to facilitate safe and efficient operational movements.

Physical and Psychological Demands of Work

What types of work require the most demanding forms of physical activity? This question matters not only to employers but also to employees, labor unions, insurance companies, and ergonomists. For employers, rating the physical demands of a job poses an ongoing challenge. To help meet this challenge, the office of Health and Safety Ontario (2011) provides employers in Canada with a "physical demands analysis" that they can use to describe how often and in what ways workers must move, as well as the postures they must assume. Ergonomists often use such analyses to evaluate workplaces in order to determine the physical demands placed on workers.

From a total health perspective, ergonomists are also interested in the psychological stresses that a job places on workers. But the most *psychologically* stressful jobs? For one answer to this question, we can turn to the *Jobs Rated Almanac* (Kranz & Lee, 2015), which ranks jobs in terms of several variables, one of which is stress. These rankings are based on such considerations as whether the worker faces deadlines, competes with others, encounters physical hazards or uncomfortable environmental conditions, is expected to perform precise movements, or is responsible for the lives of others through his or her actions. The four jobs rated least stressful were tenured university professor, audiologist, medical records technician, and jeweler. The five jobs rated highest in stress were police officer, airline pilot, military general, firefighter, and military member.

All of these jobs require physical activity, and strenuous activity may be required in the case of military personnel and firefighters. Although managers, supervisors, and professional workers do not always experience the highest levels of stress, the types of employment most likely to raise stress levels tend to be those requiring low levels of physical activity. Individuals subjected to this double whammy of high stress and low physical activity may be especially vulnerable to heart attack, hypertension, and stroke unless they pursue exercise outside of work.

This concern has led many communities to impose fitness-test requirements for law enforcement officers and firefighters. In 2015, for example, the FBI instituted physical fitness tests for its 13,500 agents. The testing is designed primarily to ensure that agents can move quickly during a mass shooting, chase suspects, and restrain them if they resist arrest. An agent's test results become part of his or her permanent record (Schmidt, 2015).

Similarly, many major companies have provided discounts for employees who join commercial fitness centers, and some companies have even built facilities and hired staff to lead employee programs in physical activity, fitness, and wellness. According to some reports, every dollar invested in such programs can bring as much as $5.81 in return. Companies see reductions in sick leave, and participants claim that they work harder and are more productive in their jobs. Unfortunately, 34 percent of workers refuse to participate in such programs ("Increasing Employee Participation," 2013). Even so, employers who recognize the practical importance of such programs have maintained their vitality even during economic downturns.

KEY POINT

Industrialization and technological advances have led to the creation of jobs that impose higher levels of stress while involving lower levels of physical activity. In response, some businesses sponsor exercise and sport programs for employees.

Sphere of Education

Education plays a crucial role both in preserving cultural traditions and in providing the knowledge and skills that enable societies to progress. All phases of education involve physical activity—for example, the eye movements required in reading; the wrist, finger, and arm movements required in writing; and the more expansive forms of physical activity required in learning how to play a musical instrument or operate a power saw. In almost every form of educational program, the aim is to change the behavior of those being instructed; for our purposes here, that means changing their physical activity patterns.

KEY POINT

The education sphere includes that aspect of life in which we seek to learn new skills or knowledge; in this process, physical activity usually plays an important role.

Although we are most interested in educational programs in sport, exercise, and recreation, they form only the tip of the instructional iceberg. Look around and you will see physical activity instruction occurring everywhere! Physical therapists teach patients how to walk; faculty members at dental and medical schools teach students how to perform the intricate movements required in surgery; fathers and mothers teach their children how to dribble a soccer ball; ergonomists train employees in new assembly techniques; American Red Cross staff teach would-be lifeguards how to rescue

Kinesiology Colleagues

Photo courtesy of the University of North Carolina

Tom Martinek

Tom Martinek is a professor of kinesiology at the University of North Carolina at Greensboro. During his 40-year tenure, he has become one of the foremost scholars studying the social and psychological dynamics of teaching and youth development. For the past 22 years, he has directed Project Effort, which offers after-school sport and leadership programs for underserved youth. He has published four books and more than 100 articles and has been honored with the Salvation Army Boys and Girls Club Youth Development Award and the University Bullard Award for Service.

When asked to identify the five most important features of a high-quality physical education program, Martinek said, "It should be an inclusive program where the needs of all children are addressed. It should be holistic in nature so that social, psychological, and cognitive skills are developed along with physical skills. It must be recognized as an integral part of the school's education mission. Parents must be informed and involved. And, finally, teachers must be passionate about providing high-quality learning experiences for children."

swimmers; and senior automobile mechanics train new mechanics to use tools. Anywhere physical activity is important, you will find some form of instruction, whether formal or informal. Indeed, instruction in physical activity may be as universal as physical activity itself.

Instruction in Sport and Exercise

Not so long ago, instruction in sport and exercise was limited largely to public school and college physical education programs, municipal recreation programs, and the military. Today, however, the ongoing explosion of interest in learning about sport and exercise has led to an expansion of instruction beyond the walls of schools and colleges. Sport and fitness instruction is now offered at gyms, sport and fitness centers, resorts and commercial spas, hospitals, ice rinks, riding stables, dance studios, martial arts centers, and private clubs (e.g., tennis, golf, swimming). In addition, millions of children are taught how to play sports by their parents or older siblings and by the coaches of youth sport teams. Physical activity instruction is also provided by companies that specialize in working with young athletes, summer sport camps offered by famous coaches, and age-specific swimming and gymnastics schools (for more on these options, see chapter 15).

Even adults, who as a group used to shy away from sport instruction, now seek it out in considerable numbers, and they pay as much as $100 per hour for lessons with top specialists in activities such as skiing, scuba diving, golf, and fitness training. In addition, many large corporations now provide their employees with recreation programs, which often include basic instruction in exercise and various sports. Instruction offered through the media has also become a huge business. For example, the Golf Channel Academy offers instruction 24 hours a day, how-to books and videos are widely available for every conceivable sport, and sport-focused specialty publications include regular instructional features.

This interest in learning how to participate in sport and exercise is coupled with a growing demand for knowledge about how to live a healthful lifestyle. At the same time, corporate managers, keenly aware of the economic advantages of having a healthy workforce, are making instructional programs available at worksites, often in collaboration with their health maintenance organizations or health insurers. Typically, these programs use a variety of methods to integrate theoretical information about exercise, nutrition, and body mechanics with practical experiences in the exercise room.

Physical Education

Although instruction in sport, fitness, and exercise now takes place in myriad formal and informal settings, it is still most visible and accessible in the physical education classes offered as part of the school curriculum in most countries (see chapter 15). The U.S. educational system is based on the belief that a democracy thrives best when its citizens are provided with a free and accessible program of education. Moreover, instruction in sport, exercise, and fundamental movements has long been considered an integral component of the educational enterprise.

In this section, we give brief attention to the life experience that you probably knew as "gym class," which is referred to as *physical education* by physical activity professionals. Because such programs hold enormous potential to influence the physical activity patterns of large segments of the population, they have attracted the attention of all physical activity professionals, as well as public health officials. For this reason, you as a future physical activity professional should appreciate

When School Districts Ignore Laws Mandating Physical Education

A study of 5th graders in 55 California school districts between 2004 and 2006 found that half of the districts failed to satisfy a state law requiring at least 200 minutes of physical education for every 10 school days (Sanchez-Vaznaugh, O'Sullivan, & Egerter, 2013). Of the nearly 100,000 students covered in the study, more than 80 percent attended schools in districts that failed to meet the law; Latino and African American students were more likely to attend such schools than were white students. Moreover, in districts that did comply with the law, 64 percent of the students were classified as meeting fitness standards, whereas the figure was only 57 percent in noncompliant districts. In 2013, a lawsuit brought by a parent and an interest group called Cal200 resulted in a settlement with 37 school districts to enforce the state code regarding physical education in public schools.

the potential benefits and pitfalls of school-based physical activity programs. More information on the latest national standards for physical education is presented in chapter 14.

KEY POINT

Physical education is the only near-universal program of sport and exercise instruction available to young people. For this reason, it should be of the highest quality possible.

The material emphasized by physical education teachers in their classes is influenced by trends and forces in society. This influence makes sense because public education must be responsive to the needs of society. For example, in the early 20th century in the United States, two societal developments—the growing popularity of sport and the burgeoning interest among educators in the phenomenon of play as an educational experience—contributed to the move by schools to begin using sport in the physical education curriculum, replacing the formal systems of exercise and gymnastics that had been typical at the close of the 19th century. Today, in contrast, fitness has assumed a renewed importance, and many schools are including fitness activities in their programs.

Physical educators pursue many objectives but have traditionally given priority in the curriculum to exercise and sport. Why are these objectives important to public education? Let's briefly examine each one.

Teaching Physical Education for Physical Fitness

Promoting physical health through exercise has been a key objective of physical education from its inception. The extent to which this objective is emphasized in public schools has varied according to pressure exerted by society. Today, fitness ranks at the top of the agenda, primarily because public health officials are alarmed at the generally poor fitness and sedentary lifestyles of many children.

As you read this text, you will become painfully aware that overweight and obesity, taken together, have been declared by many health professionals as the number one health problem in the United States. How bad is it? The CDC (2015b) reports that the percentage of U.S. children aged 6 to 11 years who are classified as obese increased from 7 percent in 1980 to nearly 18 percent in 2012. Over the same period, the percentage of adolescents (aged 12 to 19 years) who were obese increased from 5 percent to nearly 21 percent. Overall, more than a third of children and adolescents were overweight or obese.

One major contributor to the increase in obesity among children and adolescents is the amount of time spent watching television, using a computer, or pursuing entertainment on a mobile device—all of which place little stress on the muscular and cardiorespiratory systems. This grim picture was highlighted by a survey study (Common Sense Census, 2015) of media use by children aged 8 to 18, which found that tweens (i.e., 8- to 12-year-olds) and teens spend an astonishing amount of time—6 and 9 hours per day, respectively—using media for enjoyment. The amounts do not include time spent using a computer to do schoolwork.

These patterns of sedentary behavior carry ominous implications for health. For example, a large international study of 300,000 children and adolescents found that watching 1 to 3 hours of television per day led to an increase of 10 percent to 27 percent in one's risk for obesity (Braithwaite et al., 2013). Although documented evidence indicates a decrease in TV watching among young people over the past decade (Michigan Medicine, 2010), the explosion of media choices seems simply to have diverted their attention to other equally sedentary activities rather than motivating them to go outside and play.

One might think that the dangers posed by this state of affairs would be recognized by public school systems and that schools would therefore increase the number of required hours devoted to physical education. After all, as Lawrence Locke (1996), professor emeritus at the University of Massachusetts, has pointed out,

> School physical education is an experience through which almost all of our population passes, at least in some form, and it is the single place where most of them encounter formal sport and exercise fitness activities before adulthood. Those encounters also take place at precisely the time when lifestyles are being examined and adopted, and when possibilities for the self are being accepted or rejected. (p. 429)

The logic of Professor Locke's argument is impeccable, yet the evidence suggests that many school boards have yet to commit seriously to sponsoring regular physical education for students. For example, only half of adolescents in the United States attend physical education class at least once a week, and the rate decreases across the high school years.

Only 30 percent of high school students report participating in physical education on five days a week. Between 43 percent and 47 percent of schools require physical education from kindergarten to the fifth grade; the percentages decrease from that point on so that between the 11th and 12th grades, only 9 percent of schools require participation in physical education (United States Report Card on Physical Activity for Children and Youth, 2016).

Moreover, access to physical education class is related to socioeconomic status (SES). As of 2007, only 49 percent of low-SES students went to schools where physical education was required, as compared with 59 percent of high-SES students. This disparity is even more concerning in light of the fact that obesity disproportionately affects students of lower SES. Contrary to popular opinion, varsity sports do not make up for the relatively low participation rates in physical education; only 33 percent of girls and 37 percent of boys participate in varsity programs (Johnson, Delva, & O'Malley, 2007).

Recognizing the potential of childhood obesity to affect state expenditures on health care, some state legislatures have mandated daily physical education for all students, thus reversing the trend of the past few decades. Still, in order for physical education classes to yield their intended health benefits, some of them will have to be conducted in different ways. For one thing, some physical education classes include very little physical activity at the moderate or vigorous level. For example, a study of California public schools (California School Boards Association, 2009) found that students were sedentary most of the time during physical education classes; indeed, only 4 minutes per half hour were spent in what scientists would classify as vigorous activity. Students enrolled in larger classes were appreciably less active than students in classes with fewer than 45 students. Thus, it appears that mandating physical education in schools does not in itself produce the intended effects. Clearly, more attention needs to be given to increasing both the amount and the level of physical activity in which children engage during class.

Even so, some encouraging signs indicate that the United States and other countries are beginning to take the childhood obesity problem seriously. The CDC called attention to the sedentary ways of the U.S. population in 1996 with the publication of *Physical Activity and Health: A Report of the Surgeon General* (USDHHS, 1996). This document had been preceded by the USDHHS (1990) publication *Healthy People 2000*, which formulated health goals for the year 2000. After seeing only limited success in achieving these goals, the USDHHS modified and restated them in *Healthy People 2010* (USDHHS, 2000); it has since published *Healthy People 2020* (USDHHS, 2012), which sets even higher expectations. All of these reports give a great deal of attention to the problem of physical inactivity among young people and recommend increasing the percentage of schools that require physical education.

Television, the Internet, and Gaming: Formidable Challenges to Physical Educators

The uphill battle faced by physical education teachers is underscored by a study tracking the physical activity level, body mass index, and television viewing habits of nearly 12,000 Canadian young people aged 5 to 19 (Tudor-Locke, Craig, Cameron, & Griffiths, 2011). The study found that 17 percent of the individuals were overweight and 11 percent were obese. After adjusting for each child's age, sex, and parental education level, the investigators found that the odds of a child's being obese decreased by 20 percent for every 3,000 steps walked or run per day but increased by 21 percent for every 30 minutes of television watched per day.

The researchers drew the following conclusions: "Television viewing is the more prominent factor in terms of predicting overweight, and it contributes to obesity, but steps/day attenuates the association between television viewing and obesity and therefore can be considered protective against obesity" (p. 66). However, many children and adults now binge-watch television shows streamed via the Internet and play numerous hours of computer games. These and other sedentary activities reduce their opportunities for healthy physical activity.

Teaching Physical Education for Sport Skill Development

Developing physically fit students is an important objective, but most teachers also reach for objectives beyond fitness. Some of these objectives can be achieved through sport education—for example, developing responsible personal and social behavior and providing opportunities for enjoyment, challenge, and self-expression. Of course, sport can also serve as a way to attain and maintain physical fitness. However, if students do not learn to incorporate moderate to intense forms of physical activity into their lives, then being fit while they are in school will achieve little. What good will we have done if we create a population of fit young people who slip into sedentary living once they graduate?

Part of the answer is to help students develop competence in lifetime sport skills. People who have confidence in their ability to engage in lifetime sports—such as swimming, squash, handball, in-line skating, and golf—are more likely to engage in these activities throughout their lives. However, teachers have little chance to develop skill competencies in their students unless physical education classes are conducted daily. Therefore, high school physical education programs should include a heavy concentration of these lifetime activities rather than focusing on team sports, which people are less likely to play after graduation. Yet, despite the fact that health and education leaders have emphasized the importance of lifetime sports, available data suggest that the physical education curriculum in U.S. public high schools continues to be dominated by team sports—particularly basketball, baseball, volleyball, and soccer. In contrast, lifetime activities such as swimming, jogging, tennis, racquetball, and hiking are much less likely to be taught.

For example, a report (CDC, 2014) profiling physical education practices at selected sites across the United States in 2012 revealed that most programs still emphasize team sports. Specifically, only 13 percent of the schools included backpacking or hiking in the curriculum, 13 percent wall climbing, 64 percent tennis, and 10 percent swimming. In contrast, 97 percent of the classes included basketball, 94 percent football, 92 percent soccer, 98 percent volleyball, and 86 percent kickball—all of which involve skills that are unlikely to be carried over into adulthood. In addition, the typical mode of instruction in these team sports means that students usually spend a great deal of time being physically inactive. The same survey showed that teachers report devoting 75 percent or more of class time to physical activity in only 58 percent of physical education classes.

Make no mistake about it: What is taught in physical education classes can greatly affect the activities that people choose to engage in after graduation. Data from the Physical Activity Council (2016) reveals that people 18 years and older who had physical education in high school are two to three times more likely to be involved in outdoor activities such as cycling, running, jogging, and team sports—even though some of these activities (i.e., outdoor activities and cycling) are given scant attention in physical education programs. Of those who had physical education in school, 80 percent reported being more active than did students not attending physical education, and 39 percent reported being active more than 150 times per year. On the other hand, only 61 percent of those who did *not* attend physical education in high school reported being active, and only about 25 percent reported being active more than 150 times per year. Clearly, then, much work needs to be done if physical education classes are ever to realize their potential for increasing the levels of physical activity of the general population.

KEY POINT

One way to counteract the epidemic of obesity is to increase the time allotted to physical education programs in school. However, simply requiring physical education may not reduce being overweight or increase physical activity among youth.

Sphere of Leisure

Physical activity also plays a prominent role in leisure, which is not the same thing as free time (discussed later), though most laypeople think of it that way. What do you do in your free time—play sports, exercise, read, attend concerts, watch television, hike? The term *leisure pursuits* is often used synonymously with *play* or *recreation*. However, leisure is more a state of mind than simply free time. Sometimes we choose to work in our free time. Have you ever given up an opportunity to spend a summer weekend (free time) camping or at the beach in order to work an extra shift at your job? If so, was your work time "free time"?

As you can see, the terms *leisure* and *free time* are a bit more ambiguous than they first appear. Can

Any activity that allows us to feel at peace with the world and brings us a sense of contentment can be labeled a leisure activity.

you really work in your leisure time? You may have noticed that some people play softball or soccer with an intensity that seems to be anything but leisurely. Conversely, can people play (i.e., be at leisure) while they work? Perhaps. For example, artists, novelists, and professors sometimes seem to approach their work almost as if it were a leisure experience. Thus, a good place to start in our examination of the leisure sphere of physical activity experience is to clarify what we mean by *free time* and *leisure*.

Distinguishing Leisure From Free Time

When we disengage from our everyday lives and carve out minutes and hours to do what we want to do, we often refer to this period as **free time**. At such times, we are free from our everyday routines and free to do what we want to do—when, where, and with whom we choose.

KEY POINT

Leisure is a state of being, and free-time activities can help us attain this state. For example, large-muscle physical activities (e.g., sport, exercise) can help us nourish and maintain a leisure disposition.

In contrast, theorists describe **leisure** not as free time but as "a state of being" (De Grazia, 1962). Philosophers have struggled to describe this state at least as far back as the ancient Greeks, but none have done so with complete success. In its purest state, leisure is a feeling that all is well with the world—a feeling of supreme contentment. It can provoke celebration and wonder, creativity and discovery, excitement and reflection. Pieper (1952) described leisure as the "basis of culture" and as the psychological and spiritual disposition that is the fundamental prerequisite for the great works of art, music, theater, and philosophy.

How do we achieve this state of being? Sometimes we attain it while engaged in sedentary activities, such as sitting beside a gurgling stream, reading a novel, or listening to music. At other times, we achieve it through large-muscle activities, such as running, hiking, surfing, or skiing. Precisely what this state of being is or how it comes about may never be explained fully, but one thing seems clear: Leisure clarifies, for ourselves and others, who we truly are when we are free from the sense of obligation, anxiety, and pressure that can confront us in our everyday lives. When we experience leisure, we look forward to experiencing it again and again.

In order to use our free time to achieve a state of leisure, we must divorce ourselves psychologically from other aspects of life, which is a challenging task for many people, especially when smartphones keep them constantly connected with work. Moreover, we must choose an activity that is conducive to a state of leisure; for instance, though many of us choose to watch television during our free time, a study of teenagers showed that they do not find it to be a satisfying experience (Csikszentmihalyi, 1990b). This and other hollow free-time experiences that merely fill our idle hours are not leisure experiences. They do not challenge us, stimulate our imaginations, or help us reveal our identities in the way that true leisure experiences do. Thus, if we are to convert free time into leisure time, we must use it for participating in activities that tend to nourish and maintain the state of being known as leisure.

Physical Activity as Leisure Activity

The entire range of leisure pursuits—from sedentary activities such as chess and reading to large-muscle activities such as water skiing and softball—constitutes the subject of study in the discipline known as **leisure studies** or **recreation**. Kinesiology, in turn, concerns itself primarily with the large-muscle forms of leisure pursuits that require substantial physical activity. Thus, the interests of kinesiologists overlap with those of leisure professionals. Faculty and students in departments of leisure studies and departments of kinesiology both view sport and exercise as legitimate parts of their fields. Although this blurring of boundaries might seem problematic, it has existed for several decades and has not deterred the continuing expansion and refinement of either of these exciting areas of study.

In fact, in the early history of higher education, physical education and recreation were typically housed in the same department. While some schools continue to combine the two majors in one department, it is more common, especially in larger universities, for each to be housed in its own organizational entity. Regardless of administrative structure, the two fields share many common interests. For this reason, physical activity experts from kinesiology often collaborate with recreation professionals in the National Intramural-Recreational Sports Association (NIRSA) and the National Recreation and Park Association (NRPA).

Physical activities such as golf, folk dancing, softball, hiking, boating, and long runs in the park hold great potential for putting us at leisure. Yet we can also approach these activities in ways that make a leisurely disposition impossible. For example, a leisure state is incompatible with actions such as cheating, bellyaching about an official's call, intentionally injuring an opponent, and exercising under the compulsive stimulus of anorexia nervosa. The context in which the activity takes place also matters. For instance, is it possible for an athlete to attain a state of leisure in a stadium where 80,000 spectators are sending up a deafening roar?

In the final analysis, whether a given sport or form of exercise constitutes a leisure pursuit depends on its nature, on the context in which it is pursued, and on the motivations and attitudes that the participant brings to the activity. In and of themselves, the free-time activities of sport and exercise offer us only the *potential* for achieving the state of leisure. The challenge of the physical activity professions is to teach people to participate in free-time physical activity in ways that nourish the disposition known as leisure.

Because your future as a physical activity professional will likely be affected by participation trends in large-muscle **leisure activity**, it is worth taking the time to examine trends of participation in this important sphere. Remember, kinesiologists are particularly interested in leisure-time pursuits that involve relatively large amounts of physical activity. Here, the data are somewhat encouraging. In one study, for example, adults were asked the following question: "During the past month, other than your regular job, did you participate in any physical activities or exercises such as running, calisthenics, golf, gardening, or walking for exercise?" The proportion who responded no declined from 31 percent in 1989 to 25 percent in 2002 and remained at that level until the end of the study in 2010 (CDC, 2010a).

As you might expect, aging tends to suppress physical activity behaviors. According to the Physical Activity Council (2016), only 18 percent of people in Generation Z (those born in or after 2000) report being physically *inactive* in their leisure time, but the figure is 24 percent for Generation Y (born in 1980 to 1999), 26 percent for Generation X (born in 1965 to 1979), and a whopping 39 percent for baby boomers (born in 1945 to 1964). This pattern makes it wise to be mindful that as people age and take on the responsibilities of job and family, they tend to become more sedentary. Lay the foundation now for a lifetime of activity in order to avoid becoming a couch potato in your golden years!

As older people increasingly recognize the benefits of leading a healthy lifestyle throughout their lives, we will see larger numbers of older adults seeking physically active leisure pursuits. Working with this population provides an exciting professional opportunity for students who major in kinesiology.

What kinds of physical activity do people perform in their leisure time? For the moment, think specifically about noncompetitive, large-muscle forms of leisure activity. (Though competitive sport clearly fits under the umbrella of leisure activity, we look at its popularity a bit later, in our discussion of the competition sphere.) Some leisure pursuits do not involve increased physical activity—for example, watching television or using a mobile digital device—but many leisure activities do involve getting physical. In this regard, the Physical Activity Council (2016) reported moderately encouraging results from a study of participation rates for individuals aged 6 years or older, especially in terms of participation in fitness activities; for the details, see figure 2.1.

Sphere of Health

Attending to personal and community health needs can consume a large part of our daily experience. For example, we dispose of our garbage, ensure that we drink clean water and eat proper foods, and visit doctors when we are ill—all of which enables us to be productive citizens and enjoy life. To varying degrees, these and other health-related endeavors require physical activity. However, physical activity also intersects with this sphere in a much more direct way. We now possess scientific evidence showing that when physical activity is performed in the right amount and with sufficient frequency, it contributes to our health in many important ways.

We know, for example, that moderate to vigorous physical activity performed regularly and at safe levels almost always results in health benefits. We also know that the payoffs do not necessarily depend on performing painful exercise routines or long runs. In fact, it has been known for decades that most health and longevity benefits can be obtained through moderate and intermittent physical activity—for example, through sport participation and physical activity performed as part of work. It is equally clear that the safest, most effective, and most efficient way to attain health benefits from physical activity is to participate in carefully designed programs supervised by kinesiology professionals who are well versed in the science of physical activity. As this fact continues to become better understood by the general public, the demand for highly trained fitness leaders and consultants will continue to increase.

Physical Activity, Health, and the Public Interest

Why is it important for individuals to maintain and improve their health through physical activity? For one thing, healthy people are vital to the national and international economy. In 2014, U.S. health care spending increased by more than 5 percent (on top of 3 percent growth in 2013) to reach a total of $3 trillion, or nearly $10,000 per person; the total cost is expected to exceed $4.5 trillion by 2020 (Centers for Medicare and Medicaid Services, 2014).

A sick population not only incurs medical expenses but also drains the economy through lost productivity. For example, in the European Union,

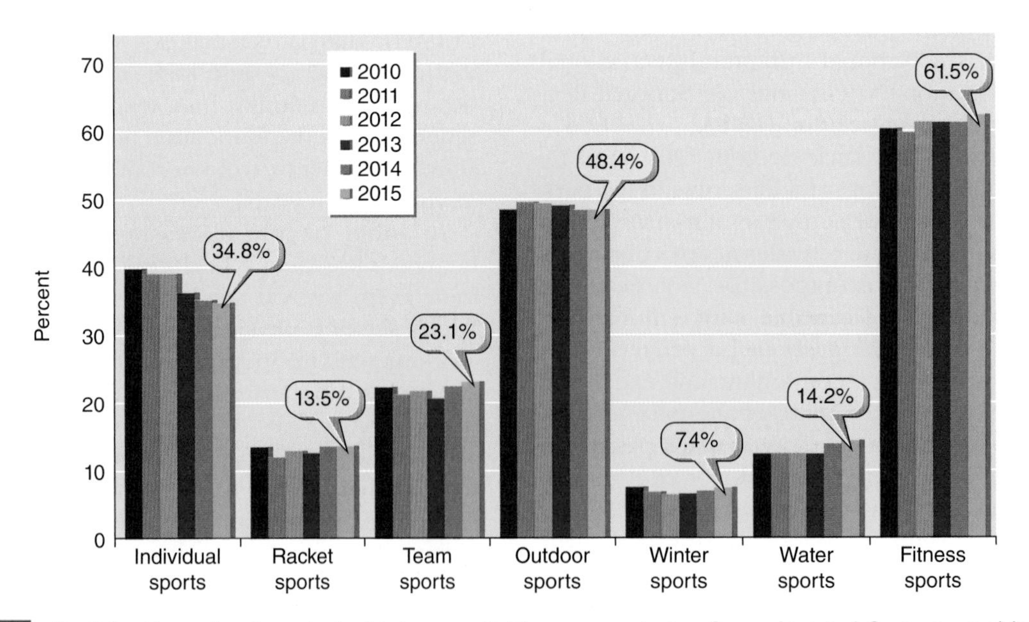

FIGURE 2.1 Participation rates in selected leisure activities; percentage of people aged 6 years or older (winter sports based on season).

Reprinted, by permission, from Physical Activity Council. Available: http://www.physicalactivitycouncil.com/pdfs/current.pdf, page 7

National Physical Activity Plan

In 2006, a group of U.S. agencies and nongovernmental organizations developed what is called the National Physical Activity Plan. Its purpose is to advocate for a set of policies, programs, and initiatives to increase physical activity in all segments of the population by coordinating efforts from business and industry; the health care sector; the education world; the mass media; parks, recreation, fitness, and sport leaders; public health leaders; the transportation sector; experts in land use and community design; and volunteer and nonprofit agencies. More than 300 experts from these sectors worked with the coordinating committee to draft an initial plan, which was created as a "living document" to be regularly updated. The current version of the plan can be downloaded from the National Physical Activity Plan website.

where 60 percent of people above age 15 either rarely or never exercise or play a sport, and a high proportion sit for more than 4 hours per day, the World Health Organization (2016) has estimated that eliminating physical inactivity would help the economy, increase life expectancy, and reduce greenhouse gas emissions by emphasizing walking and biking.

As a taxpayer, you have a large stake in this issue. As health care costs increase in the private sector, the amounts spent on health care costs by state and federal governments also increase. These expenditures put a substantial drain on state and federal budgets, thereby reducing the funding available for education and other social programs. The massive spending also poses a potentially significant problem for future generations who will have to foot the bill.

None of this, of course, measures the personal effects of poor health. Each incidence of illness brought on by lack of physical activity represents one person who has lost his or her potential for living an active, productive, enjoyable life. In the United States, three such illnesses are heart disease, stroke, and high blood pressure. The leading cause of death in the United States is heart disease, which kills more than 375,000 people per year; among women, it accounts for more deaths per year than all forms of cancer combined. Another 140,000 Americans are killed each year by stroke; the risk of having a stroke more than doubles with each decade after age 55. High blood pressure was a primary or contributing cause in the deaths of 410,000 Americans in 2014 (CDC, 2016). As a future kinesiologist, you should know that early morbidity (illness) and mortality can be mitigated by regular exercise and healthy eating.

Many of these diseases relate to obesity, which is a problem not only of childhood but also of adulthood. In turn, one major contributor to obesity is sedentary living. The U.S. Bureau of Labor Statistics (2012) estimated that, in 2011, people older than age 15 watched television for nearly 3 hours per day, which constituted about half of their total leisure time. In addition, this amount of time was more than three times the amount spent on the next most popular leisure activity—socializing with friends and family. According to the CDC (2010b), 34 percent of the U.S. population older than 20 can be classified as overweight, another 34 percent as obese, and more than 5 percent as extremely obese.

There is a strong association between obesity and socioeconomic status. People who are poor and uneducated are more likely to be obese than are those with higher levels of income and education. People with lower levels of income and education are also less likely to engage in physical activity or eat a well-balanced, nutritious diet.

Physical activity figures so prominently in U.S. health objectives because it has been shown to prevent and help rehabilitate these costly and deadly diseases. Specifically, performing regular physical activity at moderate to high levels of intensity serves as a front line of defense against not only obesity but also heart disease, stroke, and diabetes (which is strongly linked to obesity). For example, in an exhaustive review of 28 studies, researchers found that higher levels of leisure-time physical activity were associated with lower levels of type 2 diabetes (Smith, Crippa, Woodcock, & Brage, 2016). In addition, as noted earlier in this chapter, regular physical activity may enable elderly persons to live independently for longer and avoid falls and other injuries. More generally, regular exercise lowers the risk of mortality for both older and younger adults.

It is not unreasonable to view the increase in obesity and other hypokinetic diseases in the United States and other nations as an epidemic. As a result, the need to get people moving is now a top priority among health care providers. In the

United States, one of the goals of *Healthy People 2020* (USDHHS, 2012) is to reduce the proportion of the population that engages in no leisure-time physical activity to 33 percent (from the current 36 percent), increase adult participation in regular physical activity of at least moderate intensity (for at least 150 minutes per week) to 48 percent (from 44 percent), and increase the proportion of adults who engage in muscle-strengthening exercises on 2 or more days per week to 24 percent (from the current 22 percent).

A similar commitment to increase physical activity levels was seen in the Vienna Declaration issued by the World Health Organization in the European Region at a conference in 2013. The group set the following as one of its goals: "scaling up healthy eating and physical activity schemes in people-centered primary health care and ensuring an appropriate continuum of nutrition and physical activity ranging from health promotion to prevention and care throughout the life-course" (p. 3). Clearly, the profession of kinesiology is well positioned to make a significant contribution to the attainment of national and international health goals.

The Darker Side of Physical Activity

At this point, it may be helpful to do a reality check. Obviously, physical activity is a valuable adjunct to healthy living; it does not, however, provide unqualified benefit. The fact that some physical activity is good does not necessarily mean that more is always better; to the contrary, some research has shown that excessive amounts of exercise may be damaging to one's health. For example, long-term endurance athletes often suffer from diminished function of the right ventricle; moreover, immediately following a race, some have evidenced increases in cardiac enzymes that may play a role in clotting. Vigorous physical activity usually comes at some risk of injury. Figure 2.2 illustrates the numbers of acute injuries in young athletes worldwide.

Overexercise can also occur in other types of activity, including weight training, aerobics, and repetitive motions in work settings. It can result in stress fractures, strained muscles, inflamed tendons, and psychological staleness. Sometimes the damage brought on by long-term wear and tear, although more insidious, can be more traumatizing. Repeated injuries to the knees, fingers, wrists, hips, and backs can take their toll on athletes, not only in the weeks and months following an injury but also in later years. For instance, years of pitching a baseball can lead to rotator cuff injuries, and stress fractures are common among gymnasts. The most hazardous activities, however, are collision sports in which players deliberately or accidentally make contact with each other, such as American football.

Of course, football is not the only sport that produces injuries; in fact, most sports involve some degree of risk. According to Safe Kids Worldwide, 40 children died from playground accidents between 2001 and 2008. More than two million children aged 19 and under visit emergency rooms in hospitals each year for treatment of recreation and sport-related injuries. In 2013, more than 700,000 of these were for injuries sustained playing football and basketball (Safe Kids Worldwide, 2015).

Physical activity can also become problematic when the urge to engage in it is compulsive or when a person develops an unhealthy emotional dependence on exercise. Individuals with a physical activity addiction focus so single-mindedly on

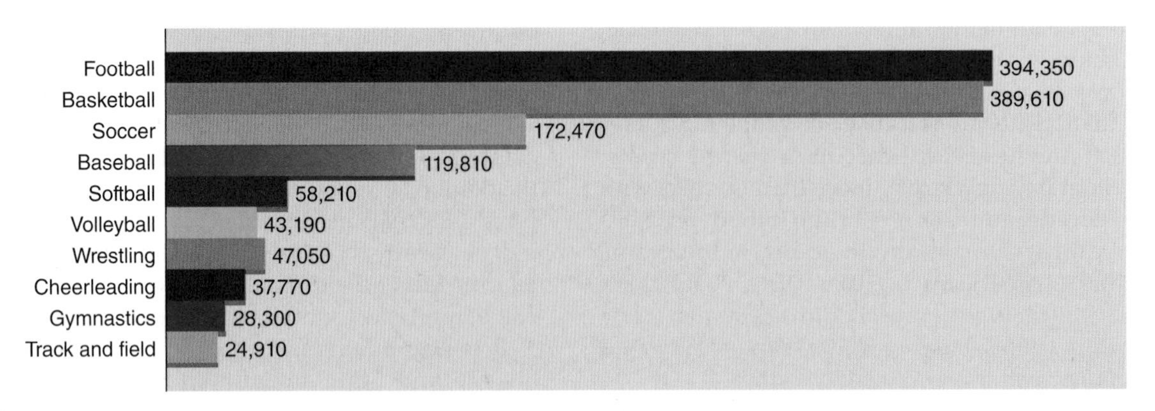

FIGURE 2.2 Number of injuries among athletes aged 19 or under (estimates based on reports from hospital emergency rooms worldwide).

Data from Safe Kids Worldwide 2015.

Injuries Plague Football Players

Few sports are more popular, more dangerous, or more likely to capture the imagination of North Americans than American football. Indeed, this sport illustrates the fact that it is sometimes difficult to objectively examine those aspects of our culture with which we are most familiar. According to the National Center for Catastrophic Sport Injury Research (Kucera, et al., 2016), American football is associated with the highest number of injuries—and the most catastrophic injuries—of any sport. In 2015, football was directly responsible for seven fatalities, all of them among high school athletes. Although some observers may consider seven fatalities out of more than a million participants to be a tolerable number, such statistics provide little consolation for the families and friends of the athletes who were fatally injured.

Moreover, recent media accounts have highlighted the terrible, lifelong damage incurred by professional football players. Sport pages are filled with first-person accounts provided by former football players in their late 30s who are unable to walk without assistance and are saddled with elbows that don't work or fingers that have been broken so many times that they can no longer hold a cup of coffee. Perhaps the greatest concern, however, involves the scourge of concussions suffered by football players. The disorienting effects of a hard collision—often jokingly described by sportscasters as "having your bell rung"—occur when the brain collides with the interior of the skull, thus injuring vital nerve tissue. The effects of repeated head trauma (often glorified by players as "head banging") are cumulative, and individuals who have spent many years playing the game sometimes suffer permanent disability—for example, dementia pugilistica, a brain disorder more commonly associated with boxers. In addition, postmortems have revealed that 87 of 91 deceased NFL players whose bodies were examined had chronic traumatic encephalopathy (CTE), a disease associated with repeated traumatic blows to the head.

The problem is not limited to the professional ranks. According to the National Collegiate Athletic Association (NCAA, 2010), concussions account for more than 7 percent of injuries to college football players. In addition, increasing attention is being paid to high school football, where, because the athletes' brains are still developing, catastrophic head injuries can have even more serious and long-lasting effects. More generally, every 3 minutes, a child is seen in the emergency room for a sport-related concussion (Safe Kids Worldwide, 2015), and the long-term effects of repeated "minor" head injuries are difficult to assess because the symptoms may not appear for years.

1. Is American football a body-threatening sport? Is it a violent sport?
2. If so, what factors contribute to the violence?
3. What recommendations would you make to reduce the injury rate in football?
4. What implications do these issues carry for kinesiology majors who plan a career in athletic training or sports medicine? In coaching? Sport management? Sport psychology?

an activity that they neglect other daily obligations that are important for a healthy and well-balanced life. When does devotion to an activity, such as running or golf, become addictive and harmful? No clear boundaries separate healthy from unhealthy exercise, but concern may arise when physical activity interferes with work or harms important personal relationships. Another signal is the onset of an emotional disorder when the opportunity to engage in the chosen activity is taken away. Many avid runners, for example, experience acute bouts of depression when they are injured and unable to run. High school wrestlers who must "make weight" in order to compete within a specific weight category often engage in harmful practices. Similarly, young women interested in gymnastics or figure skating can become preoccupied with weight loss and continue to exercise even when their body mass is already minimal. In summary, as with most other activities in our daily lives, the best prescription for a happy, healthful, and balanced life is to pursue physical activity with a sense of dedication and commitment—but in moderation.

KEY POINT

When pursued in moderation, physical activity is desirable and health promoting. It may be undesirable, however, when it puts the integrity of the body at risk or induces questionable behavior patterns and psychological states.

The response of kinesiology professionals to the problem of injury during physical activity has been vigorous and long-standing. Kinesiology departments have cooperated in establishing excellent programs of athletic training and sports medicine and in overseeing the tremendous growth of the National Athletic Trainers' Association. As you will see in chapter 13, the work of preventing and treating athletic and exercise injuries constitutes one of the fastest-growing specializations in the field of kinesiology.

Sphere of Competition

Do you like to compete against others? Is your interest piqued by the opportunity to compete with classmates for the highest grade, with co-workers for the employee-of-the-month award, or with other athletes for a league championship? It is likely that competition plays an important role in some part of your life. In fact, it is difficult for anyone to avoid competition completely, and the chances are good that physical activity figures prominently in most forms of competition that you pursue.

Competition itself is not an activity but an organizing principle for activity. It can either add to or detract from the enjoyment of activity, but it usually increases one's level of performance. For example, skipping stones on a pond can be an exciting experience for children; when they decide to see who can skip a stone the farthest, the activity becomes even more exciting. As they become more intent in their efforts to outskip their playmates, their stone-skipping performances may also improve. Similarly, some people thoroughly enjoy shooting baskets alone on the playground or playing golf by themselves or in leisurely social contexts with friends, whereas others seek the peculiar kind of excitement that comes only from competition.

Nature of Competition

The word *competition* is derived from a Latin word meaning "to strive together," but most of us think of it as striving against. We compete *against* opponents. However, if you think about it, competition between teams or individuals also requires an element of cooperation, or striving *together*. In fact, cooperation is necessary to hold the game together. Competition at its best has been defined by sport philosopher Robert Simon (2004) as "a mutual quest for excellence in the intelligent and directed use of athletic skills in the face of challenge" (p. 24). Opponents who play their best help us to play our best. In contrast, when players become spoilsports, give up, or cheat, the game falls apart. Do you think sport in the United States has lost this respect and appreciation for the opponent? Explain your answer.

Professional Issues in Kinesiology

Equality: Are Women Athletic Trainers Getting a Fair Shake?

When the National Athletic Trainers' Association (NATA) was founded in 1950, its entire membership consisted of men. This state of affairs persisted for many years until the landmark Title IX legislation began to make a difference and girls' and women's sports began to receive more support. Still, by 1994, although NATA had more than 3,000 members, only a relative few were women (less than 1 percent). Women's membership did grow, however, and in 2000 NATA elected its first woman president. Nevertheless, only 25 percent of women hold positions on athletic training staffs in Division I schools and only 16 percent hold positions as head athletic trainers (Martin, 2013). What do you think is responsible for this inequity, and what steps might be taken to correct it?

Citation style: APA

Martin, L. (2013). The role of women in athletic training: a review of the literature *Skyline 1*(1), 13. Retrieved from http://skyline.bigskyconf.com/cgi/viewcontent.cgi?article=1012&context=journal.

Identifying Good and Bad Competition

When competition goes well, sport is a beautiful thing. When competition goes awry, sport can be ugly. On the basis of your experience in playing, watching, or reading about sport, describe how each of the following factors contributes to or detracts from the potential beauty of sport.

- Players' conduct after losing a game
- Players' conduct after winning a game
- Players' conduct before a game
- Coaches' conduct before, during, and after a game
- How a coach goes about recruiting a star athlete
- The crowd's reaction to an obviously incorrect call by a game official
- The crowd's reaction when the home team wins the World Series or Super Bowl
- A parent's reaction to his or her child's failures in Little League baseball
- A university's policy for awarding scholarships to students

Competition is involved in many aspects of modern culture, but when most people think of competition they think of physical activity in general and of sport in particular. More specifically, they think of physical activities in which a premium is placed on skill, strength, endurance, or other physical qualities and in which one's performance is assessed in relation to either a standard or the performance of others. Unlike physical activity in the other spheres of experience, most sports depend on the organizing principle of competition. Remove competition from most sports and you are left with aimless physical activity. It is for this reason that we have designated a separate sphere for competition.

Competition can be added to physical activity in various ways. In **side-by-side competitive activities** (e.g., golf, swimming, running, bowling, shot put, cycling), competitors do not interact directly and do not interfere with each other's chances of succeeding. In fact, doing so is against the rules. In contrast, in **face-to-face noncontact activities** (e.g., volleyball, baseball, tennis, racquetball), competitors do interact, trying to maximize their own chances and decreasing their opponents' chances of winning. In this form of competition, players deliberately attempt to thwart the efforts of their opponents and take advantage of their competitive weaknesses, but the rules prohibit direct contact between opposing players. In **face-to-face contact activities** (e.g., football, rugby, hockey, wrestling, and, increasingly, soccer and basketball), players try to thwart the efforts of opponents by means of direct physical manipulation, such as blocks, tackles, and checks. Finally, in **impersonal competition**, participants compete against records set by themselves or others—for example, trying to set a new record time for swimming the English Channel or trying to climb a mountain by a route never before taken. In such activities, one's competitors (e.g., holders of the current record) are not present; in fact, they may not even be alive!

Problems arise with competition when it becomes more than an organizing principle that increases our enjoyment or pushes us to achieve better performances. Often, competition becomes an end in itself. When coaches and players elevate the goal of winning a contest above the sheer enjoyment of the activity or the basic values of friendship, caring, and cooperation, they convert competition into an end in itself rather than a means to a greater end. Mihaly Csikszentmihalyi, a psychologist who has taught us much about sport, said, "The challenges of competition can be stimulating and enjoyable. However, when beating the opponent takes precedence in the mind over performing as well as possible, enjoyment tends to disappear. Competition is enjoyable only when it is a means to perfect one's skills; when it becomes an end in itself, it ceases to be fun" (Csikszentmihalyi, 1990a, p. 46).

Winning Takes a Backseat to Good Sporting Behavior

In 2008, two women's softball teams in NCAA Division II met in a do-or-die contest for both teams. Central Washington and Western Oregon held first and second place, respectively, in their conference. When 5-foot-2-inch Sara Tucholsky came to the plate for Oregon, the game was scoreless with two runners on base. Given her diminutive stature, observers expected that, at best, she might single down the line or perhaps hit a sacrifice fly. After all, she had gotten only 4 hits in 34 at-bats during the entire season. However, Tucholsky found the pitch she wanted and hit a long ball over the fence. As she was rounding first base, she missed the bag and quickly doubled back to touch it—a turning maneuver that tore the ACL in her right knee. As she lay crumpled on the field, her coaches discussed their options. Under the rules, a home run hitter must touch each base and cannot be touched by teammates, trainers, or coaches. The decision was made to bring in a substitute who would be required to take first base, which is permitted under the rules (the two runs that Tucholsky had batted in would still be counted).

That's when Central Washington veteran Mallory Holtman approached the Oregon coach to ask if she and teammate Liz Wallace could carry Tucholsky around the bases. And that's what they did, stopping at each bag so that she could touch it with her left foot. Fans in the stadium gave the Washington players a standing ovation, and media around the country pointed to their action as an example of what we need to see more of in sport. Holtman seemed surprised by the fuss and told reporters that she only did what anyone else would have done in her position.

1. Would you have done what Holtman and Wallace did? Why, or why not?
2. Would your decision depend on whether your team was ahead at the time? Why, or why not?
3. Would your decision depend on whether the game would decide the conference championship? Why, or why not?

Effects of Competition

Competition is an ingredient that we add to play in order to make it more enjoyable, just as poetry is added to words, and music is added to sound, in order to increase our enjoyment of them. Shooting a ball at a hoop in the gym is fun, but shooting in the context of a game makes it even more fun because it enables you to test your skills against others. Competition also creates the drama that plays out in major sport arenas. If winning or losing wasn't at stake, few people would watch. Another reason often heard for competition is that it improves performance. The idea is that in trying to prove yourself better than your competitors, you achieve higher levels of performance than you would otherwise. However, is this always the case?

Some critics, dating back at least as far as anthropologist Margaret Mead in the 1930s, have said that competition is problematic. For example, Kohn (1992) claims that competition unnecessarily increases anxiety and reduces motivation among those who think they have no chance of winning. Other critics say that competition fosters aggression toward others. They are especially critical of zero-sum competitions, in which winning by one individual or team comes at the expense of losing by another. As for the notion that competition improves performance, these critics point to an exhaustive analysis of 75 studies regarding the effect of competition on the performance of motor skills (Stanne, Johnson, & Johnson, 2009). This investigation found that performance of motor tasks improved when participants cooperated more than they competed; the investigators did note that the strength of the effects depended on the type of competition used.

Perhaps the best approach for coaches and others whose work centers on competition is to realize that it can be a double-edged sword. On one hand, it can be fun and can help players develop both physically and emotionally. At the same time, whenever teams or individuals vie for superiority, the activity involves delicate interpersonal interaction. Therefore, care must be taken to keep the contest in proper perspective.

If you were the coach of a Little League baseball team, what rules would you enforce to help ensure that competition improved performance, was socially rewarding, and promoted positive character development? What rules would you enforce for parents? How would you ensure that competition did not tempt players to cheat? For example, would you tell players to inform the umpire if, when rounding second base, they failed to touch the bag? Why, or why not?

Wrap-Up

We have covered a lot of ground in this chapter. We began by recognizing that although kinesiology focuses primarily on exercise and sport, kinesiologists often study a much broader spectrum of physical activity. By now, you realize that physical activity is important not only to our recreational lives but also to virtually everything we do. You also know about the various dimensions or spheres of daily living in which physical activity plays a vital role. We have seen, for example, that physical activity is central to carrying out the daily chores through which we remain self-sufficient. It also provides the means by which we express ourselves and do our work. Thus, physical activity plays a significant role in education, leisure, health, and competitive pursuits.

When we examine the spheres of physical activity experience, it is apparent that we rely on physical activity every day, not only to survive but also to lead full and rich lives. Surely, then, physical activity offers many practical benefits, though we do not seek these benefits through physical activity as much as we should. Having said this, it would be misleading to leave you with the impression that physical activity is important only because it is useful in achieving some end. To the contrary, it may be as important to us for the special types of personal experience that it offers as for any concrete benefits that we derive from participating. To put it simply, our lives are enriched when we immerse ourselves regularly in a variety of physical activity experiences. This personal or subjective side of physical activity is explored further in chapter 4. First, however, let's use chapter 3 to help you more fully appreciate the vital importance of physical activity and learn how physical activity professionals use it to serve their students, clients, and patients.

REVIEW QUESTIONS

1. Why are ADLs and IADLs important to kinesiologists who work with people who are elderly or who have disability?
2. What type of physical activity professional is likely to be involved in treating an individual with carpal tunnel syndrome? What professional is likely to be involved in redesigning the workplace to reduce the risk of carpal tunnel syndrome?
3. What are gestures, and what purposes do they serve in our daily living?
4. Why is physical education important in public schools? What objectives do physical education teachers pursue?
5. List three health benefits of regular physical activity.
6. What does the element of competition add to physical activity? When is it helpful, and when might it be harmful?
7. Describe a situation in which physical activity might help nourish and maintain a state of leisure. Describe a situation in which physical activity might reduce the possibility of attaining a state of leisure.

Go online to complete the web study guide activities assigned by your instructor.

3

The Importance of Physical Activity Experiences

Duane V. Knudson

The author acknowledges the contributions of Shirl J. Hoffman to this chapter.

CHAPTER OBJECTIVES

In this chapter, we will

- help you appreciate physical activity as a signature of humanity;
- help you appreciate the ways in which experience can expand the human potential for physical activity;
- help you develop insight into the factors that influence people's perceptions of physical activity and decisions to engage in it;
- preview kinesiology concepts that should be taken into account in prescribing physical activity experiences;
- describe the importance of subjective experience of physical activity and spectatorship of physical activity;
- explain factors that influence our personal experiences of physical activity; and
- discuss how individual preferences, tastes, dispositions, activity histories, and personal circumstances contribute to our subjective experiences of physical activity.

Surfing and sailboarding are physical activities in which many people participate passionately for most of their life span. What is it about these activities that makes them so attractive to so many people? Certainly, it is exciting to balance on a board and control one's body while interacting with an unpredictable ocean wave! Beyond these purely physical demands, however, there is something thrilling about the experience, yet it can also be peaceful and spiritual.

We will see in this chapter that the experience of physical activity varies across individuals and influences their activity choices. In the case of surfing, some people relish the danger and physical challenge of riding the biggest waves in competition, whereas others are more attracted to recreational surfing with gentler waves. As these cases illustrate, there is something special about the experience of physical activity that attracts people. It is often hard to describe, however, without experiential knowledge of a given activity—that is, knowledge obtained by engaging in the activity and learning from the experience itself.

Eyewire/Getty Images

As described in chapter 2, human physical activity is experienced in a wide variety of contexts. This chapter, in turn, explores the importance of each person's experiences of physical activity; more specifically, it explores two major domains of physical activity experience. The first domain involves **direct experience of physical activity**, where *experience* can be defined as any activity that includes training, observation of practice, and personal participation. Activity experience includes performing physical activity as well as observing it. The second domain involves the inward, **subjective experience of physical activity**, where *experience* can be defined as one's individual reaction to events, feelings, or other stimuli. In this sense, we may experience sadness, joy, beauty, and a host of other emotions through physical activity.

This chapter explores why some people have a passion for physical activity whereas others just don't seem to "get it." As you move through this chapter, you will gain appreciation for the importance of both domains of physical activity experience in developing your own capacity to perform physical activity and encourage physical activity in others.

KEY POINT

Our physical activity experience consists of our history of participation, training, practice, or observation of any particular physical activity. It differs from subjective physical activity experience, which involves our reactions, feelings, and thoughts about physical activity.

Physical Activity as a Signature of Humanity

Although it is tempting to think of physical activity experience as being inferior to mental activity, such a limited view does not do justice to this important and far-reaching concept or to the integrated nature of the human mind and body. Physical activity draws on aspects of humanity that help define us as the highest order of living creatures on earth. Without this connection, physical activity experience could not work its marvelous effects. Let's explore how physical activity helps define our humanity and, in turn, how our innately human capacities enable us to develop our physical activity capacity through experience.

What characteristics set humans apart from the "lower" animals and from machines? Your first thought might be that we possess souls or a moral sense, that we are able to engage in complex forms of social communication, or that we possess a capacity for high-level reasoning that animals and machines can never replicate. A good case can be made for each of these arguments. You may not have thought, however, about another defining human characteristic—namely, our extraordinary capacity to translate complex mental concepts into precise and creative physical actions.

KEY POINT

A unique capacity for performing physical activity is one of the major features that contribute to the distinctive character of the human race.

But what about other animals? Don't they have a capacity to link movements with intentions and plans for moving, and if so can't we say that they perform physical activity? Certainly they engage in goal-oriented, voluntary movements. For example, chimpanzees use a variety of tools to probe termite mounds and even use weapons to hunt small mammals. They also use small stones to crack palm nuts using the same techniques that they have seen used by local villagers. In a less well-known example, a small ground finch can uncover food by using its feet to push aside stones that are 14 times heavier than its body weight—which is equivalent to a 200-pound (90-kilogram) human moving a 3,000-pound (1,360-kilogram) rock—by bracing its head against a large rock for leverage (Gill, 1989). Yet, although these and other animals can perform goal-oriented movements, the human capacity for physical activity is of an entirely different order.

Recall that physical activity involves human movement that is intentional, voluntary, and directed toward achieving an identifiable goal. Now consider four aspects in which physical activity may be considered unique to the human species: intelligence-based physical activity, ethically and aesthetically based physical activity, flexibility and adaptability of physical activity, and ability to improve performance through planned experience.

Intelligence-Based Physical Activity

First, because humans are big-brained, highly intelligent creatures, our physical activity tends to be rooted in more intricate plans and directed toward more sophisticated goals than is the case with other animals. We use motor plans or mental images as dynamic maps to guide our movements when we perform skills as simple as a standing long jump and as complex as catching a flying disc, piloting an airplane, or performing brain surgery. Other animals, by comparison, entertain relatively simple plans. For example, although a kangaroo has a capacity for cardiovascular endurance that humans can only dream of—hopping in 15- to 20-foot (4.5- to 6-meter) strides at an all-out pace for up to 20 miles (32 kilometers) at a time—it cannot use this capacity to play the rule-bound game of soccer. Similarly, a cheetah can easily outrun a human in a contest in which the goal is simply to run as fast as possible. However, unable to formulate a clear understanding of goals and constraints imposed by rules, a cheetah cannot adapt its extraordinary running skill to the complexities of an Olympic relay race.

Ethically and Aesthetically Based Physical Activity

Second, because humans are essentially spiritual creatures who possess unique moral and aesthetic senses, we can use our movements to express our imagination; our moral reasoning; and deep and complex moods such as joy, wonder, and appreciation of beauty. Indeed, physical activity is integral to rituals that form an important part of our religious and civic lives. This is not to deny that other animals have emotional lives, too. Chimps, for example, can become so depressed when their mothers die that they sometimes die, too (Heltne, 1989). However, although some chimps can use sign language to express emotions and can daub paint on paper, they cannot translate emotions through muscular actions into symbolic works of art or other elaborate expressions of sorrow, joy, or wonder. If thousands of chimpanzees were each given a hammer and chisel, perhaps, by mere chance, one of them might create a recognizable work of art, but never something as profound as Michelangelo's *David* or Rodin's *The Thinker*. Similarly, wading birds may engage in elaborate mating dances, but the choreography is the product of instinct—not an intentional expression of mood through movement such as, for example, *Swan Lake*.

Flexibility and Adaptability of Physical Activity

Third, human physical activity is further distinguished from that of other animals by virtue of the unique combinations of movements permitted by human anatomy. At first, this may seem to be an exaggeration. After all, elephants are equipped for performing far more forceful movements than humans are, greyhounds can surpass us in speed, dolphins swim better than we do, and monkeys are more agile. In what ways, then, can we say that humans hold a movement advantage?

The answer lies in the fact that whereas other animals may be endowed with considerable specific talents for movement, humans are the best equipped for performing a wide range of activities. This distinction hinges on two properties of human anatomy. First, with our upright posture and bipedal gait, we are the only animals whose forelimbs have been totally freed from assisting with walking, flying, brachiating (arm swinging), or swimming. The human body has a foot specially constructed to bear weight and give leverage to the leg, a pelvis specifically designed for attaching the strong muscles needed to help maintain bipedal balance, and thighbones that permit long strides in walking. Indeed, our ability to walk on two feet has been described as "the most spectacular physical trait of human beings" (LaBarre, 1963, p. 73).

The second property is the human capacity for dexterity of movement, which is made possible by a unique complex of hands, arms, shoulders, and stereoscopic vision. The human hand possesses a true opposable thumb—a much more beneficial construction than that of the typical primate hand, whose five fingers all operate more or less in the same plane. The human hand arrangement allows us to perform delicate grasping, manipulating, and adjusting movements that are not possible for other animals. However, this is not the only advantage. Our hands are also positioned at the end of a series of long arm bones, which are joined to an amazingly movable shoulder girdle that a dog or cat could

Our ability to practice and improve our performance sets us apart from lower animals.

only dream of having. (Can you imagine a dog, cat, or horse being able to scratch its back with its forelimb?) That movable shoulder girdle enables us to position our hands through an enormously large range in space. This upper-arm advantage is complemented by a facility for stereoscopic vision (lacking in many animals) that not only gives us advantages in depth perception but also allows us to perform most of our movements within our field of vision.

Ability to Improve Performance Through Planned Experience

The fourth—and perhaps the most significant—distinguishing characteristic of humans is our ability to improve our capacity for physical activity through planned, systematic practice and training. Only humans possess the intelligence to use physical activity in planned, systematic, and scientifically verifiable ways either as a means of achieving physical rehabilitation or as a means of improving their health, performance, or skill. Of course, the cardiorespiratory efficiency of a young lion or eagle improves as its hunting range expands, but the driving force of its activity is hunger and survival, not a systematic conditioning plan to improve the physiological functioning of its body. Some species do teach their young how to hunt and fish, but the methods are relatively primitive and lack the sophistication of human plans.

Factors That Influence Kinds of Experience in Physical Activity

Because our potential for physical activity helps us express what is unique about our humanity, it might seem that each of us would naturally maximize opportunities to engage in physical activity and incorporate it into our lives. However, that isn't the case. Although it is natural for us to engage in physical activity, relatively few of us are inclined to do it regularly and in its more vigorous forms. In fact, an enormous disconnect exists between what we know about the benefits of vigorous physical activity and our behavior patterns. Likewise, few of us are inclined to tap our full potential for developing skill in a variety of physical activities. We know that when people are competent in a sport, they enjoy engaging in it during their leisure time, yet relatively few people devote themselves to the considerable practice necessary to reach even a moderate level of competence.

KEY POINT

Although we possess a unique facility for performing, planning, and implementing physical activity experiences in order to improve our performance, most of us are not inclined to explore this potential to the fullest.

Why this paradox? Why do most of us fail to explore to its fullest potential something so fundamental to our human nature and so beneficial to us? In contrast, why do some people engage in physical activity throughout their lives? This is one of the most pressing questions confronting the physical activity professions today, and, as you will see when you study chapter 8, it is an area of study that is central to exercise psychology. Unfortunately, there are no simple answers. Many factors in our lives encourage us to explore physical activity, and many factors discourage us from doing so. The barriers that keep us from living physically active lives include lack of time, lack of access to an exercise or sport facility, and lack of a safe environment in which to be physically active. However, these aren't the only factors to consider. Let's look briefly at how two other major factors may affect the type and extent of our physical activity experiences: our social environments and our personal circumstances (see figure 3.1).

Social Environment

The people with whom you regularly interact can affect the types and amounts of physical activity experiences that you pursue, both as a child and as an adult. As a child, you may have received substantial social support for engaging in certain kinds of physical activity, thus causing you to be physically active in your younger years. Similarly, among both young and middle-aged men and women, the amount of vigorous exercise that a person undertakes is influenced strongly by friendships and social alliances (King et al., 1992; Sallis & Hovell, 1990). For example, if your girlfriend or boyfriend has a physically active lifestyle, you probably feel encouraged to do the same. Thus, although each of us is ultimately responsible for our decisions about physical activity, our dispositions regarding it can be affected by the people with whom we associate.

Parents

When you were a child, did your parents lead a physically active life, or were they largely sedentary? If they modeled an active lifestyle—for example, taking you with them to their workouts

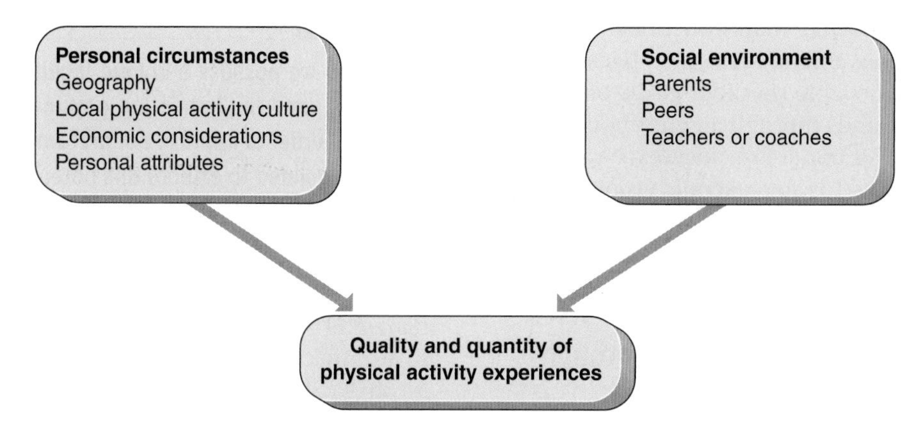

FIGURE 3.1 Many factors influence our decisions about whether to engage and persist in physical activity experiences.

at an exercise facility, taking a daily walk or run in your neighborhood, or avidly playing tennis or squash—then perhaps they can take some credit for the fact that you are an active person today. If, on the other hand, they were "couch potatoes" who avoided sport, exercise, and other physical activity, then you may have had to resist their adverse influence. From a research standpoint, however, it remains unclear how strongly children's levels of physical activity are affected by their parents' levels. Researchers have found, for example, that the activity levels of children who were participating in an obesity treatment program bore a remarkable correspondence to the activity levels of their parents (Kalakanis, Goldfield, Paluch, & Epstein, 2001). Yet other researchers have found no correlation between the levels of engagement of parents and children in physical activity (Jago, Fox, Page, Brockman, & Thompson, 2010).

Even if parents' participation in physical activity does not exert a strong influence on the physical activity levels of their children, encouraging them—especially preschoolers and adolescents—to be physically active is likely to make a difference. This is a good point for future parents to keep in mind. More specifically, merely showing support for children's physical activity by transporting them to physical activity settings or helping organize an

Do Inactive Parents Produce Inactive Children?

Researchers at the University of Bristol wanted to know whether physically active parents produce physically active children. To find out, they conducted a study that examined physical activity patterns among sixth graders and their parents from 40 primary schools (Jago et al., 2010). Specifically, they assessed levels of parent and child physical activity, as well as inactive time, through the use of accelerometers, which are small devices worn by subjects while they go about their daily activities; the devices provide reasonably reliable measures of physical activity. The investigators found no relationship between parents' and children's physical activity levels, but they did discover something very significant. The more parents watched TV, the more their children watched as well. For girls, the relative probability of watching more than 4 hours per day was nearly four times higher if their parents watched 2 to 4 hours per day as compared with girls whose parents watched fewer than 2 hours per day. For boys, the difference was even more pronounced. The chance of watching more than 4 hours per day for boys was more than 10 times higher if the parents watched more than 4 hours per day as compared with boys whose parents watched fewer than 2 hours per day.

These findings show that although parents may not directly influence their children's physical activity patterns, they can exert a direct influence on their children's sedentary behavior. This finding should concern kinesiologists because research has shown that men who report more than 23 hours per week of sedentary activity are 64 percent more likely to die from heart disease than are those who are sedentary for fewer than 11 hours per week, even though many of these more sedentary men exercised regularly (Warren et al., 2010).

activity (e.g., serving as a Little League coach or cheerleading sponsor) may increase the probability that they will be physically active (U.S. Department of Health and Human Services, 1996).

Researchers also have yet to find a strong link between one's level of physical activity in younger years and one's level later in life. For example, participating in school or college sport does not appear to influence the amount of physical activity that one engages in as an adult (Brill, Burkhaulter, Kohl, Blair, & Goodyear, 1989; Cauley, Donfield, LaPorte, & Warhaftig, 1991; Dishman & Sallis, 1994). In fact, you may be familiar with someone who was a star athlete in college yet became physically inactive following graduation, just as you may know someone who did not lead a physically active life as a child but became an avid adherent of exercise and sport in later years. Obviously, we have much to learn about this potential connection. Until a more complete picture emerges, however, it seems best to ensure that constructive physical activity experiences are integrated into the early years of development.

Peers

Your physical activity decisions may also be influenced by the extent of your friends' involvement in physical activity. Do you hang around people who like to be physically active? If so, you are likely to be physically active as well. In fact, some researchers believe that the physical activity patterns of peers may be as important as those of family in determining whether a person becomes involved in sport (Kirby, Levin, & Inchley, 2011).

As you grew older, did your peers become more physically active or more inactive? Unfortunately, the statistical probability is that your peer group became less active. For example, in 2011 only 29 percent of high school students surveyed had participated in 60 minutes of physical activity on all of the 7 days just prior to the survey (Centers for Disease Control and Prevention [CDC], 2012).

For example, if your peer group is immersed in online social networking, playing video games, or other sedentary pursuits, then you are likely to lead an inactive lifestyle as well, rather than using vigorous physical activities as the nucleus of your social life. As you grow older, peer groups may continue to affect your physical activity decisions. For example, if you are married, your spouse and your spouse's friends are likely to influence your decisions about whether to participate or not participate in physical activity (Loy, McPherson, & Kenyon, 1978). More generally, the predispositions of your relatives, friends, and social groups regarding physical activity will probably continue, indirectly but predictably, to influence your decisions throughout your life.

Teachers and Coaches

If you were an athlete in school, then your coaches—of youth sport teams, junior high school teams, and high school varsity teams—probably had a significant effect on your decisions about physical activity. Athletes tend to associate their participation in sport with the earlier influence of coaches, especially during adolescence (Ebihara, Ideda, & Myiashita, 1983). Physical education teachers and coaches are influential because they are positioned to either confirm or disconfirm a young person's sense of competence in an activity. In other words, a parent's remark that a child performed well in a soccer game may encourage the child to continue developing his or her skills, but a similar remark from a coach is likely to exert more influence.

KEY POINT

The people closest to you—including your parents, peers, coaches, and teachers—influence the kinds and amounts of experience you have with a particular physical activity, especially when you are young.

What Factors Have Influenced Your Physical Activity Patterns?

What leisure-time or competitive physical activities did you participate in as a child? Did you play soccer, football, or Little League baseball? Did you take private dance or gymnastics lessons? Did you ride a bike, climb trees, or roller-skate with your friends in the neighborhood? Did physical education motivate you to engage in physical activity outside of class? Did you participate in community-based sport and recreation programs? Do you think that these early physical activities influenced your engagement in physical activity as an adult? Do you see similarities in the general types of physical activities that you pursued as a young child and those that you pursue now? What factors have contributed to your decision to continue (or discontinue) engaging in these activities?

Of course, physical education teachers and coaches can also create social environments that *discourage* young people from seeking out physical activity experiences. In this regard, teachers and coaches may act as gatekeepers to the physically active life for countless children each year. Thus, it is no exaggeration to say that the behavior of a teacher or coach may exert profound and lifelong effects on a student's physical activity experience.

Personal Circumstances

The amount and kind of physical activity that a person experiences can be affected not only by his or her social environment but also by personal circumstances, such as the availability and accessibility of facilities and play spaces (Garcia et al. 1995; Zakarian, Hovell, Hofstettere, Sallis, & Keating, 1994). One important factor is geography, which is explored in the sidebar titled Geographical Hot Spots for Producing Professional Athletes. In addition, the weather can dictate the availability of appropriate sport environments. For instance, people who live in colder climates are more likely to develop competency in skiing, skating, hunting, hiking, and fishing than in swimming or golf. On the other hand, locations characterized by warm, sunny days tend to encourage jogging, in-line skating, and walking.

In some cases, certain sports are also emphasized by local cultural traditions. In the United States, for instance, high school football is uniquely important in western Pennsylvania, Florida, and Texas, whereas boys' wrestling and girls' basketball are popular in Iowa. In many inner cities, basketball is a cultural tradition, owing at least partly to the fact that space isn't available for sports such as baseball and football. In other cities, however, the most popular sport may be soccer, or "football" as it is called in most of the world.

KEY POINT

Often, our physical activity experiences—and, consequently, the activities in which we develop proficiency—are determined by factors that lie somewhat outside of our control, such as climate, regional culture, and financial considerations.

Activity preferences can also be dictated by financial considerations. Generally, people in higher income brackets tend to be more active than do their poorer counterparts (see figure 3.2). Of course, this disparity results in part from the fact that cost often limits opportunities for participation in physical activities to people with higher income levels. The difference is especially noticeable in sports that require expensive equipment, high admission fees, or (in the case of swimming, for example) transportation or admission to fee-based

Kinesiology Colleagues

Dann Baker

Dann Baker is a world-renowned teacher and student of karate. Dann served in the Air Force before earning his bachelor's and master's degrees in kinesiology. Through hard and consistent practice, he has earned second-, third-, and ninth-degree black belts in several styles of karate and has won numerous national and international championships. Now 69 years old, he continues to teach university students and karate instructors all over the world. Learning through experience and repetition with a skilled mentor is a time-honored tradition in martial arts instruction.

Photo courtesy of Dann Baker

Geographical Hot Spots for Producing Professional Athletes

Your decisions about playing sport can be affected by where you were born and raised. To illustrate this fact, the production of athletes by particular U.S. states and counties was quantified by researcher Christopher Storm (2005) using a database of the hometowns of every U.S.-born professional athlete in one of the four recognized major leagues: Major League Baseball, the National Basketball Association, the National Football League, and the National Hockey League. Some of Storm's findings are presented in the following list.

- *Overall:* The regions with the highest per-capita production (ages 20 through 44) were the Southeast (Texas to South Carolina), the Northern Plains (Utah to the Dakotas and Iowa), California, and Ohio. The least productive region was New England.
- *Baseball:* The most productive regions per capita tended to be rural—specifically, in the Pacific Coast area and the Southeast, along with Ohio, the Dakotas, and Alaska.
- *Basketball:* The most productive areas per capita were the Great Lakes states, the Pacific Coast, and the Southeast; in addition, basketball players came mostly from urban counties.
- *Football:* The counties that produced the most football players per capita were located in the Southeast; the counties that produced the fewest were found in New York and New England.
- *Hockey:* Predictably, the most professional hockey players per capita were produced in the upper Midwest and the Northeast, along with Utah and Colorado.

private clubs. The effect of income on participation is also particularly noticeable in skiing and golf; in fact, 50 times more skiers and 12 times more golfers come from the upper income bracket than from the lowest bracket (U.S. Department of Commerce, 2012). Those who exercise with equipment (probably at clubs that require dues) are more than 5 times as likely to be from the highest bracket than from the lowest bracket. On the other hand, given that walking for exercise is essentially free and

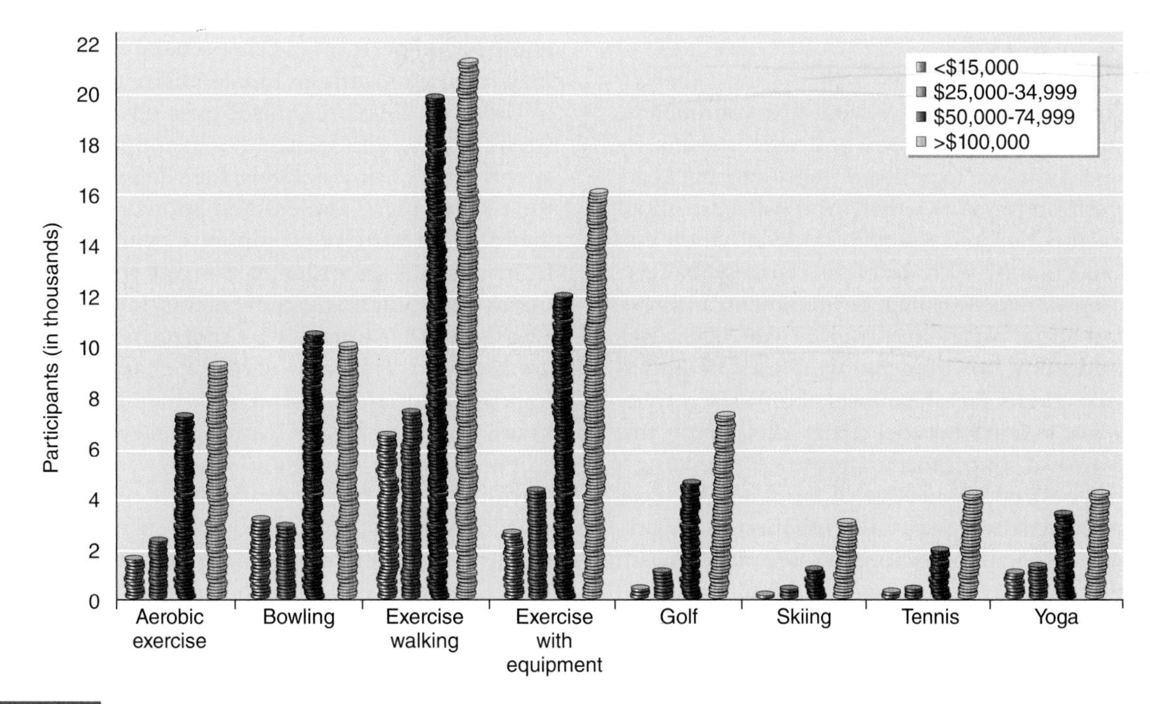

FIGURE 3.2 Influence of income on participation in selected sport and exercise activities.

Data from U.S. Department of Commerce, U.S. Census Bureau 2012.

readily available to all, financial differences may not account for the fact that 3 times more people walk for exercise in the highest income bracket than in the lowest bracket.

At the same time, our decisions about whether to become involved in certain physical activities do not always lie outside of our control. To the contrary, our physical activity patterns can also be affected by our individual attitudes and personal attributes. In fact, our own perceptions, feelings, and decisions can play a major part in the process. For example, a decision to prepare yourself for a marathon or to train for a bodybuilding competition may have to do with your personal attraction to certain unique characteristics of these activities. Similarly, if you participate in hang gliding, your decision to do so may have something to do with your willingness to take risks, or with the exhilaration you feel at high heights, or your perceived ability to keep the glider in the air.

KEY POINT

Beyond the key factors in our social and ecological environments, our decisions about physical activity are also affected by factors such as self-perception, competence in a given activity, and characteristics of the activity itself.

Ways in Which Experience Can Affect Physical Activity

As a kinesiology major, it is vital that you become familiar with the processes and mechanisms by which performance experience improves our capacity to perform physical activity. You will learn about relevant theories and research in detail when you enroll in courses such as biomechanics; exercise physiology; motor learning, development, and control; and sport psychology. Before you begin this advanced study, however, you will find it helpful to step back and look at the effects of physical activity experience in much broader terms. This perspective will give you a conceptual framework for building a more specialized knowledge of the scientific basis of physical activity. You can also use this framework, and the associated disciplinary terminology, to understand and appreciate past and future physical activity experiences, as well as activity courses that are often required of kinesiology majors.

Think back to a form of physical activity in which you had a special interest. How did your experience with this activity affect your capacity to perform it on successive occasions? If, for example, the activity involved volleyball skills, then the effects of your accumulated experiences were probably most evident in the form of changes in your ability to perform the movement sequences associated with serving, setting, or spiking. If the activity was cardiorespiratory training, then the effects of your accumulated experiences may have manifested themselves not so much in changes in the nature of your movements but in your improved capacity to perform the activity for longer periods at greater intensity. These examples illustrate the two fundamental effects of physical activity experience: the development of skill through practice and the development of physical capacity.

These effects of physical activity experience can be conceptualized on a continuum (see figure 3.3). All physical activity experiences contribute to both skill and physical capacity, but the particular mix of the two depends on many factors that can be modified by the performer or by a kinesiology professional. Movement experiences that emphasize training result in relatively greater conditioning, which improves physical capacity. In contrast, movement experiences that emphasize practice result in relatively greater learning of motor skills or coordination.

Learning, Practice, and Skill

Although physical activity is fundamental to human life, we do not come into the world readily equipped to perform it. Our prenatal development may endow us with an innate ability to move, but in order to develop control over our movements we must have experience. For example, newborns spend a lot of time kicking their legs and waving their arms; these aimless and spontaneous movements occur without prompting or encouragement from parents. In order to convert these spontaneous and automatic movements into conscious *physical activity,* much experience and maturation are required. This section of the chapter explores improvement in physical activities through the kinesiology concepts of learning, practice, and skill.

Think back to a time when you were first learning some physical activity. It is likely that your movements were jerky and inefficient. With practice, however, they became smoother, allowing you to be not only more accurate but also more forceful. This process usually involves an important change in the sequence of movements that we use. To see this principle in action, carefully examine the two photos presented in figure 3.4. In the first photo, the youngster locks shoulders, elbows,

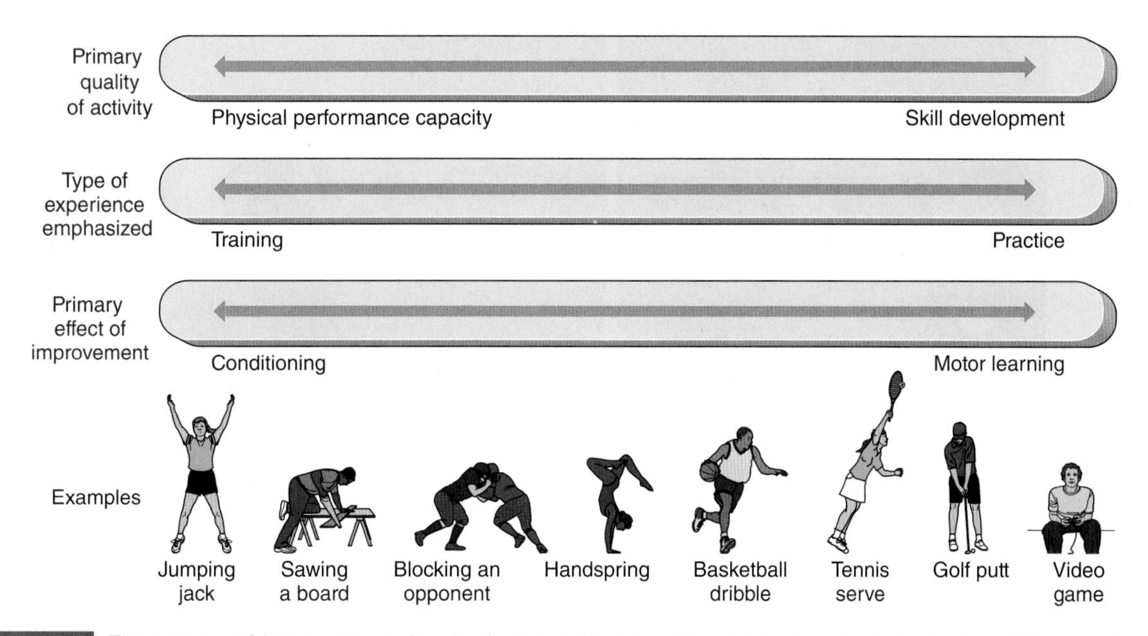

FIGURE 3.3 The nature of improvement in physical activity depends on the types of physical activity experience and the relative importance of skill and physical capacity.

and wrists in an effort to control the movement, but this approach limits how much power can be generated. In contrast, the more advanced player, shown in the second photo, rotates the hips, trunk, shoulders, elbows, and wrists independently, thus allowing her to generate much more force.

Most people who engage in physical activity want to improve their performance, whether that means performing faster, more forcefully, for a longer time, or more skillfully. The only way to make such improvements is to expose oneself systematically to appropriate physical activity experiences. Such experiences enable people enrolled in fitness programs to become stronger or to increase their cardiorespiratory endurance. They also help athletes become more skillful, dancers more graceful, and factory workers and military personnel more efficient.

Throughout life, we continue to confront the challenge of assembling sequences of movements and body positions in order to accomplish specific tasks. Activities that place a premium on efficient, coordinated motor responses are called **motor skills.** Although activities such as threading a needle, performing a cartwheel, pitching a baseball, and typing a letter are all motor skills, each one places distinctly different demands on our perceptual and motor processing systems. The general quality that underlies the performance of voluntary movements is called **skill**, which is reflected in the efficiency and accuracy with which an individual is able to attain the goal of a motor skill.

Becoming a skilled mover is a never-ending process of gaining more and more control over one's motor system by gradually refining the neuromuscular systems through performance experience. The type of performance experience required to do this is known as **practice**. Laypersons tend to think of practice as merely an endless repetition of the same movements by rote, but that isn't how it works. Practice of motor skills, especially during early learning, involves a lot of cognition. That is, it involves higher brain function and integration of spinal neural activity as well as muscular action; in fact, muscular action is merely the result of these higher cognitive processes. Practice is a deliberate effort to "get it right" by modifying the erroneous movements we have made in earlier trials and revising our cognitive plans for moving in the next trial. With each practice trial, we vary the level of force or direction of our movements, adopt new strategies, and analyze feedback from our movements so that we can, for example, throw a ball farther or more accurately. This process is why one scientist referred to practice as "a particular type of repetition without repetition" (Bernstein, 1967, p. 134).

KEY POINT

The type of physical activity experience that brings about changes in skill is called *practice*. The relatively permanent effect of practice on performance is called *learning*.

FIGURE 3.4 *(a)* Physical activity experience leads to *(b)* more accurate and faster movements.

The importance of cognitive processes in early skill learning should not blind us to the fact that learning can also proceed on an unconscious or tacit level. For example, an individual may be able to improve his or her performance (i.e., reaction time) in responding rapidly to stimuli that appear in a complex temporal pattern yet be unable to explain what that pattern is. Similarly, catching a fly ball in baseball is an extraordinarily complex skill, and many components of this action are achieved at an implicit (unconscious) level. Indeed, a young child can learn, implicitly, the movements that allow her to catch the ball. In another example, people who have had strokes that limit or eliminate their short-term memory may not remember having practiced a skill previously but may still show improvement from practice to practice (Meehan, Bubblepreet, Wessel, & Boyd, 2011).

The refinements in the nervous system that result from practice are referred to as **learning**. Although much has yet to be discovered about learning, scientists agree that it is primarily a central phenomenon rather than a peripheral one. This is another way of saying that learning results from rebuilding neural pathways in the brain and spinal cord rather than making changes in peripheral organs such as bone and muscle. (Even though we often hear the term *muscle memory*, the remembering is done not by the muscles but by the brain.) We also know that these changes tend to be relatively permanent. If, after having played many years of tennis, you do not return to the court for several years, the level of your play will deteriorate, but with a little practice you will quickly regain it. Amazing, isn't it?

When you learn a motor skill, you remap the neural routes in your central nervous system. As you know from your own experience, this remapping doesn't happen overnight. Depending on the skill involved, it may take weeks, years, or even

Other Uses of the Term *Skill*

In American football, quarterbacks (who throw passes), wide receivers (who catch passes), and halfbacks (who run and catch the ball) are often said to play the "skill positions" on the offensive team. This title is not given to offensive players (i.e., linemen, such as tackles, guards, and ends) whose primary responsibilities are to block and tackle, even though their actions also require motor skills and performance capacity. Why do you think the term "skill position" is used in this sport? Do you think it is useful? Does this terminology inadvertently distort the kinesiology concepts of skill and performance capacity? Do linemen require skill to block and counter evasive moves of defenders?

decades to master. Of course, the extent to which changes in skill become permanent depends a great deal on how much you practice the skill.

Training, Conditioning, and Physical Performance Capacity

The second type of change brought about by experience is much different from learning. Improvements in some physical activities hinge less on improving the accuracy or timing of your movements and more on improving strength, developing the capacity to exert greater force, or increasing the range of motion at joints. These factors are not elements of skill so much as they are elements of **physical performance capacity**. To put it simply, strength and flexibility are less learned than they are developed.

Consider the case of two tennis players, Samuel and Jose, both of whom have devoted long hours of practice to the sport and played in competitive leagues. They have developed about the same level of skill in tennis, as is evidenced by their equal proficiency in serving, hitting backhand and forehand strokes, and playing at the net. Then Jose, a perceptive kinesiology major, begins to supplement his tennis practice with a rigorous training program designed to increase his physical capacity through development of agility, muscular endurance, cardiorespiratory endurance, and flexibility—all of which he believes to be almost as important as skill in the sport of tennis. As a result, he becomes a more proficient tennis player in competition than Samuel. Although both players have learned the *skills* of tennis to the same degree, Jose ends up as the better player because he has supplemented his skill development with improvement of other performance qualities that are important in the sport.

KEY POINT

Physical activity experiences known as *training* are used to develop such performance qualities as muscle strength and endurance, cardiorespiratory endurance, and flexibility. The state of having developed these qualities is known as *conditioning*.

The supplemental types of experiences that Jose has engaged in include weightlifting, cardio drills, stretching, and running—all of which are kinds of physical activity experiences known as training. The changes brought about by training are referred to as **conditioning**. Training consists of physical activity experiences designed to improve muscular strength and endurance, cardiorespiratory endurance, flexibility, and other aspects of performance that condition the performer. Training increases the capacity to move one's body more forcefully (i.e., strength) over longer durations (endurance) and with more range of motion (flexibility). Generally, kinesiology majors study the experiences of training that lead to conditioning in courses on the physiology of physical activity, whereas they study the experiences of practice that lead to learning in courses on motor behavior.

Practice Versus Training

The differences between practice and training can be summed up this way: Practice primarily affects memory, cognition, perception, and other central nervous system processes that are associated with problem solving. Training, in contrast, primarily produces effects that are largely peripheral to the central nervous system and that usually affect muscle, bone, soft tissue, and the cardiorespiratory system. Training is most effective when done systematically under the direction of a professional exercise instructor, but training effects often occur as a default or unintended by-product of physical activity. For example, one chops wood or shovels snow in order to accomplish ends that have nothing to do with training per se; however, if you perform these activities enough, you will become conditioned both for them and for related tasks. Quite simply, if you repeat movements that appropriately stress muscles, tendons, and bones; if you elevate the heart rate to certain levels; or if you move your joints through large ranges of motion—then some level of conditioning will occur. On the other hand, mindlessly performing a skill, no matter how many times, is not likely to lead to high levels of learning.

So, is practice or training the most appropriate physical activity experience for improving your performance in a given activity? Most large-muscle activities require some degree of both practice and conditioning, but usually one or the other is more important. The key to choosing which result to emphasize is to ask whether success in the activity depends primarily on skill or on physical performance capacity. If the primary quality is skill, then the most important experience is practice; if the primary quality is physical performance capacity, then the most important experience is training. If skill and performance capacity are equally important, then practice and training are equally important as well.

Activities in which success depends primarily on learning precise and coordinated sequences of movements are located at the right end of the continuum shown in figure 3.3. Improvement in these activities results largely from practice. At the extreme left end of the continuum are activities that require little in the way of skill but depend heavily on factors such as strength, endurance, and flexibility. We improve our performance in these activities primarily through training experiences. Notice that activities can be distributed along the training-to-practice continuum.

Because most physical activities incorporate both skill elements and physical capacity, improvement often requires both practice and training experiences. For example, rehabilitation programs designed to help individuals learn to walk after a stroke focus on practice because much of the problem relates to neurological impairment. However, because the loss of muscular strength may also impinge on efforts to support the body, training may also be required. On the other hand, as shown in figure 3.3, some activities (e.g., playing video games) rely almost entirely on practice for improvement, whereas others (e.g., performing jumping jacks) rely almost exclusively on training.

KEY POINT

Generally, engaging in practice without training, or engaging in training without practice, is an incomplete formula for developing excellence in sport.

If you set a goal of making the varsity soccer team, much of your preparation will involve practicing such skills as dribbling, tackling, and passing the ball. At the same time, because the game requires you to run continuously, you will need to run sprints and long distances in order to develop cardiorespiratory endurance. You might also train with weights, especially with your legs, in order to increase your kicking force. Like many sports, then, soccer requires attention to both practice and training experiences.

Experience in Physical Activity and Physical Fitness

Practice and training experiences play an important role not only in developing sport proficiency and rehabilitating injured body parts but also in

Definition of Physical Fitness

We all have a general idea of what is meant by *physical fitness*, but the term has been defined in many ways. Here are a few examples:

- Centers for Disease Control and Prevention (2011): Physical fitness consists of a set of personal attributes related to the ability to perform physical activities that require aerobic fitness, endurance, strength, or flexibility; it is determined by a combination of regular activity and genetically inherited ability.
- Public Health Agency of Canada (2012): Physical fitness consists of a set of attributes related to either health or performance (i.e., skill). As its name suggests, health-related fitness includes the components of fitness related to health status. Performance- or skill-related fitness includes the components that enable optimal work or sport performance.
- *Journal of the American Medical Association* (Torpy, Lynm, & Glass, 2005): The five components of fitness are cardiorespiratory (i.e., heart and lung) endurance, muscular strength, muscular endurance, body composition, and flexibility. Cardiorespiratory endurance is the ability to perform sustained physical activity, such as walking, swimming, or running. Muscular strength and muscular endurance are linked to each other and are improved through the use of weight-bearing exercise, such as weightlifting and use of resistance bands. Body composition is determined by the proportions of muscle, fat, and water. Flexibility relates to range of motion and is improved by gently and consistently stretching the muscles and the connective tissues surrounding them.

How are these definitions similar, and how do they differ? Which definition do you think comes closest to the one discussed in the preceding paragraphs?

increasing a person's **physical fitness**—that is, his or her overall capacity to safely perform activities of daily living. A person's physical fitness depends on a complex blend of genetics and physical activity, and it involves a variety of components of interest to kinesiology professionals. One of the key early discoveries in kinesiology revealed that there is not just one domain of athletic ability or physical fitness; rather, there are numerous specific components. Among other things, this discovery produced the principle of specificity, which will be an important topic in your major.

In general, a physically fit person is one who has used physical activity to develop the capacity to perform the essential activities of daily living at a high level, possesses sufficient energy remaining to pursue an active leisure life, and is still able to meet unexpected physical demands imposed by emergencies. Because this type of fitness is reflected in a capacity to *perform* physical activities, it is called **motor performance fitness**, which has been defined succinctly as "the ability to perform daily activities with vigor" (Pate, 1988, p. 177). Your motor performance fitness is reflected, for example, in how many sit-ups or push-ups you can do and how long it takes you to run or walk a given distance. We know that this type of performance depends on unique components of fitness, such as strength, endurance, and flexibility. Because the performance of daily activities also requires other fitness components—such as balance, agility, coordination, and quickness of response—some kinesiologists include these components as additional indicators of motor performance fitness.

A second type of physical fitness, generally regarded by those in the health-related professions as more significant than motor performance fitness, is **health-related fitness**. This type of physical fitness involves using physical activity experience to develop the traits and capacities normally associated with a healthy body, specifically in relation to diseases known to result from a physically inactive lifestyle. These conditions, sometimes called **hypokinetic diseases**, include heart disease, high cholesterol, high blood pressure, diabetes, and obesity. Health-related fitness can be assessed directly by using various technologies without actually assessing motor performance. One example is the use of skinfold calipers to measure percentage of body fat. In other cases, health-related fitness can be assessed while the subject engages in motor performance. For instance, recording heart activity (via electrocardiography), blood pressure, and heart and respiration rates while a person exercises can give kinesiologists a fairly clear picture of his or her health in terms of cardiovascular-related hypokinetic disease.

Figure 3.5 illustrates the various components assessed by motor- and health-related fitness tests. As you can see, some of the motor fitness measures bear little relation to the health of the individual. Could an individual with a diseased cardiovascular system score high on a balance, strength, or agility test? Such a result is quite likely. On the other hand, some components of fitness are often assessed by both motor- and health-related tests. Flexibility tests, for example, measure motor performance capacity but also reflect something about the health-related range of motion of joints.

KEY POINT

The term *fitness activities* refers to training experiences that improve our general capacity to perform daily activities and help prevent disease processes associated with low levels of physical activity.

Identifying Critical Components of Physical Activity Experiences

How do kinesiology professionals zero in on the critical components of physical fitness and develop prescriptions for physical activity? In many (perhaps most) instances, professionals use their ability, developed over many years of practice, to conduct informal, intuitive task analyses. Like veteran physicians diagnosing their patients' ailments, experienced physical activity professionals can often

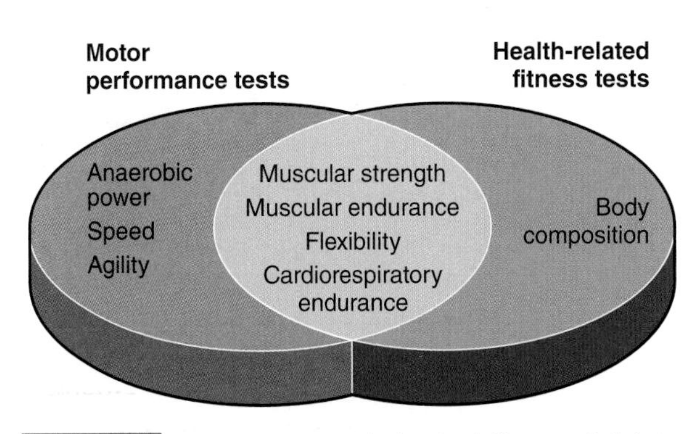

FIGURE 3.5 Components of physical fitness fall into two categories: motor performance and health-related fitness. Some components meet requirements for both categories.

Adapted from Pate 1988.

quickly identify the critical fitness components and performance elements of a task. In other cases, professionals use a checklist or other formal system to identify important components for prescribing physical activity. The following discussion provides some examples of frameworks for conducting task analyses of physical activities.

The first step is to identify the quality of experience that will bring about improvement. If the activity is located more toward the skill development (right-hand) end of the continuum displayed in figure 3.3, then the experiences chosen should be likely to improve skill. Thus, the goal of the analysis is to identify critical aspects of the performance that can be learned through carefully constructed practices. If, on the other hand, the activity is located near the physical performance capacity (left-hand) end of the continuum, then the focus should be on training experiences that promote conditioning appropriate for the activity.

Determining Skill Components Critical to Learning

If the physical activity in question falls near the skill development end of the continuum (figure 3.3), then the analyst's first task is to identify components of the activity that must be practiced in order

Depth or Breadth of Experience: Generalists Versus Specialists

Do you have in-depth experience in a particular physical activity (that is, are you a specialist)? Or do you have a broad range of experiences in a number of activities (that is, are you a generalist)? If you are like most people, you are more of a generalist. Limiting practice and training to one particular activity allows a person to develop depth of capacity in that activity. In contrast, exposing oneself to a broad range of training and practice experiences increases one's capacity for different types of skills and activities.

Physical Activity Generalist

Physical activity generalists are individuals with experience in a broad range of skills and activities. A person who has developed low-average to above-average competency in, say, rock climbing, wrestling, ice skating, football, and baseball is a generalist. The advantage that accrues to generalists is the enjoyment and satisfaction that come from being able to engage in a variety of activities. The disadvantage is that competence is not highly developed in any single activity, thus depriving the individual of the experiences that come from demonstrating excellence.

Physical Activity Specialist

Physical activity specialists devote themselves to developing depth of capacity in a single activity or a narrow range of activities. For example, a young girl who wants to become an Olympic gymnast must commit herself to training, practice, and competitive schedules; therefore, she is unlikely to be found on the tennis court or in the swimming pool on a regular basis. The advantages that accrue to specialists include the pride and satisfaction that come from being able to do one activity or a small number of activities at an above-average level. The disadvantage is that the individual misses out on opportunities to engage in a number of different activities. Concentrating one's efforts on a single physical activity can result in remarkable proficiency, as anyone who watches national- or international-class athletes can appreciate. These in-depth experiences can result in some amazing capacities. Here are a few examples:

- As a result of long hours of training, Bob Natoli was able to lift a total of 51,640 pounds (23,424 kilograms) in 1 hour of upright barbell rowing.
- Nancy Schubring trained herself to run a half marathon in 1 hour, 30 minutes, and 51 seconds while pushing a youngster in a baby carriage.
- Chris Gibson developed the capacity to perform 3,025 consecutive somersaults on a trampoline.

These examples give striking testimony to the fact that in-depth training and practice experiences can improve the sophisticated nerve, muscle, and cardiorespiratory systems that allow us to perform physical activity.

to maximize learning. In making such important decisions, physical activity professionals often consult **motor skill taxonomies**, which are classification systems that categorize skills according to their common critical elements. Obviously, motor skills can be classified in many ways. We could, for example, classify them according to whether they require large movements or small movements, whether they are performed in the water or on land, whether they involve fast movements or slow movements, or whether or not they require equipment. However, it is doubtful that any of these factors would be considered critical, because they do not appear to carry direct implications for how to plan practice experiences. Remember, if a taxonomy is to be useful, it must at least supply us with hints about how to design practice experiences.

One of the most illuminating classification systems designed for analyzing motor skills in the past 40 years consists of a simple scheme (shown in figure 3.6) that locates skills on a continuum from **closed skills to open skills** (Gentile, 1972; Poulton, 1957). The system is based on the presumption that one critical component of all skills is the predictability of the environmental events to which performers must adapt their movements. In this view, performing a motor skill is essentially a matter of adapting one's movements to relevant objects, persons, or environments. For example, to execute the simple task of picking up a ball, performers must move toward the ball, reach their hands in the direction of the ball, and position their hands and fingers in such a way as to grasp it. Other movements, such as moving the arms or hands away from the ball or keeping the fingers in a stiff and extended position, will not allow the performer to accomplish the goal. In this sense, then, the movements of the performer are controlled or regulated by the shape of the ball and by its location in space at any given time.

If the ball to be picked up is resting on a table, then its position is fully predictable before and during the movement. In this circumstance, the movements that the performer must execute in order to achieve the goal are also predictable and may be planned before the trial begins. Motor skills that are performed in such highly predictable environments and require highly predictable movements are called *closed skills*. The appropriate practice experience, therefore, focuses on stereotypical and consistent arm, hand, and finger movements (i.e., "grooving in" the movement). Also, because the ball is stationary, the performer need not be concerned about coordinating the timing of the movements in

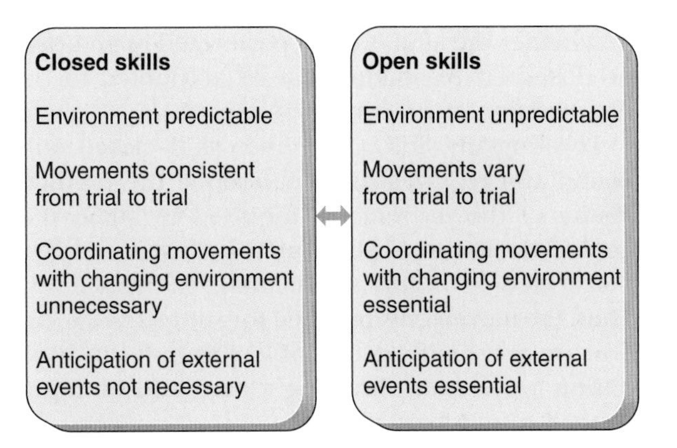

FIGURE 3.6 The open skill–closed skill continuum and possible locations of several sport skills.

order to coincide with changes in the environment, as, for example, must be done by a baseball batter. In other words, the performer here is free to execute the movements in his or her own time.

If the task goal is changed by requiring the performer to pick up the ball as it is rolled at various speeds across the table, then the performer's motor response must be variable rather than stereotypical. In addition, the performer cannot plan and execute the responses in his or her own time, because the start and duration of the movements is controlled by the speed at which the ball is rolled. Therefore, picking up the ball, which was originally a closed skill, becomes an open skill. It is open in the sense that it requires the performer to monitor a constantly changing environment and adjust his or her movements accordingly.

When the skill is "opened up," the ball not only controls the direction and location of the performer's movements, as was the case when it was stationary, but also determines when the performer must initiate movements. As a result, the performer can make errors not only by being inaccurate in the direction of movements but also by being early or late in making them. If the ball's velocity *and* direction change on each trial, then the performer faces an even more complex challenge. Because the ball might roll at either high speed or low speed and in any direction, it could well topple over the edge of the table before the performer has time to decide on a response and execute it.

Thus, activities that require open skills present the performer with unpredictable environments. This simple change in the structure of the task—creating a moving rather than a stationary environment to which the performer must adapt his or her movements—fundamentally changes the critical components of the task. Therefore, the question of whether the skill is to be performed in an open or closed environment must be accounted for in the planning of practice experiences.

Locating any skill on the open skill–closed skill continuum requires us to determine the predictability of the movements required to attain the goal. Remember, skills near the closed end are performed in highly predictable environments. Thus, the movements required for successful execution are also highly predictable. Examples include hitting a golf ball, throwing a dart, and hitting a softball from a batting tee.

In contrast, activities at the extreme open end of the continuum require movements that are highly unpredictable because the locations of the environmental objects to which the performer must adapt can change in irregular ways. For example, the activity of returning table tennis serves requires the performer to react to balls directed at multiple angles on the table and hit with enormous spin, which can affect the ball's action. In contrast, a golf drive is performed in a stable environment where no reaction is required; thus it is considered a closed skill. Skiing also meets the general definition of a closed skill, but because the body moves swiftly against the terrain it takes on some of the characteristics of an open skill as the performer must constantly adjust to the changing landscape. Dancing with a partner also requires constant adjustment in position and steps, but as these elements are learned the changes in the position of one's partner become more predictable.

KEY POINT

Skills can be located on an open–closed continuum. In closed skills, performers must adapt their movements to fixed and highly predictable environments. In open skills, performers must adapt to changing and highly unpredictable environments.

Determining Practice Experiences Critical to Improving Skills

The types of practice experiences essential for improving skills vary depending on the location of the skill on the open–closed continuum. Skills near the closed end may be practiced best in situations in which the environment is structured in the same way for each trial. In learning to hit a golf ball, for example, a player would focus on grooving a theoretically correct technique that could be repeated in trial after trial. In contrast, practice regimens for skills near the open end of the continuum are more likely to involve structuring the environment in various ways so that, for instance, the direction and velocity of relevant moving objects or persons change with each trial. Thus, the goal of practicing for open skills is to develop a flexible technique that allows one to adapt to a variety of environmental stimuli.

If a high school coach had her softball team spend a lot of time hitting balls from a tee (a closed skill) rather than hitting pitched balls (an open skill), this prescription would not focus on the critical aspect of the task: coordinating the swing with the direction and speed of the ball. As a result, players would waste their time and effort and might even reinforce inappropriate techniques. Again, practice for open skills typically concentrates on developing strategies for anticipating likely changes in the environment. This approach provides performers with more time to plan and execute their responses. Obviously, skills located near the center of the continuum require appropriate application of all of these practice strategies.

Remember that the open skill–closed skill continuum focuses on only one critical component of motor skills. Other critical components may be present as well. Chapter 7 introduces you to many other elements that must be taken into account in planning practice experiences for skills.

Determining Experiences Appropriate for Training

We have considered how to determine the types of *practice* experiences essential for *learning* skills. The question now is how can we determine the types of *training* experiences likely to lead to the *conditioning* essential for improving physical capacity. Remember, unlike skill, physical performance capacity depends on training experiences that involve such critical components as muscular strength and endurance, cardiorespiratory endurance, and flexibility. Knowing the critical components that underlie a particular activity permits you to plan appropriate training strategies for developing them.

How would you determine which types of physical activity experiences are essential for improving competitive performance in bicycling and in golf? To start, you must determine the critical components of bicycling and golf. You might approach the problem by referring to the continuum depicted in figure 3.3. You should recognize immediately that the key to successful performance in golf is learning to perform a coordinated sequence of movements and that these movements must be acquired through practice experiences. This is not to say that strength, endurance, and flexibility are unimportant in golf—only that they are less important than skill. You will also realize that bicycling, though dependent on the development of an efficient posture and a coordinated sequence of efficient movements, is determined primarily by the contribution of such performance components as cardiorespiratory endurance, muscular endurance, strength, flexibility, and balance. Given the dominant importance of physical capacity elements in bicycling and their secondary importance in golf, how might we identify the specific critical physical capacity elements around which training programs should be developed?

This problem can be approached in two ways. One is to consult experts. If we asked a panel of exercise physiologists or strength and conditioning experts to use a scale of 0 to 3 to rate the relative contributions of five critical performance components in golf and bicycling, we could draw up a chart like the one presented in table 3.1. (This table depicts the total ratings assigned by seven experts to five elements of physical capacity for each of ten physical activities.) Examining the table, we can see that four of the five physical capacity components were judged critical to bicycling, which suggests that any training regimen should focus on each of them. None of the components, however, figured prominently in golf performance, something that probably doesn't surprise you. Although strength and endurance training may improve golf performance to some degree, golf is predominately a game of skill; hence one's golf game is improved more effectively through practice. Softball also received relatively low ratings, whereas, swimming and jogging, which depend primarily on cardiorespiratory endurance, were rated high in that critical component.

A second approach to determining critical performance components is to seek out the kinesiology research on the activity of interest. Considerable research has been reported that can confirm or refute the logic of expert experience. Kinesiology professionals should seek out this research because it ranks higher than expert opinion in the hierarchy of evidence. For example, a biomechanics study of male golfers reported that total lower-extremity power (which combines strength and the velocity of its expression) accounted for about 35 percent of club speed (McNally, Yoontz, & Chaudhari, 2014). Even more useful for prescribing training of relevant components of performance are prospective studies that train these strength variables and document the change in club speed with golfers (e.g., Fletcher & Hartwell, 2004).

KEY POINT

The critical performance components underlying a physical activity can be ranked by asking experts to rate their relative importance. Kinesiology research is beginning to validate (or invalidate) such task analyses with data.

Table 3.1 Ratings by Experts of Physical Capacity Elements

Physical fitness	Jogging	Bicycling	Swimming	Handball and squash	Skiing (alpine)	Basketball	Tennis	Calisthenics	Golf	Softball
Cardiorespiratory endurance (stamina)	21	19	21	19	16	19	16	10	8	6
Muscular endurance	20	18	20	18	18	17	16	16	8	8
Muscular strength	17	16	14	15	15	15	14	14	9	7
Flexibility	9	9	15	15	14	13	14	14	9	9
Balance	17	18	17	17	21	16	16	16	8	7

Combined ratings of seven experts as to the contribution of cardiorespiratory endurance, muscular endurance, muscular strength, flexibility, and balance on a scale of 0 to 3. A maximum score is 21.

Adapted, by permission, from D.J. Anspaugh, M.H. Hamrick, and R.D. Rosato, 1991, *Wellness* (St. Louis: Mosby Year Book), 165. © The McGraw-Hill Companies, Inc.

Heredity and Experience

Although physical experience is important in determining how well we are able to perform skills, it is not the only factor. Think about the incredible performance of Simone Biles at the 2016 Rio Olympics. Can we say that the only thing that separates her from most other gymnasts is her extensive experience? Probably not. For one thing, she has a small, strong body that lends itself to generating dynamic movement and rotating in the air. In addition, her strong lower body helps her maintain balance, and her genetically endowed nervous system may give her unusual control over her movements. At the same time, she has had, since her earliest years, an unflagging commitment to becoming a world-class gymnast. This raises the question of how much heredity contributes to athletic performance. As we have seen, athletes cannot move to elite status unless they have spent years gaining experience in an activity.

Still, heredity certainly makes a contribution. For example, you and a friend may have had about the same amount of competitive basketball practice, and both of you may have adhered to rigorous conditioning programs to improve your jumping ability and overall strength. Yet you may be a much better player than your friend. Why? Perhaps you were more dedicated and more highly motivated to succeed. Perhaps you were also lucky enough to have parents whose stature and abilities were passed on to you. You may be taller, faster, or more agile, or you may have what seems to be a natural facility for changing direction, jumping, or coordinating the movements of your arms and legs—all of which are important to success in basketball. Practice and training can modify most of these characteristics to some degree, but the genetic contribution of your parents plays a significant role in your abilities and in the trainability of those abilities. Let's briefly consider the ways in which heredity might modify the effects of physical activity experience.

Abilities as Building Blocks of Experience

In research on motor learning and control, genetic predispositions that offer advantages or disadvantages for particular activities are usually termed **abilities**. Although physical activity scientists still do not completely understand abilities, researchers believe that abilities are genetically endowed perceptual, cognitive, motor, physiological, and personality traits that underlie human performance. For example, hitting a baseball requires special visual abilities, surfing requires certain balance abilities, and orienteering requires spatial ability. People who possess great amounts of the unique abilities required for a particular activity have greater potential for success in that activity than those who did not inherit such abilities. Notice that we said that these individuals have greater *potential*, which by itself is usually insufficient for achieving high levels of skill or performance. The key to exploiting and eventually realizing potential is to engage in appropriate physical activity experiences.

KEY POINT

Ultimately, how well we are able to perform an activity is determined by our physical activity experiences and by our abilities, which are influenced by genetics.

Think of abilities as the foundational building blocks on which we construct our experiences. People who possess greater amounts of the abilities required for a given activity have the potential for higher achievement, but a person will not realize that potential unless he or she also capitalizes on opportunities to improve performance through practice and training. Thus, the totality of your abilities and your practice and training experiences determines the highest level of competence that you are able to achieve in a particular activity. Sometimes, people inherit the abilities required by an activity but fail to exploit those abilities through practice and training; such individuals are sometimes viewed as underachievers. Others appear to have little natural ability for an activity but compensate by engaging in an unusually ambitious practice and training schedule; these individuals are sometimes viewed as overachievers.

Genetic factors also influence our body proportions, skeletal size, bone mass, limb length, limb circumference, and distribution of muscle and fat, all of which influence our capacity to perform various physical activities. If you inherited a small frame, you probably won't excel in American football, rugby, or sumo wrestling, no matter how much you practice and train. If you inherited a very large frame, don't expect to make the Olympic team in gymnastics, ice skating, or diving—sports in which a compact body configuration is essential to rapid twisting and turning. Figure 3.7 shows some theoretically ideal body proportions for selected sport activities. Keep in mind that although inherited body characteristics may be a limiting factor for those who aspire to elite status as highly specialized performers, this consideration is not important for most people. Indeed, individuals who hope to achieve moderate levels of performance that enable them to participate at a recreational or social level often find that appropriate training and practice enable them to achieve their goals.

Interaction of Experience and Abilities

The contribution of inherited abilities to the performance of physical activity is a much more complicated issue than most people imagine. Geneticists are just beginning to probe the ways in which heredity contributes to performance, and their explorations so far have led them to be cautious in generalizing about the topic. One method that scientists have used involves measuring performance on a battery of tests by people with common genetic backgrounds. For example, if we correlate the performances of children with those of their parents on a reaction-time test, we can presumably shed some light on the contribution made by hereditary factors. A correlation of 1.0 is a perfect correlation (rarely found), which suggests that almost all of performance can be explained on the basis of heredity. In contrast, a correlation of 0.5 suggests that approximately 25 percent of the performance (0.5^2) is accounted for by common hereditary factors. Table 3.2 shows that the correlations between various relationships and motor performance are generally not large.

The issue is complicated by the fact that genetics not only may supply us with the abilities required to perform certain activities but also may determine how our bodies respond to experience. In this way, we might say that heredity interacts with experience. As a case in point, consider long-distance running, in which performance is largely determined by the runner's *maximal oxygen uptake*, or ability to deliver oxygen to muscles (and use it) and to eliminate carbon dioxide. Endurance training generally increases runners' maximal oxygen uptake, but it does not affect everyone in the same way. A standard bout of conditioning may improve endurance performance by 16 percent for some individuals and by 97 percent for others (Lortie et al., 1984).

Why this wide variability in the physiological response to training? One reason may be the presence of genetic differences in the capacity for oxygen uptake. In other words, this kind of experience (training) appears to interact with various genetic traits. Thus, individuals who are lucky enough to have inherited this training-response ability probably show greater improvement from a standard bout of training than do those who lack it. Heredity may also play a role in heart-rate response to training, stroke volume, and cardiac output during submaximal exercise. In addition, genetic differences have been shown to explain some performance variables, such as adaptation to endurance exercise, muscle strength, and vertical jump height as a measure of dynamic movement.

Research about genetic contributions to athletic performance remains in its infancy, and we have much to do in order to clarify this important research area. Most genetic research in kinesiology focuses on muscle physiology, trainability, and pathologies. One group of researchers who exhaustively examined the possible contribution of genetic variants to elite athletic performance summarized the state of the art as follows: "There is still no evidence that any of these (genetic) variants have any substantial

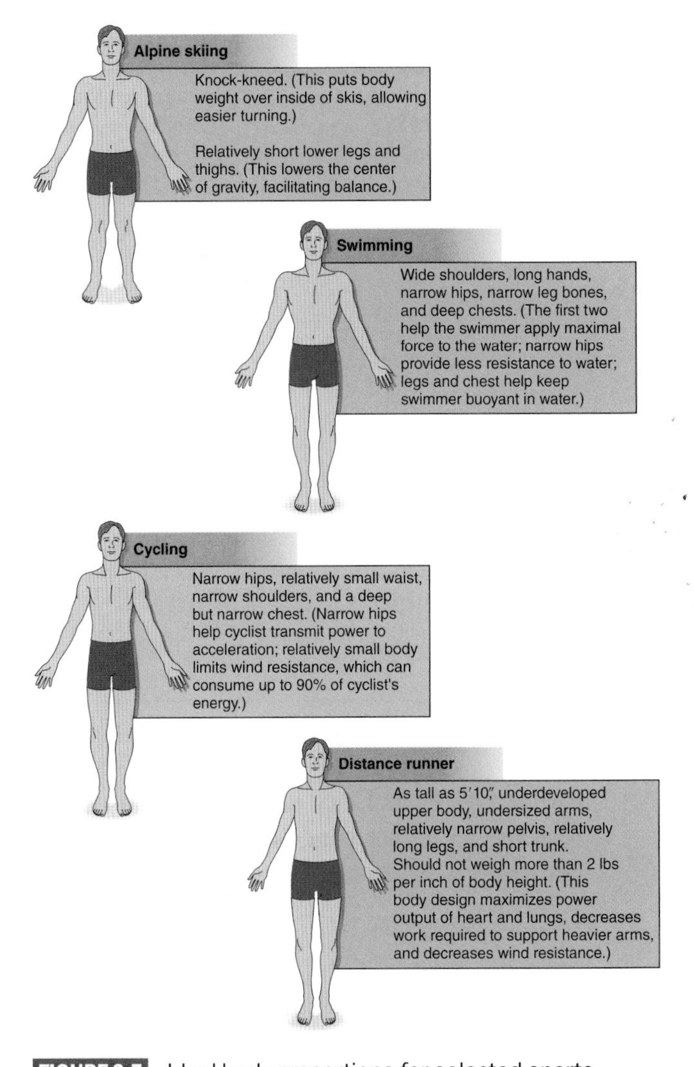

FIGURE 3.7 Ideal body proportions for selected sports.

Adapted, by permission, from R.B. Arnot and C.L. Gaines, 1984, *Sportselection* (New York: Viking Press).

Table 3.2 Correlations for Selected Aspects of Motor Performance and Selected Relatives

Relatives	Reaction time	Movement time	Eye–hand coordination	Strength (push-up)	Strength (grip)	Endurance (sit-up)
Parent and child	0.13	0.09	0.04	0.25	0.20	0.24
Siblings	0.37	0.33	0.32	0.38	0.29	0.34
Uncle or aunt and nephew or niece	0.20	0.18	0.23	0.51	−0.12	0.54

Adapted from Bouchard, Malina, and Perusse 1997.

predictive value for prospectively identifying potential elite athletes" (MacArthur & North, 2005, p. 336). Advances in genetics have recently accelerated, but for now it remains unknown whether we will someday be able to leverage genetic tests of fitness abilities and their trainability in order to identify and train elite athletes.

KEY POINT

The limits of learning and conditioning are determined by both physical activity experiences and inherited *abilities*. We control only experience; therefore, the best strategy for achieving excellence is to seek the best physical activity experiences.

Like the effects of training on conditioning, the effects of practice on learning may be modified by inherited abilities, although the process appears to operate somewhat differently in this case. Theoretically, each motor skill is supported by certain motor abilities, which are distributed unevenly among the population. For example, if you are fortunate enough to have inherited the speed-of-response ability, then you are likely to perform well in tasks that require it, such as sprinting and batting. Similarly, if you inherited an ability for mechanical reasoning, then you may do well in motor tasks that require this ability, such as the operation of complex equipment.

The issue becomes more complicated, however, when we look at the contribution of abilities to skills when they are practiced for long periods. Practicing skills usually increases levels of performance, but scientists have discovered that the abilities required to succeed at low levels of performance may differ from those required to perform at very high levels. In some classes of skills, cognitive abilities such as mechanical reasoning and spatial ability may be important during the initial stages of learning but not during later stages. In contrast, motor abilities such as speed of movement and reaction time may not be important in early stages but may be critical at higher stages of learning (Fleishman & Hempel, 1955). Thus, the effects of practice, like the effects of training, interact with the influence of inherited traits.

Even so, the average person should not decide to engage (or not engage) in a particular activity solely based on whether he or she seems to possess large amounts of requisite abilities for that activity. Genetically endowed ability may be a legitimate concern for a young person who envisions competing in the Olympics someday, but for most of us our collection of inherited abilities—whatever they may be—coupled with practice and instruction will enable us to play most sports at a level adequate to bring us enjoyment. Besides, none of us can ever know for sure what our abilities are. The best advice is to engage in the physical activities that you enjoy, realizing that regardless of your abilities, the only avenue to achievement will be through experience. In most cases, the more high-quality experience you accumulate in an activity, the better you will perform.

Subjective Experience of Physical Activity

Given the variety of external factors that influence physical activity, as well as the obvious physical aspects of experiencing an activity, one can easily forget about the importance of our inner or subjective experience of physical activity. Nonetheless, unobservable emotions are at least as important as the physical, hands-on experience of our favorite physical activities. If we are to become enthusiastic about a particular form of physical activity—be it basketball, aerobics, hiking, weight training, taekwondo, or anything else—then something in the activity must contact our innermost selves in a way that causes us to repeatedly return to it. This relationship between physical activity and our inner lives matters greatly to kinesiology professionals. Indeed, how could we promote physical activity, for

Professional Issues in Kinesiology

Subjective Perception of Overload

The kinesiology principle of overload holds that the intensity of physical training needs to be higher than the performer's current level of fitness in order to produce training effects. Developing an understanding of how much intensity is too little, how much is desirable, and how much is too much constitutes much of the learning that occurs through physical activity experience. Thus, first-hand experience is useful to kinesiology professionals and their clients, as is research about the perception of physical activity intensity, which often uses heart rate or rating of perceived exertion (RPE). This is particularly true for water activities where the person's perception of intensity or overload does not match the actual biomechanical and physiological load on the body (Alberton et al., 2016).

Citation style: APA

Alberton, C.L., Pinto, C.C., Gorski, T., Antunes, A.H., Finatto, P., Cadore, E.L., . . . Kruel, L.F.M. (2016). Rating of perceived exertion in maximal incremental tests during head-out water-based aerobic exercises. *Journal of Sports Sciences, 34*, 1691-1698. doi:10.1080/02640414.2015.113480

ourselves or others, if we could not find the motivation or internal joy associated with such activity?

To consider an example, the experience of running is not limited to putting one foot ahead of the other in a mindless repetition of movements; nor is it simply a matter of having your muscles contract and your heart pound in a mechanical fashion. The experience also involves feeling your racing pulse and the warmth of your body, the sense of energy being released in your muscles, and the pain of fatigue. In addition, it includes the memories that the experience impresses on you, as well as changes in how you think and feel about the act of running and about yourself. In this way, the physical features of activity merely set the stage for our subjective experiences. The late George Sheehan, a runner and a cardiologist who wrote extensively about the spiritual side of running, described it this way: "The first half hour of my run is for my body. The last half hour is for my soul. In the beginning the road is a miracle of solitude and escape. In the end it is a miracle of discovery and joy" (1978, p. 225).

KEY POINT

The realm of subjective experience involves the interior and sometimes mysterious aspect of our lives, in which we collect, recall, and reflect on the feelings and meanings that physical activity evokes for us.

Physical Activity Experiences as Subjective and Unique

Whether physical activity is associated with work, education, or some other sphere of experience, the

Human Desire: The Incalculable Ingredient of Elite Performance

Summing up an exhaustive analysis of the role of genetics in athletic performance, researchers Elaine Ostrander and her colleagues wondered whether the really important variable in attaining athletic success is something called "heart":

> All of the body's critical organs including lungs, muscles, and heart will, when taxed beyond comfort, send signals to the brain ordering the body to cease. Yet in elite athletes we find an extraordinary ability to disobey those signals. . . . When watching the extraordinary performances of some of the athletes mentioned in this article, we perceive an intangible quality that, for lack of a better term, is often called "heart." In every elite athlete there seems to be the ability to "dig down deep" and find the extra energy that is needed to train the extra lap, finish first, break the record, or capture the medal. (Ostrander, Huson, & Ostrander, 2009, p. 429)

feelings we have when we move our bodies differ from the feelings we have when we are sitting or lying motionless. Moreover, physical activities themselves do not all stimulate the same subjective experiences. For example, executing a series of push-ups conjures a distinctly different subjective experience than does hitting a golf ball. Similarly, the feelings associated with playing racquetball differ from those associated with running on a treadmill, and both of these feeling sets differ from the feelings associated with typing on a keyboard, even though all three endeavors require physical activity.

KEY POINT

One of the main reasons we seek out physical activity is that it supplies us with unique forms of human experience unavailable to us in other areas of everyday life.

Subjective Experiences in Physical Activity Often Overlooked

Sometimes we engage in physical activity without pausing to ask ourselves what thoughts and feelings it generates or how it fits into the larger scheme of life. For instance, even if our subjective experiences constitute the underlying reason for which we engage in an activity, we may lose sight of this important fact if we become too involved in competition or in achieving our goals.

Physical Activity Best When Personally Meaningful

Unless we are attracted to the subjective aspects of a given physical activity—that is, unless we discover something inherently enjoyable in it—then the activity is unlikely to become personally meaningful to us. In other words, the physical, tangible side of physical activity provides the raw material for the experience, but the subjective experience keeps us coming back for more. As noted in the chapter opener, some people surf primarily for subjective enjoyment and for an almost spiritual joy in the activity. In sport, and in many other kinds of physical activity, unless you are able to invest the activity with the deepest resources of your mind and will, participation can be hollow and lack meaning.

If you have ever ice-skated in solitude on a frozen pond or cycled along a country road on a brisk fall morning, then you have probably enjoyed the subjective pleasures offered by physical activity. The same pleasures may come to you when you're struggling through a difficult wrestling practice in a hot, humid room or when your lungs are bursting in the final stages of a tough aerobics workout. Engaging in physical activity can touch us emotionally, mentally, and spiritually. When we allow this to happen, our physical activity experiences are more meaningful.

Why Subjective Experiences Are Important to Kinesiology Professionals

Undergraduate kinesiology students need to learn about subjective experiences in physical activity for several reasons. For one thing, subjective experiences can help you clarify the basis of your career choices. For instance, you might plan a career as an athletic trainer because you want others to have the same subjective experiences that you had when you were in rehabilitation for a torn ligament. Or you might decide to be an exercise instructor because you want to help others feel the self-confidence that comes from physical fitness. Or perhaps you will decide to become a physical education teacher or coach because of the good feelings that you experienced when participating in sport in high school.

Listening to Your Body

As noted, we sometimes don't pay attention to the subjective experiences of physical activity. Here, however, is a much different point of view from philosopher Heather Reid: "In contrast to such sedentary activities as taking tests or watching TV in which we try to ignore our bodies, sport forces us to *listen to our bodies*. . . . [Athletes] savor the exhaustion that follows a well-played game or a race well run, even as they welcome the pain-numbing endorphins that miraculously quiet their bodies' screams midway through a hard effort. . . . What most athletes don't know, but philosophical athletes do know, is just how much can be learned from listening to their bodies. Listen closely enough to your body and you might discover a window to your soul" (2002, p. 52).

Learning about subjective experiences can also help you develop your skills as a kinesiology professional. For example, physical educators who have experienced the feelings associated with the clumsy beginning trials of a skill can better serve their students; public health professionals who experience environmental barriers to physical activity can better relate to community resistance; and personal trainers and fitness leaders who have suffered through the painful early stages of a training regimen are better able to give their clients assurance and encouragement. Such knowledge can be useful in designing programs and interventions that help people understand the subjective experiences and meaning associated with their physical activity.

But the most important reason to study and learn about subjective experiences lies in the fact that how we feel and what we think—before, during, and after we engage in a physical activity—largely determine whether we will make that activity an ongoing part of life. Being attracted to mountain climbing rather than canoeing, to aerobics rather than running, or to baseball rather than track and field may be traced in part to the subjective experiences that these activities have evoked in us. Here, we explore how physical activity can affect us in ways that cannot be measured by stopwatches, strength gauges, or competitive points; in other words, we focus here not on external performance but on internal dynamics.

The Nature of Subjective Experiences

Imagine that you have just run a half marathon in your best time ever. Now imagine that after an extremely tough football practice you are sitting on a bench in the locker room, tired beyond belief but also surprised at how good it felt to endure the discomfort and pressure of that hot August practice. Now imagine yourself reflecting on a mountain hike that you took two summers ago. You recall the dense forests, the musty smell of damp leaves and earth, the sounds of birds chirping around you, the steep and winding hills you climbed, the treacherous descents, and the exhilaration of having exerted your body to its limits.

These three situations would probably stimulate distinctly different feelings in you. In the first, your thoughts and feelings would likely be intense and sharply focused on your subjective experiences while you were running the race. The sensations of a heightened pulse rate, near-cramping legs, and burning lungs might all be indelibly impressed on your mind.

In the second scenario, you might feel pride in having gutted it out and finished the practice, but you might also think that you learned a valuable lesson about yourself and your capacity to endure discomfort. Although these thoughts and feelings are less immediate than those evoked by the first scenario, they still constitute, in part, a direct response to the activity just performed.

In the third scenario, your thoughts and feelings might be nostalgic, mellow, and far removed from the experience of performing any specific activity. No doubt you would have had many other experiences in the years since you took the hike. Of course, you could not experience this new sense of meaning without having been deeply involved in the activity in the first place, but neither could you discover this additional layer of significance without the mellowing and integration of memories over the intervening years.

These scenarios illustrate how multifaceted subjective experiences can be. They can be intense and immediate, or they can involve past experiences that you relive repeatedly. These two general kinds of experiences deserve special attention.

Immediate Subjective Experiences

When we engage in physical activity, our movements create immediate emotional and cognitive impressions. Our bodies are equipped with movement sensors called *proprioceptors*—sensory devices in tendons, ligaments, muscles, and the inner ear that are stimulated by physical actions. Proprioceptors provide us with information about the body's movements and its position in space. This information is at least as important as the unending array of visual and auditory information that often impinges on us when we move. Another sensory apparatus, this one in the circulatory system, picks up biochemical changes in the blood that affect our perceptions of fatigue and effort. Thus, each time we engage in physical activity, complex sensory signals flood our nervous system. These signals are vital to our performance and safety, and they provide raw data that gets transformed into perceptions, feelings, and knowledge. Of course, we can choose to ignore many of these sensations, just as we can ignore the vast array of subtle colors in a beautiful painting or the nuances of sound in a symphony recording. However, remaining open to the sensory information that accompanies physical activity is the first step toward appreciating its subjective dimensions.

KEY POINT

All physical activity is accompanied by sensations that we can convert to perceptions, emotions, and knowledge. To fully experience a physical activity, we must be open to the emotional and cognitive impressions that it evokes.

Replayed Subjective Experiences

Although an attention-demanding activity may sometimes yield little in the way of subjective experiences while it is being performed, one may discover much later that it has produced a lasting effect. Subjective experiences can endure in memory for months, years, or even a lifetime. In one of the scenarios that opened this section, you reflected on a hike taken two summers ago and appreciated in a new and exciting way what it had meant to you. A person might experience the same feelings when she thinks about the day that she earned a fitness award or about the early-morning high school exercise program that taught her how to control her weight and eat nutritious meals. The process by which we reexperience the subjectiveness of physical activity is called **self-reflection**, which, in simple terms, involves reliving a past experience. In this process, it is as though we replay a recording that includes not only visual but also kinesthetic, auditory, and other impressions from an earlier time. Because of the passage of time, self-reflection often enables us to place specific subjective experiences within a framework that makes them more meaningful.

KEY POINT

The subjective experiences that accompany physical activity may profoundly affect us during and immediately after our performances. Reflecting on these experiences later can help us put them into more comprehensive and meaningful frames of reference.

Components of Subjective Experience

When exploring a territory as broad and expansive as subjective experience, it is helpful to break it into its various components. To that end, the following sections address sensations, perceptions, emotions, and various ways in which we gain knowledge through subjective experience.

Sensations and Perceptions

Sensations are raw, uninterpreted data collected through sensory organs. Our experiences help us interpret these raw sensations as meaningful constructs or **perceptions**. A barrage of sensations is automatically stimulated when we engage in physical activity—for example, the feelings of contracting our muscles and moving our bodies and, perhaps, the feelings of an increased heart rate, of breathlessness, or of muscle fatigue. If we attend to such sensations, they provide us with a direct report of our inner states. Organized into meaningful information by the process of perception, these sensations allow us to make ongoing corrections in our physical activity. For instance, perceptions of fatigue or discomfort usually form the basis of our decision about when to stop an exercise. Similarly, perceptions inform our judgment about how forcefully we are moving, how much resistance our movements are encountering, or, in the case of divers and gymnasts, how our bodies are oriented to the ground. Sensations from outside of our bodies also become part of our subjective experiences. What we see or hear when we perform, and even what we smell, all coalesces into a cohesive subjective experience.

Emotions

Perceptions during physical activity can elicit various internal reactions. For example, physical activity may either increase or dampen our level of excitement and motivation. We may become angry, annoyed, surprised, pleased, disappointed, or enthusiastic. Such reactions can stem from the feelings elicited by an activity, by our impressions of the quality of our performance, or by the outcome of a particular event. These subjective reactions are called **emotions**, and they differ depending on the person and the situation. For instance, achieving a personal goal or performing well in an activity may result in the emotion of elation or joy, whereas failing to achieve a goal or performing poorly may prompt disappointment or shame. Psychologists have identified more than 200 emotions but have yet to agree on precisely how to define them. It seems clear, however, that emotions involve a complex combination of psychological processes, including perception, memory, reasoning, and action (Willis & Campbell, 1992).

The feelings associated with physical activity may be just as important as its physical benefits.
Dragan Trifunovic/fotolia.com

KEY POINT

Physical activity is always accompanied by internal and external sensations of human movement. These sensations can be organized into perceptions, which can give rise to emotions that we associate with particular physical activities.

Knowledge as an Element of Subjective Experience

When a person says that she has gained knowledge, we usually interpret this statement to mean that she has learned a fact or perhaps something like a mathematical formula or a story theme from a great work of literature. This type of knowledge is called **rational knowledge**, and you will rely on it as you read this book and as you study the theoretical aspects of physical activity, such as exercise physiology, biomechanics, and sport history.

A different type of knowledge, which is equally important, derives from our subjective experiences in physical activity. This **intuitive knowledge** results from a process by which we come to know something without conscious reasoning. It can be difficult to describe this type of knowledge or to know just how we master it. Intuitive knowledge gained from subjective experiences is usually personal in that it concerns ourselves rather than an exercise, a skill, or another person. Philosopher Drew Hyland (1990) identified three types of intuitive or personal self-knowledge associated with participating in physical activity: psychoanalytic self-knowledge, mystical knowledge, and Socratic self-knowledge. Because knowledge gained through subjective experience is closely associated with our motivations for participating in physical activity, let us now discuss each type in some detail.

Psychoanalytic Self-Knowledge

Psychoanalytic self-knowledge concerns our deep-seated desires, motivations, and behavior. This form of self-knowledge relates primarily to the types of activities in which we choose to participate and the manner in which we pursue them. For example, when a young woman with an impoverished, disadvantaged background chooses to participate in a sport such as polo, golf, or squash, her choice may constitute an effort to transcend social barriers. When a young man chooses contact sports over noncontact sports, his choice may reveal a love of competition and physical contact—or it could represent "a precariously controlled desire to physically dominate or even hurt others" (Hyland, 1990, p. 74). Many active sportspersons probably do not access this type of knowledge, but that doesn't mean such knowledge is unavailable to them. For instance, it has been hypothesized that, for some people, participation in high-risk sports (e.g., hang gliding, scuba diving, motor-car racing) may be "propelled by personal needs to seek god-like experiences, which can activate an inflation of the ego" (Heyman, 1994, p. 194).

Psychoanalytic self-knowledge may also be revealed by the manner in which we choose to participate. Some exercisers, for example, are so intent on performance improvement that they cannot enjoy their participation. As a result, they may develop a narrow, single-minded pattern of behavior that does not serve them well in ordinary life, where they must meet many different responsibilities. Similarly, a softball player who is a poor loser may have invested an inordinate amount of her self-worth in the game; the same may be true of a winner who gloats. Thus, we can learn a great deal about ourselves by carefully examining both the activities in which we choose to engage and our styles of involvement.

Mystical Knowledge

Mystical knowledge, termed *Zen self-knowledge* by Hyland (1990), consists of subjective experiences that are available to experienced performers in rare and special circumstances. It touches on a dimension not experienced in ordinary life. The experiences that give rise to this type of knowledge are sometimes characterized as *transcendent*, thus suggesting that they take performers out of the real world. These subjective experiences can be so powerful that memories of them remain with performers for years, perhaps for a lifetime. The most frequently cited type of mystical experience is called **peak experience**. Research suggests that peak experiences tend to come involuntarily and unexpectedly (Ravizza, 1984). In these experiences, time usually seems to slow down, and specific features of the environment sometimes stand out in sharp contrast to the background. In addition, performers usually report being totally absorbed in the activity.

You might also have heard people speak of experiencing "flow" or "being in the zone." Both are quite similar to peak experience. Psychologists Susan Jackson and Mihaly Csikszentmihalyi interviewed many elite athletes and asked them to describe their flow experiences. For example, Simon, an elite cyclist, talked about his last day in the Tour de France:

> My body felt great. Nothing, you feel like just nothing can go wrong and there's nothing that will be able to stop you or get in your way. And you're ready to tackle anything, and you don't fear any possibility happening, and it's just exhilarating. Afterward I couldn't come down, I was on a high. I felt like I wanted to go ride, up that hill again. (Jackson & Csikszentmihalyi, 1999, p. 9)

Performers describe such experiences as opening a window to a more extensive and inclusive world. This type of experience may include feelings of awe and reverence, to which a performer may attach a religious or philosophical interpretation. In other cases, performers may describe gaining new insights into themselves as beings connected to an all-encompassing universe.

Socratic Self-Knowledge

Socratic self-knowledge involves understanding the difference between what we know and what we don't know—or, in the realm of exercise and sport, what we can and cannot do. Knowing our performance limits can help us operate within the boundaries of our skills and abilities. Coaches give testimony to the importance of Socratic self-knowledge when they say that athletes should "play within themselves."

Ignoring personal limits can lead to disastrous performances and physical harm. Such is the case for weightlifters who try to lift beyond the amounts recommended for their weight class and for ski jumpers who challenge heights for which they have no preparation. The purpose of practice and training is to expand the range of our performance abilities systematically and gradually, thereby not only improving our performance quality but also reorienting our perceptions of our own limitations in light of newly developed skill or performance capacity. In this way, performers can set realistic goals that allow them to test their performance limits both safely and productively. Helping them do so is one of the important roles played by personal trainers, coaches, physical education teachers, physical therapists, and exercise leaders.

KEY POINT

The performance of physical activities can be a source of knowledge about our motivations for engaging in activity, about different dimensions of reality, and about our personal performance capabilities.

Talking About Subjective Experiences

Since subjective experiences are unique, you cannot know precisely what the person riding the stationary bicycle next to you is experiencing; neither can that person know exactly what you are experiencing. Because subjective experiences are inherently interior, the only way to compare our experiences with those of others is to talk about them. Some people describe their experiences quite eloquently. For instance, Yiannis Kouros, a legend in the world of ultra running, once explained what he felt when he ran:

> During the ultras I come to a point where my body is almost dead. My mind has to take leadership. When it is very hard there is a war going on between the body and the mind. If my body wins, I will have to give up; if my mind wins, I will continue. At that time . . . I see my body in front of me; my mind commands and my body follows. This is a very special feeling, which I like very much. . . . It is a very beautiful feeling and the only time I experience my personality separate from my body, as two different things. (Kouros 1990, p. 19)

We have no reason to suspect that Kouros was exaggerating or fabricating his experiences, but it is impossible to validate his accounts. We can know for certain only what we experience; we cannot have the same certainty about the experiences of others.

If you've never before talked about your subjective experiences in physical activity, you may find it difficult at first. Most people are hesitant to share their deepest feelings—about physical activity or anything personal—especially with others whom they don't know well. You may worry that what you say about your subjective experiences will be misunderstood or seem silly to others. Such feelings are normal, because most of us have had little practice in sharing such information, and we have not been encouraged to do so. Unfortunately, the subjective aspects of human performance often get less attention than do the physical aspects from coaches, personal trainers, aerobics instructors, and physical education teachers. Thus, these individuals rarely encourage students to talk about the subjective aspects.

Even when we overcome our natural reluctance to talk about our subjective experiences, we often cannot find the right words to describe our feelings. In fact, Scott Kretchmar has suggested that words may simply be inadequate to describe what we feel about physical activity. He pointed out that "the 'medium of exchange' in sport is 'feel'" and noted that "any meaningful distinctions in this realm typically outrun any verbal ability to refer to them" (Kretchmar, 1985, p. 101).

KEY POINT

It is often difficult to find the words to communicate our subjective experiences in sport and exercise. Doing so matters, however, because it helps us understand the meanings that we create in physical activity.

Nevertheless, words remain our primary medium for communicating our subjective experiences. Furthermore, struggling to put our feelings into words may help us come to grips with the deeper meanings that physical activity holds for us. Sy Kleinman, a philosopher who is particularly sensitive to the meanings of physical movements, would agree: "Engagement in game, sport, or art and *description* of this kind of engagement enable us to know what game, sport, or art is on a level that adds another dimension to our knowing" (italics added; 1968, p. 31). Therefore, although you may feel inadequate to describe your physical activity experiences, such descriptions can provide an important means for clarifying to yourself why you engage in activity and what it means to you.

Factors Affecting Our Enjoyment of Physical Activity

It is not altogether clear how internalizing physical activity differs from learning to enjoy it. Clearly, we are unlikely to internalize an activity that we don't enjoy. If you were asked why you work out regularly, go in-line skating, or play lacrosse, you might answer that you find the activity enjoyable. This answer may sound simple, or even simplistic, but the concept of enjoyment is complex. Let's focus briefly on this concept in order to better understand how enjoyment attracts us to physical activity.

What is it about physical activity that causes people to enjoy it? Although many factors are probably involved, we focus here on three general categories: factors related to the activity, factors related to the performer, and factors related to the social context in which the activity is performed.

Factors Related to the Activity

If you reflect on the physical activities that you enjoy and ask yourself why you enjoy them, you probably think of specific characteristics of the activities. For instance, you might enjoy tennis because you prefer individual sports to team sports and because you like a game that makes you concentrate. You might prefer aerobics to running because you enjoy moving your body to music. In both examples, the enjoyment comes from a specific quality of the activity itself. Csikszentmihalyi (1990a) identified several characteristics of activities that make them enjoyable; here, we discuss the following three: (1) balance between challenge and ability, (2) clear goals and feedback, and (3) competition.

Evenly Matched Challenges

Imagine that two students want to develop their strength by using free weights. The first student selects a weight that is very easy to lift (say, 30 pounds, or 14 kilograms), and each day he lifts this weight 10 times, never increasing the weight or the number of repetitions. Soon, he becomes bored, realizes that he's not enjoying the activity, and quits. The second student decides to begin his weight training program by trying to lift 200 pounds (91 kilograms). Each day he comes to the gym and pulls with all of his might but is unable to lift the weight off of the ground. Soon, he becomes frustrated and, like the first student, realizes that he's not enjoying the activity; he also quits.

As shown in figure 3.8, when our skills and abilities go far beyond the challenges of the activity, we usually experience boredom. In contrast, when success requires skills and abilities that we lack, we often experience frustration or anxiety. As Csikszentmihalyi put it, "Enjoyment appears at the boundary between boredom and anxiety" (1990a), which is another way of saying that enjoyment requires a delicate balance between the challenges of the activity and the skills and abilities of the performer.

Clear Goals and Feedback

Part of the fun we have in sport derives from testing our skills and abilities against the challenges of the game. For example, making a golf shot is something that we try to accomplish by executing a series of coordinated movements. The enjoyment stems not so much from the fact of the goal being accomplished—that is, the ball being in the cup—as from our having *tried and succeeded* at putting the ball into the cup. Thus, it is not surprising that the enjoyment we experience in reaching a goal is greatest when the goal is difficult to attain. If we attained the goal on every attempt, our enjoyment would soon vanish.

If trying to attain a goal is central to our enjoyment of an activity, then it makes sense that lacking a clear goal can detract from our enjoyment. This situation can occur when we find ourselves in a game—say, cricket—in which we don't understand the rules or purpose. Similarly, exercising on a piece of equipment without knowing much about it may be unenjoyable because we lack a clear sense of what we are supposed to be doing.

Our enjoyment of physical activity also depends on receiving some degree of feedback. Imagine yourself shooting free throws. Now imagine that each time the ball leaves your hands, the lights go out and you never know whether the ball goes through the hoop. At first, the novelty of the situation might make it fun, but soon it would become dull. Thus, even if we have a clear idea of our goal for an activity, our enjoyment also depends on knowing how we are doing in relation to that goal. Of course, activities vary in the degree of feedback they provide. For example, table tennis and squash provide a good deal of immediate feedback. In marathon running, on the other hand, the only outcome of the contest is decided more than 2 hours after the starting gun is fired; thus, this activity provides much less feedback.

Learners also receive important feedback from teachers, coaches, and trainers. After each trial or group of trials, a teacher usually comments on any major mistakes made by performers and provides recommendations for correcting those mistakes. This information, when coupled with encouragement, can affect performers' subjective experiences. For example, age-group swimmers are more likely to enjoy swimming when they perceive their coach as one who gives both informative feedback and encouragement after an unsuccessful performance (Black & Weiss, 1992).

KEY POINT

We are more likely to enjoy a physical activity when its challenges match our abilities, when it has clear goals and is followed by feedback, and when it is arranged in a competitive framework.

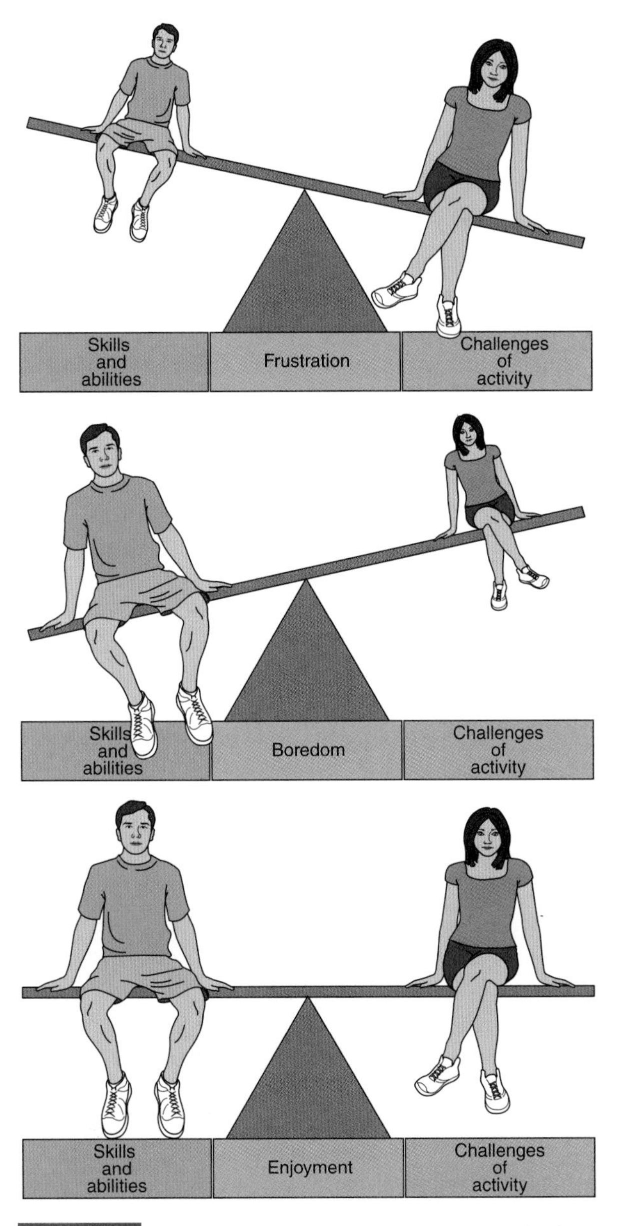

FIGURE 3.8 We are more likely to enjoy a physical activity when its challenges are balanced by our skills and abilities.

Competition

As noted in chapter 2, enjoyment of physical activity is often heightened by competition (Csikszentmihalyi, 1990b). **Competition** is not an activity per se but an organizing principle that frames physical activity within a larger purpose. Usually, it is a way of relating to other people by comparing one's performance with theirs. It may also involve comparing one's performances with a standard (e.g., golf par), a personal best, or an event record. Just as we can add the dimension of poetry to language or the dimension of harmony to music to make them more enjoyable, we can add competition to physical activity to make it more enjoyable.

On the other hand, adding competition to an activity can become a problem if competition becomes an end in itself. When coaches or players value winning above enjoyment of the activity—or above the basic individual and social values of friendship, caring, and cooperation—then they elevate competition to the status of an end rather than a means. The issue is summarized as follows by Csikszentmihalyi:

> The challenges of competition can be stimulating and enjoyable. But when beating the opponent takes precedence in the mind over performing as well as possible, enjoyment tends to disappear. Competition is enjoyable only when it is a means to perfect one's skills; when it becomes an end in itself, it ceases to be fun. (1990b, p. 50)

Factors Related to the Performer

Our enjoyment of physical activity is also influenced by factors that lie within us—specifically, our dispositions and attitudes. **Dispositions** are short-term, highly variable psychological states that may be affected by a host of external factors. **Attitudes** are relatively stable mind-sets about things that may be favorable or unfavorable.

Disposition

Csikszentmihalyi (1990a) identified three dispositions affecting enjoyment that are particularly relevant to physical activity: (1) how competent we feel in performing the activity, (2) the extent to which we are able to become absorbed in the activity, and (3) how much control we feel we have over the activity. Here is a brief description of each.

› **Perceived competence:** As a rule, we enjoy activities in which we do well more than those in which we feel incompetent. Psychologists use the term *self-efficacy* to refer to how adequate we feel to perform a task. Research has shown, for example, that people who feel competent in exercise are more likely to adhere to a rehabilitative exercise program (Dishman & Sallis, 1994), are more likely to engage in higher levels of intensity when participating in physical activities (Sallis et al., 1986), and are more faithful in attending exercise programs (McAuley & Jacobson, 1991). In another study, aerobic dancers with higher ratings of self-efficacy reported enjoying the activity more and exerted more effort than did those with low ratings of self-efficacy (McAuley, Wraith, & Duncan, 1991).

› **Absorption:** We all remember times when we were deeply engrossed in an exercise or sport activity, as well as other times when we simply went through the motions. When we surrender to an activity and become absorbed in it, we lose consciousness of ourselves as distinct entities apart from the activity. Csikszentmihalyi (1990a) found that when we lose a sense of self while performing an activity, our feelings of enjoyment increase, partly because when we aren't self-conscious, we aren't as likely to judge our own performance. The reason lies in the fact that most of us set higher expectations for our performances than we are able to achieve; therefore, self-evaluation usually results in negative feedback, which is not enjoyable.

› **Perceived control:** It is inherently enjoyable to experience a sense of mastery or control over our environment. Knowing that we can master a tough cross country course, control the basketball (when dribbling, passing, or shooting), control our opponent in a wrestling match, or control our bodies in a difficult gymnastics skill adds to our sense of enjoyment of those activities.

KEY POINT

Temporary dispositions affect our enjoyment of physical activity. Enjoyment tends to increase when we feel competent in an activity, become absorbed or lost in it, and feel that we control our body and the environment.

At first glance, this last principle might seem to be contradicted by people who enjoy **sensation-seeking activities** (e.g., hang gliding, parachute jumping, ski jumping, rock climbing), in which so many factors are uncontrollable. Strange as it may seem, however, a sense of control is one of the factors that draws people to such high-risk sports (McIntyre, 1992). In these cases, enjoyment appears to stem not from the presence of threat-

Have you ever put the value of winning above the enjoyment of the activity itself?

ening forces but from feelings of being able to control these forces through elaborate preparation, training, and painstaking adherence to safety procedures. After all, if we had complete control of our bodies and the surrounding environment, sport of all kinds would be boring because we could always accomplish our goals. Thus, some degree of risk and uncertainty is fundamental to our enjoyment.

Attitudes

Along with these temporary dispositional factors, more stable and enduring attitudes also can affect our enjoyment of physical activity. Just as individual preferences and attitudes vary widely in regard to food, movies, clothes, friends, cars, and the like, the same is true about physical activities. It is largely inexplicable why we vary so much in our tastes regarding physical activity, but there is no question that we do. In fact, a scale for assessing attitudes toward physical activity was designed in the late 1960s by Gerald Kenyon (1968) and his associates. They used the scale in surveys of individual preferences regarding a wide range of activities—in relation to which people could be either participants or observers—and identified six categories of attitudes toward physical activity. Each category represents a distinct preference for a particular form of physical activity based on what are called "sources of satisfaction," which accrue to performers or observers: social experience, health and fitness, pursuit of vertigo, aesthetic experience, cathartic experience, and ascetic experience.

Physical Activity as a Social Experience Some forms of physical activity, such as team sports and group exercises, are intrinsically social events; that is, they normally take place in active social environments and involve interaction with other people. The most social sport activities are team sports, as well as team aerobics or other exercises done in health clubs where others are present. Some other forms of physical activity (e.g., rock climbing, surfing, long-distance running) incorporate less social interaction, but even those activities may provide opportunities for social engagement before, during, and after participation. Generally, participants consistently characterize the social aspects of physical activity as positive experiences (Neulinger & Raps, 1972). More specifically, both youngsters engaged in youth sport (Wankel & Krissel, 1985) and elite ice skaters (Scanlan, Stein, & Ravizza, 1988) have indicated that developing friendships and experiencing social relationships are important sources of enjoyment. Thus, individuals who enjoy high levels of social interaction are more likely to seek out physical activities that maximize such experiences.

Physical Activity for Health and Fitness Some people value physical activity for the contribution that it makes to their health and fitness. When a person feels "in shape" and able to meet whatever physical demands may be faced in the course of a day, he or she enjoys a sense of confidence and well-being that adds immensely to daily quality of life. However, as we have seen, many people simply enjoy engaging in activities that happen to lead to fitness. The unique sensations that accompany the physiological response to vigorous and sustained exercise cause many people to return again and again to running trails and gymnasiums.

Physical Activity as the Pursuit of Vertigo Some people are attracted to a certain category of physical activity that presents an element of risk or thrill, usually through the medium of "speed, acceleration, sudden change of direction, or exposure to dangerous situations, with the participant usually remaining in control" (Kenyon, 1968, p. 100). The thrill that comes from disorientation of the body in such activities is called **vertigo**. (This experience differs from the kind of vertigo caused by certain medical conditions, in which a person can be quite nauseated, dizzy, and disoriented even when motionless.) Some people pursue vertigo through sensation-seeking activities that reorient the body with respect to gravity, as in amusement park rides, free-fall parachuting, bungee jumping, downhill skiing, ski jumping—or any other activity that creates a sense of danger, thrill, and intense excitement.

We do not know why some individuals seek out such activities whereas others just as deliberately avoid them. Petrie (1967) suggested that those who pursue vertigo may tend habitually to reduce perceptual input so that they experience a specific event at a lower level of intensity than do people who habitually augment (exaggerate) perceptual input. Involvement in vertiginous activities may also be an attempt to compensate for tedious experiences in the workplace (Martin & Berry, 1974), a form of stimulus addiction (Ogilvie, 1973), or the result of complex sociocultural factors (Donnelly, 1977). Whatever the cause, the level of interest in such activities, whether as participant or as spectator, shows no sign of declining.

Physical Activity as an Aesthetic Experience We normally associate **aesthetic experience** in physical activity with dance, but some individuals, as either participants or spectators, perceive sport—and to a lesser extent, exercise—as providing certain artistic or aesthetic experiences (Thomas, 1983). In sport, the aesthetic element is especially prevalent in activities such as gymnastics, diving, and ice skating, in which grace and beauty of movement constitute primary considerations in awarding scores to competitors. The aesthetic element is less apparent in sports in which the outcome is usually viewed as more important than the manner in which the athlete moves—for example, football, basketball, and baseball. Nevertheless, it is common to hear sportscasters refer to a "well-choreographed" football play, speak of the dancelike characteristics of a basketball player leaping through the air, or even describe the movements of a tennis player with a phrase such as "art in motion."

Exercise seems less likely to be a source of this type of subjective experience, although dance aerobics clearly contains a strong aesthetic component. Aesthetic experience in physical activity can also be facilitated by outdoor activities (e.g., hiking, kayaking, mountain climbing) that bring participants into contact with the beauty of the natural environment. Women are more likely than men to engage in physical activities for aesthetic experiences, whereas men are more likely to value activities for their vertiginous, ascetic, or cathartic elements (Smoll & Schutz, 1980; Zaichowsky, 1975).

KEY POINT

People who find enjoyment in the aesthetic experience of physical activity may engage in gymnastics, diving, skating, and other movement forms that emphasize grace and beauty of motion.

Physical Activity as a Cathartic Experience The notion of catharsis posits a purging or venting of pent-up hostility, either through aggressively attacking an enemy or inanimate surrogate (i.e., object) or through watching an aggressive event. This hypothesis owes a debt to the work of behavioral physiologist Konrad Lorenz (1966), who wrote in his best-selling book *On Aggression* that sport furnishes a healthy safety valve for "militant enthusiasm," which he viewed as the most dangerous form of aggression. Scientists have since questioned this "hydraulic hypothesis" of aggression, both generally (Berkowitz, 1969) and in its application to sport. In fact, it is no longer viewed as a credible hypothesis for explaining our attraction to sport. For example, if watching sport lowered levels of aggression in spectators, we would expect to find lower levels of aggression in spectators following their attendance at a game than before the game began, but research has shown just the opposite to be the case (Goldstein & Arms, 1971).

However, physical activity can be viewed as cathartic in another way; specifically, scientists have discovered that a vigorous bout of physical activity can lower anxiety (Herring, Jacob, Suveg, Rishman, & O'Connor, 2011), thereby inducing a sense of relaxation and calm (but not necessarily reducing pent-up hostility). Leisure theorists have long pointed to the value of leisure activities, including sport, in promoting feelings of rest and relaxation. Moreover, sport and exercise are novel activities that provide us with a change of pace, which may in itself recharge our batteries by shifting our attention away from problems and worries. Thus, the noontime racquetball player may derive a sense of enjoyment, relaxation, and calm not because he purges hostility but simply because he directs his energies toward a new task and becomes absorbed in its features.

KEY POINT

Participating in sport and exercise does not purge us of hostility or aggression, but it may calm, relax, and refresh us.

Physical Activity as an Ascetic Experience Physical fitness and training programs often require us to delay gratification and undergo pain, sacrifice, and self-denial. These experiences are sometimes referred to as **ascetic experiences**, and elite athletes are accustomed to them in the form of torturous training regimens intended to improve their capacity for performance. Ascetic experiences are also familiar to recreational athletes. Occasionally, for instance, you may see a runner who is (thanks to her attire and running style) obviously a novice and who is wheezing, groaning, and, from all appearances, suffering. In fact, fitness clubs specialize in ascetic experiences; patrons with pained faces struggle to complete their exercise routines while a personal trainer shouts encouragement as they work through the pain.

You may be surprised to learn that not everybody interprets the ascetic experiences of breathlessness, fatigue, and muscle soreness as unpleasant. For example, in one study, 12- and 13-year-old soccer players told investigators that they enjoyed not only winning, learning new skills, and playing with teammates but also "working hard" and "feeling tired after practice" (Shi & Ewing, 1993). In another study (Wankel, 1985), exercisers who did *not* drop out of an exercise program reported experiencing physical discomfort more frequently than those who *did* drop out. In this study, middle-aged men who reported experiencing fatigue because of vigorous exercise also reported *decreased* feelings of distress and *increased* feelings of psychological well-being because of the exercise. Thus, for some people, the discomfort of physical activity is attractive rather than unappealing.

KEY POINT

Participating in exercise or sport can be painful and can lead to uncomfortable subjective experiences. Yet participants do not always interpret these sensations negatively; indeed, the discomfort may be a source of attraction or achievement.

Factors Related to the Social Context

Our enjoyment of physical activity can also be affected by the social context in which the activity occurs. Have you ever competed in an important sport contest with a large crowd looking on? How did that experience compare with playing the same sport on the playground or in an intramural setting? Do you feel different when you work out in a fitness center with many onlookers than you do when you work out alone? What about exercising with music playing in the background? Most people find these experiences to differ profoundly from each other. One isn't any better than another; they are simply different.

The way you feel when you are engaged in physical activity may depend on the social conditions surrounding the activity as much as or more than on the activity itself. Specifically, your experience of physical activity can be affected by factors such as the presence of others (e.g., parents, friends, strangers), the hype preceding a contest, your feelings toward or relationship with an opponent, and the way in which media members or coaches interpret a particular game. For example, running on a treadmill offers a different subjective experience than running through a park. Running when it is cold, dark, and raining offers a different subjective experience than running on a bright, warm, summer day. Adding people to your exercise group changes the subjective experience of exercising, just as adding competitors can change the social climate and subjective experience of playing a game. Whether these changes enhance or detract from your enjoyment depends on your individual preferences.

The social atmosphere created by an exercise leader can exert a powerful influence on class participants' enjoyment of the experience. Fox, Rejeski, and Gauvin (2000) reported on a study that tested the effects of leadership style and group environment on the enjoyment experienced by members of a step aerobics class. The instructor, who was highly trained in aerobics, adopted either an "enriched style" that included high levels of

Does Music Improve Exercise Performance?

Are you one of those people who plug into their iPod when they are on the treadmill or exercise bike? Most people who listen to music while exercising do so because it takes their mind off of the pain that usually accompanies a hard workout. However, does listening to music improve performance?

Atkinson (2004) examined whether background music improved starting, finishing, or overall power during a 10-kilometer cycling time trial. The study also evaluated whether music affected heart rate and subjective response (i.e., perceived effort). In the study, 16 subjects performed two 10-kilometer time trials on a Cybex cycle ergometer—one with and one without "trance" dance music (a combination of forms of electronic music) being played in the background. The cyclists also completed a music rating inventory after the time trial in the music condition. When music was played, cyclists completed the time trial in an average of 1,030 seconds; when music was not played, the average time was 1,052 seconds—a difference that is significant both statistically and practically. Music brought about the largest increases in both time trial performance and heart rate during the first 3 kilometers of the trial.

Participants rated the tempo and rhythm aspects of the music as more motivating than the harmony and melody aspects. Although the cyclists performed faster in the music condition, their perceptions of the effort required to cycle were also higher under the music. This finding suggests that listening to music did not make the exercise *seem* easier.

interaction with class members (e.g., addressing them by name, providing positive reinforcement, orally rewarding effort) or a "bland style" (e.g., not addressing participants by name, avoiding conversation, not providing positive reinforcement, not rewarding effort).

The investigators manipulated the group environment by planting undergraduate students in the experimental sessions. The planted students established an "enriched" environment in one exercise group and a "bland" environment in another. In the rich environment, the students introduced themselves to all participants, initiated casual conversation, were compliant with the instructor's wishes, and made positive remarks about the instructor. In the bland environment, the students did not introduce themselves to other members, did not initiate conversation, and were compliant but not enthusiastic or orally supportive of the instructor. Thus, the study established two leadership styles and two group environment conditions. The investigators examined the effects of each condition, as well as combinations of the conditions, on participants' enjoyment. They also examined the effect of these variables on participants' intentions to take another exercise class.

The investigators found that both leadership style and social environment affected the participants' enjoyment and their intentions regarding whether to enroll in another class. Ratings of enjoyment were higher for the enriched leadership style than for the bland leadership style; they also were higher for the enriched group environment than for the bland group environment. The most powerful effect, however, was found for the condition in which participants were exposed to an enriched leadership style *and* an enriched group environment. On average, ratings of enjoyment were 22 percent higher for this condition than for any of the other conditions. Intention to enroll in another class was also highest in this condition (about 16 percent higher than in the other conditions). The conclusion is easy to draw: Both the style adopted by the exercise leader and the social environment in which the exercise occurs can exert a major effect on participants' enjoyment.

Another way in which the social context can affect our enjoyment is by influencing our sense of perceived freedom. As you learned in chapter 2, leisure theorists have long known that we enjoy activities more when we are free to choose them than when we feel obligated to do them. In fact, Bart Giamatti, former president of Yale University and, at the time of his death, commissioner of Major League Baseball, described leisure activity as "that form of nonwork activity felt to be chosen, not imposed" (1989, p. 22). This sense of freedom to participate is not simply a matter of being free from actual coercion; it also involves feeling free from any subtle coercive mental force that might instill a sense of obligation. For example, getting up early on Saturday morning to lift weights with a friend because she has been begging you for weeks to join her robs you of your sense of freedom; as a result, you may have a far different subjective experience in this situation than when lifting weights because you simply want to.

KEY POINT

We never perform physical activities in a vacuum, and the social context that surrounds them can affect our sense of enjoyment.

Watching Sport as a Subjective Experience

One of the most influential people in the development of modern physical education theory was Jesse Feiring Williams. He was one of many medical doctors who became professors and devoted their lives to health promotion by preparing generations of physical education teachers, coaches, and professors. Like many scholars of his generation, Williams (1964; originally published in 1927) was skeptical of sport watching because he believed that it demanded only "simple sensory responses" and did not require "expressive, cooperative skill activity" that could be related to the purposes of education. The rise in opportunities for sport spectatorship that worried Williams has continued into the 21st century and, for the most part, shows no sign of slowing down as we discover more sports and other physical activities to watch—and more ways to watch them. Whether or not all of this sedentary sport watching has been beneficial to the public, it clearly acts as a source of vivid subjective experience for wide segments of the human community.

Because sport watching is connected to the physical activity professions, we summarize here some of the factors that affect our attraction to, and enjoyment of, watching sport. Although this discussion focuses on sport examples, many people also enjoy watching other physical activities that may lack explicit competitive elements—for example, dance and fishing.

Ways of Watching Sport

People engage in sport spectatorship in many ways, ranging from watching a contest while walking by a playground, to watching friends or family members participate in a tournament, to watching live or televised **sport spectacles**. We generally reserve the term *sport spectacle* for collegiate, professional, or international events that attract large numbers of spectators and involve supporting casts of cheerleaders, bands, officials, and members of the media. In such spectacles, both players and spectators fill defined roles. Players, of course, take on various roles associated with the competition per se, whereas spectators play various roles as consumers, supporters, or detractors. Thus, when spectators cheer, jeer, do the wave, or sing an alma mater, they are participating in a spectacle, of which the sporting contest itself is only one aspect.

Vicarious Participation

Vicarious sport participation is a form of watching in which observers participate in a contest through the power of imagination. If, while watching a sport event, you notice yourself tensing your muscles or adjusting your body position in accordance with an athlete's movements, then you probably are participating vicariously in the activity. Vicarious participants are usually fans who identify with a particular player or team and thus have a stake in the contest's outcome. Indeed, the word *fan* is rooted in *fanum,* the Latin word for "temple," which underscores a metaphoric link between religious zeal and the enthusiasm that fans often feel for their favorite teams and players. When their team wins, it is as though the fans themselves have won; when their team loses, they may experience profound disappointment.

Vicarious participation in sport has expanded because of greater access to fantasy sport and sport gambling. Sport fans can now select their own fantasy teams and compete in virtual leagues with friends or strangers. This world overlaps with that of sport gambling when people heighten their participatory excitement by betting on the outcomes not only of sport contests but also of fantasy sport competitions.

KEY POINT

Vicarious participation in sport occurs when spectators engage emotionally with a team or athlete or imagine themselves performing the same activities as the athlete or team they are watching.

Disinterested Sport Spectating

Sometimes we watch sport events without great emotional investment—a phenomenon known as **disinterested spectating**. We are likely to watch a contest in this manner when we care little about the outcome. For example, we may engage in this sort of spectatorship when watching a contest between two teams with whom we are unfamiliar or a game that we have attended only because we were invited by a friend. People also often watch in a disinterested fashion when watching sport is part of their job, as is the case for reporters, referees, hot dog vendors, and sometimes coaches.

Nevertheless, it would be a mistake to assume that disinterested spectators derive no pleasure from watching contests. For example, coaches who watch competitions between teams other than their own tend to observe athletes' performances in a detached, objective manner, yet they may also enjoy and admire those performances. In these cases, spectators watch in much the same way that an audience watches a symphony or ballet; enjoyment comes not from the drama of competition but from observing the skillful actions of the performers.

Fantasy Sport and Sport Gambling

Internet, cable, and smartphone access to college and professional sport means that many people now have virtually constant access to sport competition and commentary. The opportunities for spectator activities have been increased further by the expansion of professional sport into fantasy sport. Do you think that betting on sport competition should be legal? If so, should limits be placed on sport gambling, either in order to protect the integrity of sport or to protect fans from financial ruin? Do virtual sports such as fantasy sport constitute gambling? Or are they virtual games of skill, somewhat like video games?

Factors That Affect Enjoyment of Watching Sport

If someone asked whether you like to watch sporting events, you might answer sometimes yes, sometimes no. Your answer might depend on a number of factors. Few people enjoy watching their favorite team get trounced, and fewer yet are likely to enjoy watching a game that they know little about. In addition, Americans are unlikely to get excited about watching cricket or Australian rules football for the same reason that Australians and citizens of the United Kingdom are unlikely to enjoy American baseball. In other words, our attitude toward a given sport is often influenced by our level of familiarity with it, as well as its importance in our culture. To put it in a nutshell, the level of enjoyment that we experience when watching a sport event is determined largely by three major factors: our knowledge of the game being played; our feelings toward the competing teams and players; and the extent to which the game entails a sense of drama, suspense, and uncertainty.

Game Knowledge

Think about the sports you most like to watch; they are probably the ones that you know the most about. Indeed, knowledge of the game—what sport philosopher Scott Kretchmar (2005) referred to as **game spectator knowledge**—often determines our enjoyment as spectators. Game spectator knowledge involves knowing about the game, including the players, strategies, and competitive tactics. Only when we possess comprehensive knowledge of the activity can we fully appreciate the quality and significance of the athletes' performances. Such knowledge may come from participating in the activity or from watching it or reading about it. Thus, it is not necessary to play a sport in order to enjoy watching it; for example, an individual who knows little about the game of lacrosse may enjoy watching a game without knowing the sport's history, strategies, rules, or personalities. It is unlikely, however, that this person's enjoyment of the game will be as robust as that of a fan who understands precisely what the players and teams are trying to do at every turn of the game.

Feelings Toward Competing Teams and Players

Recall the last time you watched a contest in which one of your favorite teams was competing against an opponent that, for one reason or another, you had come to dislike. What turn of events added to your enjoyment of the experience? Perhaps you enjoyed the experience more when good things happened to your favorite team (it played well) and when bad things happened to the team that you disliked (it played poorly).

KEY POINT

Possessing comprehensive knowledge of a sport's players, rules, and competitive strategies adds to our enjoyment of watching it, as do the feelings that we harbor toward the participating teams.

Connections between enjoyment and one's feelings toward the competing teams serve as the focal point of the dispositional theory of enjoyment. This theory has been the subject of considerable research (Zillman, Bryant, & Sapolsky, 1979) involving spectators of football, tennis, and basketball. For example, the researchers found that spectators applauded the failed plays of the disliked team almost as much as they did the successful plays of the favored team. In addition, by comparing reports from spectators who watched the Minnesota Vikings defeat the St. Louis Rams in a professional football game, the investigators found that those who both liked the Vikings and disliked the Rams reported maximal enjoyment, whereas those who both disliked the Vikings and liked the Rams reported maximal disappointment (Zillman & Cantor, 1976).

Other research has shown that the enjoyment brought about when a fan's favored team wins can lead to increases in self-esteem and self-confidence. In contrast, diminished self-esteem and self-confidence have been associated with losses by one's favored team. Our enjoyment of sport spectating also tends to increase when our favorite team defeats an opponent of high quality rather than low quality. In addition, we tend to enjoy a game more when our favorite team plays much better than we expected (Madrigal, 1995).

Human Drama of Sport Competition

The sense of drama, suspense, and uncertainty that often accompanies sport contests also enhances our enjoyment of them. This sense usually requires that the competing teams be equally matched in talent and ability, though some of the best drama in sport occurs when an underdog overcomes great odds to defeat a heavily favored opponent. Even if the teams are evenly matched, games tend to be less enjoyable when one team far outplays the other, as has been the case in many Super Bowls.

Enjoyment also relates to the extent to which spectators perceive a contest to involve human conflict. Research by Zillman, Bryant, & Sapolsky (1979) showed that when spectators viewed identical sport telecasts, they experienced more enjoyment when the broadcasters' commentary described the game as involving bitter rivals than when the game was described as involving friendly opponents. Enjoyment has also been found to increase when watching rough and aggressive play (e.g., more minutes spent in the penalty box by hockey players), presumably because it demonstrates that the players are sincere in their efforts to win.

These findings demonstrate that in spite of the enormity of the sport spectator business, we know relatively little about the subjective effects of sport watching on the collective psyche. Is watching sport a constructive use of our time? When the game is over, do we feel good about having watched it? Is it possible that sport spectating is a culturally hollow experience—little more than junk food for the human spirit? Is it sometimes even a debasing experience? What changes might make sport spectating a more ennobling human experience? These important questions offer fertile ground for research by future kinesiologists.

Wrap-Up

This chapter begins by making the obvious observation that the key to improving physical activity performance is to accumulate physical activity experience. Clearly, then, we as physical activity professionals should engage ourselves and those we serve in experiences of the right quality and quantity. Therefore, our expertise must include knowing the specific types of experiences that are essential for developing physical fitness or for learning skills or improving physical capacity in a given activity.

Of course, you cannot fully appreciate all of the variables that come into play in prescribing physical activity experiences for your clients until you have completed your study of kinesiology. However, this chapter introduces you to a number of general factors that you must take into account. These factors include considering whether the activity consists of skills that must be learned through practice or physical capacities that must be conditioned through training. In addition, in order to improve performance, any practice or training regimen must focus on the critical elements of the activity and provide in-depth experiences. Finally, experience interacts with inherited traits to bring about performance changes.

When we move our bodies, unique sensations elicited by internal and external sources give rise to emotions, thoughts, and other inner states. These sensations and our interpretations of their meanings constitute the subjective experience of physical activity, which helps determine our activity preferences and constitutes our primary reason for engaging in activity in the first place. It also helps us learn about ourselves in physical activity environments.

The goal of most physical activity professionals is to help people develop an intrinsic orientation toward sport and exercise, though in most cases a participant's extrinsic orientation also plays a part. We can think about intrinsic orientation as enjoyment; anything that we value for its intrinsic qualities usually constitutes an enjoyable experience for us. Many factors can make sport and exercise more or less enjoyable, including those that lie within the structure of an activity itself, those related to the dispositions and attitudes of the performers, and those associated with the social context in which physical activity is performed.

This chapter also examines briefly the subjective experiences associated with watching sport, as well as the fact that people have different ways of watching. Our subjective experience is affected by our knowledge of the game, our feelings about the competitors, and the drama of the competition, including its description by sportscasters. Because significant parts of our profession—sport management, athletic coaching, and administration—are geared toward athletic displays for spectators, we need to carefully monitor the long-term effects on spectators.

This concludes our study of what it means to experience physical activity through performance and spectatorship. You are now ready to proceed with in-depth study of the research knowledge about physical activity. As you become engrossed in the basic concepts of the scholarly subdisciplines of our field, remember that performing physical activity is an important source of our knowledge about it.

REVIEW QUESTIONS

1. List four unique characteristics of human physical activity.
2. What factors influence our decisions about what physical activities to engage in and how physically active to be?
3. What type of physical activity can we improve most through practice? Through training?
4. What is an ability? How might ability level limit the proficiency that we can develop in a given physical activity?
5. Give an example of how an individual might internalize a daily run through the park.
6. Describe three types of knowledge available to us from subjective experiences in physical activity.
7. What evidence exists to refute the notion that people do *not* like the sensations that accompany hard physical effort of exercising vigorously?
8. What types of physical activities might be chosen by a person who values physical activity as an aesthetic experience? As an ascetic experience? As a social experience?
9. What are the various ways in which one can watch sport? In your opinion, which ways add the most to your subjective experience of physical activity?

WWW **Go online to complete the web study guide activities assigned by your instructor.**

PART II
Scholarly Study of Physical Activity

Although it is essential for kinesiologists to learn through their own physical activity experiences, that alone is not enough. Professionals in the field are also defined by their mastery of a complex body of theoretical knowledge about physical activity. Some people who have not formally studied kinesiology may think they know a good deal about physical activity, but their information is often incomplete, out of date, or simply untrue. As a kinesiology major, you—more than the typical person on the street or your friends in other disciplines—need to possess a solid base of knowledge about physical activity.

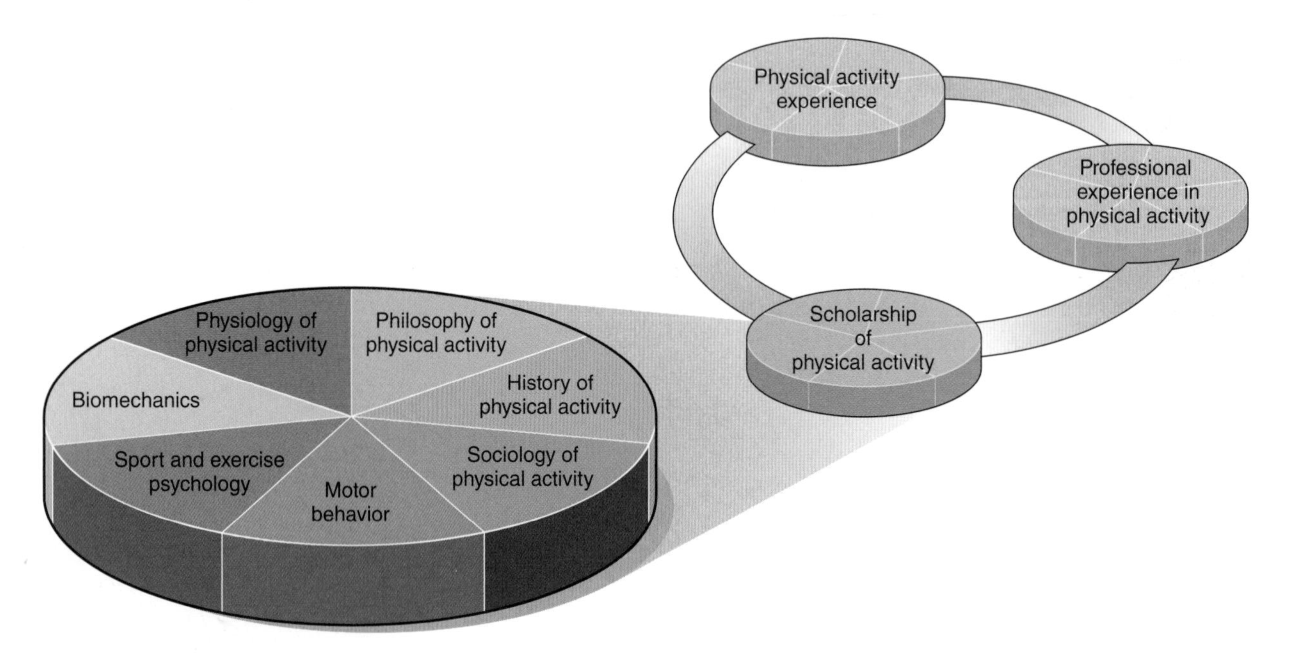

To that end, part II of the book introduces you to the scholarly study of physical activity through seven subdisciplines of kinesiology. Although the field also includes other subdisciplines, as well as various professional areas, these seven subdisciplines have gained recognition as the major categories of kinesiology knowledge. Chapter 4 focuses on philosophy of physical activity, asking and answering questions about values, meaning, and ethics in sport and exercise. Chapter 5 is devoted to the history of sport and exercise; it provides you with a glimpse of how the field of kinesiology has evolved. In chapter 6, you will examine relationships between sport and exercise, as well as some of the pressing social issues of our day, such as racial and gender discrimination. Chapter 7 directs your attention to the performance of skilled movement and factors affecting it, including those related to human development.

In chapter 8, you will study the psychological aspects of physical activity and learn how this important dimension of our lives both affects and is affected by sport and exercise. Chapter 9 introduces you to the fascinating world of biomechanics, which provides insights into the ways in which forces affect and are affected by physical activity. Finally, chapter 10 addresses basic physiological aspects of physical activity, including the effects of exercise on muscular and cardiovascular systems and the physiological basis for training.

Even though these chapters are jam-packed with information, they provide a "once over lightly" preview of these scholarly subdisciplines. Thus, when you finish reading and studying them, you will not have mastered the content or the holistic body of knowledge that constitutes kinesiology. However, you will have taken a glimpse at what lies ahead of you in your program of studies, in which you will delve into each of these subdisciplines in greater depth.

4

Philosophy of Physical Activity

Scott Kretchmar and Cesar R. Torres

CHAPTER OBJECTIVES

In this chapter, we will

- examine the nature of philosophical thinking;
- describe how philosophy fits into the field of kinesiology;
- review the history of the philosophy of physical activity;
- describe research methods used in the philosophy of physical activity; and
- provide an overview of knowledge in this domain by focusing on three issues: the nature of the person, the nature and value of play, and the ethics of sport.

Suppose that you have been successfully coaching track and field in a public high school for almost 25 years. You are recognized as a caring, fair, and committed coach who seeks to help young athletes succeed both on and off the field. Your efforts, accomplishments, and reputation have been rewarded twice with state and national coaching awards.

Your school district wants to develop a policy on transgender athletes, and, given your distinguished coaching career, you have been invited to serve on a committee to do so. You agree, but you also realize that you know little about the subject, apart from the fact that transgender people are those whose gender identity does not match the sex they were assigned at birth. The research you complete in preparation for the committee's first meeting indicates not only that transgender athletes face prejudice and discrimination but also that participation, particularly by transgender females, is controversial. Concerns have been raised about competitive advantage and about physical risks to nontransgender females. Further doubts have been raised by national precedents. Whereas some states allow transgender athletes to participate according to their gender identity, others require that their participation be determined by the sex indicated on their birth certificate, and at least one state requires transgender athletes to undergo hormonal treatment in order to participate in athletics. On the other hand, many states have no policy at all. You want to do what is right for all student-athletes, but this array of information has left you at a loss. What should you recommend?

If sport and coaching are not your thing, here is another ethical dilemma. You are the administrator of a health and recreation program for a mid-sized city. One of your responsibilities is to promote physical activity. You are well aware that this responsibility can be met by supporting traditional activities such as tennis, swimming, hiking, gardening, yoga, Pilates, weight training, and golf. Therefore, you make sure that your town maintains the requisite facilities for these activities—including, among others, swimming pools, jogging trails, and golf courses. You also conscientiously advertise these kinds of sport and exercise programs for all ages. However, you have noticed a decline in participation and have become aware that many people in your community, particularly children and young adults, are attracted to alternative sports and adventure activities, such as X Games sports, white-water rafting, wind surfing, hang gliding, obstacle course races, and ultra endurance events. You would like to bring your own program up to date and provide a more attractive array of activities. At the same time, you know that most of these alternate exercise and sport forms carry higher degrees of risk; in fact, for participants, the danger and extreme challenges contribute to the activities' attractiveness.

All of your training as a kinesiology professional has taught you that safety must always be the top priority. Moreover, you know that most school and recreation programs have eliminated activities that carry unnecessary risks, such as boxing and trampoline jumping. Some schools are even debating the future of their football programs, given recent research findings about head injuries. Thus, you feel caught: How much risk is tolerable? At what age should individuals be able to decide for themselves what sporting risks they take? You wonder if school and recreation programs have erred on the side of safety and fear of liability. What should you do?

Sport and physical activity organizations have struggled when dealing with sex and gender issues; in fact, some would argue that they have not done particularly well in this regard. They are not alone—religious organizations, schools, the military, and other social institutions have a similar track record. This difficulty may stem in part from the fact that attitudes toward sex (biological status) and gender (the characteristics that society associates with sex) are culturally shaped and emotionally charged.

In addressing such issues, philosophy can be useful because it aims to provide a clear and objective perspective from which to untangle complex considerations. Philosophic methods would have us see things both as they are and as they should be. In the case of sex and gender as they relate to sport, philosophy would have us take a step back and attempt to answer the following kinds of questions: Are the empirical claims about "unfair advantage" and "likelihood of causing injury" factual? If the facts point in the direction of "slight advantage" or "small increased risk," are these differences tolerable? Are they in line with other advantages and risks that are part and parcel of sport? What are the rights of transgender individuals? Is self-identification a valid basis for making important policy decisions? Does gender identification carry important values for people's lives?

Our second scenario—the one about high-risk activities—pits two goods against one another. The first involves health, safety, and longevity; the second involves meaning, significance, and happiness. These two goods acknowledge two truths: Life itself is precious, and we are built to pursue meaning or significance. A short life is typically regarded as unfortunate if not tragic, and many would also argue that a life with no purpose or joy is not worth living.

Some individuals find meaning and significance in extreme challenges and high-risk activities even when at play. Here, enculturation and emotion can cloud our vision. In a society that deifies health and safety—and one that is highly litigious and values work over play—high-risk leisure activities may seem irrational and even immoral. This cluster of factors has the potential to result in an existence that is safe but insipid. Thus, philosophers would also want to ask about the individual right to choose one's own life course—both at work and at play. Doesn't the freedom to choose carry some weight? In addition to life itself, doesn't the *quality* of life matter? Do we own our own bodies, and should we have a say in body-related decisions?

When faced with such questions, philosophy may not be able to provide "proofs" or "slam-dunk" arguments on one side or another. However, this does not mean that philosophy cannot produce persuasive evidence. Moreover, even partial or tentative answers are often more valuable than no answers at all.

Why Use Philosophical Thinking?

The short answer to this question is that philosophical thinking is distinctive and valuable. Philosophy, in fact, has been called the queen of the sciences, an honorific title that points to its special qualities. In this introductory section, we discuss the nature and value of philosophy, as well as its role as part of a cross-disciplinary endeavor.

Nature and Value of Philosophy

Philosophical thinking involves reflecting more than it requires testing, measuring, or examining things with our senses. In contrast, physiologists test physiological responses, for example, by putting people on treadmills to find out how they react to various conditions. Biomechanists measure the forces generated by various movement techniques and kinds of equipment. Psychologists examine subjective reports from individuals who have been interviewed, and historians analyze various documents for information about our past.

Goals of Philosophy of Physical Activity

- To understand the nature of health and physical activity, particularly in the forms of exercise, sport, games, play, and dance
- To understand how confident we can be about our claims in kinesiology
- To understand the most important values of physical activity, as well as its contribution to good living
- To learn how we ought to behave in sport and in our professional lives as kinesiologists

Philosophers also learn a great deal by living in the world and experiencing it. Thus, in some ways they also test, measure, and examine. They use this information from daily living for later reflections. In other words, the tools they use relate to different ways of mulling things over, using logic, speculating, imagining—or simply thinking! You will have a chance to learn about and practice three of these ways of reflecting in a later section of this chapter.

Here, we need to appreciate two important clarifications. First, philosophical methods typically do *not* include gathering data from controlled experiments; instead, they use various types of reflection. Second, contrary to what some believe (and perhaps to doubts of your own), philosophy can produce results that are as valid and reliable as anything discovered by the physical sciences. Of course, this description does not apply to all philosophy, but that does not mean that everything in philosophy is unreliable or consists of little more than guesswork, speculation, or personal opinion.

Let's take a look at four types of philosophical claims. They range from personal opinion, characterized by little supporting evidence, to claims that seem patently true because they come with much supporting evidence.

- **Personal opinion:** This type of claim indicates individual taste or preference. This is the lowest level of truth claim because it is difficult to rally persuasive arguments to support such judgments. Behind these judgments lie individual differences, diverse life experiences, and contradictory but equally valid attitudes. Thus, we might say that personal opinions are true for a given person but not necessarily for others. For instance, my claim that vanilla is the best flavor of ice cream is a reasonable one, but your own preference for, say, coffee-flavored ice cream is reasonable, too. Thus, we agree to disagree without any hope of solving the matter. You are entitled, as the common expression goes, to your own opinion, as long as it is a reasonable one.
- **Speculation:** Although speculations are not merely opinions, they are still very hard to support with evidence. When someone tells us why he or she believes that a speculation is true, we may believe that the claim is plausible, but typically we also retain reasonable doubts. For example, philosophical speculations such as the claim that God exists or that love is the preeminent value could be true, but we may never know for sure. Also, it is unlikely that evidence will ever come along to support or refute such claims conclusively.
- **Probable assertion:** This is the kind of claim around which a considerable amount of evidence can be gathered. This evidence may speak to personal experience, logic, immediate insight or intuition, or objective observations of the world and human behavior. When we hear evidence for a probable assertion, we are likely to conclude that it is likely to be true or at least partly true. When we argue that all people are created equal, or that all human beings should enjoy certain rights, a considerable amount of persuasive evidence can be brought to bear on these claims. These claims were viewed as self-evident truths by the framers of the U.S. Constitution; perhaps that was an overstatement, but the claims still possess a high degree of appeal and probable truth value.
- **Truth assertion:** Some have argued that philosophy can never provide sufficient evidence for truth claims, but the same has been said in relationship to much research in science. For our purposes here, we regard truth assertions as those claims that are very difficult to question. When we claim that it is wrong, with rare, if any, exceptions, to torture people, we think we are making an ethical claim that is objectively and universally accurate. When we argue that compassion is not the same thing as hatred, it would be hard to dispute this powerful distinction known to all of us. Even if some characteristics are similar (e.g., compassion and hatred are both emotions), other aspects of these two things provide a conceptual foundation for telling them apart and never confusing them.

KEY POINT

Philosophers reflect more than they measure, and their reflections range from informed personal opinions and highly speculative assertions to claims that are likely to be true.

Behind all of this activity lies what many have called a love of wisdom; in fact, the word *philosophy* means precisely that. The pursuit of wisdom requires that we ask different questions and use different tools of research than those in other parts of our colleges and universities. Philosophical questions probe the very core of our discipline and professions: Why should you be a kinesiologist? What kind of good do kinesiologists bring into the world? Is a profession that focuses on health and physical activity as important as a profession that focuses more on specific knowledge, such as the teaching of writing or mathematics? Is it ethically acceptable to spend more time on a needy client and neglect one who seems to be in better health?

Answers to such questions cannot be found merely by using skin calipers, microscopes, petri dishes, or even surveys. Instead, such questions require that we reflect, examine with the mind's eye, use logic, and discern value. In short, philosophy asks the big, meaning-of-life questions and uses research tools that transcend those of other disciplines.

KEY POINT

Philosophy of physical activity involves reflecting on the nature and value of both tangible and intangible objects.

Philosophers and Scientists Working Together

The methods and products of philosophy are typically thought of as complementing those of the natural and social sciences; similarly, we are used to thinking of kinesiology not as cross-disciplinary but as multidisciplinary. This conception implies that kinesiology's component subdisciplines and spheres—such as physiology, biomechanics, and philosophy—are relatively independent. In this view, investigators in each area conduct their research on their own and then join hands after the fact in order to solve larger problems. This process is much like putting together the pieces of a jigsaw puzzle. In other words, philosophy produces its own puzzle piece, then takes it to a meeting where all of the diverse pieces are put together. Thus, philosophers and scientists can work together, and one way to do so is to contribute valuable puzzle pieces for the purpose of producing a more complete picture of things.

But what if philosophy and other disciplines were to realize that they are actually interdependent? What if philosophy cannot stand on its own but rather needs insights from cell biology, anthropology, and other areas of study in order to ask the right questions and proceed with useful reflections? What if this is also true of such fields as physiology, biomechanics, and sport psychology? What if they need the insights of literature, history, and philosophy in order to ask the right questions and design their empirical studies most effectively? If this is true, then we can argue that philosophers and scientists have a moral obligation to work together; that is, they *should* work together. If this is the case, the validity of both scientific and philosophic discoveries depends on such upfront collaboration.

If our research areas are interdependent, then we need a holistic representation of the spheres of physical activity—one that differs a bit from the traditional representation we use in other parts of the book (see the part II opener figure). In this view, although we can identify different areas of study, which are listed in the big circle of figure 4.1, no given area is entirely distinct. Thus, no lines divide the areas; they are not separate puzzle pieces. Rather, they fade off and run into one another—from philosophy to history to sociology and so on. The satellites attached to the big circle depict the ways in which scholarship and

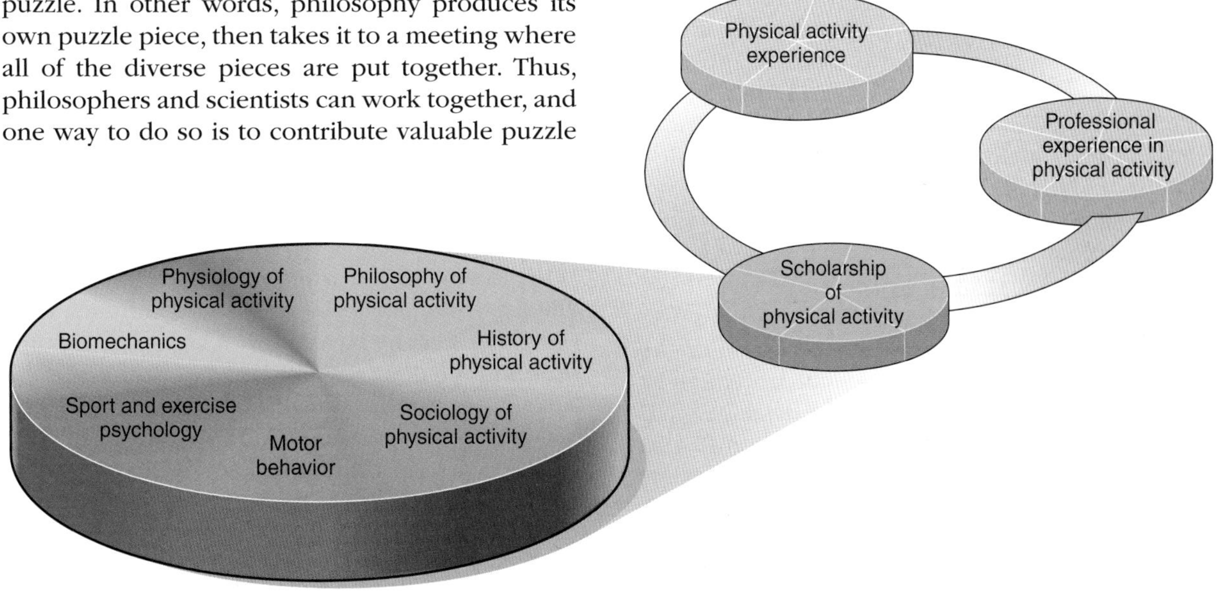

FIGURE 4.1 Holistic kinesiology encourages cross-disciplinary research, professional practice, and performance, as illustrated by the removal of hard lines and the opening of doors between the subdisciplines.

application are related. The sphere nearest the big circle represents the sum benefit of our cross-disciplinary field of study. A reciprocal relationship then exists between the two other parts. One leads to participation in physical activity—in work, exercise, games, play, and sport—whereas the other leads to the workplace and the practical problems faced there. Interaction between the spheres is two-way, thus suggesting that theory informs practice and, conversely, practice informs theory.

In the big circle, no single discipline rules the roost. We are mutually dependent on one another, and we are the mutual beneficiaries of one another's work. Likewise, for the satellites to the right, no circle dominates the others. Physical activity experience, scholarship, and professional experience depend on one another and benefit from each other's perspectives. This conception produces the holistic, democratic, and egalitarian notion of the discipline of kinesiology as described in chapter 1.

What Do Philosophers Do?

Philosophers pursue the same goal as other members of the cross-disciplinary kinesiology team. In the broadest terms, this goal is simply to better understand the world and our lives in it. For our field in particular, the objective is to better understand human movement or, as it is identified in this text, physical activity. Philosophic thinking is needed in order to address a number of issues that are important for kinesiologists. Here are a few.

› What is the scope of our field? Is kinesiology responsible for all forms of movement or only skilled movement? Should we focus on physical fitness, or is our domain that of overall health? For people interested in cultural activities, which ones should be included under the umbrella of kinesiology? Sport? Games? Play? Exercise? Dance? To make such judgments, we must lay some groundwork in an area called **metaphysics**, which is the branch of philosophy that distinguishes one thing from another. The work done here is largely descriptive. Before we can decide, for instance, whether the scope of our field should include all movement or only skilled movement, we need to describe what these two things are. Thus, in metaphysics we ask: What is movement? What is skilled movement? What is health?

› How confident can we be as kinesiologists about the objectivity and validity of our attitudes and perceptions about health, sport, obesity, and related matters? Are we confident enough in our understandings to export our conclusions to other cultures? Is obesity really an epidemic, or does that depend on one's perspective? Should obesity be conceptualized as inherently dangerous, or does that too depend, for example, on how it is seen in different cultures? To make judgments about such questions, we must again lay some groundwork, this time in an area called **epistemology**. This branch of philosophy examines how we know things and with what assurance we can claim that something is true. Thus, it has to do with bias, perspective, objectivity, socialization or enculturation, and other issues related to the strength of the conclusions that we reach. In epistemology, we ask questions such as the following: Am I aware of all (or even most) of the attitudes and potentially skewed perspectives that I bring to my reflections? Can I become entirely objective, or at least objective enough to trust my conclusions?

› What really matters in kinesiology? Is it health? Pleasure? Knowledge? Skill? How does kinesiology contribute to the good life? Can a person live a meaningful life without spending time outdoors or without possessing a variety of recreational skills? Should everyone be able to swim? Is kinesiology largely an auxiliary profession in which we teach skills, fix broken bodies, and promote health? Or is it a liberal profession in which we celebrate play and other forms of freedom in movement? Should we promote life or quality of life? To make judgments about these questions, we must lay some groundwork in an area called **axiology**, which is the branch of philosophy that examines the value of things. It requires that we make judgments about the relative worth of things in the world (such as health) and experiences (such as pain-free living). In axiology, we ask questions such as the following: What kind of life should we pursue? What components go into high-quality living? How do we rank values that provide a means to other good things against values that constitute ends in themselves?

› How should we behave as kinesiologists? What exactly do we owe our employers, patients, clients, or students? When we sign a contract, what obligations do we have? What if we wake up some morning and don't feel like going to work? If we are not sick but are emotionally exhausted, would it be acceptable to take the day off? What if we see one of our colleagues doing something inappropriate or even illegal? Do we have a moral obligation to blow the whistle? What if blowing the whistle carries some risk to ourselves? What do we do then? Judgments about such questions fall within an area called **ethics**, which is the branch of philosophy that examines how we ought to behave. It differs

Kinesiology Colleagues

Angela Schneider

Photo courtesy of John P. Metras Museum, London, Ontario

One of the most pressing ethical issues in sport involves the prohibition of some forms of performance enhancement, or, more specifically, the elimination of doping. This issue is well known by Angela Schneider, professor of kinesiology at the University of Western Ontario and former silver medalist in rowing at the 1984 Olympics, who co-chaired and then became the first woman director of the Ethics and Education Committee of the World Anti-Doping Agency (WADA). She and her committee—which included other sport philosophers as well as sport scientists, coaches, and international-level athletic administrators—reviewed policy and enforcement techniques. Based on contributions from Schneider's committee and others, as well stakeholder groups, WADA directors wrote the World Anti-Doping Code. Schneider's committee then developed an educational strategy and helped WADA launch an advertising campaign to promote clean sport and fair competition.

During this process, it was important to have philosophers at the table and to draw on the kind of leadership shown by Schneider. Among the questions debated were the following: What counts as doping? (This is both a metaphysical and a scientific question.) What is an unfair advantage? (This is a question of metaphysics and ethics.) What is good about drug-free competition? (This is a question of ethics and axiology.) What right should informed, adult athletes have in making decisions about performance enhancement? (This is a legal and ethical question.) How can we persuade athletes to become or remain drug free? (This is a question of policy and education.) What kind of evidence exists to support the values of clean competition? (This is an educational and epistemological question.) In short, Schneider's theoretical expertise in philosophy allowed her to rise to a position of considerable visibility and influence in international sport.

from historical, cultural, and so-called descriptive ethics, which describe attitudes and how people *do* behave. Ethics in this sense focuses on right and wrong, good actions and bad actions, what morally responsible people do, and what irresponsible or selfish people do—in short, how we *should* behave.

History of Philosophy of Physical Activity

By the 1960s, U.S. physical educators studying philosophy of physical activity were pursuing new directions by distancing themselves from educational philosophy and from the single profession of physical education. Instead, they began using philosophical techniques to produce new insights about physical activity itself—especially sport. This approach attracted a few scholars in other disciplines as well. The two most important physical educators in this vanguard were Eleanor Metheny (*Connotations of Movement in Sport and Dance,* 1965; *Movement and Meaning,* 1968) and Howard Slusher (*Man, Sport, and Existence: A Critical Analysis,* 1967), both of whom were active in doctoral education at the University of Southern

California. Also at the forefront of this interest in sport was philosopher Paul Weiss (*Sport: A Philosophic Inquiry,* 1969), then at Yale. In a nutshell, their contributions were as follows: Metheny conceptualized movement as a source of insight and meaning for individuals; Slusher used existential perspectives to examine how sport enhances our authenticity as human beings, as well as our freedom and responsibility; and Weiss celebrated the pursuit of excellence in sport.

In the middle of the country, physical educator Earle Zeigler mentored a group of graduate students at the University of Illinois, and another important center of doctoral education developed at Ohio State University under the leadership of Seymour Kleinman. Zeigler wrote about philosophical schools of thought and prominent philosophers while extrapolating implications of their work for the field of physical education. Kleinman focused on philosophical analyses of movement, dance, Eastern movement forms, and the bodily dimensions of being human.

In 1972, discussions between Weiss, Warren Fraleigh (State University of New York at Brockport), and German philosopher Hans Lenk at the Olympic Scientific Congress in Munich led to the formation of the Philosophic Society for the Study of Sport (PSSS), now known as the International Association for the Philosophy of Sport (IAPS). Fraleigh was the key organizer and continued to be centrally involved as the organization developed. The PSSS almost immediately set about sponsoring a scholarly publication for the subdiscipline, and the first issue of the *Journal of the Philosophy of Sport* (JPS) appeared in 1974 with Bob Osterhoudt as its editor.

Philosophy of sport grew slowly for at least three reasons. First, both philosophers and philosophy-of-sport classes were few and far between because many departments of physical education and kinesiology emphasized the physical sciences. In fact, some schools that had multiple physiologists, biomechanists, and motor control specialists did not have a single sport philosopher. Second, the cognate area of philosophy is uniquely conservative. As a result, many philosophers who resided in philosophy departments preferred not to be associated with a group that studied sport, and they stayed away from meetings addressing philosophy of sport and chose not to publish in the *Journal of the Philosophy of Sport*. Third, early philosophers of sport worked hard to separate themselves from education and other applied fields. As a result, they produced very abstract literature that was often difficult to read and was perceived as having little utility.

Over the past two decades, however, changes have occurred that favor philosophy of sport. First, appreciation has grown for cross-disciplinary study and the role that philosophy, history, and the social sciences can play in inter-disciplinary research. For example, guidelines from a number of funding agencies indicate that single-discipline research is often constrained and ineffectual and that collaboration across disciplines is to be encouraged. Second, philosophers in parent disciplines are beginning to show up more regularly at meetings about the philosophy of sport, and the literature is improving considerably in both quantity and quality. The last three editors of the *Journal of the Philosophy of Sport* have been parent-discipline philosophers, and journal contributions are now more balanced between sport-study and philosophy scholars. In addition, philosophy of sport now has a second refereed journal thanks to the debut in 2007 of *Sport, Ethics and Philosophy*.

Finally, we have seen philosophy of sport spread to universities around the world. Whereas activity in the 1960s and 1970s was confined largely to North America, groups now exist in South and Central America, Australia, Great Britain, Scandinavia, Middle Europe, South Korea, and Japan. The philosophy of sport is also growing, albeit slowly, in other parts of the world. For instance, one 2010 issue of the *Journal of the Philosophy of Sport* (volume 37, issue 2) was devoted to philosophy of sport in non-English-speaking countries.

This positive activity notwithstanding, the fate of sport philosophy in North American kinesiology departments remains uncertain. Due to various factors—including the dominance of the biological sciences, the emphasis on large external grants, and the push for scientific specialization—relatively few doctoral programs in philosophy of sport remain in the United States. Thus, it is unclear where the philosophical leaders and researchers related to kinesiology will be produced in the years to come.

Research Methods for Philosophy of Physical Activity

Now that you know about the goals of philosophy of physical activity, as well as a bit of its history, let's look at how scholars produce knowledge in this subdiscipline. As you have seen, philosophers use various techniques of reflection rather than measurement. Like all scholars, they ask questions.

However, unlike many other researchers, philosophers try to answer those questions by reflecting on experience, using logic, and trying out their conclusions in everyday life to see how well they work. Thomas Nagel (1987) summarized the philosophical process by noting that philosophy "is done just by asking questions, arguing, trying out ideas and thinking of possible arguments against them, and wondering how our concepts really work" (p. 4).

Thus, in order to do philosophy, you don't need to be able to conduct scientific experiments or work through mathematical proofs. You do, however, need to be able to think carefully and precisely, and you can improve this skill with practice, just as people improve writing and athletic skills by working on them. For some of us, reflection may be a relatively new experience. This is so because in our fast-paced, rapidly changing world, finding time for contemplation can be difficult. Moreover, in our modern science-dominated world, measurement tends to be trusted more than reflection. As one academic humorist put it, "In God we trust. All others bring data!" Because of these sentiments and biases toward measurement, you may not have had sufficient opportunity to practice your skills of reflection.

In truth, the gap between those who measure and those who reflect is not as great as it may seem. Most philosophers think that good empirical experimentation is crucial and that the gains made by the physical sciences have been both impressive and useful. In addition, philosophers use scientific findings to put parameters on their own work. For instance, in the philosophy of mind, philosophers are very much aware of advances made in neurophysiology, and philosophical claims about the nature of mind need to be consistent with the findings of brain science. It could be the case that intelligence is the product of an incredibly complex computer-like brain. Or perhaps it is better to model thinking on chaos theory and dynamic systems. Perhaps intelligence is a single generic capability. Or maybe we have multiple intelligences that have been developed across the history of evolution in specific environments to solve specific problems. It is possible, in other words, that we have multiple IQs—one for spatial problems and others for motor problems, music problems, mathematical issues, language dilemmas, and so on—rather than a homogeneous aptitude that allows us to call ourselves "smart." Philosophers need to know such things and use them to constrain and direct their philosophical reflections on the so-called theory of mind.

The reverse is also true. That is, scientists necessarily lean on philosophical methods and on intangibles such as meaning. They necessarily reflect on various matters throughout the inductive scientific process, even though some are unaware that they are doing so. First, scientists employ values to determine what is worth looking at and what is not. They settle on issues that are "important," "significant," "of potential value for humankind," or "crucial for the advancement of science." Second, they reflect during the process of formulating their research problem, coming up with potential hypotheses, and developing a research design. Then, in deciding to test in one way or another, they imagine different possibilities and use logic as well as empirical data. Ultimately, as the data begin to come in, they reflect on their findings. They ask questions such the following: What do these data mean? Could they mean two or more different things? What are their implications for future study? In short, data do not interpret themselves.

KEY POINT

The gap between those who reflect and those who measure is narrower than it may seem. Both philosophers and scientists are affected by physical realities and by the force of ideas and meanings.

When scientists are good at making value judgments and reflecting on meanings, their work goes better. If they are not, their research never gets off the ground, and their findings are never published in reputable journals. In short, scientists who are unable to reflect skillfully and are not good at uncovering the meanings associated with their data are unlikely to be very good scientists.

Even though scientific and philosophical methods overlap in some ways, they are not the same. Reflective reasoning processes are *central* to philosophical research, and three of them are highlighted in the following sections (see figure 4.2) (Kretchmar 1994, 2011):

- Inductive reasoning
- Deductive reasoning
- Descriptive reasoning

Inductive Reasoning

Inductive reasoning moves from examination of a limited number of specific examples to broad, general conclusions. Most scientific research is

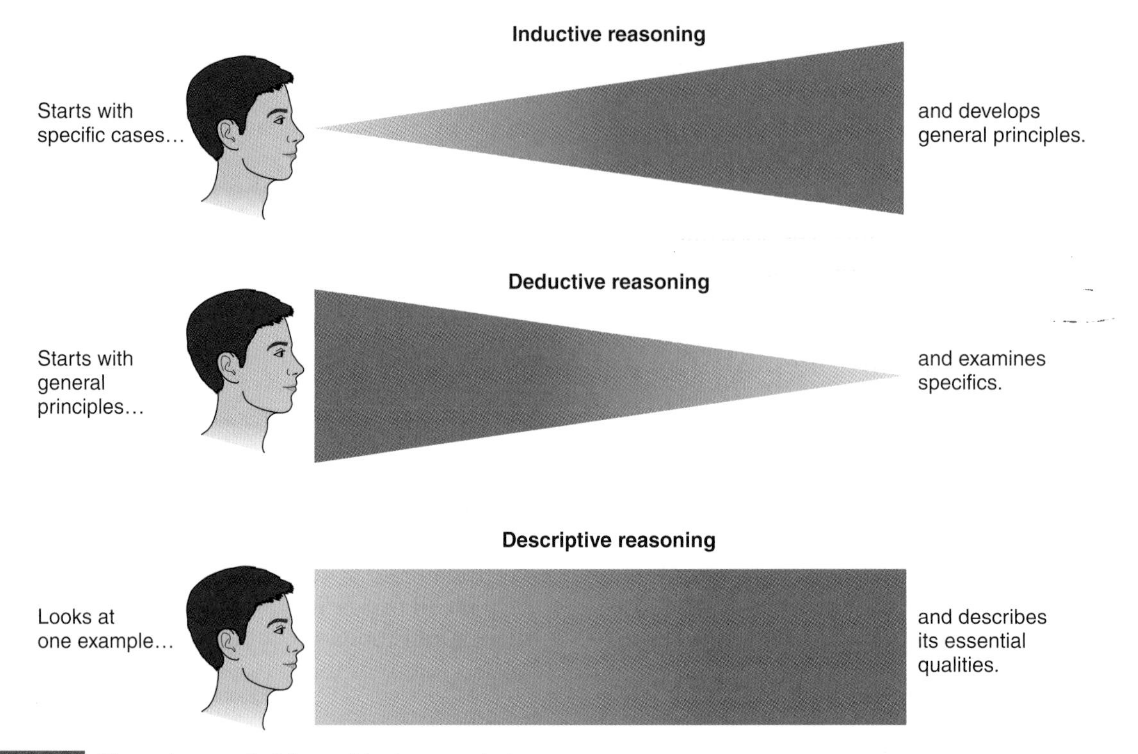

FIGURE 4.2 Three types of philosophical reasoning.

inductive, because it measures a limited number of specific examples or conducts a manageable number of trials and then draws larger conclusions based on those measures. Induction is also used by philosophers. For instance, you could use inductive reasoning to examine a metaphysical question about the nature of exercise. You could start by developing a list of examples of exercise activities, such as participating in rhythmic aerobics classes, training for strength to achieve better performance in football, lifting weights to develop muscles for a bodybuilding contest, mowing the lawn, walking to work for lack of a car, and taking a daily walk at lunch to improve your health. You could then decide whether all of the items are genuine examples of exercise, eliminate any that are questionable, and then identify the general or common characteristics of the remaining items—in particular those features that distinguish exercise from activities such as dance, sport, and play.

Once again, you would need to ensure that your examples were noncontroversial, or "safe," and thus unlikely to contaminate your inductive conclusions. You would also need to ensure that your generalizations were neither too narrow (e.g., exercise requires the use of equipment) nor too broad (e.g., exercise requires a person who moves). When conclusions are too narrow, they are typically inaccurate (it is true that many types of exercise require equipment, but some do not). When conclusions are too broad, they are unhelpful (a performer is also required for sport, dance, or play, so this characteristic does not tell us much about what is unique to exercise).

Deductive Reasoning

Deductive reasoning starts with one or more broad premises and attempts to identify conclusions that follow from it or them. For example, you might use deductive reasoning to determine whether a particular activity you happen to be observing is a sport. Let's say that you are attending your first chess tournament and are curious about whether chess should be identified as a sport. You develop a major premise and a minor premise as follows: If it is true that sport requires the use of physical skills to solve its game problems (major premise), and if it is true that chess does not require the use of physical skills to solve its problems (minor premise), then it necessarily follows that chess is not sport. Of course, the validity of this reasoning depends on the accuracy of the two premises, and someone might take issue with the first premise by asserting that physical skills should not be viewed as a requirement for sport. This person might then argue further that because chess is a competitive game, it deserves to be identified as a sport.

Descriptive Reasoning

Descriptive reasoning involves looking at one example of an event and describing its essential qualities. For example, you might imagine slightly altered forms of basketball and then, through a process of descriptive reasoning, determine whether they constitute the same game. What do you think about the following variations on the sport of basketball? Would each one still be basketball?

- Changing the goal of the game from winning to achieving a tied score at the end of the contest
- Eliminating all restrictions on the means of scoring a basket, thus making it legal, for example, to use a stepladder under the basket or stand on a teammate's shoulders
- Playing on an outdoor court rather than in an indoor facility

A negative answer for any of these variations should help you zero in on what is central to the activity of basketball. If, on the other hand, a given variation does not eliminate basketball, then the variation in question is probably not crucial to the identity or integrity of the game.

One important research technique in which descriptive reasoning is often used is **phenomenology**. Phenomenologists, as Edmund Husserl (1900/1970: 252) noted, are guided by the requirement that they "return to the things themselves." This means that they should get back in touch with their experiences and examine them in order to determine what exactly they mean. As we saw with the basketball example, variations help phenomenologists distinguish meanings that are central from those that are peripheral. For example, playing outdoors rather than indoors does little to change the phenomenon of basketball, but if people use ladders to score baskets then something more central or essential to the game has been lost. The essence of basketball, in other words, is tethered to the unique nature of the basketball problem, not to situational variables such as indoor versus outdoor courts or wood-floored gymnasiums versus asphalt playgrounds.

KEY POINT

Inductive reasoning begins with specific cases and develops broad conclusions. Deductive reasoning begins with broad premises and determines specific conclusions. Descriptive reasoning begins with one example and varies it to see how dramatically the phenomenon changes.

Overview of Knowledge in Philosophy of Physical Activity

It is now time to look at several important topics in the philosophy of physical activity. Although scholars have produced a complex, broad literature related to human movement, we examine here only a few examples of the insights produced by this fascinating subdiscipline. Specifically, we examine the following:

- Who our clients are and what a person is
- Why sport, competition, and play are so popular
- What values are connected to physical activity
- What sport ethics requires

Who Are Our Clients?

How are we to understand the nature of the people we serve? Some sport philosophers believe that you should begin your work on understanding human movement with the adjective (i.e., *human*) before you get to the noun (i.e., *movement*). We follow that recommendation here. That is, we look at who moves before we get to the nature and value of movement itself in forms such as play, games, sport, and competition.

Good reasons exist for using this sequence in dealing with philosophical knowledge. It can be argued that we must understand the person before we can understand how and why he or she is attracted to games and play. For practitioners, the same logic holds. If you are going to use play or exercise as an intervention to promote *human* welfare, you must first know what it is to be human. In short, the intervention must be appropriate for the recipient; therefore, it is with the recipient that we should begin our analysis.

Philosophers have taken many positions on what human beings are. Human beings have been described as social animals, symbolic creatures, individuals controlled by subconscious drives, products of economics, composites of two elements (mind and body), composites of three elements (mind, body, and spirit), and mere machines (albeit complex ones) and thus little more than collections of atoms. We have also been described both as free and idiosyncratic and as controlled and fully predictable. Although all of these positions may merit some attention, we focus here on three that are labeled as follows: materialism, dualism, and holism. If you understand these three, you will have a reasonably good grasp of fundamental positions that can be taken on the nature of the human being. You should also have a good grasp of the implications of these positions for kinesiology and the movement professions.

Materialism

Perhaps the simplest and most straightforward position on human nature is captured by **materialism**. Philosophers call it a monistic position because it posits the person as being made of only one thing—namely, atoms. In this view, all that truly exists in the world are atoms and void. The human being, including the body, the blood, the wiring, the brain—all of the human being—is nothing more than a complex machine. In this view, the brain can be likened to an incredibly sophisticated computer, and the body is cast as a combination of very complex lever systems, motors, wiring, and plumbing.

Consciousness for materialists is often thought of as a kind of sideshow. Although these philosophers may acknowledge that subjective experiences are real, materialists do not believe that these experiences have any power or efficacy. Instead, they view ideas and values as products of brain states; moreover, they assert that physical brain states are what make things happen.

This is a no-nonsense kind of philosophical position. What you see is what you get! The person appears to be an impressive material object . . . and is! Nothing more, nothing less. None of us need to rely on spooky substances like mind or spirit to account for human behavior. Neither should we discount the considerable capabilities of this powerful machine. Once we gain the ability to measure and understand all parts of the organism, all human capabilities will be accounted for in terms of atoms and void.

Materialistic accounts of persons were popular in the time of Plato and even before, but they became ever more prominent during the period of the Scientific Revolution beginning in the 16th century. The careful skepticism inherent in scientific inquiry, the emphasis on measurement, the onset of the Copernican revolution and the realization that the Earth is not the center of it all, the rise of Newtonian physics and the increased appreciation of causal relationships, the invention of the microscope and the clock—all of these developments and others encouraged human beings to think of themselves as simply one kind of physical object among others.

The assumptions and perspectives associated with materialism were adopted by many physical educators and exercise scientists, who began to study the person as a complex, mobile machine. More specifically, physiologists, biomechanists, motor control and motor learning experts, sport psychologists, and even some sociologists and philosophers of sport began looking for "underlying mechanisms" (the holy grail of materialism)—that is, the multiple physical causes that, when pasted together, would explain the whole.

Dualism

Some philosophers appreciate the simplicity and honesty of materialism yet doubt that consciousness can be cast aside so quickly and easily. In other words, they doubt that human behavior can be accounted for by looking only for underlying mechanisms or only for material causes. While acknowledging that human beings have a body that works very much like a machine, these philosophers argue that people are also endowed with a mind, with consciousness, with a steady stream of very concrete emotions and ideas. Moreover, these subjective states have power. Ideas of love and hope and justice, for example, can (and do) change our behavior. In addition, in this view, it is the ideas that do the work—not just the physical brain states.

When these philosophers are asked how many parts go into the production of a human being, their answer is two: mind and body. Thus, instead of being monists like the materialists, they are identified as dualists. René Descartes, perhaps the most articulate and forceful of the modern dualists, said that a person was composed of two substances—one that is extended in space (body) and one whose nature it is to think (mind).

Dualism is an attractive philosophical position because our intuitions and daily experiences suggest that our thoughts must count for something. We have an idea, we draw up a plan reflectively, and then we act on it. We have an exciting idea, and everything from our heart rate to our outlook on life is affected. In short, dualism can be attractive because it takes ideas seriously.

Many physical educators and exercise scientists are dualists because they believe that consciousness is real and that nonphysical ideas somehow invade the world of motor behavior—of swimming across pools and catching baseballs and performing pirouettes. Many have expressed their respect for both of our human elements and the close interrelationship between them, as the Roman poet Juvenal did when he argued for "a sound mind in a sound body" in the poem *Satire X* and as Jesse Feiring Williams did when he proposed an idealistic "education through the physical" (Williams, 1930: 279).

Holism

More recently, a third position—holism—has received a considerable amount of attention. In some ways, it can be considered a middle position between monistic materialism and mind–body dualism. Holism takes seriously both the mechanistic and the thoughtful sides of our human nature, but it refuses to separate them. In other words, it argues that our physical nature, all the way from our anatomy to our genes, is always shaped and influenced by emotions and ideas. Likewise, our subjective experiences are always shaped and influenced by our physicality. In short, we never get one without the other. Thus, neither the physical nor the thoughtful side of our nature has any independence. We are thoroughly ambiguous—whether one is looking at an injured knee or a child playing soccer. Each of these phenomena is a composite of "bodymind"!

The central tenet of holism shows its diametric opposition to materialism. Holists argue that the whole is more than the sum of its parts. In this view, no matter how sophisticated our measurement techniques become, underlying mechanisms will never give us a complete picture of human behavior.

From a dualistic point of view, our ability to complete physical tasks results from our ability to put thoughts into action.

KEY POINT

How we deal with people (e.g., cure, educate, or teach them) is affected by what we think a person is. If we carry false impressions of human nature, we risk using inappropriate, ineffectual, or dangerous interventions.

Philosophers who are frequently identified as champions of holism include William James, John Dewey, and Maurice Merleau-Ponty. They all claim that it is unhelpful to think of a person in terms of parts or dichotomies such as mind and body. They all also believe that intelligence works on many levels, from mere nonverbal reflex behavior to creative action to reflection and verbalization. Thus, behavior can be more or less mechanistic or more or less creative. However, no behavior is simply mechanical, and no actions result from an independent mind giving instructions to a dependent body. Once again, all behavior is ambiguous.

Kinesiologists and physical educators have seen the relevance of such a position for their own work. Among other things, it has expanded the significance of physical activity as something that counts as creative, intelligent behavior in its own right, without the addition of theory or other propositions. Holism has increased the significance of the body in understanding human nature and behavior. Holists, unlike materialists and dualists, do not say that we *have* a body. Rather, they argue that we *are* our bodies; or, more accurately, as some Zen Buddhists say, we are simply bodymind! Finally, holism has expanded the kinds of interventions designed to improve health or repair injury. Rather than working on the plumbing (cardiovascular

Kinesiology Colleagues

Jesús Ilundáin-Agurruza

Photo courtesy of Linfield College

Dr. Jesús Ilundáin-Agurruza teaches a philosophy-of-mind course unlike any other in the U.S. curriculum. His course should allay the worries of Richard Shusterman, pioneer of somaesthetics (a cultural-philosophical appreciation of embodiment), that perhaps somaesthetics cannot be incorporated into the standard philosophical curriculum found in Europe or the United States. Fortunately, Ilundáin-Agurruza's classroom is anything but typical because it richly incorporates disciplined yet playful movement.

Contemporary philosophy of mind is notoriously technical and centered on both metaphysical and neuroscientific issues. Ilundáin-Agurruza's course implements a radically holistic pedagogy that integrates both theory and practice, both thinking and moving, and both Asian and Western philosophies. It includes weekly swordplay practice that combines medieval European and Japanese techniques. Wielding wooden swords, students come to learn about and connect the more technical theoretical aspects with *felt* movement. They cut and parry their way through recent and cutting-edge philosophical analyses, such as radical enactivism and Asian notions of consciousness. The course also instills emotional intelligence and temperance. That is, it teaches meditative techniques and philosophical principles meant for use in duels and in stressful situations such as exams—all while conducting a rigorous philosophical examination of the mind.

system), or fixing a hinge (e.g., knee), or manipulating the computer (i.e., brain), we can intervene at multiple levels—from our embodied ideas and aspirations to our idea-affected or idea-shaped cells. Holistic medicine is now taken seriously in many quarters. A "bad back," for instance, can be addressed by any number of interrelated therapies, such as meditation, yoga, chiropractic manipulations, pharmaceuticals, and surgery.

Why Are Sport, Competition, and Play So Popular?

Sport occupies a prominent place in modern society, not only in the United States but virtually everywhere in the world. From the professional sport ranks to friendly pickup games to modifications used in physical education classes and summer camps, the different modalities of sport touch the lives of a vast number of people. What makes sport so attractive? Where does its value reside? To answer these questions, let us consider the structure of sport as a unique kind of game that can be experienced both competitively and playfully.

KEY POINT

Sport's popularity presents us with a philosophical question: What accounts for the allure of competition, play, and games?

Sport Grounded in Artificial Problems

Games are peculiar activities. They are created by a set of rules specifying a goal to be achieved and limiting the means that participants can use to reach that goal. The rules exist to create the challenges of the game. In contrast, imposing such rules would be absurd in ordinary life (Suits, 1978), where it would be senseless to make our workaday activities more difficult than they already are.

Games, therefore, are constructed or artificial problems (Kretchmar 2004; Morgan 1994). The built-in inefficiencies imposed by game rules differ from the inefficiencies encountered in other realms of life—for example, in education, politics, and business—where limits on efficiency come mainly from ethical constraints. For instance, we cannot make writing easier by copying someone else's essay and calling it our own. That would be an act of plagiarism, which constitutes a serious breach of ethics. Game rules, however, constrain us in gratuitous or unnecessary ways that are not occasioned by ethics.

Almost any activity can be turned into a game by imposing gratuitous difficulties. Think about the ordinary activity of studying for a test. You could turn it into a game by putting time limits on your reading, attempting to read backward, or reading exclusively in a dimly lit room. These methods are not the most efficient ways to study for a test, but they do turn the activity into a game. In this way, nearly any activity can be converted into a game, which, as an artificial problem, invites and challenge us to solve it. This artificiality constitutes one dimension of sport's popularity.

Sport as Showcase for Motor Skills

In light of this definition of games, we can describe sport as a game in which motor skills are required to reach its goal. In soccer, for example, the objective is to get the ball over the goal line, between the goalposts, and under the crossbar. Furthermore, the rules prohibit certain actions that might otherwise increase participants' ability to accomplish the testing objective (in this case, getting the ball into the goal). Here are some examples of such restrictions, most of them designed to test certain physical skills:

- Each team may have only 11 players on the field at one time.
- With some exceptions, field players may not use their hands to control the ball.
- Players taking a penalty kick must kick the ball forward.
- Players may not prevent goalkeepers from releasing the ball from their hands.
- During free kicks, all opponents must be at least 10 yards from the ball.
- Goalkeepers may not touch the ball with their hands after it has been deliberately kicked to them by a teammate.

The restrictions imposed by the rules of soccer provide the sport with its distinctive artificiality, which is recognized by and has captured the attention of people around the world. More generally, the gratuitous invitations offered by various sports to test particular sets of motor skills represent another dimension of sport's popularity. Clearly, sport is a game meant primarily to test our motor skills.

Sport and the Creation of Excellence and Meaning

Rules are game cues that allow us to formalize the artificial problem of a given game. They clarify what should be accomplished and how it should be accomplished. Thus, it is within the rules of a game that the skills meant to be tested become relevant and therefore valuable. Otherwise, what would people think of jumping to shoot a ball into a hoop or hitting a ball with a solid piece of wood and then running around between padded cushions? These skills, known as jump shooting and batting, are recognized as components of basketball and baseball, respectively, because of the rules of these sports. More broadly, rules allow sport to be practiced by diverse people and open the door for the creation of excellence and the emergence of meaning, both in sport and in our life narratives—all of which helps explain why sport has become a worthwhile human project.

Thus, engaging in acts that would be considered inconsequential in ordinary life also facilitates the exploration of our capabilities in a protected environment. To keep that exploration appealing, we often modify the rules of sport so that the artificial problem does not become too easy or too difficult to solve. People seem attracted to sport when accomplishing the goal is challenging yet lies within the reach of their capacities. Neither foregone conclusions nor unworkable feats are good conditions for preserving sport's meaning.

Just as rules are indispensable in crafting and preserving the distinctive artificiality of sport, the deployment of motor skills is indispensable in instantiating sport. For instance, successful participation in soccer requires proficiency in kicking, passing, running, and tackling, among other motor skills. Indeed, soccer expertise involves a blend of these motor skills. Rather than being generic in nature, soccer skills are highly specialized. This specialization emerges both from the test established by soccer's rulebook—which sets the goal to be accomplished and the means allowed (and prohibited) in pursuing it—and from the players' ingenuity in negotiating it.

This is the case not only for soccer but for all sports, because the rulebook of each sport indicates the set of motor skills that it is designed to test (Torres, 2000). Thus, a sport's distinguishing set of motor skills provides another component that differentiates it from other sports and may well account for a given sport's popularity. Moreover, motor skills represent the standards of excellence by which players evaluate their performance. Aspiring to reach and expand a sport's standard of excellence also provides meaning and value for our involvement in the sport, and more generally for our lives. This significance is part of what we find appealing about sport.

Professional Issues in Kinesiology

Equality and Fair Play

What is required for fair play in athletic competition? The principle of equality demands that contestants adhere to the rulebook in order to guarantee that everybody has the same formal opportunity to advance in competition. Many philosophers also argue that contestants should abide by the "spirit" of the rulebook, which is typically articulated as requiring respect for opponents and the promotion of legitimate comparisons of the contestants' motor skills.

But is this all? What about access to coaching and facilities? Or the ability to pay for travel, equipment, and specialized health care, among other areas of potential inequality brought about by differences in socioeconomic status? Doesn't fair play demand that all who so desire have equivalent opportunities to participate in sport and develop their sporting capabilities? How meaningful are victories if they mainly reflect socioeconomic inequalities rather than talent, merit, and effort? Promoting fair play is more challenging than we tend to think. For a solid analysis of fair play and the challenges it poses, see Sigmund Loland's *Fair Play: A Moral Norm System* (2002).

Citation style: APA

Loland, S. (2002). *Fair play: A moral norm system.* London: Routledge.

Sport and Fair Competition

So far, we have not focused on competition. This is so because sport is not necessarily competitive. Any sport, or any game for that matter, can be organized both competitively and noncompetitively. Noncompetitive sport requires participants to engage fully in trying to meet the challenge set by the rules. Think about a person who uses a wheelchair and tests herself by seeing how fast she can cover the sprint distance of 400 meters. Facing such a challenge can be fun and meaningful even though it involves no comparison, no ranking, and no competition. By way of contrast, competitive sport requires not only engaging fully in trying to meet the challenge set by the rules but also attempting to establish athletic superiority. After a contest, for instance, a competitor in wheelchair racing learns how fast she can complete the sprint distance of 400 meters *and* how she fares in comparison with her opponents. Similarly, in a full-fledged competitive game, a rugby team learns two things—how well it does on the test provided by the sport per se *and* how good it is in comparison with its rival.

Competition requires that contestants share the same test and that they commit themselves to surpassing each other's performance (Kretchmar, 1975). When individuals enter the world of competitive sport, they enter a community that defines itself in terms of the unique test they share. Here, testing evaluations and competitive comparisons form a decisive unity. Although competition determines winners and losers, something more valuable may be at stake: the attempt to challenge each other to achieve excellence (Simon, Torres, & Hager, 2015). This view of competition highlights the value of performance and cherishes the process as much as it values the outcomes that result from contestants' performances (Torres & McLaughlin, 2003). That is, excellent performance transcends mere winning and losing and can even become the central value of competitive sport. Thus, the pursuit of athletic superiority, doing better than one's opponent, stands as another possibility for explaining the widespread popularity of sport.

Sport as Delightful Play

We have seen that sport is a goal-oriented activity structured by rules that impose limitations on how we can go about attaining the goal. This is done primarily to test our proficiency in using a given set of motor skills. We have also seen that this activity can be organized either just to evaluate our proficiency in solving the test or, in addition, to compare our skills with those of an opponent. Another distinction matters as well: Sport may be encountered as play. Here, we explore the differences between playful and dutiful modes of experiencing sport and physical activity—both of which help explain their popularity.

We have all tasted the delight and mesmeric quality of play, yet it is a difficult phenomenon to conceptualize. Although play is found during engagement in an activity, it does not consist of the activity itself. Rather, play is the manner in which we approach, embrace, and experience the activity (Meier, 1980). Play requires an **autotelic attitude**, in contrast with all forms of instrumental or utilitarian orientation toward the world. This distinction means that while at play, we are interested not in any payoffs that might come from participating in a given activity but rather in the activity itself. In other words, when we are at play, the doing supplies its own reward. For example, a playful encounter with biking is experienced as utterly absorbing and fulfilling. Even if extrinsic rewards and goals were of concern when the activity started, they recede into the background as we take great pleasure in experiencing our biking skills. In such moments, biking is lived primarily as an activity that we like simply for what it is and for the quality of experience it provides.

When sport or physical activity, or any activity for that matter, is lived in a playful mode, it becomes something dear to us. In a sense, it becomes ingrained in who we are. At play, we forget about time, obligations, and the usual roles we fulfill in life; instead, we surrender to the magnetism of the playground and experience the doing as valuable in itself (Kretchmar, 2005; Torres, 2002). Playful sport and physical activity are characterized by what could be described as an exuberance of being; they involve feelings of freedom, spontaneity, creativity, joy, and meaning. To put it in a more pedestrian fashion, play feels good! We also know that this extraordinary way of embracing and experiencing the world is fragile and temporary. However, even though play cannot be guaranteed by command, it is more likely to appear if we predispose ourselves to find it by being patient, respectful, ingenious, and caring with the activity we are engaged in and where it is taking place.

As professionals, we can develop strategies to help our students and clients find play in a variety of potential playgrounds. The challenge for us in this capacity is to facilitate growth in the reciprocal relationship between players and playgrounds. The playground shapes and affects the player, who

Physical activity that brings us joy and pleasure is delightful play.
Bananastock

becomes, for instance, "basketballed," "bicycled," or "table tennised." In turn, the player affects the playground, which is shaped by the player's personality, goals, and personal history. Thus, basketball becomes "Mary-ed," "John-ed," or "Susie-ed."

Play can be fostered in both competitive and noncompetitive sports. Many athletes, both casual and professional, are clearly drawn not only by the activity's instrumental value but also to the process of doing the activity—that is, to an internal focus. Similarly, play does not recognize differences in age, gender, race, religion, socioeconomic status, or motor skill development. It is not the exclusive property of any particular group, which means that it does not have to be restricted to children and retirees, as is often the case with conceptions of play in contemporary society. We are always at risk of finding play, or being found by it, or both (Kretchmar, 1994). The key is to be aware of and embrace such risks! The magnetism of play, and the way we live an activity while at play, make play an essential component of our lives. No wonder some people have recommended that we live our lives by fostering as much play as possible (Lasch, 1979; Morgan, 1982; Pieper, 1952), and no wonder that play is so often associated with sport and physical activity. Don't you think this association helps us understand people's attraction to sport?

Sport as Beneficial Duty

A second mode of experiencing sport and physical activity also helps explain sport's attractiveness. It differs markedly from the play model because it focuses not on the process but on the payoffs of participation in sport and physical activity. This model supports many of the most common arguments used to include physical activity in education and in personal exercise programs: the power of sport and physical activity to bring about beneficial effects. That is, sport and physical activity are presented as valuable because they help us improve ourselves in concrete and desirable ways, such as preserving our health, teaching civil and moral values, fostering national pride, and combating sedentary living and obesity. This account is clearly based on utility, which has been emphasized for years as a rationale by many professionals in the field. When sport and physical activity are formulated in this way, we are likely to approach them as means to an end, as pursuits that we must engage in. In short, they become a form of duty, a way of participating that provides a clear contrast to playful sport and physical activity.

Sport and physical activity are more likely to be encountered as a form of duty when the focus is placed on external rewards. In contrast, playlike sport and physical activity are self-contained. As a way to live these activities, play accentuates their intrinsic values and ends. At play, sport and physical activity are experienced as delightful, absorbing, and meaningful; thus, they allow us to show our human potential.

There is nothing inherently wrong in regarding sport and physical activity as pursuits that produce external rewards. Doing so, however, tends to diminish their intrinsic worth. The more we remind ourselves that our participation in sport and physical activity produces good results, the less we are prone to relish their intrinsic meaning and value. In other words, the more we talk about

How Movement Playgrounds Are Created

Given that play serves as a valuable ally in promoting movement across the life span, it behooves you to become good at "growing playgrounds" for your students and clients. To that end, take a few minutes now to develop a list of dos and don'ts as a professional playmaker. In doing so, consider what you have learned in this unit about the attractiveness of sport and physical activity, the nature of games, the values of competition, and the qualities of play.

dutiful sport and physical activity, the less we talk about their potential to foster play. One case in point is the popular and well-intended Exercise is Medicine program. Medicine typically does not taste very good, even if we need it to make us well; thus, it is shortsighted to equate physical activity with medicine.

In summary, there is much to be said about promoting good outcomes through the practice of sport and physical activity, but we should not do so at the expense of play by emphasizing only the expected results of a *workout*. In addition, even the most committed and cautious among us would welcome the addition of some delight and sensuousness to our training regimen. Tedious duty can take us only so far. The beneficial effects produced by dutiful sport and physical activity are important, and they illuminate the social acceptance of these pursuits. However, their limitations should not be overlooked. After all, when sport and physical activity became just *workloads*, we often cannot wait to escape them, in spite of their esteemed payoffs, in order to pursue more intrinsically rewarding projects.

Meaningful Combinations of Sport, Play, and Competition

To begin with, sport offers an intriguing physical problem. It invites us to give it a try. It asks us to find an answer to its riddles. The physicality of the test provides further reasons for skillful, embodied individuals like us to enter its domain. Sport is a sensuous problem. When sport is experienced playfully, the testing becomes intrinsically meaningful and even more delightful. This combination of sport and play makes us more likely to continue our relationship with sport and physical activity.

If a sporting test is appealing in itself, many of us encounter yet another layer of meaning when competition is added to the mix. This combination of sport and competition offers not only the uncertainty and tension of taking our chances in the sport but also the ambiguity and drama of learning how we will fare in comparison with our opponents—and the tighter the contest, the more attractive the comparison. This is not surprising given people's interest in competition, whether athletic or of some other type.

An even more powerful combination occurs when a sporting test is shared competitively *and* lived in a playful mode. What a formidable triad this is! When competitive sport is experienced playfully, we focus on and thoroughly enjoy the doing while aiming for excellence—not just victory—among fellow sportspeople who are collaborating in a common project. If sport as a test is attractive, and if competitive sport is doubly attractive, then competitive sport experienced as play stands as a triple delight. In this form, sport stands as a project of problem solving that has been deeply enhanced.

What Values Are Related to Physical Activity?

Given past and current scandals, what do you think about rewarding players or teams for being good sports by giving extra points at the end of games based on referees' judgments of players' behavior? In a different vein, have you ever wondered what your responsibility would be if you discovered that a teammate had been cheating? Should you tell your coach? League officials? Would it be acceptable to just keep quiet about it? Regarding yet another issue, do strong reasons exist for eliminating dodgeball from school physical education programs? More generally, will physical activity become antiquated if science and technology solve most of our health-related problems? The search for answers to such questions requires you to examine your values and develop persuasive arguments to defend them. In doing so, you can benefit from using the techniques of reflection discussed earlier in the chapter.

All of us use values—our conceptions regarding the importance of things—to make decisions in both personal and professional matters. Values can be either moral or nonmoral. Moral values relate to our character and how we ought to behave, whereas nonmoral values relate to objects of desire such as happiness, ice cream cones, and good health.

Several values are central to physical activity. Once again, these values relate to things or conditions that people desire, or things that make life go better. Kretchmar (2005) identified four of them:

1. Health-related physical fitness
2. Knowledge of the human body, physical activity, and health practices
3. Motor skill
4. Pleasure or fun related to physical activity

KEY POINT

Each of the four values promoted by physical activity programs—fitness, knowledge, skill, and pleasure—would lead to a different sort of program if used as the central guide or priority for planning and intervention.

Each of these values supports a different approach to developing physical activity programs. Physical fitness, for instance, provides direction for those who are most interested in health, functioning, and appearance. From this perspective, physical activity professionals should be concerned primarily about such things as assessing fitness, measuring fitness changes that result from physical activity, and promoting physically active living.

In contrast, knowledge is the primary value for those who believe that information is a precursor to improved behavior—for example, for staying healthy, looking good, and performing physical skills more proficiently. This view would encourage more research in scholarly subdisciplines of kinesiology. Thus, physical activity professionals should become well-grounded in these subdisciplines on the assumption that possessing more of this knowledge is good for its own sake and will also make them better practitioners.

Motor skill is valued by those who like to move well and who know the joy of proficiency. Some might rank this goal highest because the pleasures of skilled movement may encourage people to stay physically active and thus improve their fitness. In other words, intrinsic motivations lead to extrinsic gains. For those who place high value on motor skill, physical activity professionals should concentrate on teaching appropriate movement techniques. For example, in school and college physical activity classes, professionals would focus on particular sports such as golf or softball.

Valuing pleasure resonates well with people's conceptions of the good life. Fun and enjoyment are important for just about everyone. Enjoyment grows when a situation is complex enough to present stimulating challenges but simple enough so as not to overwhelm people's capabilities (Csikszentmihalyi, 1990). We become bored if things are too easy but anxious if the situation is too difficult. Physical activity professionals who see the world from this perspective should make sure that games are simpler for beginners and more complex for advanced players and that the games are suited to the motor-skill level of the players. In addition, as we have seen, the complexity of a game is increased by competition, which probably also increases the fun as long as the challenge does not overwhelm the players. One way of modulating the complexity is to ensure that opponents are evenly matched.

The four values discussed here—fitness, knowledge, skill, and pleasure—do not exist in isolation from one another. All are important in physical activity, and each one influences the others (Kretchmar, 2005). Consequently, the main task we face is not to pick one or two values to the exclusion of the others but to weigh their relative merits and rank them. We can use such rankings to guide us in setting priorities related to physical activity, whether for experiencing it as performers or spectators, for studying it, or for engaging in professional practice centered on it.

In terms of this latter dimension—engaging in professional practice centered on physical activity—professionals might use value rankings to prioritize topics for discussion with people who control funding (e.g., city government officials, school board members, representatives of private foundations that donate to worthy causes) in order to encourage more spending for physical activity programs. For instance, if they emphasize fitness, then the profession is cast largely as a means to an end, or, as Kretchmar (2005) argued, an "auxiliary profession," albeit a very important one. If, in contrast, values such as skill and knowledge are emphasized, then the field takes on a liberal arts character. In this formulation, movement contributes directly to the good life by freeing us to move well and to understand its significance.

For his part, Kretchmar clusters a kind of knowledge (wisdom- or encounter-understanding) with skill as his own choice for the two highest-priority values, whereas fun, fitness, and theoretical knowledge rank lower. This profile presents physical education as a liberal arts subject matter aimed more at improving the quality of one's life than at meeting various day-to-day health and recreation needs. Kretchmar's rationale for his rankings depends in part on his thoughts about the intrinsic and extrinsic worth associated with each value. He cites three criteria:

- Intrinsic values are better than extrinsic values because intrinsic values hold immediate and direct worth, whereas extrinsic values lead only indirectly to an intrinsic value.
- Among intrinsic values, those that involve broader satisfaction and contentment are better than those that involve more limited pleasure.
- Among intrinsic values involving broader satisfaction, those that involve long-term meaning and coherence in our lives are better than those that involve short-term or immediate pleasure.

What Ethical Obligations Do Sportspeople Have?

Sport headlines often tell of cheating, illegal performance enhancers, gamespersonship, and various forms of corruption. It is tempting to think that such behaviors affect mainly professional sport, but they often appear in intercollegiate sport as well. Even organized youth sport, as well as recreational sport, can be affected by various forms of dishonest and troublesome conduct. This state of affairs has been attributed to multiple and complex causes, including, among others, overemphasis on winning, the transformation of sport into a commodity, the rise of individualism, and strategic reasoning. Whatever the causes, it is clear that many athletes, coaches, administrators, parents, and spectators are tempted to overlook questions of morality. This reality highlights the fact that sport tests, and invites us to exercise, our capacities for moral reasoning. If the moral status of sport is to be improved, then we need to engage seriously in this kind of reasoning.

As noted earlier, ethics involves determining what is right, what is wrong, and what ought or ought not to be done. To put it another way, ethics is concerned with questions about how we should live our lives; in sport, this endeavor involves formulating defensible standards of behavior. While thinking ethically, you must make an effort to be impartial, consistent, and critical. As you know, our thinking can be biased by our emotions, our assumptions, and our human tendency to give ourselves a break. The moral point of view is universal, one in which you do not count more than anyone else. Thus, it is difficult to justify bending the rules simply because "everyone does it" or intentionally injuring the best opposing player in a soccer match in order to send her to the sidelines for the rest of the game. In the context of universality, we would not want to be the recipient of such actions, and, because no one counts as more important than anyone else, it would not be right for us to prohibit rule bending or intentional harm in general while exempting ourselves from these restrictions. In short, societal understandings about right and wrong and the force of self-interest do not provide solid foundations for ethics.

In sport, decisions about right and wrong actions require that we keep in mind our conceptual analysis of competition. The more we understand the characteristics of competitive sport, the clearer and stronger our deliberations will be about what actions—by athletes, coaches, and administrators—promote its flourishing. These deliberations provide principles about how these folks ought to behave. Sometimes ethical considerations in sport also include spectators, parents of young athletes, and other people related to sport. The goal is to produce the best possible world of sport for everyone.

When we consider competitive sport as a reciprocal challenge in the pursuit of excellence—and

Winning Strategies

Using principles of universality, what kinds of strategies do you believe should be acceptable in sport? What about flopping in soccer to draw a foul call, intentionally fouling an opponent in basketball to stop the clock, or intimidating a batter by intentionally throwing a pitch toward his or her head? What principles can you develop to distinguish between acceptable and unacceptable sporting actions? What ideas can you generate to discourage the use of morally unacceptable strategies? For example, should we levy harsher penalties? Use incentives? Provide more education? Do something else?

when we use the moral guideline of creating the greatest good for the most people—we can generate the following behavioral guidelines.

- Follow the rules of the sport not simply because they are the rules but because they serve as the foundation for the artificial problem that you regard as special. Avoid cheating, which alters, and even destroys, the sport and vitiates the legitimacy of results.
- Respect your opponent as someone who not only facilitates the contest but also makes possible the creation of athletic excellence. View your opponent as a partner who belongs to the same community of contestants and shares your interest and passion.
- Strive to bring out the best performance in one another. This approach honors each person's motor skills and the sport's standards of excellence.
- Recognize and celebrate athletic excellence—your own as well as that of your opponent.
- Seek opponents who are close to you in ability and who compete with personal resolve to win without slacking off during the contest.
- Care about your opponent's well-being as much as your own. Your opponent is integral to the contest, and a victory is fully meaningful when your opponent is at his or her best.
- Remember that how you play says as much about you as an athlete as the scoreboard does.

The first scenario presented at the beginning of the chapter poses a moral dilemma related to high school sport. What should you recommend to the committee charged with developing the school district's policy regarding transgender athletes? This issue relates to the eligibility and admission of athletes for competition (Meier, 1985). The task of determining who is eligible for and admitted to competition does not fall within the logic of sport as an artificial problem meant to test motor skills; nor does it regulate actual contests themselves. It does, however, bear on fairness and respect. More specifically, it bears on the requirements under which athletes are allowed to participate in sport and the preservation of conditions of equality among athletes.

Ethics and Transgender Competitors

Ethical reflection in sport typically requires that we critically examine societal understandings about right and wrong, as well as tendencies that favor self-interest. When evaluating the best course of action in a given situation, we must also consider what is good for sport and for the interests of everyone involved. As you know, one of the main concerns in high school athletics involves the possibility that, because of the supposedly unfair advantage conferred by their physiology, transgender females might negatively alter the condition of equality among athletes. However, scholars have argued that transgender athletes possess no such advantage and consequently do not disrupt competitive fairness (Karkazis, Jordan-Young, Davis, & Camporesi, 2012; Teetzel, 2006; Wahlert & Fiester, 2012). Thus, there is no reason to exclude transgender athletes from competition or force them to compete under the sex identified and assigned to them at birth.

Another argument in favor of including transgender athletes in competition against members of their self-identified gender does not hinge on physiological equivalency but on the relevance of gender in individuals' lives and identities. Sport participation can be a powerful way for athletes to write a meaningful gendered narrative (Gleaves & Lehrbach, 2016). Transgender athletes have the same right as any other athlete to write such narratives and should be able to self-select the category in which they wish to compete. This view suggests that transgender athletes' inclusion in sport should not be conditioned on hormone therapy, surgery, or any other intervention meant to conform them to the stereotypical male–female binary. In short, self-identified gender identity should be enough for inclusion in sport because it implies sensitive and respectful treatment of all athletes. Given these arguments, you now may have a good idea of what you will suggest to the committee.

What Ethical Guidelines Exist for Questions of Performance Enhancement?

Our capacities for moral reasoning are severely tested in sport when questions of performance enhancement are raised. Efforts to improve performance have ranged from high-tech shoes to improved diets to sport psychology—and, of course, drugs. How and on what basis do we permit many enhancers but outlaw others?

Although athletes have used performance-enhancing drugs for a long time, the practice became more prevalent in the late 1950s, when anabolic steroids came onto the scene. Their use has been on the rise ever since (Todd, 1987). Although precise figures are difficult to find, data from the National Football League show a large number of suspensions since 1982, when the league implemented a policy of suspending players who tested positive for drugs (Boeck & Staimer, 1996; Mosher & Atkins, 2007). In this section, we look at the use of performance enhancers—drugs and other things—from ethical perspectives (Osterhoudt, 1991; Simon et al., 2015).

The main ethical problem is that performance enhancers give athletes an unfair advantage over opponents. You might ask, then, which substances improve performance unfairly, and which substances improve it fairly? Your first answer might be that natural substances (e.g., food, testosterone) are fair, whereas unnatural substances (e.g., steroids, amphetamines) are not. However, the distinction between *natural* and *unnatural* is problematic. For example, unnatural performance enhancers include technological breakthroughs in facilities and equipment, such as long-body tennis rackets, artificial track surfaces, sophisticated basketball shoes, and better-engineered racing bicycles. Yet these enhancers are generally considered fair, though they are sometimes subject to limits (golf balls and baseballs, for instance, could be engineered to travel farther, but limits are placed on this attribute). In contrast, although testosterone is a natural substance, its use is typically considered unfair when it is taken in doses beyond what a person might produce in his or her own body (Brown, 1980).

One solution is to create a level playing field by giving all competitors access to whatever performance-enhancing aids they want. Another possibility is to distinguish between performance enhancers that increase a body's capabilities and those that just make it possible to use more effectively whatever capabilities the body already possesses (Perry, 1983). For a variety of reasons, however, neither of these options has proved satisfactory.

Another view suggests that performance enhancers should be prohibited if they put the health of an athlete or others at risk or if they are shown to make athletes dangerous role models for young children. However, this approach raises serious questions about the rights and freedoms of athletes to make informed choices about their own health. It raises equally complicated questions about where, when, and under what conditions adult freedoms should be curtailed for the well-being of less mature or less well-informed individuals. As for the matter of privacy and individual rights, when is drug testing acceptable, and when does it constitute an ethically unwarranted intrusion on an athlete's freedom as a human being? If we test athletes for recreational drugs or illegal drugs that don't enhance performance, some would argue that we are invading their privacy in unacceptable ways (Thompson, 1982).

Clearly, difficult ethical issues are connected with athletes' use of performance enhancers. Uncertainties remain concerning which aids are morally unacceptable, although performance enhancers that present health risks to athletes seem to be viewed as less acceptable than other performance enhancers. Additional investigations are needed to examine the ethics connected with using different kinds of aids to enhance sport performance.

At the present time, sport administrators, ethicists and others are debating the wisdom of the current bans on a wide variety of performance enhancers. This debate is becoming ever more lively in a culture allowing many kinds of chemical enhancers that produce effects ranging from improved sexual performance and lower blood pressure to increased ability to sleep or stay awake. Why, then, should we not allow enhancers that enable someone to work out longer, build muscle faster, or recover from athletic stresses more quickly? Even if some risk is involved, why not let informed adults make decisions for themselves?

What New Ethical Concerns Have Appeared on the Horizon?

Ethical issues do not stand still. In other words, as society evolves, and particularly as new technologies are developed, new ethical questions arise. One hundred years ago, nobody was concerned about the ethics of social media, the ethics of organ transplants or genetic testing, the ethics that should govern the use of drones in attacking military targets, or the ethics of global warming. The same is true in the domains of health, sport, and physical activity.

New ethical questions have emerged with the development of new technologies. Some examples are given in the following list.

- Should technology be used to improve the fairness of games by eliminating or reducing errors made by umpires or other officials? For example, should all balls and strikes in baseball be determined electronically? Should more plays in football and basketball be subject to replay reviews, even if doing so lengthens games and interrupts the natural rhythm of play?
- Should safer, high-tech equipment be required in sport in order to reduce the risk of injury? What if such equipment is unduly expensive and thus prevents all but the wealthy from participating? Should a high-tech football helmet be mandated even if it means that fewer schools can afford the sport and therefore fewer athletes get to play?
- Should newly developed, high-performance equipment be allowed? For instance, should new swimsuits be permitted? Should "hot" golf balls be legal? What does new equipment do to our ability to compare the performances of those who played many years ago (with inferior equipment) and those who play now (with more advanced equipment)? How do we determine which players rank among the best?
- Should electronically enhanced feedback be allowed to help athletes make decisions in the heat of competition? Should bikers, for instance, be allowed to wear measuring devices that provide information about heart rate and energy expenditure?
- Should athletes be permitted to use new medical procedures to enhance their natural abilities? For instance, should athletes who already enjoy 20/20 vision be allowed to have LASIK eye surgery to give them even better visual acuity? Should elective muscle or joint surgery be permitted in order to enhance strength or flexibility? Should athletes be allowed to undergo amputations and substitute high-tech "blades" for their own legs?

Some technologies affect both health and sport. One of them is genetic manipulation, or, as it is often called, "gene doping." While much of this approach remains futuristic, possibilities loom on the horizon. Genetic science is already providing much-needed tools for promoting health and preventing disease. In the not-too-distant future, it might well be possible to use genetic interventions to correct for health-endangering profiles—for instance, profiles that increase the risk of certain kinds of cancer, Alzheimer's disease, or diabetes.

Most ethicists today support such research and welcome the resulting health-promoting advances—as long as they are broadly available (not limited to those who are wealthy) and carry few if any unintended consequences. On the other hand, many ethicists are concerned about using genetic therapies to enhance people in significant ways, thus in effect changing human nature. What if we could elevate everyone's IQ by 50 points through a safe and inexpensive genetic intervention? Would we want to do that? Is it right to make that decision for one's own offspring when that individual has no say in the matter?

Such possibilities raise ethical questions about genetic interventions designed to promote superior athletes. Would it be right for parents to select the profile of a super-athlete from a genetic menu? How far is too far when trying to achieve athletic excellence?

When Overload Become Overemphasis or Overspecialization

The pitfall of going too far in the quest for athletic excellence is not limited to futuristic genetic therapies. To the contrary, some athletes today suffer from injury or burnout related to overuse, excessive training at an early age, overspecialization, or extreme parental or social pressure. On one hand, such commitment may seem to be required by one's pursuit of excellence, hope for a college scholarship, or desire to keep up with friends. Surely, it is hard to argue against the quest for excellence and a good work ethic. On the other hand, the trade-offs may be too severe in physical, emotional, or social terms. Some ethicists have argued that such extreme measures distort the person and lead ultimately to a less satisfying life. Children, they say, have a right to enjoy sport and experience it as play.

KEY POINT

Ethical reflection in sport requires critical examination of societal understandings about right and wrong, tendencies that favor self-interest, and conceptualization of what is good for sport and everybody involved in it.

Wrap-Up

Philosophy offers not just a body of knowledge but, more important, an intriguing way of thinking about reality. More specifically, philosophy of physical activity helps us reach new insights about various forms of physical activity and what it means to move and do so skillfully. Philosophers use reflective techniques to answer questions that cannot be fully addressed through testing, measurement, or other experimental methodologies. Philosophy is particularly important at a time when many ethical questions need answers and when the values of physical activity are contested. Philosophy is practical because the answers it produces help shape our professional practice. No kinesiologist can avoid establishing priorities or making value judgements and ethical decisions. Because these things cannot be avoided, they should be done well on the basis of thoughtful examination.

MORE INFORMATION ON PHILOSOPHY OF PHYSICAL ACTIVITY

Organizations

British Philosophy of Sport Association

European Association for the Philosophy of Sport

International Association for the Philosophy of Sport

Journals

International Journal of Religion and Sport

Journal of the Philosophy of Sport

Quest

Sport, Ethics and Philosophy

Introductory and Reference Texts

Kretchmar, R.S. (2005). *Practical philosophy of sport and physical activity* (2nd ed.). Champaign, IL: Human Kinetics.

Kretchmar, R.S. (Ed.). (2017). *Philosophy of sport.* Farmington Hills, MI: Cengage Learning/MacMillan.

Kretchmar, R.S., Dyreson, M., Llewellyn, M., & Gleaves, J. (2017). *History and philosophy of sport and physical activity.* Champaign, IL: Human Kinetics.

McNamee, M., & Morgan, W.J. (Eds.). (2015). *Routledge handbook of the philosophy of sport.* London: Routledge.

Reid, H. (2012). *Introduction to the philosophy of sport.* Lanham, MD: Rowman & Littlefield.

Ryall, E. (2016). *Philosophy of sport: Key questions.* London: Bloomsbury.

Torres, C.R. (Ed.). (2014). *The Bloomsbury companion to the philosophy of sport.* London: Bloomsbury.

REVIEW QUESTIONS

1. Describe the main goal of philosophical study of physical activity, as well as the four major kinds of issues it most commonly tackles.
2. Describe the major change in philosophy of physical activity that occurred in the 1960s.
3. Describe the three reasoning processes that serve as central research tools in philosophical studies of physical activity.
4. Discuss the concept of blended unity of mind and body, as well as its implications for school physical activity programs, research on physical activity, and the well-being of competitive athletes. How does this concept constitute an improvement over dualistic views of mind apart from body and over materialistic views that picture persons as complex machines?
5. Explain the relationship between rules and skills in sport. Why is this relationship relevant to competition?
6. Discuss dutylike play, sport, and physical activity and how people relate to and experience these types of activity.
7. Describe the four values promoted by the field of physical activity and their implications for designing physical activity programs.
8. How can we decide what kinds of performance enhancement are morally defensible?
9. Are ethical standards unchanging? If not, are they influenced by social, cultural, and technological factors? Are they influenced by reason or insight?

WWW Go online to complete the web study guide activities assigned by your instructor.

5

History of Physical Activity

Richard A. Swanson

CHAPTER OBJECTIVES

In this chapter, we will

- explain what a physical activity historian does,
- describe the goals of history of physical activity,
- describe how the subdiscipline of history of physical activity developed,
- explain how research is conducted in history of physical activity, and
- explain what research tells us about physical activity in U.S. society from the Industrial Revolution to the present.

A survey course in U.S. history will tell you about some rather strict rules in the Puritan society of 17th-century New England regarding the use of time away from work, behavior and practices on the Sabbath, and certain forms of play and amusement that were banned altogether. Many of these rules were made into laws in most of the U.S. states following the Revolutionary War and were still in effect in the late 19th and early 20th centuries. In contrast, by the 1870s and 1880s, churches, YMCAs, and YWCAs were sponsoring sport competitions and even building gymnasiums and playfields. You can find photographs of such facilities, both in books and online, that show children and young men and women participating, for example, in exercise classes and playing baseball on a lot next to a church. In 1891, in just such a setting—the International YMCA Training Institute in Springfield, Massachusetts—a young instructor developed the rules for the first game of basketball. By the 1920s, some religious publishing companies were producing folk and square dance records for use in church youth recreational programs.

What happened to change beliefs and practices related to sport so rapidly over a 50- to 60-year period? What societal influences played a role in changing religious behaviors so profoundly? What Biblical understandings and interpretations had to be modified to allow such an alteration in practice and teaching? Did theological differences arise over this issue within and between religious denominations and congregations? Was this evolution led by figures in various denominations, or did secular culture influence the behaviors of otherwise devout church members? Was the change positive or negative for the church? For society?

Courtesy of the Library of Congress, LC-DIG-npcc-31876

The developments presented here describe one aspect of the historical relationship between two cultural institutions found in most, if not all, societies in the world: (1) organized play, sport, and exercise, which in this text we place under the umbrella-like term *physical activity,* and (2) religion. Historians also examine relationships between physical activity and other social institutions, including, among others, those in the realms of economics, politics, business, and education. The same is true of music and art historians, labor historians, medical historians, and military historians. This is the case because the history of humankind is too complex to be understood through only one lens, such as the political development of a nation.

In this chapter, you will begin to see many ways in which physical activity has been acted on by the larger society, as well as how it has influenced other areas of communal life on the national and world levels. You will also begin to appreciate the exciting array of opportunities that exist for individuals who choose to become physical activity historians, as well as the many ways in which knowledge of our professional history can be used by kinesiologists in other subdisciplines.

Goals of History of Physical Activity

- To identify and describe patterns of change and stability in physical activity in particular societies or cultures during specific periods
- To analyze such patterns

Why Use History of Physical Activity?

Consider what life would be like if you didn't have a memory. You wouldn't be able to recall the 10 years that you spent training with a swim team, the pickup basketball games you played at your local neighborhood courts, the time you broke your leg falling off your bike and went to an exercise rehabilitation program for several months, or the fortunes of your favorite team. Memories give you insights about how things came to be the way they are. No one possesses perfect memory, and no one can predict the future with total accuracy, but our recollections can help us act intelligently by avoiding errors from the past and developing reasonable plans for the future.

History offers both broad and detailed insights that go far beyond our own memory; thus, it gives us the opportunity to develop a more extensive memory than we could ever acquire independently. History consists of a vast collection of information, mostly about events that occurred before we were born and often in geographic regions and societies different from our own. For instance, the ancient Greek Olympic Games, 18th-century peasant ball games in Europe, and 20th-century basketball in the United States have all influenced the physical activities that we take part in today. Studying history gives us windows onto the past and magnifying lenses with which to look closely at things that we find especially compelling. It helps us understand how and why our current physical activities are structured the way they are, allows us to compare them with physical activities from earlier periods, and gives us the tools to look toward the future from new vantage points.

For instance, a kinesiologist who is knowledgeable about history of physical activity would probably interpret an issue such as a fitness trend—or the popularity of spectator sports in colleges and universities or the availability of opportunity in sport across racial, ethnic, and gender lines—quite differently from someone who lacks such knowledge. Similarly, a person who has studied history of physical activity could look at the chapter-opening scenario and offer answers for the questions posed. For example, such a person would know that Protestant Christianity in the United States underwent a gradual but radical change in attitude toward recreational physical activity—a development known as "Muscular Christianity"—in the latter half of the 19th century. In the ensuing years, as organized sport and exercise were increasingly incorporated into the missions of many congregations and approved by many clergy members, physical education was increasingly accepted in public schools, and parks and recreation departments developed in cities across the United States in the early 20th century. Today, universities with strong ties to Christian churches often pursue fame and recognition through investments in intercollegiate athletics.

Although historians of physical activity are academic specialists in their own right, with specific goals that help guide their research, the fruit of their labor is often applied in a variety of other professional roles. For instance, many of the modern exercise machines found in college recreation centers, athletic training facilities, and commercial fitness centers are based on models developed by an early U.S. physical education leader—namely, Dudley Allen Sargent, who directed the Hemenway Gymnasium at Harvard University in the late 19th and early 20th centuries. Later, exercise specialists and engineers adapted Sargent's ideas and principles to contemporary knowledge and construction materials that were unavailable in his own time. In turn, some of Sargent's ideas were based on principles developed by even earlier inventors of whom he undoubtedly learned through the work of historians. In a similar fashion, coaches often create new playing strategies and training techniques by adapting those of an earlier generation of coaches whose work they have studied.

Likewise, knowing the history of sport, games, and exercise gives creative physical education teachers a tool for building interest in their lessons by helping students discover the development and evolution of an activity. In yet another application, scholars and scientists who perform research in any of the other kinesiology subdisciplines described in this book also rely on the work of earlier researchers in their own fields. They can learn about this history thanks to physical activity historians who have analyzed past work and summarized it in historical context. For example, one might examine how the first generation of scholars in a given field resolved ethical considerations related to the use of human or animal subjects and whether those issues have changed as the field has matured. Understanding past practices can help us avoid earlier mistakes, make adaptations that work within a new scheme, and gain insights that sometimes lead to entirely new ways of thinking about a specific problem.

Knowledge of history also helps us make better decisions today—for example, the decision by sport organizations to ban players, coaches, and managers from contact with gamblers. Major League Baseball was the first to adopt such a policy, doing so in the wake of the 1919 World Series, in which players from the heavily favored Chicago White Sox were accused of accepting money from gamblers in exchange for purposely losing the series to the Cincinnati Reds. Public outrage threatened to reduce confidence in the integrity of the game, which of course could have cut into ticket sales. With that lesson in mind, similar policies were adopted by other sport organizations, such as the National Football League, the National Basketball Association, the National Hockey League, the National Collegiate Athletic Association, and the U.S. Olympic Committee. Thus, the memory of the 1919 scandal has informed decisions by various organizations to fine, suspend, and dismiss players and management personnel for associating with gamblers in any manner.

It is hard to miss the phenomenon of U.S. society's current interest in exercise, health, and fitness. Exercise studios and health clubs dot the landscape, and these topics are widely addressed in videos, blogs, websites, books, and magazines. However, if you asked teenagers or people in their 20s about the origins of this trend, most would probably draw a blank. Moreover, at the same time that much of U.S. society seems absorbed in the idea and the practice of exercise, health, and fitness, the nation also faces rising obesity rates in the youngest of children and in the adult population. The origins of this perplexing issue, as well as the current conditions contributing to it, are still being sought out by medical professionals and kinesiologists, including some historians.

KEY POINT

History of physical activity teaches us about changes as well as stability in the past. This knowledge helps us understand the present and make reasonable decisions for the future.

Kinesiologists do know that interest in exercise picked up in the United States in the 1950s due to a variety of influences, including Cold War fears, President Dwight D. Eisenhower's heart attack, and evidence suggesting that U.S. children were less physically fit than their European counterparts. Kinesiologists also know about an earlier period in U.S. history when exercise was prominent—the late 19th century (Park, 1987a).

We can also point out that U.S. fascination with structured exercise has fluctuated over the years, focusing on different benefits of physical activity. This information raises important questions: Is the current, relatively high level of interest in exercise for health likely to decrease (as is perhaps suggested by the historical fluctuation)? If so, the coming decrease carries serious negative implications for the health of many Americans given the decline that would result in physical fitness for most of the U.S. population. Can we do anything to prevent this possible downward cycle? Or do we now have such strong scientific evidence confirming the health benefits of exercise that we will break the roller-coaster pattern of the past? We could not even raise these questions without understanding the history of physical activity.

Your own previous physical activity experiences, academic studies, and even observations of professional practice have given you some insight into recent history of physical activity. For example, experience in competitive team sports may have given you firsthand knowledge of the fact that basketball, volleyball, and soccer are incredibly popular in Western societies and, in recent years, have grown tremendously in Asian and African countries. If you have ever wondered how this came to be, then you have been on the verge of historical inquiry. Or perhaps you have noticed gender inequities in physical activity and wondered what historical events led to their existence. If you are a man, then perhaps at some point you have feared being teased for a lack of athletic ability. If you are a male athlete, then you may have noticed with some unease how much more financial support you received to develop your athletic talents than your female counterparts did. If you are a woman, then you may have felt strangely out of place in a competitive environment or noticed the disproportionate limitations on the financial support available to help you improve your athletic skills. Whatever your particular case, if you have ever wondered why such situations exist, then once again you were on the verge of delving into history.

More formally, your previous scholarly study has probably familiarized you with some historical information that will help you understand the history of physical activity. Much of the historical research on physical activity takes a Eurocentric perspective deriving from ancient Greece and Rome, Europe and Great Britain, and North America. Knowing something about the overall history of these societies will make the history of physical activity easier for you to learn and remember. For example, if you know that nationalism and national patriotism swept through Europe in the 19th century, then you will not be surprised to learn that in countries such as Germany and Sweden people sought to use special gymnastics systems to build a healthy citizenry that would keep their nations strong. In turn, making this connection helps you remember when, and at least one reason why, such gymnastics systems came into being. We also have much to learn about the rich physical activity histories of Asian, Native American, and other cultures.

Finally, from watching physical activity professionals in action, you may have observed some history in the making. Perhaps you know people who work as athletic trainers, sport marketers, personal trainers, or sport physical therapists. These professionals are part of an important historical trend. Not long ago, career opportunities for people in the physical activity field were limited to professorships, physical education teacher positions, and coaching jobs. In recent years, however, professional opportunities have expanded greatly, and this is a historical phenomenon.

What Do Historians of Physical Activity Do?

As outlined at the beginning of the chapter, physical activity historians pursue two main goals. The first goal is to identify and describe patterns of change and stability in physical activity in particular societies or cultures during specific periods. The second goal is to analyze such patterns. In terms of the first goal, historians provide enormous factual detail to give readers a thick, robust description of, for example, what it was like to be involved in a particular physical activity, as well as the values and attitudes that people had toward the activity, the people's broader lives, the general societal values and attitudes prevalent at the time, how people's physical activities fit into their overall lives, and how their physical activities changed.

goals

Comparing Sport: Then and Now

Many movies show people playing sport during different historical periods. Here are just a few examples:

42 (set during mid-1940s; racial integration of professional baseball)
A League of Their Own (set during 1943; female professional baseball)
Breaking Away (set during late 1970s; bicycling)
Chariots of Fire (set during 1924; Olympic track and field)
Cinderella Man (set during 1930s; boxing)
Field of Dreams (set during 1980s with throwback to 1919; baseball)
Hoosiers (set during 1950s; basketball)
Kansas City Bomber (set during 1970s; roller derby)
Leatherheads (set during 1925; American football)
Race (set during 1936; Olympic track and field)
Seabiscuit (set during 1930s; horse racing)

There are many others. Think about film footage that you have seen of people playing a sport during different historical time periods. Compare it with how the sport is played today or how contemporary sport is presented in movies.

Sometimes changes occur slowly, giving us the impression that things have stayed the same. For example, the rules, strategies, styles of play, and equipment in many of our sports changed considerably during the 20th century; however, because these shifts occurred relatively slowly, the games remained recognizable. At other times, changes take place much more rapidly and dramatically. For example, improvements in bicycle design in the United States more than a hundred years ago contributed to a rapid rise in bicycling and enabled the bicycle craze of the 1890s (Hardy, 1982).

The second main goal pursued by physical activity historians is to *analyze* the patterns of change and stability that they find in physical activity in particular societies or cultures during specific periods. That is, most historians go beyond description in order to demonstrate relationships and influences and thereby explain why certain things occurred. To do so, they look for individuals, groups, events, and ideas that helped bring about changes or maintain stability. For example, Hardy's (1982) *How Boston Played: Sport, Recreation, and Community 1865–1915* focuses on the city-building process that took place in Boston after the Civil War. During that time, the population of the city mushroomed as European immigrants and people from the U.S. countryside arrived to take jobs created by the Industrial Revolution. The growing population led to major problems, such as overcrowded tenements, poor health, neighborhood destruction, crime, ethnic conflict, and political and moral disorder—all of which threatened traditional ways of life. Hardy believed that Bostonians were looking for ways to revitalize their sense of connectedness to one another and that sport and recreation played an important part in that process.

Looking at what happened in Boston during this period, Hardy posited three categories of ways in which Bostonians responded to the problems in their city: escape from the city and its problems, reform of the city and its problems, and accommodation to new forms of city life through developing a renewed sense of group identity (1982, pp. 197–201). Hardy used these categories to examine the complex ways in which people sought community by building new playgrounds and parks, expanding sport and exercise programs in schools and universities, joining community-based sport clubs, riding bicycles, and conferring hero status on top athletes. Understanding how this earlier generation of Bostonians used physical activity to benefit the entire city could well prompt modern-day community leaders to work with physical educators, recreation specialists, exercise consultants, landscape architects, land developers, and urban planners to create new and exciting opportunities for current and future generations of citizens.

Historians who study physical activity are usually college or university faculty members, but a handful are librarians, journalists, consultants for book publishing companies, archivists in charge of special collections of documents, or museum curators. We focus here on faculty members, whose typical activities include teaching, research, and professional service. They often teach broad survey courses on history of physical activity, and some of their classes focus more narrowly on topics such as the ancient Olympic Games, physical activity in colonial America, and sport and ethnic relations in the United States. Examples of professional service include giving a community presentation about the history of a region's minor league baseball team, participating in an educational television panel to discuss the history of gender relations in American sport, and helping a city's nonprofit historical society put together a museum display about the history of rehabilitative exercise.

KEY POINT

Most physical activity historians are college and university faculty members. They pursue the two main goals of the subdiscipline—identifying and describing key patterns and analyzing those patterns—through their research, teaching, and service.

Libraries are perhaps the most important research tools for historians. For example, if you want to study the physical activities of African American slaves in the United States, you could use a library to locate sources such as interviews with former slaves, published accounts from white slaveholders written during the time of slavery, and accounts from Northern whites who visited plantations where slaves worked (Wiggins, 1980). Because libraries are so important, most historians take great care to develop solid working relationships with librarians. In some cases, library staff act almost as gatekeepers who decide how much access you should have to rare and valuable materials.

Historians of physical activity sometimes apply for research money, either from their own institution or from outside organizations that specialize in supporting historical research. Historians often use such funding to travel to libraries in distant cities or to pay research assistants to help collect information. When asking for research money, historians must write clear, persuasive proposals. Funding organizations usually require applicants to describe such things as the topic they want to investigate, why it is important, the sort of information they will collect, the sources where they will find this information, what makes them competent to conduct the study, how much money they need, and what they will buy with that money.

History of Kinesiology and Physical Activity

A handful of U.S. scholars studied history of physical activity in the early 20th century, and even in the late 19th century at least one physical educator viewed the subject as important. Even so, it wasn't until the 1960s that the subdiscipline began to develop a recognizable identity in North American kinesiology.

One of the earliest U.S. reports to emphasize history of physical activity was Edward Hartwell's *On Physical Training* (1899), which included information about ancient Greece, Europe and Great Britain, and the United States (Gerber, 1971). Hartwell believed that professionals who would be charting a course for the future should know about the past, and as early as 1893, he pointed to the need for a textbook on history of physical education. Decades would pass, however, before such texts appeared.

In 1917, historian Frederic Paxson wrote a classic article titled "The Rise of Sport" (Paxson, 1917). Drawing from the well-known frontier thesis developed by his graduate school mentor Frederick Jackson Turner, Paxson argued that the disappearance of the Western frontier spurred an increase in the popularity of sport because sport served some of the same functions of escape or release that the frontier had enabled (Pope, 1997; Struna 1997). This safety valve function and Paxson's notion of the gradual, inevitable social evolution of sport served as central ideas in the research of many sport history scholars for decades to come.

In the 1920s and 1930s, physical educators published several textbooks on the history of physical education, as well as articles in professional

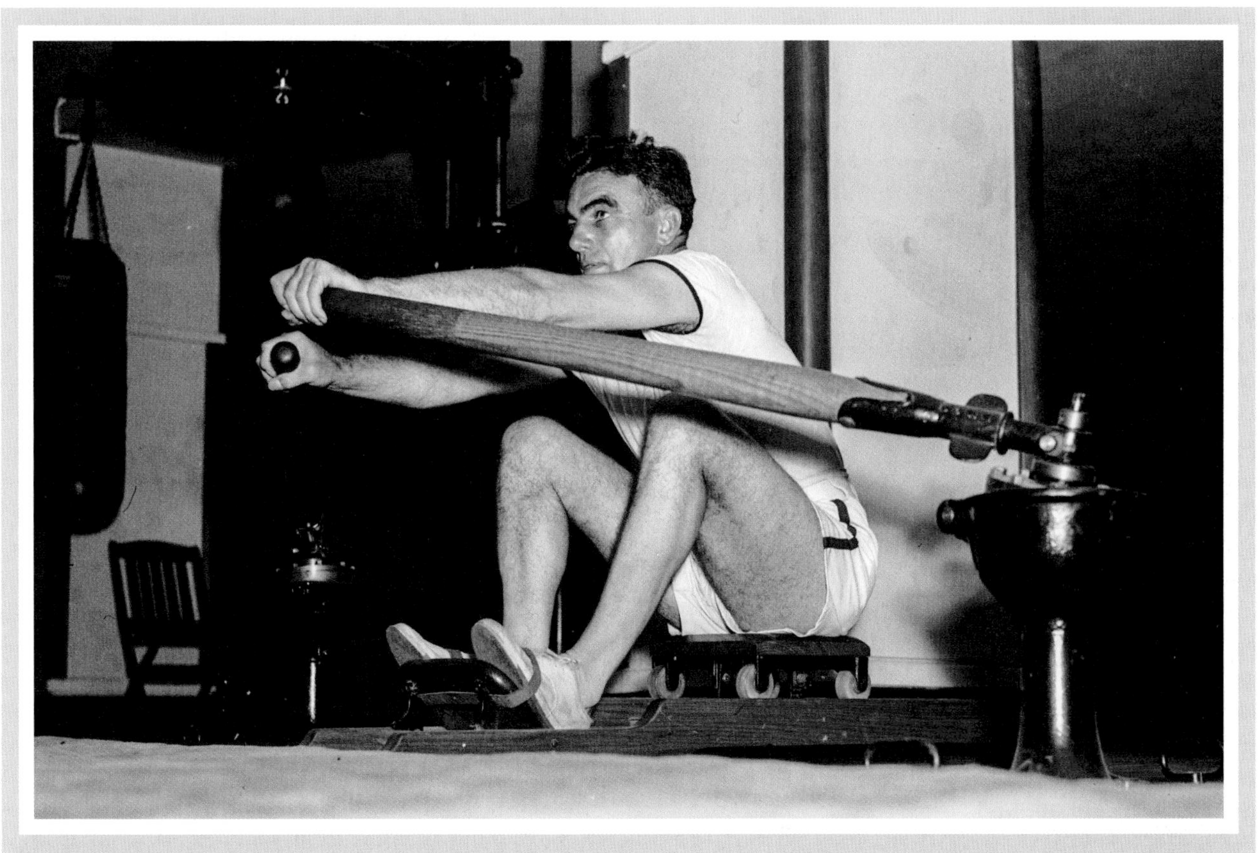

Sport historians look at where we've come from to help us see where we're going.

Courtesy of the Library of Congress, LC-DIG-hec-23026

journals on the development of physical education programs in schools and colleges. These books and articles probably helped lead physical education departments at colleges and universities to recognize the importance of history of physical activity.

Seward Staley (1937), professor of physical education at the University of Illinois, opened the door wider when he suggested that courses on the history of physical education should include a focus on the history of sport. Under his direction, graduate students would begin studying sport history at the University of Illinois after World War II (Struna, 1997).

In 1940, historian Foster Rhea Dulles published his widely read *America Learns to Play: A History of Popular Recreation.* Dulles went beyond Paxson's safety valve and evolutionary perspectives to point out more positively that popular 20th-century recreational pursuits—including sport—were well liked because they were satisfying to the U.S. public (Struna, 1997). Furthermore, a broad range of people had access to them, indicating to Dulles that U.S. democracy had come of age (Pope, 1997).

After World War II, historian John Betts completed an important doctoral dissertation titled "Organized Sport in Industrial America" (1952). He stressed that sport played an active role in society by helping bind people together, and he highlighted the ways in which late 19th-century entrepreneurs had harnessed a variety of new manufacturing processes, new forms of transportation, and new forms of communication to shape sport into a profitable commercial enterprise (Pope, 1997; Struna, 1997). In sharp contrast to Betts' active model, a passive model of sport was offered a decade later by journalist Robert Boyle in *Sport—Mirror of American Life* (1963). As the title suggests, Boyle viewed sport as a mirror that simply reflected society. Both Betts' and Boyle's work exerted considerable influence on the work of other sport historians.

The ranks of sport historians began to grow in the 1960s, but no scholarly association yet existed to help the subdiscipline advance. The first attempt to remedy this lack occurred in 1962 when Marvin Eyler, Seward Staley, and Earle Zeigler worked to develop a section for sport history in the College Physical Education Association, which was soon renamed as the National College Physical Education Association for Men (NCPEAM). Because this was the only formal organization for sport historians at the time, the NCPEAM *Proceedings* became an important published record of their work (Berryman, 1973; Struna, 1997).

Scholars in the discipline of history became more interested in studying sport when Eugen

Kinesiology Colleagues

Photo courtesy of Jennifer Wicks/The California State University

Allison Wrynn

Allison Wrynn, a professor in the department of kinesiology at California State University, Fullerton, specializes in the history of sport and kinesiology. She is a widely published historian with a major focus on women and girls in sport and physical activity. She is also a voracious reader who, unsurprisingly, was drawn to biographies as a child. Her professional training began at Springfield College, the birthplace of basketball and an early leader in the field of kinesiology, where she majored in physical education and was surrounded by the history of the field on a daily basis. She also trained at UC Berkeley with one of the leading sport history scholars in the world, Dr. Roberta Park. Wrynn has traveled all over the world to conduct research and attend conferences. As she says, "It's gratifying to see the discipline of sport history grow and mature." In addition, as the first (and thus far only) female editor of the *Journal of Sport History*, she has encouraged women, young scholars, and those doing research who might be considered less "traditional" to submit their work in order to advance the field.

Weber, a prominent historian, presented a paper titled *Gymnastics and Sports in Fin-de-Siècle France: Opium of the Classes?* at the 1970 meeting of the American Historical Association (AHA) and published it the following year in *American Historical Review* (Weber, 1970; Berryman, 1973). The program of the 1971 AHA meeting included an entire section on sport history.

In another development, the North American Society for Sport History (NASSH) was established in 1972, held its first annual conference in May 1973, and was soon recognized as the main scholarly association for sport historians in the United States and Canada. NASSH was designed to attract a broader scholarly membership, including women, Canadians, scholars beyond North America, and sport historians from disciplines beyond physical education. It began publishing *Journal of Sport History* in 1974, and a report issued in 1985 identified this publication as the seventh most frequently cited history journal among the hundreds in existence (Struna, 1997). During this same period, the academic study of the history of sport and physical activity was advancing throughout the world. The International Society for the History of Physical Education and Sport (ISHPES) was formed in 1989 from the merger of two existing organizations: The International Committee for the History of Sport and Physical Education (founded in 1967) and the International Association for the History of Physical Education and Sport (founded in 1973). The British Society of Sports History, the Australian Society for Sports History, and the European Committee for Sports History are among the many national and regional scholarly organizations devoted to this vibrant area of study.

Research Methods for History of Physical Activity

In the 1970s and 1980s, research in sport history expanded in several directions (Pope, 1997; Struna, 1997). For one thing, scholars began using two new analytical frameworks, or sets of general concepts or ideas, to make sense of the historical information they collected. The first framework, **modernization theory**, emphasized that the rise of modern sport occurred during the Industrial Revolution as U.S. society shifted away from agricultural and local economies toward city-based industries rooted in science and technology. During this process, sports changed from relatively unspecialized games to highly organized contests involving many rules and specialized playing positions. The use of modernization theory was pioneered by Allen Guttmann in his book *From Ritual to Record: The Nature of Modern Sports* (1978).

The second conceptual framework emphasized **human agency** and suggested that people were actively involved in developing or constructing their own sports. Research in this vein focused on the details of how this construction took place (Pope, 1997; Struna, 1997). One of the earliest works to use this approach was Stephen Hardy's (1982) work, mentioned earlier, on sport and recreation in Boston after the Civil War. Hardy pointed to many local struggles among middle-class and working-class groups—often tinged with ethnic distinctions—that occurred as Bostonians went about structuring their sport and recreational activities.

Other scholars expanded the focus on human agency to look at gender differences. For example, in *Cheap Amusements: Working Women and Leisure in Turn-of-the-Century New York,* Kathy Peiss (1986) examined working women's culture between 1880 and 1920. She emphasized their newfound recreational activities outside of the home in dance halls, nickelodeons, social clubs, and amusement parks.

Another important research direction, inaugurated in the 1980s, dealt with exercise and health (Struna, 1997). This trend was sparked both by the U.S. public's growing interest in exercise and physical fitness in the late 20th century and by the expansion of advanced course work on exercise in college and university programs of kinesiology and physical education. One of the earliest studies of this type was James Whorton's *Crusaders for Fitness: The History of American Health Reformers* (1982). Whorton showed the role of exercise among the many practices advocated in 19th- and early-20th-century reform movements aimed at improving the health of the U.S. public.

In the 1990s, physical activity scholars continued to use the approaches to historical research that had been pioneered in the 1970s and 1980s. In addition, they increasingly linked their work to research in other social science disciplines, especially anthropology, economics, and sociology (Struna, 1997). This approach has continued into the second decade of the 21st century.

Thus, building on a small base of scholarly knowledge gathered earlier in the 20th century, the subdiscipline of the history of physical activity grew in size, sophistication, and scope from the 1960s onward. Scholars initially focused on sport,

Thinking From a Historical Perspective

Before the 1970s, scholars in the discipline of history had little interest in studying sport history. Why did these historians ignore sport for so long? What might have been their concerns? What challenges or roadblocks might they have faced in conducting historical research in the field? What characteristics of sport itself might have made historians slow to identify sport history as a viable research topic?

but since the 1980s have also given increased attention to exercise and physical fitness.

Now that you've learned about the goals of history of physical activity, as well as the chronological development of the subdiscipline, let's look at how scholars produce this knowledge. How do they find out what happened, who was involved, when and where things took place, and why? In some ways, the process resembles what detectives and attorneys do when they hunt for evidence to reconstruct important details of a crime. Specifically, the process involves locating evidence; critiquing and examining it; and piecing it together into a coherent, insightful framework that explains how and why things occurred. We can divide this process into three stages: finding sources of evidence; critiquing the sources; and examining, analyzing, and synthesizing the evidence.

Finding Sources of Evidence

Think about the many physical activities that are common in current society. If historians 200 years in the future wanted to study these activities, what sorts of items would be useful? All kinds of things may come to mind: exercise videos; websites, books, articles, and magazines about exercise and physical fitness; business records of professional sport franchises; videos of televised sporting events; curriculum guides for school physical education; videos of dance performances; policy statements from the National Collegiate Athletic Association; and the 1996 report from the U.S. Department of Health and Human Services titled *Physical Activity and Health: A Report of the Surgeon General.*

Of course, the list could go on and on. For instance, the digital age has brought us countless websites devoted to health, fitness, exercise, and organized sport ranging from youth activities through professional and masters-level competition. Sport fans and people seeking information about health and exercise can now blog about their interests, instantly share items, and express their own reactions to what they read. These digital sources will serve as a rich resource for historical researchers in the future. As with traditional primary and secondary resources, their hardest task will be to test the validity and reliability of the data. In addition, the sheer volume of data available online presents its own set of challenges in reading, sorting through, and organizing the countless resources for further analysis and interpretation.

The items just discussed constitute primary sources of information, meaning that they were produced in the society and period being studied. However, where would scholars find them? We can only guess where such things might be stored in the future, but we know where historians look today—libraries, archives (storage facilities for documents from people and organizations deemed important), and private collections (Struna, 1996a). In addition, researchers sometimes interview people to tap their memories of life in earlier times.

Relevant materials are scattered throughout the United States. For example, the University of Illinois is home to the Avery Brundage Collection. (Brundage was president of the International Olympic Committee from 1952 to 1972.) The collection contains items such as letters to and from Brundage, books, pamphlets, scrapbooks, Brundage's notes on various topics, speeches given by Brundage, and policies from the International Olympic Committee and other sport governing bodies (Guttmann, 1984). In another example, copies of *Boston Medical and Surgical Journal,* from the 19th century, are located both at the National Library of Medicine in Bethesda, Maryland, and at a library at the University of Chicago (Struna, 1996a). Meanwhile, the Archives and Special Collections of Springfield College house original documents from early physical education and YMCA organizations, as well as Olympic and other amateur sports. In addition, local historical societies often provide rich sources of information, and the same is true for the archives located in the library at your own college or university.

Historical reports written by people in later periods—referred to as secondary sources—can be useful in finding primary sources. Examples of secondary sources include a journal article written in the 1980s about African Americans who participated in intercollegiate athletics in the 1930s and

1940s, as well as a book-length report produced in the 1990s on physical activity in the American colonies. Authors of such sources may include detailed notes pointing to the nature and location of specific information (i.e., primary sources and other secondary sources) used in their investigations. In addition, most libraries have electronic databases to help you locate sources, and most of them are accessible online.

Critiquing the Sources

After you locate primary sources, you must scrutinize them carefully for authenticity and credibility. Sources that are authentic and credible make your historical research believable. Consider the case of written documents such as newspaper articles, books, and letters. Determining the authenticity of such sources involves identifying who wrote a given piece and when. You might learn that a document was written many years after the events it describes or that the author listed on the document was not the person who wrote it. For example, if a letter dated as being from the 17th century uses the phrase "didn't get to first base," then you can be sure that it was written much later because this term comes from the game of baseball, which didn't exist until much later (Shafer, 1980). Similarly, if the ink used to publish a certain report wasn't manufactured until several years after the indicated date, then you know that the piece was actually written later. (Such a determination would require a technical specialist.)

Determining the credibility of written documents helps you interpret them. Specifically, the *rule of context* encourages you to make sense of a document's language in relation to what the words meant to people in the society in which it was produced. The *rule of perspective* requires you to examine an author's relationship to the events that he or she describes and how the author obtained the information. The *rule of omission* or *free editing* reminds you that you always have partial records of events—not the complete events themselves; therefore, it is important to locate multiple sources to round out your evidence (Struna, 1996a).

Examining, Analyzing, and Synthesizing the Evidence

After you have located authentic and credible sources, you should examine the evidence to find information that addresses the tentative hypotheses or research questions that you established at the beginning of your investigation. The goal is to describe events in detail and then analyze them in order to learn how and why things took place. Historians accomplish this goal by placing events in an analytical framework that uses either trends and relationships in the events themselves or theoretical models from the social sciences. You have already seen three examples of analytical frameworks—the binding together in community framework (Betts, 1952), the modernization framework (Guttmann, 1978), and the human agency framework (Hardy, 1982; Pope, 1997; Struna, 1996a, 1996b)—and there are probably as many approaches to analysis and synthesis as there are researchers. The main goal is to piece together the evidence in order to gain a detailed understanding of what happened, as well as how and why (Shafer, 1980; Struna, 1996a).

KEY POINT

Historical research involves finding sources that contain evidence about past events, critiquing the sources for authenticity and credibility, and analyzing the data contained in the sources in order to learn how and why things happened.

Overview of Knowledge in History of Physical Activity

Now that you know what historians of physical activity do, let's look at some of the fascinating information they've discovered. The body of knowledge is large and complex; it deals with many societies worldwide and covers many periods. Here, we touch on a few highlights to provide a well-rounded perspective of kinesiology and to stimulate your interest in taking a sport history course.

If you were a boy living in the Greek city-state of Sparta during the 6th century BCE, you would receive more extensive and harsher physical activity training than would a boy in the city-state of Athens, who would receive a more balanced physical and intellectual education. In both places, however, the physical activity instruction would pursue the goal of preparing you for military service. Although the ancient Greeks considered the Olympic Games similar to warfare in competitiveness, this prestigious event for full-fledged citizens was distinct from the brutal gladiators' contests—which involved mostly slaves and convicts—that were staged several hundred years later in the Roman Empire. In turn, the Roman contests differed greatly from the informal, rough team ball games played by European peasants in the Middle Ages and from the high-profile women's gymnastics performed in the Olympics of the late 20th century. We haven't even mentioned other forms of physical activity, such as dance and exercise! Thus, it is indeed a monumental task to understand the ways in which physical activities have fit into societies around the world since the beginning of history.

In North America, much is known about the history of physical activity in Western civilizations—especially ancient Greece and Rome, Europe, Canada, and the United States. We must recognize, however, that rich physical activity traditions have also been developed by people in Africa, Asia, Australia and New Zealand, the Indian subcontinent, Central and South America, the Caribbean region, and the Pacific Islands. These traditions include a variety of martial arts in Asian societies, sumo wrestling in Japan, ball games played in high-walled courts in native Central American civilizations, an African martial art brought to the Atlantic and Caribbean coasts of South America, and a sport in Afghanistan involving hundreds of mounted horsemen who simultaneously try to gain possession of a goat or calf carcass. Historical information exists about all of these activities. In this text, however, we focus on the history of physical activity in North America.

After an earlier trickle of travelers, British and European colonists began arriving in North America in greater numbers in the early 1600s. Initially, people worked hard to establish themselves in unfamiliar and often harsh environments. These hardships, coupled in some places with strict religious prohibitions against idleness and amusement, meant that participation in sport and games was limited. Nevertheless, people sometimes engaged in their own versions of traditional British and European sport and recreational activities. Participation increased as the colonists became more established and religious sanctions were lifted. Activities included horse racing, fishing, hunting, sailing, boat racing, team ball games, sleigh races, skating, footracing, boxing, wrestling, animal baiting (in which a wild animal such as a bear was tied by a chain and attacked, often by dogs or rats), cockfighting, and billiards. Some of these activities were commercialized and yielded profits for promoters; gambling was also common (Baker, 1988; Lucas & Smith, 1978; Struna, 1996b).

Next, we'll look more closely at physical activity in North America during three periods:

- 1840–1900: North American society was characterized by industrialization, growth of science and technology, immigration, urbanization, democratization, and westward expansion. The period also included the American Civil War.
- 1900–1950: North American society was characterized by the growth of consumerism, immigration, and democratization. The period also included a major economic depression and two world wars.
- 1950–present: North American society was characterized by the growth of electronic communication, the growth of global trade, immigration, and democratization. The period also included the Cold War, the Korean War, the Vietnam War, the first Iraq War, and the Afghanistan and Iraq wars of 2001 to 2012.

For each of these periods, we'll examine the following aspects:

- Participation in physical activity
- Physical activity professions
- Scholarly knowledge about physical activity

1840–1900: Industrialization and Westward Expansion

From the mid-1800s to about 1900, the economy, population, and geography of the United States and Canada expanded. These developments manifested themselves in diverse ways. The growth of business and industry was fueled by new developments in science and technology. Meanwhile, as city populations mushroomed, people's health was threatened by major sanitation problems, crowded living conditions, lack of space for physical exercise, and unsafe factories. At the same time, the advent of more leisure time sparked interest in establishing city parks and playgrounds, as well as in developing more commercial entertainment and amusements, such as spectator sports. In addition, many new schools and colleges opened, offering both physical training and student-run intercollegiate sport programs. The citizenry also started buying more products in response to glimpses of the "good life" seen in newspapers, magazines, and eventually movies. As westward expansion continued, highlighted by the completion of the transcontinental railroad in the United States, massive numbers of European and British immigrants arrived, along with Asians on the West Coast. The Civil War (1861 to 1865) exacted a terrible toll of death, destruction, and disruption, but it also led to legal affirmation of the ideal of racial equality—an important goal in U.S. society ever since.

Participation in Physical Activity

Liberal religious and philosophical currents had begun to flow together in the early decades of the

19th century, and they emphasized the view that a human being's body, mind, and soul formed an integrated whole. This perspective promoted the importance of keeping one's body healthy in order to achieve peak mental functioning and moral reasoning. Some even believed that good health practices indicated a morally righteous person in the eyes of God. This thinking, combined with squalid conditions in overcrowded cities and a nationalistic desire to make the United States self-sufficient, led to widespread efforts throughout most of the 19th century to improve people's health. The recommended practices included proper diet and exercise (Berryman, 1989; Whorton, 1982).

Calls for more vigorous physical activity—especially for boys and men—were issued by physicians, writers, and teachers, who addressed primarily the middle class and social elites. These authorities coupled their notions of the importance of health reform with ideas from England about the value of sport for developing moral character and manliness. As the century progressed, physical activity grew increasingly attractive and popular.

Many professionals also recommended moderate exercises for improving the health of girls and women. However, straining hard in exercise or competitive sport was usually considered unladylike and even dangerous for women's physical well-being and childbearing capability (Vertinsky, 1990). Even so, by the end of the 19th century, some girls and women were involved in vigorous physical pursuits.

During this time, European systems of gymnastics, especially those originating in Germany and Sweden, gained an important foothold in the United States. German cultural societies called *Turnvereins* eventually offered well-equipped gymnastics facilities in many North American cities. By 1867, 148 of these facilities existed in the United States, serving more than 10,000 members (Swanson & Spears, 1995). The programs centered on exercise performed on heavy equipment, such as parallel bars, vaulting horses, and hanging rings. In contrast, the Swedish system involved calisthenics and exercises performed with lighter equipment, such as dumbbells, wooden pins, long rods, and weights. Elements from both systems, as well as other exercise programs developed by Americans, grew increasingly popular in YMCAs and YWCAs and at city parks and playgrounds.

Public spaces, such as skating ponds in New York City's Central Park, provided residents of crowded cities with places to participate in healthy activities.

Image from *Harper's Weekly*, 1860, Skating on The Ladies' Skating Pond

Sports were popular with many of the new immigrants who arrived in the United States and Canada from 1840 onward. They played baseball as well as American and Canadian football, which evolved from rugby and soccer later in the century. Youngsters often learned American sports and games at local playgrounds, and many civic leaders hoped that immigrants' participation would speed their assimilation into U.S. society. Immigrants also sought to use physical activities to keep the cultures of their homelands alive. Many belonged to special clubs in which sport, exercise, and other forms of recreation were prominent. Examples in the city of Boston included the Boston *Turnverein*, which featured gymnastics and catered to German immigrants; the Caledonian Club, devoted to Scottish culture and offering an array of sports and games, such as caber toss, racing, and pole vault; the Irish Athletic Club, the Irish American Athletic Club, the Ancient Order of Hibernians, the Boston Hurling Club, and the Shamrock Hurling Club, all of which focused on Irish sports and games, including oarsmen's regattas, stone throwing, and hurling; and cricket clubs formed by West Indian immigrants (Hardy, 1982).

Other immigrants took part in sports and amusements that, at the time, were considered less than respectable by members of mainstream society. These activities included prizefighting, billiards, bowling, cycling, and wrestling. Prizefighting was dominated by the Irish for several decades; later in the century, boxing gyms could be found in African American neighborhoods (Gorn, 1986). After the Civil War, African Americans found their way into mainstream sports—professional baseball and horse racing, for example—but by 1890, racial prejudice forced most African Americans into segregated leagues and organizations.

In school and college physical activity classes, gymnastics exercise was preeminent. The Swedish and German systems were commonly used, as were exercise programs developed by Americans that centered on calisthenics and lightweight equipment. Such programs focused primarily on achieving health through fitness and were promulgated by physical education leaders such as Edward Hitchcock (Amherst College), Dudley Allen Sargent (Harvard University), and Delphine Hannah (Oberlin College). Sport was not usually included in these classes but was practiced in extracurricular settings. Originally taking the form of low-key intramural activities, sport served as a rich supplement to the often-stodgy exercise classes (Rudolph, 1962; Smith, 1988). Student-run intercollegiate competition appeared about midcentury and gathered steam in the 1870s. The most common intercollegiate sports were American football, crew, baseball, and track and field.

College and university administrators believed that intercollegiate sport could help recruit new students, attract alumni contributions, and promote a sense of community in an increasingly diverse student body. According to thinking that arrived on North American shores from England, sport could also be used to develop leadership, vigorous manliness, and upstanding moral character, which were considered to be desirable qualities for college men. For all of these reasons, athletics were considered too important to leave in the hands of students; as a result, campus administrators, faculty, and coaches gradually took control.

During this era, fewer sport opportunities were available for college women, although they did engage in occasional intercollegiate competition and somewhat more frequent intramural competition. In the 1870s and 1880s, however, many female physical educators began including sport activities in their classes (Swanson & Spears, 1995). Still, women who wanted to participate in competition faced societal concerns about supposed negative effects of strenuous and highly competitive physical activities on women and girls. Thus, they avoided sports in which injuries were more likely (e.g., football) in favor of sports such as basketball, tennis, golf, and baseball. In addition, women's sport was usually conducted in secluded locations with only a few spectators present. In these ways, a delicate balance was struck between making competition available to college women and keeping the activities acceptable to faculty and campus administrators (Park, 1987b).

KEY POINT

Throughout the 19th century in the United States, interest increased in physical activities, including gymnastics exercises, baseball, football, crew, track and field, horse racing, boxing, bicycling, and less well-known sports tied to people's ethnic origins.

Beyond schools and colleges, many amateur baseball teams were on their way to becoming professionalized in the 1860s as spectators were charged admission fees and players were paid. Baseball developed an avid following, and newspapers gave it the most coverage of any sport (Adelman, 1986). By the 1880s, the most popular sports were baseball, horse racing, and prizefighting.

U.S. athletes also competed in a new amateur event developed by Europeans and first held in

Athens, Greece, in 1896—the modern Olympic Games. The United States fielded a team of 10 male track-and-field competitors, all of whom were collegians or ex-collegians. These competitors chalked up nine victories in twelve events, along with five second-place finishes. Three other American athletes—two pistol shooters and one swimmer—also competed in the Games and garnered two victories in the sharpshooting events.

Physical Activity Professions

An identifiable physical activity profession did not appear in the United States until the late 19th century. Nevertheless, a variety of practitioners in earlier decades focused at least some of their work on physical activity. These practitioners included physicians, athletes, journalists, educators, ministers, health reform advocates, entrepreneurs, and a handful of European gymnastics specialists who immigrated to the United States. Some of them wrote about the physical, intellectual, and moral benefits of exercise and sport. Some developed exercise programs in schools and colleges. Others worked to establish various European gymnastics systems on U.S. soil. Some wrote popular self-help manuals about healthy living that included how-to information on exercise routines. A few became professional athletes in sports such as horse racing, pedestrianism (long-distance walking), boxing, and baseball. Others tried to sort out the best training techniques for athletic success, became coaches, or bankrolled professional sports.

By the 1880s, a climate of intense interest in exercise and sport had developed in the United States. More generally, numerous professions were beginning to organize in response to the growing need for people with specialized knowledge and skills in technologically based businesses and industries. As a result, efforts were launched to develop more stringent educational standards for new practitioners.

In this atmosphere, the physical education teaching profession took root (Park, 1987a), thus becoming the first recognizable physical activity profession in the United States. In 1885, some 60 people who wanted to promote the new profession formed the American Association for the Advancement of Physical Education. Even though university programs of physical education remained segregated for decades, physical education was a progressive field that involved professional collaboration between men and women on the faculty. After several name changes and adjustments in its mission, the AAAPE is known today as SHAPE America, or the Society of Health and Physical Educators.

Aaron Molyneaux Hewlett was an early professional in the physical activity world who served as the first director of the Harvard College Gymnasium.

HUP Hewlett, A. Molyneaux (3a), olvwork173667, Harvard University Archives

KEY POINT

The earliest identifiable U.S. physical activity profession—teaching physical education—was established in the late 19th century during a period of high interest in physical activity for health among college students and the general public.

Several teacher training programs were inaugurated in physical education, primarily in the northeastern United States, and they varied in length from a few weeks to about 2 years. Some of the most famous were the Sanatory Gymnasium (opened near Harvard University in 1881 by Dudley Sargent and

Curriculum Changes in Physical Education

In the late 19th century, physical education programs emphasized health and activities focusing on calisthenics; sport was not usually included in men's physical education classes, although it was an integral part of the curriculum in many women's colleges. What do you think was the reason for this emphasis? What types of activities have been included in your physical education classes in high school and college? What do you think caused the change between what was offered in the 1890s and what is offered today?

renamed the Normal School of Physical Training in 1894), the Harvard Summer School of Physical Education (opened in 1887 by Dudley Sargent), and the Boston Normal School of Gymnastics (opened in 1889 by Mary Hemenway and Amy Morris Homans). Around the turn of the century, several 4-year bachelor's degree programs were initiated for the preparation of physical education teachers.

Scholarly Knowledge About Physical Activity

Scientific discoveries in the 19th century produced much new information about human anatomy and physiology. By the 1840s, improved microscopes permitted more detailed studies of phenomena such as oxygen transport in blood, energy transformation in muscles, and the anatomy and functioning of the nervous system (Park, 1987a). Such information was used later in the century by scholars working to better understand the biological effects of physical activity on the human body.

This knowledge was also incorporated by some educators into physical education teacher training programs. For example, Edward Hartwell, an associate in physical training and director of the gymnasium at Johns Hopkins University, published an article in 1887 titled "On the Physiology of Exercise." George Fitz, an instructor and researcher in physiology and hygiene at Harvard University, established a 4-year undergraduate program called Anatomy, Physiology, and Physical Training. Not far from Harvard, the Boston Normal School of Gymnastics—a 2-year institution providing physical education teacher training that attracted mostly female students—offered a curriculum focused on Swedish gymnastics that also included a variety of basic science courses taught by well-known scholars from major universities around Boston. These excursions into the biological sciences by early leaders in physical education demonstrate the value that they placed on developing a scientific base for the emerging profession of physical education teaching (Gerber, 1971; Park, 1987b).

KEY POINT

Though scholarly knowledge about physical activity became important in a few teaching training curriculums in the late 19th century, it more often took a backseat to learning physical activities and practical knowledge about teaching them.

Despite this new interest in the sciences, scientific curriculums were outnumbered by professional programs that focused on learning physical activities and how to teach them. By the end of the 19th century, the emphasis was clearly placed on the professional programs and on the positive social values that students could learn through participating in both play and sport. Investigations into the biological mechanisms underlying physical activity were not completely curtailed, but it was not until the 1960s that scientific information again moved to center stage in college and university physical education curriculums (Park, 1987a).

1900–1950: Consumerism, Immigration, Democratization

In the first half of the 20th century, U.S. industry perfected mass production and churned out a wide variety of consumer goods, ranging from automobiles to refrigerators to canned food. With readily available products, entrepreneurs stepped up their advertising to encourage members of the public to become enthusiastic consumers. The rise of consumer culture also benefited big-time intercollegiate sport and professional baseball and football leagues, which became major parts of the growing entertainment industry. Meanwhile, many new immigrants, especially from southern Europe, arrived on American shores, thus adding more ethnic diversity to an already multifaceted population. At the same time, U.S. society made progress toward the democratic ideal of equal opportunity and social justice, as women won the right to vote and African Americans were hired in greater numbers for federally funded jobs. Full achievement of social equity, however, remained distant and elusive. In addition, the severe economic depression that began in 1929 left personal, lifelong scars on massive numbers of Americans who suffered through it. Many other horrible and unforgettable disruptions also occurred during World War I (1914 to 1918) and World War II (1939 to 1945).

Participation in Physical Activity

During the first half of the 20th century, competitive sport was in the limelight. In particular, professional baseball, professional boxing, horse racing, and collegiate American football—all men's sports—were especially popular with fans. In professional baseball, the American League and National League joined forces in 1903 to keep player salaries in check and reduce competition for fan loyalty. Professional basketball and American football gained a degree of prominence by the 1930s, although at this point they were still fledgling enterprises. World War II temporarily put a damper on men's professional athletics because most healthy young men entered military service. This development spurred Chicago Cubs owner and chewing-gum magnate Philip K. Wrigley to organize the All-American Girls Professional Baseball League, which played in the Midwest from 1943 to 1954.

Intercollegiate sport during this era was dominated by American football. Major college games received widespread news coverage and served as occasions for lively partying and revelry. By 1903, Harvard University was home to a stadium seating 40,000 fans, and by 1930 seven of the "concrete giants" on college campuses could hold more than 70,000 people (Rader, 1990; Smith 1988).

For a time, competitive sports for boys were common in high schools and elementary schools, on playgrounds, and in youth service agencies such as YMCAs. In the 1930s, elementary school competition began to decline as physical education teachers complained about the high stress of competition. However, many people still wanted competition for youngsters, and the void was filled by new programs designed by city recreation departments and organizations such as Little League Baseball (formed in 1939) and Biddy Basketball (formed in 1950).

Girls' and women's sports in this era were subdued in comparison with the major boys' and men's events and their accompanying publicity and hoopla. A few women did achieve national and international success in sports such as tennis, basketball, golf, long-distance swimming, and track and field. For the most part, women trained and competed outside the educational system, in settings such as private country clubs and industry-sponsored leagues. For example, in 1930, Mildred "Babe" Didrikson was offered a job as a stenographer at Employers Casualty Insurance Company in Dallas so that she could play for the company's Golden Cyclones basketball team. The company also sponsored her participation in the Amateur Athletic Union's national track-and-field championships in 1932. Her phenomenal success led her to the 1932 Olympics, where she won two gold medals and a silver (Guttmann, 1991).

In schools and colleges during the 1920s, female physical educators turned toward intramural sports and low-key extramural competition for girls and women. They wanted to avoid the stress of high-level competition and to encourage all girls and women to participate. Although a few institutions continued to sponsor elite sport for females, the goal at most schools and colleges was "fun and games," which greatly curtailed the development of elite female athletes in educational settings. Things didn't begin to change until the late 1950s.

Public interest in competitive sport was so pronounced in the early decades of the 20th century that sport replaced gymnastics exercise as the centerpiece of most school and college physical education curriculums. A large number of physical education teachers believed that student participation in sport would help develop high moral character and other qualities needed by a good citizen. Students seemed more interested in physical activities that were spiced with competitive excitement than in the tedium of repetitious, traditional gymnastics exercises. Still, exercise did not completely disappear from physical education classes. Teachers sometimes replaced the old gymnastics systems with expressive movements designed to communicate ideas and feelings or with sport-related exercises such as shooting baskets or batting baseballs. In addition, by the 1920s, "corrective" physical education classes offered special exercises to students who faced challenges with posture, fitness, or health (Van Dalen & Bennett, 1971).

The U.S. military recognized the need for physical training during World War I, when approximately one-third of U.S. draftees were initially declared unfit to serve. Moreover, recruits with musculoskeletal incapacities such as flat feet and backaches deluged overseas hospitals. Recreational sport—organized by about 345 military athletic directors—was widespread and popular with the

The Golden Age of Sport: The 1920s

The 1920s, sometimes referred to as the Roaring Twenties, were a decade of change and excess. Whether one measures the change in terms of clothing fashion, political opportunity, technology, or popular culture, there is no doubt that a seismic shift took placed in U.S. society. Change was apparent as women exercised their newly acquired right to vote, as well as the freedom to raise the hemlines of their skirts. Excess was manifested in the form of stock market and real estate speculation and in increased personal spending as the United States (now free of the constraints of the First World War) increasingly became a consumer nation.

Moreover, sport became one of the products consumed by a people now hungry for entertainment—be it on the stage, on the silver screen, on the radio, or in the stadium. As a result, whereas childhood heroes and role models had previously consisted of the titans of business and industry (e.g., Thomas Edison, Henry Ford, John D. Rockefeller), they now included movie stars such as Charlie Chaplin and Mary Pickford and sport figures such as Babe Ruth and Gertrude Ederle. In a nutshell, the middle class was growing, times were good for larger numbers of Americans, and more people now had more leisure time to watch and follow their favorite teams and individual sport stars.

Whereas earlier decades had celebrated a few athletic heroes—such as boxer John L. Sullivan in the 1880s and baseball pitcher Cy Young in the 1890s and early 1900s—the 1920s produced heroes in abundance across a wide array of sports. Representatives of this sudden explosion of athletic excellence and public interest in sport included Babe Ruth and Lou Gehrig (baseball), Bill Tilden and Helen Wills (tennis), Bobby Jones (golf), Gertrude Ederle and Johnny Weissmuller (swimming), Red Grange (American football), and Jack Dempsey and Gene Tunney (boxing).

To accommodate spectator demand, larger and grander stadiums were constructed by both professional baseball clubs and college football programs. New York's Yankee Stadium was opened in 1923 with more than 50,000 seats to meet the demands of fans flocking to see the great Ruth hit his astounding home runs. New college football stadiums exceeding 60,000 seats were built at institutions including Ohio State University (1922) and the University of Illinois (1923) to handle the masses of people who arrived from near and far by train and private automobile.

This phenomenon was fueled by a media industry just beginning to realize the potential revenue available through the exploitation of commercial sport. Newspaper owners had begun to recognize this opportunity around the turn of the 20th century, when the sports page had made its debut in many large cities to meet the growing demand for coverage of Major League Baseball. In the 1920s, many of these papers began hiring talented writers such as Grantland Rice *(New York Herald Tribune),* John Drebinger *(New York Times),* Shirley Povich *(Washington Post),* and Paul Gallico *(New York Daily News),* who further popularized the stars of the sports world through their game coverage and feature stories.

Sports coverage expanded into a whole new dimension with the introduction of commercial radio in 1924. By the end of the decade, announcers such as Graham McNamee were re-creating football and baseball games for listeners in millions of homes across the nation. Contrary to the initial fears of college administrators and baseball club owners, radio coverage brought in countless new fans whose interest led them to purchase tickets to see in person what they had heard over the airwaves.

In the mid-1930s, writer Paul Gallico, looking back on the tremendous growth of interest in sport during the 1920s, dubbed it the "Golden Age of Sport." Today we might refer to it more accurately as the "*First* Golden Age of Sport," since we have certainly been in a similar "sport-crazed" era for at least the past five decades or more. In fact, given the rise of cable television and its numerous dedicated sport networks beginning in the 1980s, the publication of sport-specific magazines and almanacs, the construction of newer and ever more grandiose stadiums and arenas, and the creation of new and increasingly extreme competitive activities, it is fair to say that we live today in an unprecedented age. We can trace the roots of today's sport mania to that amazing decade some 90 years ago.

troops (Murphy, 1995; Rice, Hutchinson, & Lee, 1969). On the home front, physical education teachers added military training, such as marching drills and calisthenics, to school and college physical education programs. Recognition of the poor physical condition of the troops also served as a wake-up call after the war, spurring the expansion of school and college physical activity programs. By the end of the 1920s, many people could even exercise by following an instructor on the radio.

Racial discrimination eliminated most African Americans from the highest levels of competitive sport by 1900. In spite of this problem, a few black fighters were active in boxing, and Jack Johnson and Joe Louis became heavyweight champions. There were also other exceptions. For instance, a small number of African Americans participated in intercollegiate athletics at predominantly white schools, as well as at predominantly black colleges. A few also competed in the National Football League until 1933, when racial barriers were raised, and a few took part in the Olympic Games (track star Jesse Owens won four gold medals at the 1936 Berlin Games). Meanwhile, African American baseball players had no choice but to play for African American teams until Jackie Robinson signed with the Brooklyn Dodgers organization in 1945. After playing on a minor league team in 1946, he moved up to the major league Dodgers club in 1947.

For people in many racial and ethnic minority groups, sport involvement was influenced by cultural heritage. For example, second-generation Jews participated in sports such as baseball, basketball, and boxing—often at the local Young Men's Hebrew Association (YMHA) or Jewish community center—as a sort of middle ground that allowed them to celebrate Jewish heritage while assimilating into mainstream U.S. society through these popular activities (Levine, 1992). First-generation Japanese immigrants who arrived early in the 20th century often used traditional perspectives from their homeland to make sense of American sports—for instance, recalling samurai principles of courage and honor (Regalado, 1992).

KEY POINT

Sport was Americans' favorite activity in the first half of the 20th century. They paid less attention to exercise, though interest picked up during the two world wars in order to improve physical fitness.

Interest in exercise was renewed during World War II. As with the First World War, some young military draftees failed their induction physicals, which generated interest in improving school physical education. After the war, educators expanded many school and college physical activity curriculums in response to weaknesses observed in the fitness of wartime military recruits; thus, sport held center stage once again (Swanson & Spears, 1995).

Professional Issues in Kinesiology

Equality: Racial Integration of Professional Sport

When Jackie Robinson broke the race barrier that had existed since the mid-1880s in Major League Baseball, he not only opened doors of opportunity for baseball players of color but also contributed greatly to the racial integration of the National Football League and the newly created National Basketball Association in the late 1940s and 1950s. Ongoing efforts to achieve this goal had been made over a period of decades by individuals both within and outside of professional sport. Sportswriters in the African American press had kept up a steady drumbeat of pressure on Major League Baseball owners, as had several politicians, both black and white, in New York City and other urban centers. Another major factor in finally integrating the sport turned out to be the large numbers of African Americans who served in the military during World War II. Supporters argued effectively that those who fought for their country and its values should be able to pursue their dreams. Still, in the end, it took outside pressure and a few courageous owners and general managers to achieve the goal. Today, the issue of equality in professional sport continues to be addressed in the form of providing access for members of underrepresented groups to leadership positions in coaching, administration, and sports medicine.

Citation style: Chicago Manual of Style

Tygiel, Jules. *Baseball's Great Experiment: Jackie Robinson and His Legacy.* New York: Oxford University Press, 1983.

Physical Activity Professions

The number of bachelor's degree programs in physical education increased in the early 1900s, expanding to a total of about 135 by 1927 (Park, 1980). Master's degrees were also available just after the turn of the century, and doctoral programs appeared in the 1920s. Undergraduate programs focused on training physical education teachers, continuing the trend that had started in the late 19th century. Graduate programs offered advanced training for teachers, as well as academic preparation for becoming a college or university faculty member. The quality of these programs improved as research on physical activity picked up in the late 1920s.

Despite the popularity of sport during the first half of the 20th century, college and university physical education curriculums did not include much course work to prepare students for the occupation of coach or athletic trainer. Instead, coaches usually came from the ranks of successful athletes. If they worked in a school or college, they were often required to teach physical education classes as well, so they earned the necessary college degrees and professional teacher certifications.

The few athletic trainers on the scene during this period had little formal education in health care. In colleges and universities, they often began as gymnasium jacks-of-all-trades who also handled custodial responsibilities and other duties such as repairing equipment and facilities, laundering clothes, driving the team bus, and maintaining outdoor playing fields. They sometimes met their counterparts from other institutions at intercollegiate athletic events and shared training techniques (Smith, 1979). Because elite sports were primarily limited to boys and men, most coaches and athletic trainers were men.

This growing focus on physical activity created a market for activity-related products and services. For example, the Cramer Company, founded in 1922, identified a commercial niche and began selling liniment for sprains, as well as other products for athletic training. In 1933, Cramer inaugurated the *First Aider* newsletter, which became popular among high school coaches seeking to understand training techniques. For many years, Cramer also sponsored educational seminars for athletic trainers. The profession grew, but it wasn't until 1950 that a professional organization was founded—the National Athletic Trainers' Association.

Interest in physical therapy was sparked by the outbreak of World War I. By the end of the war, U.S. military reconstruction aides—an entirely female corps of nurses trained to help those who had been wounded—were receiving training in massage, anatomy, remedial exercise, hydrotherapy, electrotherapy, bandaging, kinesiology, ethics, and the psychological effects of injuries. In 1921, the reconstruction aides formed the American Women's Physical Therapeutic Association, which eventually became the present-day American Physical Therapy Association—the main professional association for physical therapists (Murphy, 1995). This therapeutic field borrowed corrective exercise knowledge from physical education and from the research published in its journals, such as *Research Quarterly* (now *Research Quarterly for Exercise and Sport*) and the *American Corrective Therapy Journal* (now *Clinical Kinesiology*).

Scholarly Knowledge About Physical Activity

At the beginning of the 20th century, a number of scholars were investigating topics such as neuromuscular fatigue, the vascular effects of exercise (known today as exercise physiology), kinesiology (biomechanics), body measurements and proportions (anthropometrics), the psychological aspects of play, and the history of physical education and sport (Park, 1981). However, research on physical activity did not gain much external visibility until the late 1920s. Developments included the 1927 opening of the prestigious Harvard Fatigue Laboratory, which focused on research in exercise physiology, and the 1930 launch of *Research Quarterly*, a scholarly journal devoted to physical activity. The

Sport and Ethnicity

Think of as many examples as you can of people who share an ethnic heritage getting together to do the following:

- Play or watch a sport associated with their cultural heritage.
- Play or watch a sport that is widely popular in the United States.

Is there any overlap between your two lists? Why, or why not?

journal featured topics such as the following examples from a 1934 issue: "physiology of respiration, reflex/reaction time, measurement of motor ability, effects of temperature on muscular activity, and test construction" (Park, 1980, p. 5). Physical educators often did research applicable to teaching because of the importance of teacher training in college and university physical education departments. For instance, studies of techniques for assessing motor skills could be used to help teachers develop ways to evaluate student progress.

KEY POINT

Research on physical activity started to expand in the late 1920s. Many physical educators studied topics relevant to teaching physical education, while some continued to examine other biophysical aspects of physical activity.

At the same time, some physical education faculty continued to research aspects of physical activity not centrally related to teaching. Examples include kinesiology (biomechanics), motor ability, motor capacity, physical fitness, and exercise physiology. Research on these topics continued to increase in the 1940s and 1950s but did not begin to dominate the field until the 1960s.

1950–Present: Electronic Communication and Globalization

In the second half of the 20th century, electronic communication expanded at a breathtaking pace, as illustrated by advances in television, computing, satellites, and air travel. Television—in its infancy in 1950—became a common fixture in most homes in the following two decades. Digital computers evolved from massive, room-sized arrays of vacuum tubes with limited capabilities into tiny but enormously powerful collections of memory chips and electronic displays. Satellites were placed in orbit, which greatly simplified intercontinental communication, and air travel became both faster and easier. As in past eras, each technological advance affected sport and exercise in numerous dimensions.

At the same time, global trade expanded, accompanied by greater worldwide political and economic interdependence. New immigrants continued to arrive in the United States—especially from Asia and Latin America. Meanwhile, progress was made toward the American democratic ideal of equal opportunity and social justice. Specifically, federal civil rights legislation mandated greater equality in education, business, and housing; in addition, racial integration also increased in professional and intercollegiate sport. Nevertheless, complete achievement of this goal remained elusive.

The period was also affected by war and the prospect of war. The sometimes-frightening Cold War between communist and capitalist countries shaped worldwide reality for more than 40 years. Along the way, it stimulated interest and investments in science and technology for the purpose of national defense. The Cold War also included hot, disruptive episodes in the form of the Korean War (1950 to 1953) and the Vietnam War (1955 to 1975). When communism unraveled after the fall of the Berlin Wall in 1989, vast global changes took place in political and economic relations.

Conflicts continued to erupt in the latter part of this period. In the early 1990s, Iraq invaded Kuwait, and the United States and many allies responded by driving the Iraqi armed forces back and then imposing a decade-long air blockade of Iraq. Also in the 1990s, ethnic conflict between Muslims and Christians led to civil war in the Balkan countries of the former Yugoslavia, and the city of Sarajevo, site of the 1984 Winter Olympic Games, was extensively damaged. The September 11, 2001, attack on the World Trade Center in New York City led to more than a decade of war in Afghanistan and Iraq. In 2011 and 2012, several Middle Eastern nations experienced revolts by ordinary citizens that overthrew long-standing dictators. However, tribal and religious sectarian hostilities in most of these nations have led to continuing conflict and widespread displacement of refugees seeking relief from war. These citizen movements were organized in part through modern electronic communication via e-mail and mobile texting; those same technologies, however, have also been used by terrorist organizations to spread mayhem worldwide.

Participation in Physical Activity

Since the 1950s, health-related exercise has increased dramatically among some adults in the United States. This trend has derived from a number of events. For one, the Korean War underscored the importance of physical fitness for military preparedness, and the continuing threat of communism was highlighted by the Soviet Union's 1957 launch of Sputnik—the first earth satellite. In addition, a highly publicized 1953 report revealed that more than 50 percent of U.S. children could not pass a strength and flexibility test, as compared with a failure rate of less than 10 percent among children from European countries.

Moreover, after President Dwight D. Eisenhower's heart attack in the mid-1950s, many people were well aware that his doctor prescribed exercise for recovery, which was a radical idea at the time. In the early 1960s, President John F. Kennedy continuously demonstrated an active lifestyle through his participation in touch football, tennis, swimming, sailing, horseback riding, badminton, and general exercise (Rader, 1990).

In the 1970s, adult participation increased sizably in various activities, including running, tennis, racquetball, and aerobic dance. The number of marathon races has also increased in the United States, rising from fewer than 25 in 1970 to some 1,100 in 2015 (Eitzen & Sage, 1997; Running USA, 2015). The number of recreational runners completing shorter road races also grew, increasing from 5 million in 1990 to more than 19 million in 2013 and then experiencing a slight decline over the following two years (Running USA, 2016).

Similarly, by one estimate, the number of commercial health clubs increased from 350 in 1968 to more than 7,000 in 1986 (Rader, 1990). Moreover, adult membership in health, racket, and sport clubs reportedly grew by a whopping 50 percent between 1987 and 1996, rising from about 14 million to about 21 million (Carey & Mullins, 1997). By 2016, one industry survey reported a total of more than 35,000 fitness centers in the United States and more than 6,000 in Canada (International Health, Racquet, and Sportsclub Association, 2016).

More generally, according to studies published in the late 1980s, the proportion of children and adults who said that they did something daily to keep physically fit increased from 24 percent in 1961 to 59 percent in 1984; conversely, adults who said that they had a sedentary lifestyle decreased from 41 percent in 1971 to 27 percent in 1985 (Blair, Mulder, & Kohl, 1987; Ramlow, Kriska, & LaPorte, 1987; Stephens, 1987). However, more recent surveys conducted by the Centers for Disease Control and Prevention—which used more precise criteria for vigorous physical exercise—indicate that in 2014, slightly more than 50 percent of U.S. adults were not engaging in the amount of vigorous aerobic activity recommended by 2008 federal guidelines. Moreover, engagement in vigorous physical activity in leisure time was significantly lower for females, ethnic minorities, and rural populations. In addition, adults in families with incomes above twice the poverty level were more likely to engage in regular leisure-time physical activity than were adults in lower-income families (34 percent versus about 21 percent; age-adjusted) (National Center for Health Statistics, 2015).

How might a sport historian determine what led to an increase in participation in Little League Baseball? How might one track changes in participation in other sports and recreational physical activities?

Between 1950 and 2003, the number of youngsters worldwide competing in Little League Baseball increased by a factor of 140, from about 18,000 to nearly 3 million (Little League Baseball, personal communication, April, 2004). However, from 2000 to 2009, the number of children and youth participating in baseball worldwide fell 25 percent (Futterman, 2011). As this overall trend has continued within the United States, some neighboring communities have resorted to merging their leagues in order to field enough teams, and teams often struggle to fill their rosters at various levels of youth baseball (Fisher, 2015). This decrease may be due in part to increased participation in other youth sport programs, such as basketball, American football, soccer, and lacrosse. Participation in youth tackle football increased 21 percent between 2000 and 2009; in addition, some studies have suggested that "more people now play soccer in the U.S. than baseball, and lacrosse participation . . . more than doubled" between 2000 and 2011 (Futterman, 2011). According to US Lacrosse, participation by high school youth topped 300,000 in 2015 (US Lacrosse, 2016).

Since 2010, however, various studies have shown a decline in overall participation in youth sport. One study showed that among 6- to 17- year-olds, the average number of team sports played per participant has fallen 6 percent in the last 5 years, dropping from 2.14 to 2.01 (King, 2015). Factors often cited for this decrease include athlete burnout, increased pressure to specialize in one sport, expense, travel to distant events and tournaments, and, in some sports, fear of concussion.

Elite athletic competition for girls and women reappeared in high schools and colleges beginning in the late 1950s. In 1972, the U.S. Congress approved Title IX of the Education Amendments of 1972, requiring equal opportunity for males and females in educational programs, including athletic programs. In its wake, participation in athletic competition by high school girls increased almost 11-fold between 1971 and 2007. Moreover, in the 2014–2015 school year, almost 3.3 million girls participated in high school sport in the United States (National Federation of State High School Associations, 2016). A similarly dramatic increase occurred in the number of female players at colleges and universities; for example, in the 2014–2015 academic year, more than 200,000 women participated in intercollegiate sport in the three divisions of the National Collegiate Athletic Association (2015).

In the 1950s, African Americans trickled back into elite professional athletics; in the 1960s, this thin stream grew into a fast-moving stream. In the late 1950s, African Americans constituted about 10 percent of the players in professional baseball, basketball, and American football. By 1994, nearly 80 percent of professional basketball players were African American, along with 65 percent of professional football players and nearly 20 percent of professional baseball players (Eitzen & Sage, 1997; Leonard, 1998). By 2006, however, the proportion of African American players in Major League Baseball had fallen to about 8 percent, although the number of African American front office personnel, coaches, and managers had significantly increased (Lapchick, 2007). In fact, that number increased to 10 percent in 2010, but it fell back to 8.3 percent by 2016 (Lapchick, 2016).

Many observers agree that the recent decline in baseball participation by African American youth has resulted from the growing attraction of football and basketball. To counteract this trend, Major League Baseball has begun sponsoring youth baseball programs in many U.S. inner cities. These programs have begun to show great promise, and in the 2015 Major League Baseball draft, 9 of the 36 players picked in the first round (25 percent) were African American—the largest total since 1992. As for other minority groups, in 2012, Latinos constituted nearly 30 percent of Major League Baseball players; in contrast, the number of Asian major leaguers increased steadily during the first decade of the 21st century to 2.1 percent but dropped to 1.7 percent in 2016 (Lapchick, 2016).

Increased participation in grass-roots and high-level sport, in combination with systematic training and improvements in equipment and facilities, has led to a steady increase in performance levels since the middle of the 20th century in most, if not all, Olympic sports. Also, as professionalism has been introduced into most Olympic sports, the ability of athletes to train on a full-time basis has undoubtedly contributed to this constant breaking of records and improved performances in team sports for both women and men.

One study has suggested that because "the performances of athletes are the product of genetic endowment, hard work, and, increasingly, the contribution of science," the mere willingness to work harder is no longer sufficient to break world records. They conclude: "Future limits to athletic performance will be determined less and less by the innate physiology of the athlete, and more and more by scientific and technological advances and by the still-evolving judgment on where to draw the line between what is 'natural' and what is artificially enhanced" (Lippi, Banfi, Favaloro, Rittweger, & Maffulli, 2008, p. 14).

Beginning in the 1960s, participation in outdoor recreation began a steady increase that has continued into the 21st century. For instance, bike path mileage in the United States went from almost none in 1965 to more than 15,000 miles (24,000 kilometers) in 1974 (Wilson, 1977). By 2016, the Adventure Cycling Association's Routes and Mapping Department had linked some "45,999 miles of rural roads to create low-traffic bike routes through some of the nation's most scenic and historically significant terrain"; this network, known as the TransAmerica Trail, was begun in 1976 (Adventure Cycling Association, 2016). Another indication of growing interest in outdoor activity lies in the fact that total U.S. National Park recreational visitations increased from 169,410,115 in fiscal year 1979 to 323,556,290 in 2016 (National Park Service, 1979, 2016).

In the second decade of the 21st century, participation in outdoor recreation activities continued at a high level. In a 2016 report, the Outdoor Foundation of the Outdoor Industry Association found that "nearly half of all Americans . . . participated in at least one outdoor activity in 2015" (p. 1). It also found that paddle sports continued to be a growth area: "Over the past three years, stand-up paddling was the top activity for growth, increasing participation an average of 26 percent from 2012 to 2015. Kayak fishing, whitewater kayaking, and sea/tour kayaking also saw some of the biggest participation increases during that time" (p. 1). The five most popular outdoor recreation activities among people of ages 6 through 24 were running, jogging, and trail running (20 million participants); bicycling (road, mountain, BMX; 17 million); camping (car, RV, backyard; 15 million); fishing (freshwater, saltwater, fly; 15 million); and hiking (11 million). The same five categories were most popular among adults over 25 but in a slightly different order: running, jogging, and trail running (32 million); fishing (31 million); hiking (26 million); bicycling (26 million); camping (25 million).

Americans also increased their spectator involvement between 1950 and 2016, through both in-person attendance and television (and, recently, online) viewing. For example, game attendance in Major League Baseball quadrupled between 1950 and 2006 as the number of teams increased from 16 to 30. In 2007, total MLB attendance set a record of nearly 80 million spectators. Since then, total attendance has hovered above 73 million. The National Football League expanded from 13 teams in 1950 to 32 teams in 1995. Regular season attendance grew from 14,202,747 in 1995 to 17,788,671 in 2016, although it has experienced declines each year beginning in 2010. The reasons for this decline include increasing ticket prices and improvements in television viewing capability. Nevertheless, lucrative television contracts and record-setting viewership have enriched the NFL and confirmed its popularity. Game attendance has also grown in the National Basketball Association, rising from a total of 1.3 million for regular-season games during the 1959–1960 season to 22 million during the 2015–2016 season.

Although men's soccer has been, arguably, the preeminent spectator sport throughout most of the world over the past century, it did not gain much of a foothold in the United States as a professional entertainment event until the late 1990s. The final rounds of the 1994 men's World Cup were held on U.S. soil, and in 1996 entrepreneurs developed what has become a relatively stable men's professional league—Major League Soccer—offering teams in the United States and Canada and holding special allure for Latino spectators.

With the growth of sport for girls and women at the youth, high school, and college levels, professional team sports have followed, albeit initially in a rather hit-and-miss fashion. In 1978, the Women's Professional Basketball League was launched with teams in eight cities; a year later, the name was changed to the Women's Basketball League, and twelve teams finished the season. The league disbanded in 1982 due to financial problems linked to low attendance and lack of television coverage. In 1997, the Women's National Basketball Association (WNBA) played its first season, and it has grown in popularity over the years with significant sponsorship by the NBA and television contracts. In addition, professional softball for women has been modestly successful since the 1990s in the form of a league now known as National Pro Fastpitch. There are also a number of regional professional women's football leagues with varying levels of success.

Soccer, another popular sport for women at the high school and college level, has had a more difficult time sustaining a professional league, in spite of the spectacular success of the U.S. women's national team in World Cup and Olympic competition. In 1991, the United States won the first Women's World Cup with hardly anyone in the nation noticing. After losing to China in the 1995 World Cup, the Americans won the gold medal at the 1996 Olympic Games before the largest crowd to ever witness a soccer game on U.S. soil. In 1999, the team again defeated China to win the World Cup before 100,000 spectators at the Rose Bowl, as well as a worldwide television audience numbering in the millions.

These successes led to the creation of the professional Women's United Soccer Association (WUSA), which was moderately successful but fell into financial exigencies following the 2003 season. Nevertheless, the league did much to promote women's sport, especially soccer, in North America. In 2009, a new league, Women's Professional Soccer (WPS), began play, fielding seven teams during the first two seasons and six teams for the 2011 season. The WPS disbanded in 2012 due to inadequate financing. Later in the same year, it was replaced by the National Women's Soccer League, run by the United States Soccer Federation.

KEY POINT

The enthusiasm of Americans—both as direct participants and as spectators—for a widening array of sports and exercises mushroomed in the second half of the 20th century.

Changing Attitudes Toward Exercise Among Adults

In the mid- to late-1950s, most Americans perceived regular participation in exercise and sport to be the exclusive province of young people in high school, college, or professional sport. Today, large numbers of people pursue such activities throughout much of their lives. What historical events and societal changes have influenced this shift in attitude and practice? What have you seen of this shift in your own family or circle of acquaintances?

By the early 1960s, Roone Arledge, head of sports programming at ABC, was incorporating the highly innovative and popular instant-replay technique into sport broadcasts. In 1970, he unveiled additional novel production technologies—for example, close-ups of players and fans, multiple cameras, and directional microphones—with the inauguration of ABC's *Monday Night Football* (Gorn & Goldstein, 1993). These innovations made televised sport much more exciting to watch and drew new viewers. More generally, the arrival of all-sport cable channels, beginning with ESPN in the late 1970s, wedded sport and television in a 24-7 relationship. The commercial success of sport on television—primarily men's team sport—provided vast amounts of money for professional and big-time intercollegiate athletics.

The 1990s, of course, also brought the beginnings of the Internet revolution. By 2007, anyone with a computer attached to high-speed Internet service had instantaneous access to worldwide live video broadcasts of sport events, as well as news and discussions about sport through electronic chat rooms and websites. Within a few years, the addition of smartphones, wireless tablets, and streaming services allowed fans to view events in nearly any sport.

Physical Activity Professions

At midcentury, most U.S. college students who majored in physical education—the name used almost universally at the time—became elementary and high school teachers; some also coached. Things began to change in the mid-1960s with the development of kinesiology. It soon became apparent that many careers other than teaching and coaching also centered on physical activity (they are explored in part III of this text).

As interest in exercise and physical fitness grew in the United States, physical activity programs began to be offered by health clubs, cardiopulmonary rehabilitation programs, worksite health promotion programs, and health maintenance organizations. As a result, people with specialized training in exercise began working in a variety of jobs, such as cardiopulmonary rehabilitation, personal training, aerobics instruction, and strength training. Some of these individuals advanced rapidly into management roles and business ownership.

To support the increasing number of specialists working to prevent and treat athletic injuries, the National Athletic Trainers' Association (NATA) was founded in 1950. The membership consisted almost entirely of men until 1972, when Title IX (which required equal opportunity for males and females in educational settings) became law, thus encouraging women to enter the field in greater numbers. As of 2015, there were 395 NATA-accredited degree programs, including both master's and bachelor's programs. By 2023, professional education needed to sit for national certification in athletic training will require a master's degree from a CAATE accredited program (Commission on Accreditation of Athletic Training Education, 2014–15).

At midcentury, little was offered in the way of standard educational preparation for most careers in sport management and leadership. As the sport industry grew, however, more jobs became available in sport settings, and the range of jobs widened. Job opportunities arose, for example, with intercollegiate athletic departments, professional sport teams, city stadiums, and participant-oriented nonprofit organizations, such as YMCAs and Boys & Girls Clubs. Some of these jobs didn't even require a college degree (e.g., YMCA program leader, Minor League Baseball ticket seller), whereas others required an advanced degree or extensive high-level leadership experience (e.g., city stadium manager, big-time intercollegiate athletic director).

This unevenness made it difficult to accredit degree programs or certify practitioners. In 1987, however, curriculum guidelines were published for bachelor's and master's degree programs in sport management in a joint effort by the North American Society for Sport Management and the National Association for Sport and Physical Education (affiliated with the American Alliance for Health, Physical Education, Recreation and Dance—the precursor to SHAPE America). In 1993, these guidelines were revised and published as voluntary accreditation standards. By 2016, undergraduate or graduate degrees in sport management were offered by some 482 American and 16 Canadian colleges and universities (North American Society for Sport Management, 2016).

In the 1960s, physical therapy continued to grow as a profession. The number of physical therapy jobs in the United States increased 26-fold from 1960 to 2015, rising from about 8,000 to 209,000 (U.S. Bureau of Labor Statistics, 1961, 2016). Dramatic change also took place in the way these individuals were trained. In the 1950s, physical therapy students were trained in undergraduate degree programs or in older, hospital-based certificate programs designed for people who had completed bachelor's degrees in other fields. By the late 20th century, hospital-based programs were phased out, undergraduate programs were in decline, and master's programs were on the rise. Consider the following comparison: In 1981, accredited physical therapy curriculums included 100 undergraduate degree programs, 7 certificate programs, and 8 master's programs (U.S. Bureau of Labor Statistics, 1982); in the 2015–2016 academic year, there were 233 accredited physical therapy programs in the United States.

Today, master's programs in physical therapy are not offered to new students in the United States. As of 2014, "Professional (entry-level) physical therapist education programs in the United States only offer the Doctor of Physical Therapy (DPT) degree to all new students who enroll" (American Physical Therapy Association, 2016). Raising the degree requirements from bachelor to masters to doctoral has been a central part of the profession's effort to accommodate the increase in subject matter and the expanded role of the physical therapist in the modern world of health care.

KEY POINT

In the middle of the 20th century, most physical education majors entered the same profession—physical education teaching. By the end of the century, they could choose from a wide array of physical activity careers.

Scholarly Knowledge About Physical Activity

During the final 35 years of the 20th century, scholars developed new knowledge about physical activity at a feverish pace—a trend that continues today. You will learn much more about this development in the chapters of this text that address other subdisciplines. Here, we focus on the big picture.

By the 1950s, researchers had already produced an extensive body of knowledge about physical activity and its effect on human health and performance. However, this knowledge was scattered across many research journals, and no one had taken the time to make it easily accessible. In 1960, in an attempt to remedy this situation, Warren Johnson, professor of health education and physical education at the University of Maryland, published a book titled *Science and Medicine of Exercise and Sports,* which contained 36 essays summarizing current knowledge. Another early attempt at organizing research knowledge appeared in the May 1960 *Research Quarterly* supplement, which was edited by a team of scholars headed by Raymond Weiss of New York University; each paper summarized research concerning a different aspect of physical activity (Karpovich, Morehouse, Scott, & Weiss, 1960).

Evidence-Based Practice: Balancing Risk and Safety in Sport

Efforts to document how athletes are affected by concussions have been made by kinesiology scholars in athletic training, biomechanics, and exercise physiology, as well as researchers in other medical fields. The resulting evidence can be used as the basis for making rule changes in various sports. This work is only the most recent instance of numerous endeavors to retain the excitement of sport while making it safer for both athletes and spectators. In the early 1900s, for example, the "flying wedge" formation was outlawed in American football due to player deaths and permanent injuries. Since then, various improvements in safety equipment (e.g., helmets, pads, masks) have been made in various contact sports in order to provide players with better protection. Changes have also been made in training practices. For instance, prior to the 1960s, it was common for coaches to withhold water from athletes of all ages during practice sessions for fear of inducing abdominal cramping. Medical research has since proven that thinking to be mistaken.

Future research will undoubtedly bring additional changes in both rules and practices as the search continues for balance between risk and safety in sport. Such evidence-based consideration of risk often stands in contradiction to what may be deemed beneficial in terms of entertainment and spectatorship by sport officials and for-profit teams. Historical research about such issues can help us understand why some safety innovations are quickly implemented and others are resisted.

KEY POINT

Beginning in the 1960s, the discipline of kinesiology grew rapidly and branched into numerous scholarly subdisciplines.

In the early 1960s, politicians and educators issued calls for better quality control in teacher education programs, which resulted in scrutiny and criticism of physical education curriculums, many of which emphasized the diffuse research noted in the previous paragraph. In response, physical education faculty began charting new directions for their discipline. One such person was Franklin Henry, physical education professor at the University of California at Berkeley and a top-notch scholar in exercise physiology and motor learning. Henry outlined his views on the nature of the discipline in a paper titled "Physical Education: An Academic Discipline" (1964). He envisioned a body of knowledge about physical activity that would draw from "such diverse fields as anatomy, physics and physiology, cultural anthropology, history and sociology, as well as psychology" (p. 32). He also believed that the field should be cross-disciplinary, binding together knowledge from various disciplines to form a distinct, new body of knowledge concerning human movement.

Henry's vision of the discipline spurred rapid changes during the rest of the 1960s and the 1970s. Specifically, a nucleus of physical educators who were already heavily engaged in research began pushing to make scholarship more central in university physical education departments. They continued their own research and trained a cadre of doctoral students who expected to be centrally involved in inquiry throughout their own careers. They also communicated frequently with scholars in other parent disciplines—for example, physiology, medicine, psychology, and history—who were also studying physical activity. This exchange led to rapid growth in the size and scope of the overall body of knowledge in kinesiology. Henry's cross-disciplinary or interdisciplinary vision within kinesiology departments, however, was slow to develop.

The quest for research credibility in higher education tended to make kinesiology more specialized into subdisciplinary areas, reinforcing subdisciplinary research sometimes published in "parent" discipline journals. Seven of these subdisciplines of kinesiology are introduced in chapter 1 and in part II of this book. Throughout the rest of the 20th century, scholars worked to support the growing subdisciplines by forming many new associations and working with publishers to inaugurate a host of new research journals. In the 1980s and 1990s, the amount and quality of research continued to increase, and the subdisciplines became stronger and more identifiable (Massengale & Swanson, 1997; Park, 1981, 1989). Today, research performed in kinesiology departments is high-quality work that is published in a wide variety of multidisciplinary kinesiology journals, subdisciplinary journals, and other disciplinary journals.

Wrap-Up

History extends your memory, as it were, with fascinating, evidence-based stories about how and why physical activities (e.g., exercise and sport) and the field of kinesiology came to be shaped the way they are. Knowledge of the past gives you important, broad understanding about the present that you can use to make better-informed personal and professional decisions. As demonstrated in this chapter, historical knowledge is used by kinesiology practitioners in a variety of ways across many specialties and professions. Since the 1970s, history scholars have gone beyond earlier descriptive approaches to put more emphasis on analytical frameworks that help answer questions about why things happened as they did. So far, this subdiscipline has focused mostly on sport, but since the 1980s scholars have given increasing attention to exercise.

If you are intrigued by what you have learned here, then you might enjoy venturing into sport history textbooks, taking an introductory course in sport history, looking through the book-length research studies mentioned in this chapter, or exploring major journals such as *Journal of Sport History* and *International Journal of the History of Sport*. This provocative subdiscipline can give you a new outlook that enriches your understanding of physical activity as you step into the future.

MORE INFORMATION ON HISTORY OF PHYSICAL ACTIVITY

Organizations

American Historical Association (AHA)
Australian Society for Sports History (ASSH)
British Society of Sports History (BSSH)
International Society for the History of Physical Education and Sport (ISHPES)
National Association for Kinesiology in Higher Education (NAKHE)
North American Society for Sport History (NASSH)
Organization of American Historians (OAH)

Journals

American Historical Review (AHA)
Journal of American History (OAH)
Journal of Sport History (NASSH)
Olympika: The International Journal of Olympic Studies
Quest (NAKHE)
Sport History Review
Sport in History (BSSH)
Sporting Traditions (ASSH)

REVIEW QUESTIONS

1. List and discuss the goals of history of physical activity.
2. List and discuss three ways in which a kinesiology practitioner might use knowledge of physical activity history in her or his area of specialization.
3. Describe participation in physical activity in the United States during the following three periods:
   1840–1900
   1900–1950
   1950–present
4. Describe professional practice centered on physical activity in the United States during the following three periods:
   1840–1900
   1900–1950
   1950–present
5. Describe scholarly knowledge about physical activity during the following three periods:
   1840–1900
   1900–1950
   1950–present

Go online to complete the web study guide activities assigned by your instructor.

6

Sociology of Physical Activity

Katherine M. Jamieson

The author acknowledges the contributions of Margaret Carlisle Duncan to this chapter.

CHAPTER OBJECTIVES

In this chapter, we will

- explain what a sociologist of physical activity does;
- identify the goals of sociology of physical activity;
- discuss how sociology of physical activity came into being;
- explain how research is conducted in sociology of physical activity; and
- examine what research tells us about inequitable power relations relevant to physical activity, especially gender relations, ethnic and racial relations, and socioeconomic relations.

Suppose that you're having coffee with your roommate at the local bookstore, and a couple of people at the next table are having an intense conversation. You can't help overhearing, and you're interested, because they're talking about the new plan for city parks and recreation spaces. As a longtime resident and soon-to-graduate kinesiology major, you have enjoyed lots of fun and health-related physical activity in local public spaces. It occurs to you that these public resources have received support from the taxes paid by your parents, and, more recently, by you as well. While the other folks in the bookstore seem to be fearful about who the new park spaces will invite into their neighborhood—especially with a skateboard park included—your mind explodes with all kinds of kinesiology-related questions and visions.

You begin to wonder what it would take for all residents of this city to be able to access and enjoy meaningful physical activity. Who are the major players that need to be consulted in order to reimagine the city as a place for daily, safe, enjoyable human movement? Will your own neighborhood community have different needs than those nearby? How can this plan meet such diverse needs? What types of physical activity are most engaged in on a daily basis? You consider the wastefulness of one-time, elite competition venues such as Olympic sites, and you begin to ponder how we might build such resources in a sustainable manner. Then you get really excited as you start to imagine this physical-activity-friendly city in which you may work and live, and you call city hall to volunteer for the planning committee.

© Human Kinetics

For students and emerging professionals in the field of kinesiology, it is easy to get completely enamored of the amazing capacity of the human body. In fact, fascination with human movement at all stages of life and skill development is what keeps many of us deeply engaged in the field. Who wouldn't want to teach a young person to swim for the first time, guide someone through a challenging but rewarding rehabilitative process, or co-create a health-related physical activity plan for an entire community? As kinesiology professionals, we put our skills to use in a variety of human movement arenas. Moreover, though physical performance by human bodies often lies at the core of our professional engagement, we also bear responsibility as kinesiologists for developing knowledge and skills related to the sociocultural aspects of promoting physical activity.

Imagine teaching a young relative to swim in the same backyard pool where you and other family members have begun your own joyful and skilled engagements with this activity. Now, imagine teaching an older adult who grew up in a community where public pools were racially segregated, thus depriving her of swimming experience and leaving her with a very rational fear of bodies of water. Imagine also the rehabilitation setting for another older adult who serves as the sole income provider for his family and has worked in a demanding skilled trade since high school. Is it any surprise that research indicates that men recovering from cardiac events prefer not to choose yoga as part of their rehabilitation?

Goals of Sociology of Physical Activity

- To look at physical activity with a penetrating gaze that goes beyond our common understanding of social life
- To identify and analyze patterns of change and stability in physical activity
- To critique physical activity programs in order to identify problems and recommend changes to enhance equality and human well-being

More generally, in light of these examples, can you begin to ponder the social codes and expectations related to gender, age, race, and ability that may inform even the most supposedly personal of health decisions? Beyond the crucial core of physical performance, it is through the shared experience of human movement spaces that we come to hold beliefs or develop understandings about our social world and about the value and structure of daily life. As a kinesiology professional, you will be expected to understand the swiftly changing social and cultural contours of all forms of physical activity across the life span.

Why Use Sociology of Physical Activity?

Our social lives are made up of mutually influential social practices (everyday behaviors) and shared beliefs that are so commonplace that we rarely think about them. Whether we're playing soccer, coaching a basketball team, rehabilitating a broken leg, cheering for a favorite sport team, or working out at a local gym, our social practices as athletes, coaches, exercisers, and spectators are influenced by commonly held beliefs in our society. In turn, it is through our social practices that we come to have beliefs or understandings that we share with others. Yet few of us spend time analyzing the social arrangements—social practices and shared beliefs—that underlie physical activity. When we do look more closely, though, we gain insights about ourselves and our culture. Giving us this closer look is what sociologists of physical activity do.

Although the subdiscipline of sociology of physical activity focuses primarily on sport, increasing attention is being paid to exercise and the ways in which people conceive of the human body. By exploring how we, as a culture, view our bodies, we may come to a renewed understanding of what it means to be human. Sociology of physical activity, then, involves a set of intellectual tools that one can develop in order to look at human physical activity as more than physical—that is, as cultural, too. In other words, when we intellectually unpack patterns, beliefs, values, and power relations, we reveal human physical activity as a crucial organizing component of all social systems. These intellectual tools include an understanding of theoretical perspectives, knowledge of research methods, and the ability to identify and think through social problems as they relate to social spaces of human physical activity.

Additional course work in the humanities and social sciences will enhance your understanding of principles of sociology and, more specifically, the sociology of physical activity. You do not have to know all of these disciplines in depth, but the more information you have, the more insightful you will be about the social arrangements underlying physical activity. For example, if you know a little about the general nature of socialization—the process of learning about your own society and how to get along in it—then you will find it easier to understand how members of a society become *socialized* into dominant sports or into subcultures centered on activities such as surfing, triathlon, mountain climbing, skateboarding, rugby, and exercise related to social or political causes (e.g., Race for the Cure).

KEY POINT

Sociology of physical activity focuses on the shared beliefs and social practices that constitute specific forms of physical activity (e.g., sport, exercise). Thus, sociological information enhances the breadth of knowledge of a well-educated kinesiologist.

As you get to know more about socialization, however, you will likely notice details and nuances of physical activity involvement that may lead you to ask more complex questions. For example,

Kinesiology Colleagues

© Human Kinetics

Joan Neide

Joan Neide is a professor in the Department of Kinesiology and Health Science at California State University, Sacramento. Her broad interests in physical culture lead her to combine her expertise in physical education, martial arts, and Asian studies to conduct research on play patterns among Southeast Asian children. Neide holds a master rank in Uechi-ryu karate and is currently an eighth-degree black belt. She works to bring the best of Eastern and Western social and philosophical traditions to the field of kinesiology.

rather than being content to think that socialization related to participating in physical activity flows in only one direction—from parents to children—you might become interested in the extent to which youngsters contribute to socializing their parents into participating in physical activity.

Even your observations of professional practice have given you some insight into sociology of physical activity. Perhaps, for example, you have noticed that there are few female athletic trainers for American football teams, that many people cannot afford to belong to a country club, that fitness professionals often promote achievement of a dominant body ideal, and that jobs in college and professional coaching are held mostly by white males. All of these observations are ripe for sociological study, and sociologists of physical activity have already documented and analyzed many such realities.

Let's look again at our opening scenario and consider how a well-educated kinesiologist who is versed in sociology of physical activity might help plan a physical-activity-friendly city. Kinesiologists with a background in sociology are aware of dominant values and ideas in competitive, leisure, and educative models of sport, as well as issues of access and control, especially in public settings for physical activity. Well-informed kinesiologists also recognize that varied social and cultural contexts exist for physical activity and that different forms of sport, exercise, and recreational activity may be assigned different values across time and cultural groups. Recognizing this diversity of values, a sociologically informed kinesiologist might advocate for nontraditional play spaces, for mixed-use facilities, and perhaps for unified sport leagues as he or she collaborates with other professionals and community members to create a city that is friendly to physical activity.

Sociology of physical activity in kinesiology typically advances three main goals: to look at physical activity with a penetrating gaze; to identify and analyze patterns of change and stability; and to critique physical activity programs in order to enhance equality and human well-being. The first goal is to look at physical activity with a penetrating gaze that goes beyond our common understanding of social life. This is a habit that must be developed, and acquiring it involves exercising a bit of skepticism about common assumptions. If you remember that the complex workings of social life are usually not readily observable, then it will be easier to form the habit of looking further into your own physical activity experiences, knowledge, and professional practice. For example, on the surface, it appears that sport and exercise are personal choices available to all members of society, but a closer look reveals that our physical activity choices are influenced by a host of social conditions.

To illustrate, in the United States, becoming an Olympic competitor in a sport such as figure skating, gymnastics, or speed skating often requires access to a number of resources, including societal cachet for the chosen sport; development of experts and training facilities in the sport; financial commitment of the athlete and family throughout training; a formal set of rules, guidelines, and benchmarks for elite performance in the sport; and sufficient international interest for the sport to remain an Olympic event. Kinesiologists who lack sociological understanding of physical activity might misinterpret trends in Olympic participation as indications of natural interest or willingness to work hard among particular groups (e.g., men more than women; whites more than other racial or ethnic groups). A kinesiologist who is able to perceive what some sociologists refer to as "chains of interdependence" are much better able to respond to the physical activity needs of an ever-changing, diverse societal membership.

The second goal of the subdiscipline is to identify and analyze patterns of change and stability in physical activity. Once you start looking beneath the surface of social life, your goal is to identify both changes and ongoing regularities. Returning to our example of Olympic involvement, sociological analyses indicate that the Olympic Games continue to reflect the sporting traditions of the most developed nations. Despite Pierre de Coubertin's desire to generate a modern Olympic movement involving genuine cross-cultural exchange, the Games continue to privilege large, economically advanced countries in the form of extravagant sporting displays of nationalist propaganda. Yet change has occurred. More women and more countries now participate in the Olympic Games than in earlier times (see table 6.1), new sports have been added to the Games, professional athletes are now allowed to compete in certain sports, and environmental issues are often considered part of planning discussions for Olympic venues. For sociologists,

Is your opinion of skateboarding changed when viewed with a penetrating gaze from a sociological perspective?

Galina Barskaya/fotolia.com

then, the issue is not whether you agree with the current state of affairs of the Olympic Games; rather, it is crucial to identify both shifting trends and stabilizing forces in the physical activity experiences of members of society.

Now that you have learned to see physical activity in new ways and are able to identify patterns of change and stability, you are ready to engage the third goal of sociology of physical activity: to critique physical activity programs in order to identify problems and recommend changes to enhance equality and human well-being. As you may already have guessed, sociologists view the constant stream of social interaction in any society as the most important source of societal problems, although they don't deny that our lives are also influenced by psychological and biological factors. Sociologists don't just study social interaction; they also evaluate it. That is, they pass judgment on its contributions to overall human well-being and equitable social relationships (e.g., relationships between racial and ethnic groups, men and women, or people at different socioeconomic levels). Returning to our Olympic example, sociologists have critiqued the Games as an exclusionary physical activity space that privileges societal members who happen to come from the best-resourced countries, teams, and families.

In addition, as mentioned earlier, the Olympic Games typically require host cities to use public resources, land, and human resources to build the facilities necessary to host a sporting event that lasts only a couple of weeks. Some suggest that hosting such an event will create local interest in sport and exercise, thereby increasing the physical activity of societal members. Sadly, no evidence has been found to support this claim. For kinesiologists who take a sociological perspective, limited outcomes for the public good from mega events such as the Olympic Games are considered problematic enough to warrant modifying the Games to produce more beneficial outcomes for all societal members, not only those who compete in or profit from the Olympics.

Table 6.1 Number of Women's Events and Female Athletes in Olympic Games

	Summer Games				Winter Games			
Year	# of women's events/total events	% women's events	# of female athletes	% female athletes	# of women's events/total events	% women's events	# of female athletes	% female athletes
1980	50/203	24.6	1,115	21.5	14/38	36.8	232	21.7
1984	62/221	28.1	1,566	23	15/39	38.5	274	21.5
1988	86*/237	36.2	2,194	26.1	18/46	39.1	301	21.2
1992	98*/257	38.1	2,704	28.8	25/57	43.9	488	27.1
1994	n/a	n/a	n/a	n/a	27/61	44.3	522	30
1996	108*/271	39.9	3,512	34	n/a	n/a	n/a	n/a
1998	n/a	n/a	n/a	n/a	31/68	45.6	787	36.2
2000	132*/300	44	4,069	38.2	n/a	n/a	n/a	n/a
2002	n/a	n/a	n/a	n/a	37/78	47.4	886	36.9
2004	135*/301	44.9	4,329	40.7	n/a	n/a	n/a	n/a
2006	n/a	n/a	n/a	n/a	40/84	47.6	960	38.2
2008	137*/302	45.4	4,637	42.4	n/a	n/a	n/a	n/a
2010	n/a	n/a	n/a	n/a	43/89	48.3	1,044	40.7
2012	141*/302	46.7	4,862	44.4	n/a	n/a	n/a	n/a
2014	n/a	n/a	n/a	n/a	49/98	50	1,120	40.3
2016	145/306	47.4	4,700	45	n/a	n/a	n/a	n/a

*Including mixed events. n/a = not applicable.

Adapted, by permission, from International Olympic Committee, 2016, *Women in the Olympic movement.*

KEY POINT

Sociology of physical activity in kinesiology looks beneath the surface of social life to point out social problems and advocate for changes that can enhance human well-being and promote equitable social relationships.

In keeping with this third goal, sociologists of physical activity are not always inclined to support or reinforce status quo physical activities (e.g., exercise programs, competitive sports). For example, a former Olympic athlete who becomes a coach might start an Olympic training center where there was none, or a biomechanist might design an exercise machine that meshes with the human body to improve elite performance. Both examples illustrate working within the status quo—that is, working to improve performance in activities as they are currently structured. In contrast, a sociologist might advocate changing the structure itself by questioning the desirability of staging fierce competitions between nations or treating the human body like a piece of machinery that needs to be meshed with other pieces of machinery.

What Do Sociologists of Physical Activity Do?

Sociologists of physical activity are usually faculty members in colleges or universities whose teaching, research, and service advance a scholarly and practical understanding of the complex social and cultural contexts in which societal members experience physical activity. They may teach an introductory course on sociology of physical activity and offer courses on specialized topics such as the Olympic Games; interscholastic and intercollegiate athletics; and the ties between physical activity and phenomena such as gender relations, violence and aggression, social media, and the mass media (e.g., television, magazines, the Internet).

Sometimes sociologists of physical activity combine teaching with professional service by encouraging their students to assist in physical activity

Sport sociologists investigate why some activities seem to cater more to females than to males, or vice versa.

Jon Feingersh/Blend Images/Getty Images

programs located in nearby neighborhoods. This work helps local communities and fosters a sense of civic responsibility in future kinesiology professionals. Faculty service activities may also include consulting with a community group about the role played by physical activity in agendas for bringing about community-wide change (e.g., immigrant settlement issues and culturally specific forms of physical activity; city planning and priorities regarding physical activity spaces, access, and use).

As illustrated by these examples of scholarly engagement, sociologists of physical activity cannot carry out their research in a typical laboratory. Even if it were possible to bring a "slice of life"—say, an aerobic exercise class or a group of fans watching a game—into a laboratory for study, such analyses would not be useful in most cases. Instead, they would usually produce an unrealistic result, because manipulating an event usually alters the very thing that one is trying to study. In this case, you would be studying an aerobic exercise class in a laboratory rather than in ordinary life. For this reason, sociologists usually prefer to study their subjects in the field—for example, professional golfers on the Ladies Professional Golf Association tour, a television production crew covering a downhill ski race, the political processes that underlie the preparation of elite Canadian athletes for international competition, or the preferred physical activity practices among immigrant groups.

As part of this desire to understand everyday life and physical activity engagement, sociologists of physical activity spend considerable time making contacts and gaining the confidence of the people and communities with whom they wish to collaborate for the purpose of developing new knowledge. Regardless of the field setting, the quality of such relationships bears on the quality of the research. As a result, sociologists of physical activity seldom rely on a physical laboratory space; instead, they view all physical activity settings in the world as their laboratories.

History of Sociology of Physical Activity

This subdiscipline is relatively young. It began to take shape in North America in the mid-1960s with the influence of English physical educator Peter McIntosh's 1963 book *Sport in Society.* His analysis of the social significance of sport encouraged several North American scholars to channel their careers toward this emerging area (Sage, 1997). In 1964, the International Committee for the Sociology of Sport was founded; 2 years later, it inaugurated a scholarly journal, the *International Review for the Sociology of Sport,* which was joined in 1977 by the *Journal of Sport and Social Issues.* In 1978, the North American Society for the Sociology of Sport was formed under the leadership of physical educators Susan Greendorfer and Andrew Yiannakis. The society held its first annual conference in 1980 and began publishing an academic journal, the *Sociology of Sport Journal,* in 1984. Both the society and the journal continue to serve as crucial, active scholarly spaces for scholars interested in the sociology of physical activity.

KEY POINT

Since the 1970s, sociologists of physical activity have explored increasingly wide-ranging topics, including various forms of inequality, globalization and regional and national differences, societal conceptions of the human body, and disability and ability.

Since the mid-1970s the number of scholars working in sociology of physical activity has increased, as has the range of theories and research methods used by these scholars. The hottest topics have focused on social inequities—especially those connected with gender, race, ethnicity, wealth, sexual orientation, and diverse cultures around the world. Research interests have also included globalization and regional and national differences, exercise and societal conceptions of the human body, disability and ability, and obesity.

Research Methods for Sociology of Physical Activity

Now that you know something about the subdiscipline and its historical roots, you may be wondering how sociologists of physical activity produce new knowledge. What methods do researchers use to answer important questions about the social side of physical activity? Because questions in social science tend to be complex—often involving many organizational components and layers of human interaction—social scientists rely on a variety of traditions and techniques. Like all research scientists, social scientists begin by identifying meaningful questions, which means current questions that relate to the broader knowledge base and enhance people's lives in some way. Some research questions demand a narrow, standardized focus; with others, it is best to take a descriptive approach that is more nuanced and deeply contextualized. Yet other cases call for a mixture of these approaches in order to craft a well-informed, trustworthy response.

Social science questions generally fall into three broad categories of knowledge production: landscape, descriptive, and analytic (Jamieson & Smith, 2016). Within these categories, various techniques may be identified as the best way to generate data in order to craft a meaningful response to a social science question. These techniques typically include surveys, interviews, thematic analysis, ethnography, and societal analysis.

KEY POINT

Research methods used in sociology of physical activity include surveys, interviews, thematic analysis, ethnography, and societal analysis.

Surveys

Doing survey research involves using questionnaires that are either completed directly by respondents or filled out by a researcher during brief, highly structured interviews. Questionnaires are used to collect data from a large sample of people. The largest survey project in the country is the U.S. census, which is conducted every 10 years. Another kind of survey is a political poll about voting preferences. Investigators have conducted surveys to address numerous topics related to physical activity, including youngsters' opinions about what led them to become involved in sport, former collegiate athletes' thoughts about leaving competition at the end of their 4 years of eligibility, and college athletic directors' opinions about the characteristics of successful coaches.

Interviews

Researchers use interviews when they want broader and deeper information than they can get through a questionnaire or when they want information about activities that would be difficult or impossible to observe themselves. Because interviews are time consuming, studies often focus on a small number of people; they may be conducted either one-on-one or in a group. One form of group interviewing used increasingly in our field is **focus group research**. This method is particularly valuable when the researcher wants to gain insights into people's shared understandings and learn about how individuals are influenced by others in a group situation (Gibbs, 1997). Typically, interviews are audio-recorded and later transcribed; less often, they are also video-recorded. Researchers often ask relatively open-ended questions and then probe for details. For example, one investigator may ask top athletes in high school, college, and the professional ranks to discuss their initial involvement in sport, their participation over the years, and their disengagement from sport. Another researcher may ask young people about the characteristics of their athletic heroes.

Thematic Analysis

Researchers use **thematic analysis**, sometimes referred to as *content analysis* or *textual analysis*, to investigate cultural material such as magazine and newspaper articles, photos, verbal and visual content of television programs, interview data, sporting events, and sport celebrities (Birrell & McDonald, 2000). This analytic procedure involves examining material and categorizing its content based on a strategy of thematic coding. For example, magazine photos of female and male Olympic athletes may be analyzed using the categories of physical appearance, poses, position of the body, emotional displays, camera angles, and groupings of people. In doing such analysis, researchers often identify several main categories or themes, as well as subthemes. In some cases, investigators count the number of times that each theme or subtheme occurs; in other cases, they focus on qualitative data, such as the richness and complexity of less explicit themes. Identifying themes helps a researcher organize a large mass of data into a manageable number of categories that can be analyzed further in relation to theories of interest.

Ethnography

Researchers who use **ethnography** spend many months or even years observing in a particular social setting. They "hang around" while ordinary day-to-day events occur, often taking part themselves, talking with people about what's happening, and keeping careful field notes so that they can remember details. Although this type of observation constitutes their primary source of information, they may also look at local documents or use interviews and questionnaires. Researchers analyze most of the data that they collect using thematic analyses. Ethnography has been used to study a variety of sport settings, including minor league ice hockey, baseball in the Dominican Republic, women's softball, the Olympic Games, women's professional golf, rodeo, and boys' Little League Baseball.

Societal Analysis

In societal analysis, the researcher's goal is to examine the sweep of social life, usually from the perspective of a broad social theory. This method, of course, isn't the only one tied to theory—all good social research is theoretically based. The theories used in societal analysis, however, are extremely broad because they attempt to explain the most fundamental ways in which societies operate. Examples include Marxism (and its many derivatives), modernization, structural-functionalism, figuration theory, various strands of feminism, various cultural studies frameworks, and various forms of postmodernism. These theories are extremely complex; each one is variously supported, refuted, and expanded on in massive collections of scholarly papers and books. Examples of societal analysis in sociology of physical activity include a study of modernization that focuses on sport in preindustrial and postindustrial societies (Guttmann, 1978) and an investigation (focused on social class) of societal constraints and human freedoms in sport (Gruneau, 1983).

KEY POINT

Research in sociology of physical activity involves identifying meaningful questions and generating trustworthy data through the best-fitting research traditions and techniques.

Overview of Knowledge in Sociology of Physical Activity

Sociologists of physical activity often look at human movement from the standpoint of power relationships or the ways in which physical activity choices

How does this activity and its participants fit with current cultural norms of acceptable activities for men and women?

across the life span are influenced by external, dominant cultural beliefs about social life, morals, valued behaviors, and so on. Power relations can be difficult to see in our everyday lives, but once you learn to recognize them you may begin to see them everywhere. You can then look more carefully, with a more penetrating gaze, at various fascinating social aspects of physical activity.

For sociologists of physical activity, it is crucial to focus on access and control in order to understand lifelong, human engagement in volitional movement. This focus, in turn, centers on understanding power in the context of cultural analysis. Current understandings of power reflect the presence of systems of power rather than focusing on power as something that individuals either have or don't have. For example, gender operates as a system of power or organizing system in our daily experiences of physical activity.

In fact, a dominant power structure that features particular forms of masculinity and femininity already exists in the physical education classroom, at the exercise facility, and on the sport field long before we step into those social spaces. For example, there are no signs saying that the cardio room and group exercise classes are for women, but in most fitness facilities these spaces are populated by women. When a male enters these spaces, he may be teased by some; he may also be celebrated by others, who are happy that he broke the gender barrier. Both of these responses reflect a somewhat subtle cultural knowledge that we all share at least to some degree. This "knowingness," as it is referred to by some scholars, constitutes a form of power in society because it reinforces a certain way that societal members think things *ought to be* organized (e.g., males on the free weights and females in the cardio room).

With specific reference to physical activity, power inequalities sometimes affect our participation as performers, as well as our opportunities for involvement in influential leadership roles such as health club owner, coach, athletic director, or cardiac rehabilitation program director. People with more access to power can usually get involved more easily. Those with less access to power often face social conditions that make involvement and employment more difficult. For example, African American players in the NFL have a harder time than white players do in obtaining coaching and administrative positions in the league.

Such power differences are often displayed, and thus reinforced, in movies, television programs, music, and the structure of physical activities themselves. Whether we participate directly in activities such as sport, dance, and exercise or watch others perform them, the activities express subtle information that tends to support and reinforce status quo power differences. For example, telecasts that highlight the physical appearance of female athletes—with special focus on their clothes and hairstyles—contribute to undermining, and thus trivializing, their athletic accomplishments.

It is crucial for kinesiologists to understand power in society as they identify the best ways to promote meaningful physical activity for all societal members. In fact, Jennifer Hargreaves (2001) has suggested that "stories of sport are almost exclusively stories of those in power" (p. 8). Thus, kinesiologists might ask, what are the hierarchies of bodies that emerge in physical education classes, rehabilitation settings, or school-based sport programs? What social structures—such as laws, policies, school curricula, expert knowledge—reinforce societal privilege for particular bodies? How do physical educators, athletic trainers, fitness professionals, and coaches contribute to (or resist) these processes of bodily privilege?

In the remainder of this overview, we relate sociological analyses directly to the ability of kinesiologists to promote high-quality physical activity, including sport and exercise, for all societal members. Social inequalities are underlain by power relations, which affect people's quality of life, their chances for a better life in the future, and their opportunities for engaging in physical activity. Accordingly, sociologists produce research that examines the influence of systems of power (e.g., gender, race, social class, and sexuality) on the everyday experiences of societal members. In this abbreviated review of research in the sociology of sport and physical activity, we use three material aspects of societal power—participation, leadership, and cultural expression—to illuminate key findings and spaces for change within varied sport and physical activity contexts.

Gender Relations

Gender differs from sex. As a system of power, gender is not an identity that one has (e.g., biologically male or female) but a set of norms or expectations about how we should behave, and these factors are linked to societal understandings of sexuality and procreation. At birth, babies are assigned to a sex category based on the appearance of the genitalia (Lorber, 1994). This assignment process combines with the subsequent differential treatment of youngsters, based on different sex categories, to constitute the gendering process. Thus, genders are not natural, biological categories; they are socially defined. You can't inherit a gender—you have to be assigned to it and learn it (Fausto-Sterling, 2000).

In U.S. society, men typically hold more power in a gender hierarchy than do women, although this relationship can change when race, social class, and sexual orientation are considered. For example, white middle-class women often exercise more societal power than do non-white working-class men. The situation has changed somewhat during the 20th century, but many inequalities remain. In the field of physical activity, inequalities prevail because of beliefs about the appropriateness of certain forms of physical activity for recognized gender groups.

Participation

In the United States, many girls participate in organized youth sport, and many of their older sisters play on high school and college teams. The most popular collegiate sport for women is basketball, followed by volleyball and, in a tie for third place, soccer and cross country (Acosta & Carpenter, 2014). In high school, the top five sports for girls are basketball, track and field, volleyball, softball, and cross country. This diverse sport participation by girls and women is relatively recent and can be attributed largely to the passage of Title IX in 1972, which required that women be provided with equitable opportunities to participate in education and sport.

With that legal foundation in place, girls' and women's participation in sport has increased dramatically over the last 4 decades. The average number of women's collegiate teams per school increased from 2.5 in 1970 to 8.7 in 2014 for a total of nearly 10,000—the highest numbers ever (Acosta & Carpenter, 2014; Women's Sports Foundation, 2009). In 1971, only 1 of every 27 high school girls played a varsity sport; now, that figure is 1 of every 2.4. High school girls' participation in athletics exceeded 3 million for the first time in the 2006–2007 school year and has risen to more than 3.3 million; the rate of boys' participation has also risen (National Federation of State High School Associations, 2016). While the combination of compliance with Title IX and decreased state funding for sport has resulted in changes in the kinds of sports offered to boys and girls, a recognition of the ongoing increase in rates of participation among boys is important, especially in light of the often-misleading claims made that Title IX has decreased opportunities for boys and men in athletics.

In North American societies, sport is viewed as an important avenue for exploring and confirming masculinity. More specifically, it is thought to sharpen a number of qualities traditionally considered appropriate for men and boys, such as toughness, aggressiveness, working well under pressure, and competitiveness. In reality, you probably know girls and women with some of these qualities—a fact that points out the indistinct boundaries between the two genders. Nevertheless, many people still conceptualize these qualities as masculine.

KEY POINT

Although many more females engage in physical activity today than they did several decades ago, they tend to participate in sports considered socially appropriate (e.g., involving less body contact, prominent aesthetic dimensions, and less strength development).

Although many sports attract large numbers of both males and females, some are more gender specific. For example, girls and women are less visible in weightlifting and cycling, whereas boys and men demonstrate disproportionately lower rates of participation in exercise walking and aerobic exercise. Yet, as shown in table 6.2, male and female rates of participation differ by only a narrow margin in a significant number of activities (e.g., hiking, working out at a club).

If a woman or girl you know started training to be a football player, what would you think? Many Americans would judge her action negatively. What makes it inappropriate? Although many females are now developing muscular strength and endurance through weight training, few take part in sports requiring great strength, physical domination of an opponent through bodily contact, or the manipulation of heavy objects. Furthermore, only in the last several decades have women taken up competition in strenuous, long-distance runs, including marathons and ultramarathons. One early, cutting-edge sociological study reported that the sports we judge more appropriate for females, such as gymnastics and tennis, entail little body contact and involve aesthetically pleasing movements and manipulation of light objects (Metheny, 1965).

The continued emphasis on masculinity in sport often means that women's competition is down-

Table 6.2 U.S. Participation in Selected Sports and Physical Activities by Gender

Sport or physical activity	Total (thousands)	% male	% female
Exercise walking	95,803	39	61
Exercising with equipment	55,286	47	53
Swimming	51,943	47	53
Bicycle riding	39,789	56	44
Bowling	38,980	53	47
Aerobic exercising	38,541	30	70
Hiking	37,704	51	49
Working out at a club	36,278	48	52
Running or jogging	35,524	52	48
Weightlifting	31,480	65	35

Top 10 activities listed in thousands (e.g., 95,803 = 95,803,000).
Data from NSGA 2010.

played. For example, female sports may receive fewer resources, such as travel money, coach salaries, publicity funds, media exposure, and corporate sponsorships. In some cases, they are even given team mascots or nicknames that suggest physical ineptitude, such as the Teddy Bears, Blue Chicks, Cotton Blossoms, or Wild Kittens (Coakley, 2007; Eitzen, 2006).

Leadership

Our knowledge of gender inequities in physical activity leadership pertains primarily to sport participation as an athlete, but let us briefly examine leadership in other physical activities. We know that opportunities are increasing for U.S. women in coaching, sportscasting, and officiating but remain more limited than they are for men (Acosta & Carpenter, 2004; Carpenter & Acosta, 2008; Coakley, 2007). Moreover, from the 1970s into the 1990s, sharp decreases occurred both in the percentage of female coaches of female high school and intercollegiate teams and in the proportion of female administrators heading women's intercollegiate programs. For example, in the 1977–1978 academic year, 79 percent of the coaches of female basketball teams were women; by the 1991–1992 year, that number had dropped to 64 percent (Acosta & Carpenter, 1994).

More generally, the proportion of female collegiate teams led by a female head coach decreased from 90 percent in 1972 to 43 percent in 2014. By 2014, there were 4,154 female head coaches of women's collegiate teams—the highest representation ever and an increase of 180 over the past 2 years (since 2012). Sadly, this increase in the number of women's head coaches did not address the drop in proportionality of all head coaches. Interestingly, as of 2014, only 4 percent of men's collegiate teams were coached by women, whereas 57 percent of women's teams were coached by men. Between 2000 and 2014, 2,080 new head coaching jobs appeared in women's athletics, of which about one-third were filled by females and two-thirds by males. To state the comparison another way: In 1972, nine out of ten coaches of women's collegiate teams were female; as of 2014, that figure had dropped to four out of ten (Acosta & Carpenter, 2014). Beyond amateur athletics, in 2016, only 50 percent of WNBA head coaches were women (Lapchick & Baker, 2016).

A similar pattern appears when we look at administrative positions. For instance, the percentage of female administrators heading women's U.S. intercollegiate programs plummeted from 90 percent in 1972 to 23 percent in 2014. Also in 2014, only 32 percent of head athletic trainers and 12 percent of head sports information directors were female (Acosta & Carpenter, 2014).

On a global scale, Lapchick, Davison, Grant, and Quirarte (2016) found that representation of women was abysmal in international sport federations recognized and relied on by the International Olympic Committee (IOC). More specifically, the researchers assigned a grade of D+ to the IOC and a grade of F to the collective set of national sport federations for not doing more to live up to their own rhetoric about the need for gender equality in sport leadership. In this first "report card" study of women's leadership in international sport settings, women accounted for just 6 percent of international federation presidents, 12 percent of vice presidents, and 13 percent of executive committee members. In the IOC itself, women made up a mere 24 percent of members. As you consider these leadership data with participation rates among athletes in Olympic sports, it is even more clear that structural inequalities related to gender continue to create disproportional opportunities for men and women to move into global sport leadership roles.

KEY POINT

Women occupy a relatively small proportion of coaching and leadership positions in sport.

Why are more men now coaching and administering women's teams and programs? Researchers have posited a number of reasons. To begin with, the larger number of male players constitutes a larger pool of qualified candidates. In addition, the fact that women's coaching salaries still lag behind those of men may cause some of the top female players to seek careers outside of athletics. Furthermore, men occupy most of the influential athletic leadership positions (e.g., collegiate athletic directors, Olympic leaders) that exert strong control on hiring decisions. When making decisions about hiring coaches or appointing people to important committees, people who already hold such positions likely use their convenient and trusted network of connections in athletics, often referred to as the "old boy network," to find candidates. Finally, many of us may have learned to favor applicants with stereotypical masculine qualities—that is, those who are aggressive, dominating, and physically tough—thus putting many well-qualified women at a disadvantage (Coakley, 2009).

More recently, two other factors have also gained attention as potential barriers to women's and men's involvement in collegiate coaching—namely, homophobia in collegiate athletics and incompatibility between parenting roles and coaching roles (for men and women, but especially for women). As Lapchick, Davison, et al. (2016) have pointed out, we are all still working toward full accountability to rhetorical calls for equity. It is not enough to assert the value of equality in sport; individuals and organizations must also strategize in order to recruit and retain high-quality sport leaders who can bring a collective diversity of ideas, cultural knowledge, and lived experience to the ethical governance of sport. Given these challenges, kinesiologists who develop a sociological perspective will be well prepared to lead the field toward achieving such goals.

Cultural Expression

Given that societal gender inequalities exist and that all physical activities have expressive dimensions, these inequalities are often demonstrated in our physical activities. For example, in some forms of dancing (e.g., waltz, many country-western dances), men usually move forward and lead women, who usually move backward. This technique communicates a nonverbal message about men's supposed superiority.

In sport, our most popular events—the ones we see constantly on television—celebrate a form of heterosexual manhood centered on aggressiveness, roughness, and the ability to physically dominate others. In contrast, girls and women have most often been relegated to marginal positions as spectators of boys' and men's competition or as players in their own, less popular versions of these and other sports. This contrast communicates a message that females don't matter much in "real" sport. Moreover, when coaches criticize boys or men for making weak or ineffective plays, they sometimes do so by calling them "sissies" (Coakley, 2009). Such comments, intended to spur better performance, send a message that females are inferior. Coaches communicate a similar message when they use gendered descriptions such as "throws like a girl." These examples show ways in which sport tends to reinforce ideas about the acceptability of heterosexual male dominance.

KEY POINT

Many sports serve as vehicles for exploring, celebrating, and privileging masculinity; in doing so, they express ideas that are problematic for girls and women and for boys and men who are not athletically inclined.

Girls and women are not the only ones shortchanged by the gendered attitudes that prevail in sport. In fact, mainstream society often questions boys and young men who lack athletic ability or are not interested in athletics. They may be called "sissies," and their sexual orientation may be challenged with taunts such as "gay," "fag," or "queer" (Coakley, 2009, p. 268). In high school, boys' popularity is often closely connected with being good at athletics. In one study, a group of white college graduates who had not played a varsity sport in either high school or college said that as boys they had experienced social ostracism and name-calling because of their poor athletic ability. One said, "I identified sports as a major aspect of what I was supposed to be like as a male, which oppressed me because I could not do it, no matter how hard I tried" (Stein & Hoffman, 1978, p. 148).

Gender inequity is also manifested in sport media. Have you ever thought about media coverage of men's and women's sport? About 90 percent of sport on television focuses on men's athletics; or, to state the point in reverse, women's competition gets less than 10 percent. The picture is about the same in U.S. newspapers, where women's sport gets less than 15 percent of overall sports coverage (Coakley, 2009). This disparity trivializes women's athletic accomplishments and sends an underlying message that women's events aren't worth much.

Racial and Ethnic Relations

Race is a historically, culturally, and socially defined category of social difference that is typically marked by phenotypic variance (e.g., in skin color, body shape, facial features, hair type). In other words, race is socially defined on the basis of characteristics that we select; it is not a natural, biological category or difference. Like gender, then, race is not a fixed identity that one has but rather a shifting condition of social life that one experiences every day. More generally, we live in a society that is organized around arbitrary beliefs about gender, race, and social class. Kinesiologists must understand how to work through these social conditions and inequalities in order to make meaningful physical activity available to all societal members.

Ethnicity refers to cultural heritage. People who share important and distinct cultural traditions, often developed over many generations, are classified as an ethnic group. Ethnic markers include such things as language, dialect, religion, music, art, dance, games and sports, and style of dress. Examples of ethnic groups in the United States include Latinos and Latinas, persons of Jewish descent, Inuits, Amish, and Cajuns. In some cases, ethnicity overlaps with race. For example, persons of African American descent are most often defined as a racial group, yet there are many distinct African cultural traditions, African nations, influences from various Caribbean cultures, and a variety of European colonial influences that makes racial categories limiting.

Throughout history, including current times, racial and ethnic-minority communities have lived with contingent civic representation and rights and have typically held less civic and legal power than have racial-majority communities. Typically, white European dominance—exercised in and through societal structures of governance—has contributed to inequalities ranging from subtle forms to overt forms such as Jim Crow laws in the southern United States. Such inequalities have been both highly visible and powerfully resisted in various societal and cultural settings, including those related to sport. Thus, many people can recall the breaking of the color line in Major League Baseball, but a deeper look quickly reveals that Jackie Robinson's presence also marked deep divisions in U.S. society and in its progress toward creating, in the words of the U.S. Constitution, a "more perfect union."

Although strides have been made toward equality, many difficulties remain. For its part, physical activity is not immune to such problems. We focus here mostly on sport, and we highlight rates of participation across racial-ethnic groups and demographic groupings deemed meaningful as identity politics and preferences continue to shift throughout society.

Participation

In elite team sports, we often see disproportionate representation among various racial-ethnic groups, and we often interpret this pattern in certain ways.

Masculinity and Femininity Examined

Think about physical activities you have experienced in which questions arose about a male's masculinity or a female's femininity. What do these examples suggest about our shared beliefs concerning appropriate activities for boys and men and for girls and women? What do they suggest about gender inequalities in these physical activities? What do they suggest about societal definitions of masculinity and femininity?

The high participation rates of African Americans in many professional sports probably relate mostly to U.S. social structure (Eitzen & Sage, 2009). More specifically, a history marked by limited outlets for making public contributions to society has surely influenced the U.S. preoccupation with the athletic capacity of men and women in nonwhite racial-ethnic groups. This historical legacy is joined by the mass media's tendency to push narratives of athletic achievement by members of minority racial-ethnic communities as "a way out" or one's "only option" or as a result of one's "natural ability." The combination of this legacy and these narratives limits our understanding of rates of participation among racial groups across various sport and physical activity settings.

For example, much of our research fails to account for African Americans' high participation rates in a few sports and their low involvement in most others. This situation is probably influenced both by opportunities and lack of opportunities (Eitzen & Sage, 2009). That is, African Americans tend to be more visible in sports with easy access to facilities and coaching, primarily in school and community recreation programs. In contrast, they are less likely to participate in sports that require private coaching, expensive equipment, empty land for playing fields, or facilities at private clubs—for example, tennis, golf, swimming, soccer, and gymnastics. However, it remains unclear to what extent these factors influence African American athletic involvement. In addition, the success of certain African American athletes in historically segregated sports—such as Tiger Woods in golf, Venus and Serena Williams in tennis, Bill Lester in NASCAR, Simone Biles in gymnastics, and Simone Manuel in swimming—raises new questions about access, identities, and underlying reasons for sport involvement.

In U.S. higher education, equitable access is expected to be provided to members of all social groups, but a quick review of NCAA Division I representation in leadership and athlete participation reveals disproportionate involvement in terms of both race and gender (see figure 6.1). For example, although African American males account for nearly

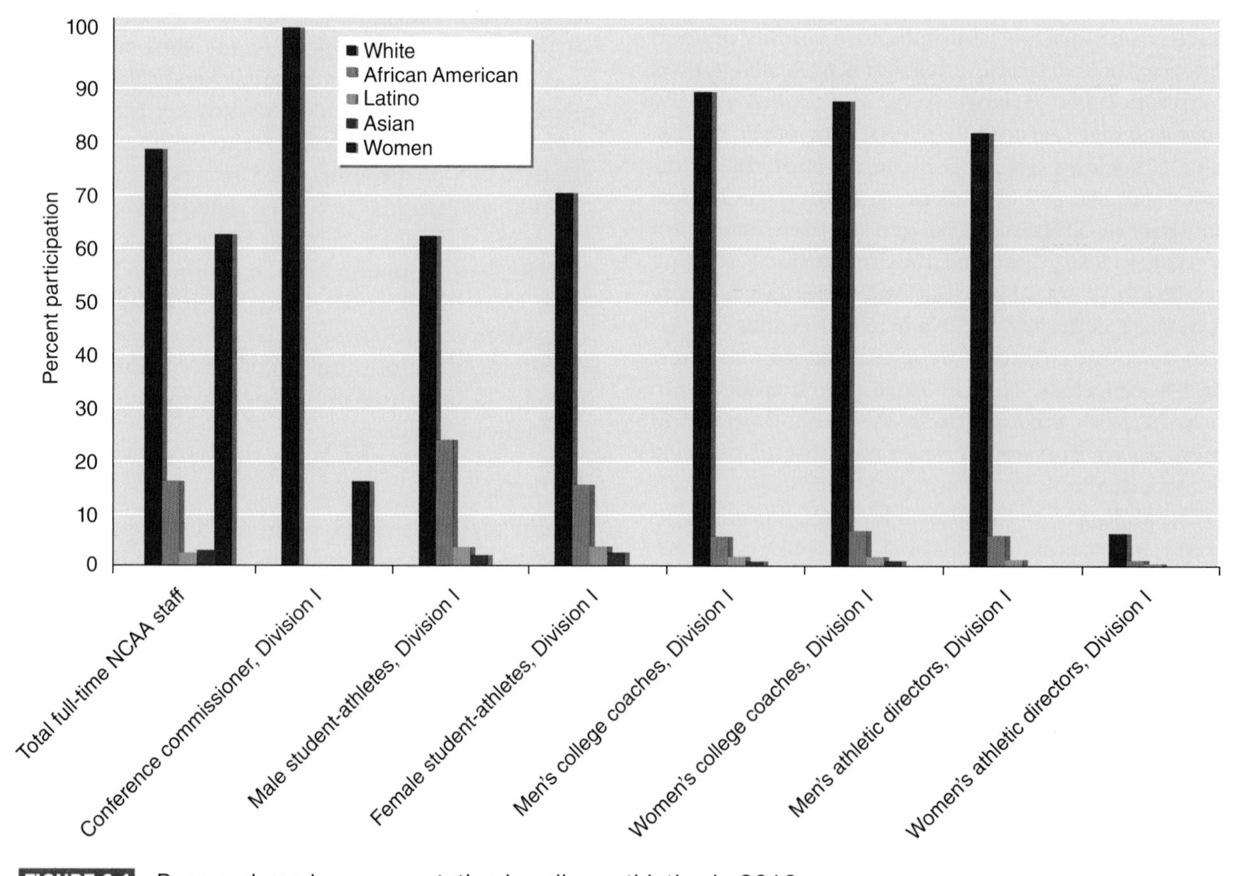

FIGURE 6.1 Race and gender representation in college athletics in 2010.

25 percent of male Division I athletes, they make up less than 7 percent of athletic directors and head coaches. In comparison, white male student athletes are overrepresented in both roles.

KEY POINT

Large numbers of African American athletes play major team sports and participate in track and field, but their representation remains disproportionately low in most other sports and in leadership roles in U.S. society.

Some of the same factors affect recreational physical activities in which competition is not essential, as well as exercise programs geared toward improving health and appearance. These types of activities—for example, surfing, backpacking, exercising at a health club, swimming, and skiing—require various combinations of expensive equipment, facilities, lessons, and travel, thus limiting involvement by individuals in lower-income families. Because poverty levels in the Unites States are strongly associated with race, racial-ethnic minorities tend to be underrepresented in more costly physical activities (U.S. Census Bureau, 2007).

More specifically, in major professional sports, African Americans are overrepresented in basketball (men's and women's) and American football but underrepresented in baseball and soccer. Latinos are underrepresented in all professional sports, especially the WNBA, NBA, and NFL. Asian women are not represented in the WNBA, whereas Asian men make up nearly 2 percent of players in MLB and the NFL but are underrepresented in the NBA and MLS. Taken together, "international players" (i.e., players from outside of the United States) account for 22 percent of players in the NBA, 42 percent in MLS, and 16 percent in the WNBA. In the case of the MLB, though many Latino players are "international players," they are not counted as such; tracking the globalization of the MLB would be interesting to observe. For more information about representation of various groups in specific sports, see table 6.3 and the reports produced by Lapchick and colleagues at the Institute for Diversity and Ethics in Sport (TIDES; Lapchick & Bullock, 2016; Lapchick, Dominguez, Haldane, Loomer, & Pelts, 2014; Lapchick, Malveaux, Davison, & Grant, 2016; Lapchick & Nelson, 2015; Lapchick & Salas, 2015).

Leadership

If U.S. society were characterized by full racial equality, then we would expect to find rather high proportions of African American coaches in sports that have high proportions of African American players—for example, American football, basketball, and baseball. However, this is not the case.

Table 6.3 Racial Representation in Professional Sport

Racial group or league	MLB %/#	NBA %/#	NFL %/#	MLS %/#	WNBA %/#
Report year	2016	2016	2016	2014	2015
White	59/443	18.3/82	27.4/618	51.1/276	24.3/37
African American	8.3/62	74.3/333	69.7/1,573	11.3/61	71.7/109
Latino	28.5/214	6.3/28	0.8/18	17/92	0.7/1
Asian	1.7/13	0.2/1	1.9/44	0.7/4	0/0
Other	2.4/18	0.9/4	0.2/4	19.8/107	3.3/5
Total people of color	—	—	—	48.9/264	—
International	—	22.3/100	—	41.9/226	16.4/25
Total players	750	448	—	—	—

Data from TIDES reports in each league, 2014, 2015, and 2016.

Between 2014 and 2016, African Americans represented only 3 percent of Major League Baseball managers, 16 percent of NFL head coaches, 27 percent of National Basketball Association (NBA) head coaches, and 25 percent of Women's National Basketball Association (WNBA) head coaches (see the sport-specific TIDES reports by Lapchick and colleagues, cited earlier). The 2007 Super Bowl was historic in this regard because it featured two African American head coaches in Tony Dungy and Lovie Smith. Some credit has been given to an NFL rule (the Rooney Rule) requiring that at least one minority candidate be interviewed when any team is searching for a new head coach. On the college level, among NCAA Division I-A programs in 2014–2015, African Americans held 8 percent of head coaching positions for football, 22 percent for men's basketball, 15 percent for women's basketball (11 percent African American women), and 1 percent for baseball (Lapchick & Baker, 2016).

These leadership inequities have been the subject of two major explanations: racist stereotypes and "old boy networks." Racist stereotypes that characterize African Americans as having inferior intellectual capabilities hinder their attempts to attain athletic leadership roles in which thinking and working under pressure are viewed as crucial. When people in charge of hiring rely on racialized views supported by institutional racism, African Americans (and other nonwhite racial groups) are likely to be shortchanged (Eitzen & Sage, 2009). Ricky Stokes, an African American former assistant basketball coach at Wake Forest University, put it this way: "As an African American assistant coach, sometimes you can be labeled as a recruiter. A lot of times people think that's all you're supposed to do. Sometimes people don't think you can do the job on the floor—coaching, teaching the fundamentals, and, taking another step, speaking to booster clubs." Although Stokes expressed optimism about landing a job as a head coach, he realized that people might perceive hiring him as taking a chance (Ross, 1996, pp. C1–C2).

KEY POINT

Few African Americans reach important sport leadership positions, even in team sports that feature many African American players.

A second reason for leadership inequities relates to networking. Athletic administrators who hire people for entry-level coaching or management positions are likely to be white. They are also likely to learn about job candidates from their professional friends in other athletic programs, most of whom are also white. Because ethnic and racial minorities are not usually involved in these networks, athletic administrators can easily overlook them. Relatively few people from marginalized ethnic and racial groups hold leadership positions in programs featuring expensive sport and exercise regimens (Coakley, 2009; Eitzen & Sage, 2009). A quick tour of sport and activity facilities—such as health clubs, soccer fields, skating rinks, ski areas, gymnastics schools, and tennis courts—would probably turn up only a handful of coaches and administrators of color.

Cultural Expression

Expressive behavior by all athletes is sometimes appreciated by audiences, sometimes regulated by league rules, and sometimes encouraged by coaches as a way to intimidate one's opponent. In the American racial hierarchy, expressive behavior by African American male athletes has been examined as a possible form of resistance to the surveillance that African American males face in U.S. society. When African American males experience restrictions on their access to education, jobs, and power, they often seek to prove their masculinity in other ways. One way is to construct an expressive "cool pose" consisting of "styles of demeanor, speech, gesture, clothing, hairstyle, walk, stance, and handshake" (Majors, 1990, p. 111). Cool pose sends a message that the person possesses a potent, intriguing lifestyle that deserves respect. It is an embodied recognition of people's right to occupy spaces that often exclude them based on racial hierarchies in conjunction with ideas about social class, gender, and, at times, sexuality.

Sport provides a ready-made arena for cool pose. Examples include celebration dances in the end zone; Muhammad Ali's boasting, poetry, and dancing in the ring; and Julius Erving's gravity-defying moves to the basket that often began at the foul line (Majors, 1990). Many current African American sport stars continue to create images rooted in the traditions of cool pose. In addition, cool pose is threaded through a variety of stylish gestures and "walks" that are embedded in the ordinary activities of African Americans and in distinctive styles of dance such as hip-hop.

Beyond the overly deterministic cool pose framing of expressiveness among African American male athletes, one might consider the following more obviously politicized moves:

- Former San Francisco 49ers' quarterback Colin Kaepernick taking a knee during the National Anthem as a way to protest what he viewed as the oppression of black people and other nonwhites in the United States.
- Grambling State Tigers football players who refused to take the field as a political statement on conditions and leadership in their program (Zirin, 2013).
- African American WNBA players and teammates wearing Black Lives Matter t-shirts during warm-ups (Pandian, 2016).

Beyond the white–black binary of racial recognition in the United States, one might consider the early retirement decision of Mexican professional golfer Lorena Ochoa and the resistance of Indian runner Dutee Chand as refusals to accept race-gender-sexuality inequities in sport governance policies (*The Guardian*, 2014).

It is important to reconsider the notion of cool pose and perhaps replace it with a more complex notion of nuanced refusal to be held captive by sport governance and a consumer culture that both organized around privileged white, middles class, masculine modes of expression. This kind of analysis of athlete expression across racial groups and with intersectional attunement will turn our attention to the ways the media, sport governance, and sport leaders may intervene, along with athletes, in disrupting the American racial hierarchy. For sport sociologists, these athlete actions are important sites of change-oriented expression.

African American boys and young men often develop their own renditions of cool pose at playgrounds and school athletic facilities. Their style is a crucial part of their sports, especially basketball (Carlston, 1983; Kochman, 1981; Majors, 1990; Wilson & Sparks, 1996). They challenge each other with their creative, flashy moves, daring other players to top them. Their goal is to heighten their reputation by combining effective game skills with a spectacular look. This kind of look can be seen in a range of media images designed to make a particular form of black masculinity profitable through black athleticism. For further reading on this subject, see Neal (2006).

Unfortunately, sport also communicates ideas that help maintain racial and ethnic inequities. For example, the fact that many top athletes are African American seems to send an upbeat message to African American boys and young men that they can make it to the pros if they work hard enough. However, this message is misleadingly positive. For example, although a number of African American high school varsity basketball and football players believe that they have a chance of playing professionally, only a minuscule number actually make it. Because constant sport practice usually gets in the way of acquiring other job skills, the combination of racial inequalities and athletic identity may limit one's ability to move beyond athletics for employment opportunities (Coakley, 2009; Eitzen & Sage, 2009). White youths in higher socioeconomic brackets who dream of playing professional sports usually end up short of their goal as well, but because they don't face racial discrimination and do come from higher-income families who have more contacts with potential employers, they are often better positioned to pursue other job opportunities.

Racial Stereotyping

Racist stereotypes are sometimes used in television portrayals of white and African American athletes. In two articles, published in 1989 and 1996, Derrick Jackson studied comments made by television sportscasters about African American and white athletes. He noticed that sportscasters tended to recognize white basketball and football players for their brains—specifically, intelligence,

Assessing the Ethnic and Racial Mix in Sport

Think about the local physical activity programs with which you are familiar in schools and community recreation centers—for example, physical education classes, varsity teams in high school and college, rhythmic aerobics classes at local YMCAs, pickup basketball games on inner-city courts, gymnastics instruction at private centers, and youth sport leagues formed by groups such as the American Youth Soccer Organization and Little League Baseball. From your observations, what ethnic and racial minorities are represented among the people who play leadership roles in these programs (e.g., supervisor, director, coach, official)? How do the proportions of ethnic and racial minorities in these positions compare with the proportions of ethnic and racial minorities among the players? What are possible reasons for the similarities or differences in these proportions?

thoughtfulness, and strategy. African Americans, on the other hand, were noted more for their brawn—that is, physical skills, moves, and muscular strength. Such reactions reinforce a prominent racist stereotype. It is heartening to learn, however, that although brains were still mentioned more often in 1996 television portrayals of white athletes, this tendency was not as strong as it was in 1989. Unfortunately, over the same period, the "brawn gap" between African Americans and whites doubled, and African Americans were twice as likely as whites to be praised for their physicality in 1996.

Jackson's research shows that our understanding of racial categories can change over time and place; therefore, as consumers of sport, we need to be aware of the racialized rhetoric used to convey sport stories. Moreover, the brains-versus-brawn manner of describing athletic performance is just one way to convey cultural messages in regard to bodily performances. The phenomenon also takes other forms as we consider how race intersects with gender, social class, sexuality, and ability (Coakley, 2007; Douglas & Jamieson, 2006; Eitzen, 2006).

KEY POINT

African American athletes who engage in "cool pose" express creativity, strength, and pride associated with masculinity; however, media members sometimes characterize them in racially stereotyped ways that strip away some of their racial identity.

In the context of this discussion of expressiveness, the case of Michael Jordan, probably the most visible athlete in the world in the 1990s, provides much food for thought. A special collection of articles devoted exclusively to this athletic superstar demonstrates the ways in which he was constructed as a media icon who transcended race (Andrews, 1996). In part, this process involved stripping away some of Jordan's "blackness"—that is, his racial identity. For example, even as advertisements accentuated his physical, athletic black body (a common media portrayal of African American athletes), the potential threat of the heightened sexuality stereotypically associated with black bodies in U.S. society was defused by frequent references to his wholesomeness and strong family ties.

These mediated images of Jordan took hold so strongly that even following his divorce, he retained a wholesome image. Such mixed messages probably contribute to his popularity among whites and African Americans alike (McDonald, 1996). Media portrayals of Jordan give us a collage of meanings, some lending positive support to a distinctive African American culture and in other cases reinforcing old ideas that help maintain racial inequalities. Sorting out which are which requires sociologically informed analysis.

Socioeconomic Relations

Wealth, education, and occupational prestige are all ingredients of socioeconomic status. People who have more money, more education, and greater occupational prestige have higher socioeconomic status and usually more access to power. To illustrate, it's not hard to recognize the many advantages of being a corporate chief executive as compared with being a custodian at corporate headquarters. The chief executive can afford a luxurious lifestyle, whereas the custodian may have to scrimp to make ends meet. The chief executive can send his or her children to the most expensive universities, whereas the children of the custodian may have to work to put themselves through college. The chief executive can make major decisions affecting future directions of the company, whereas the custodian may only be able to recommend better ways to clean the office.

Abundant evidence indicates that the income gap between rich people and poor people is widening in the United States; in fact, the pattern of income growth has changed from fast and fair to slow and distorted (Economic Policy Institute, 2011). For example, between 1947 and 1973, family income for the bottom 20 percent of households grew at a rate of 117 percent, whereas the rate for the top 20 percent of income earners grew at a rate of 89 percent. In another report, the Economic Policy Institute (Sommeiller & Price, 2015) tracked unequal or lopsided income growth on both the state and national levels. Data indicated that the richest 5 percent of U.S. households had an average income 13 times higher than the poorest 20 percent of households. This sort of lopsided income growth constitutes a long-term trend; specifically, a nearly 30-year pattern (1979 to 2007) shows the top 1 percent taking home more than half (54 percent) of the total increase in U.S. income. During this period, the average income of the bottom 99 percent of U.S. taxpayers grew by 19 percent, while the average income of the top 1 percent grew by more than 10 times as much (by 200 percent).

Social categories of race, class, and gender operate not independently but interdependently. Another Economic Policy Institute report (Wilson

Professional Issues in Kinesiology

Equality: Native American Mascot Controversy

One longtime controversy in the sporting world involves the use of American Indian symbols and rituals for team names and mascots, such as the Cleveland Indians and the associated mascot named Chief Wahoo. If you are a baseball fan, you've seen the team's baseball logo, which features the head of Chief Wahoo, whose toothy grin, hook nose, fire-engine-red complexion, and single-feather headdress serve stereotypically to identify him as an American Indian. Critics argue that such representations create false and degrading images of American Indians and contribute to the miseducation of Americans by reproducing a kind of cultural illiteracy about indigenous (native) peoples (Staurowsky, 1999). Indeed, the elements of "Indian" life featured in sport imagery—war paint, masks, costumes, drumming, chants, and ceremonies—are neither authentic nor accurate; instead, they serve as pseudo-Indian symbols (King, 2004) that feed long-standing stereotypes of Native Americans.

A number of college teams (e.g., Cornell, Stanford, Syracuse) have seen merit in these arguments and changed their names and mascots accordingly. Professional teams have not—for instance, the Washington Redskins and Atlanta Braves. Those who defend sport-team use of American Indian symbols assert that their use of such imagery constitutes a tribute and argue for their right to adopt and own the images for marketing purposes (Staurowsky, 2004). Most American Indians themselves are critical of the violent and aggressive images associated with sport team naming. Chief Wahoo, for instance, is seen as a racist caricature of an American Indian that contributes to the long history of oppression in white dealings with American Indians (King, 2004). Do you think race or ethnic-related sport mascots promote equality in sport and society? Why, or why not?

Citation style: APA

King, C.R. (2004). This is not an Indian: Situating claims about Indianness in sporting worlds. *Journal of Sport & Social Issues*, *28*, 3–10.

Staurowsky, E.J. (1999). American Indian sport imagery and the miseducation of Americans. *Quest*, *51*, 382–392.

Staurowsky, E.J. (2004). Privilege at play: On the legal and social fictions that sustain American Indian sport imagery. *Journal of Sport & Social Issues*, *28*, 11–29.

& Rodgers, 2016) found that black men made 22 percent less in average hourly wages than did white men with the same education, experience, metro status, and region of residence. Black women made 34 percent less than white men and 12 percent less than their white female counterparts. The widening wage gap has not affected everyone equally. Since 2000, young black women (those with 0 to 10 years of experience) have been hardest hit, perhaps due to a number of structural inequalities, including histories of structural inequality for family asset development, and racial inequalities in hiring practices.

This vast and growing income disparity in the United States carries important implications for many aspects of life, including education, recreation, health care, and physical activity. For example, among industrialized nations, citizens of countries with wider income gaps are generally less healthy (Siedentop, 1996). Much evidence indicates that poor people are more likely to suffer from health problems, due in large part to structural inequalities that result from increased privatization of social services and decreased opportunity to access health resources (e.g., reduction in public facilities, increases in fee-for-membership facilities).

The United States has experienced enormous growth in private health and exercise clubs that charge membership fees and cater to the upper middle class and the wealthy. Meanwhile, a decline has occurred in publicly funded programs in schools and recreation departments that are more accessible to middle- and working-class people. For example, it is increasingly common for students to pay user fees to participate in varsity athletics in the public schools and for families to pay user fees for youngsters to compete in community youth sport programs (Coakley, 2009). Thus, even though most U.S. citizens could benefit from physical activity, barriers are in place that make programs less accessible to low-income people. Table 6.4 shows how participation in certain kinds of physical activity varies by level of income. For example, participation in all physical activities increases at or above the $35,000 annual income mark, and some activities (tennis, saltwater fishing, and golf) have quite low participation rates below the $35,000 or $50,000 marker.

Participation

Research indicates that people's level of education is clearly tied to their participation in physical activity. One study showed that highly educated groups were 1.5 to 3.1 times more likely to be active than were the least educated groups (Stephens & Caspersen, 1994). In addition, salaried professionals are much more likely to participate in corporate

Table 6.4 Participation in Selected Sport Activities by Household Income: 2009 (in Thousands)

Physical activity	Number	Under $15,000	$15,000–$24,999	$25,000–$34,999	$35,000–$49,999	$50,000–$74,999	$75,000–$99,999	$100,000 and over
Total	269,988	25,568	24,659	27,297	39,689	54,549	41,485	56,740
Aerobic exercising	33,138	1,760	2,076	2,443	3,748	7,437	6,287	9,387
Backpacking	12,281	1,325	764	1,218	1,770	2,724	1,613	2,867
Baseball	11,507	573	456	1,076	1,772	2,473	2,366	2,791
Basketball	24,410	1,816	1,078	1,852	3,702	5,069	4,739	6,154
Bicycle riding	38,139	2,433	1,894	2,529	5,266	8,321	6,859	10,837
Billiards	28,172	2,763	1,624	2,300	4,397	5,961	4,863	6,265
Bowling	44,972	3,337	2,414	3,241	6,867	10,415	8,422	10,275
Camping	50,863	4,119	2,685	3,833	7,520	13,219	8,808	10,678
Exercise walking	93,359	6,855	7,061	7,911	12,813	19,961	16,814	21,944
Exercising with equipment	57,206	2,917	2,885	4,423	8,112	12,161	10,707	16,001
Fishing (net)	32,876	2,539	2,152	2,775	5,821	7,500	5,359	6,729
Fishing (freshwater)	28,996	2,327	1,869	2,635	5,350	6,552	4,429	5,834
Fishing (saltwater)	8,195	453	435	453	768	2,056	1,765	2,265
Football (tackle)	8,890	952	1,247	881	1,335	1,640	1,282	1,553
Golf	22,317	606	675	1,078	3,061	4,614	4,589	7,693
Hiking	34,013	2,294	1,858	2,126	4,234	7,587	6,726	9,189
Running or jogging	32,212	1,189	1,784	2,033	4,340	5,787	7,970	9,109
Soccer	13,578	956	539	727	1,644	2,527	2,603	4,583
Softball	11,829	1,055	536	874	2,165	2,658	2,354	2,186
Swimming	50,226	3,171	2,313	4,125	3,442	11,031	8,918	14,227
Tennis	10,818	411	436	509	1,301	2,149	1,955	4,058
Volleyball	10,733	707	651	447	1,917	2,014	1,969	3,027
Weightlifting	34,505	2,029	2,123	2,330	5,224	6,976	7,192	8,631
Yoga	15,738	1,025	1,013	1,254	1,885	3,524	2,574	4,462

Data from U.S. Census Bureau, *Statistical abstract of the United States: 2012* fitness programs than are hourly workers (Eitzen & Sage, 2009). To the extent that educational attainment is related to economic resources, these data may also reflect availability of time and resources for participation in sport (Coakley, 2009).

Income is also linked to the types of sports and recreational activities in which people take part. Wealthy adults tend to participate in individual activities such as tennis, golf, and skiing, whereas those with lower incomes tend to play team sports such as football, basketball, and baseball (see table 6.4). This difference can be accounted for by at least two explanations. First, equipment and facility costs are so high in some sports that they don't attract people with low incomes. In addition, the working hours of higher-income professionals often fluctuate, making it more difficult to schedule regular competition with a team. In contrast, blue-collar workers tend to have standard working hours, which makes it easier to schedule regular team-sport competition (Eitzen & Sage, 2009). On the other hand, expensive physical activities that lend themselves to flexible scheduling may have more appeal to higher-income people. Examples include skiing, scuba diving, mountain climbing, hang gliding, and white-water kayaking.

KEY POINT

Socioeconomic status influences the types of physical activities to which people have access; physical activities that require expensive equipment, facilities, and coaching are mostly beyond the reach of people at lower income levels.

Because we can point to many athletes who grew up in poverty and eventually acquired enormous wealth and fame as professional athletes, many people think that talented athletes have a sure path to upward social mobility. In reality, this route is traveled by only a tiny percentage of athletes, most of whom are male. In 2004, for example, the 1,035 major league professional players in football, men's and women's basketball, and baseball amounted to only 0.2 percent of the number of high school athletes playing these sports. More specifically, in 2004, about 16,000 players from colleges and universities were eligible to be drafted as rookies in the NFL, but only about 250 ended up on a team roster (Coakley, 2009).

Leadership

The people who control elite sport are generally quite wealthy. At the top stand the owners of professional sport teams, sport media moguls, and corporate executives. Each year, *SportsBusiness Journal* reports its list of the 50 most influential people in the business of sport; the list is heavily populated by media moguls, league commissioners, and corporate giants. A similar list, including 25 heavy hitters, was published in 2016 by *USA Today* (The Big Lead Staff, 2016). These two lists are not comprehensive, and they may reflect mainstream ideas about influential persons in sport; even so, they illustrate the vast white masculinity that controls sport both in the United States and in international settings.

Today, corporate heads are being hired as athletic directors in NCAA Division I college and university programs, and their salaries and other aspects of financial compensation have to be high enough to lure them away from private business. Top coaches are also well paid. In 2006, the average NCAA football coach's salary was $950,000 per year, not including benefits, perks, and incentives. By 2011, it had risen to $1.5 million, an increase of nearly 55 percent over the span of just six seasons (Brady, Upton, & Berkowitz, 2011). Also in 2011, at least 32 coaches in the NCAA's top-tier Football Bowl Subdivision earned more than $2 million; of those, 9 made more than $3 million and 3 made more than $4 million. In 2016, Alabama's Nick Saban topped the list at more than $7 million.

The top 25 coaches in NCAA Division I basketball earn between $2 million and $6 million in yearly salary alone—that is, not counting additional perks such as private jets, housing allowances, severance packages, private endorsements deals, speaking fees, and summer camp income (McKenna, 2016). In comparison, the median salary of presidents of public universities in 2010 was about $475,000, and faculty earned an average salary ranging from $20,000 to $126,000 depending on rank (McKenna, 2016). Moreover, in at least two states, college basketball coaches were the highest-paid public employees.

Some people believe that these incomes are justified because many top college coaches are in contention for jobs with professional teams that can offer huge sums of money. Taking a different track, economist Andrew Zimbalist (2001) claims that these salaries are made possible in part by the fact that collegiate athletes are amateurs—a stark difference from professional leagues, where at least half of the revenue goes to player salaries. Thus, a kinesiologist with an emergent sociological perspective could ask, "How might collegiate sport-revenue streams change if athletes were paid more than the cost of a college degree?"

Few opportunities are available for less affluent individuals to occupy important leadership roles. In the commercial enterprises of our most popular spectator sports, the goal is to make money—either for profit (as in the case of media companies and professional sport franchises) or to improve the program (as in the case of nonprofit, big-time intercollegiate sport). People who are particularly adept at doing so—for example, winning coaches; astute collegiate athletic directors; efficient professional sport franchise managers, television executives, and production personnel; and forward-looking team owners—are usually rewarded with higher incomes or more profits. Less affluent people have few opportunities to occupy important leadership roles in this high-stakes business. As an example, most elite coaches make their professional journey through the collegiate ranks, where a graduate education is often required. This is an expense that may limit this professional occupation to people who can find educational financial support.

People with average incomes have much better chances to assume leadership roles in lower-level collegiate programs, high school athletics, and community recreational programs, such as organized youth sport and adult athletic leagues. Of course, the competitive structure of these programs often mimics elite athletics. For example, youth sport teams often take the names of major professional squads. In addition, such programs often adopt procedures from elite college and professional sports, such as cutting players from teams, staging playoffs, and naming all-star teams. People at the lowest income levels don't participate in sport as frequently and are therefore unlikely to advance to positions of leadership even in lower-level programs and leagues.

KEY POINT

Wealthy people occupy influential leadership positions in our popular spectator sports and in some physical activities. Those at the lowest levels of socioeconomic status rarely find themselves in positions of leadership.

Leaders in other types of physical activity programs—for example, noncompetitive recreational activities, instructional programs, and health-related exercise programs—also range from people who are extremely wealthy to those with modest incomes. For example, owners of health club chains, ski resorts, and golf courses are probably among the most affluent. Because these are private businesses, it is difficult to obtain specific financial information about them; in addition, as in any industry, their profitability undoubtedly varies widely. Still, people who manage or teach in such facilities are likely to come from the middle and upper middle classes. In contrast, most individuals at the low end of the socioeconomic continuum are unlikely to participate enough or have access to the financing needed to make it into the leadership ranks of such programs.

Cultural Expression

Certain sports and noncompetitive activities are sometimes used to mark economic affluence; examples include golf, sailing, skiing, and scuba diving. Belonging to a private country club or indulging in vacations at fancy sport and recreational resorts lets others know that an individual has the free time and money to enjoy luxurious physical activities.

In similar fashion, lean, well-exercised bodies often serve as status symbols in white North American society. They adorn magazine and television advertisements, model the latest designer clothes, and appear in popular movies and television shows. A thin, well-contoured "hard body" tells others that an individual has the time, money, and self-discipline to shape him- or herself according to today's difficult-to-attain standards. These ideals are readily observable in almost any magazine or movie. Having a personal trainer sets the individual off as even more affluent. Of course, most people cannot afford all the body work—exercise, dieting, beauty treatments, plastic surgery, and other procedures—needed to achieve this look. As a result, the look becomes a scarce commodity that marks a person as wealthy and stylish.

KEY POINT

Well-sculpted bodies and participation in expensive sports can mark individuals as wealthy. More generally, sport can reinforce the socioeconomic status quo through, for example, the notion that winners and losers deserve what they get.

Sport can also send other messages that support socioeconomic inequities. For example, winning is the most prevalent organizing theme in sport media stories and telecasts (Kinkema & Harris, 1998). In addition, winning is usually attributed to self-discipline, talent, and hard work; therefore, if an athlete or a team doesn't win, then we may assume that the player or team was lazy or lacked talent and thus didn't deserve to win. Such beliefs underscore the American conception of merit that links hard work and talent to financial success. The flip side is that if someone fails financially, then it must mean that she or he isn't talented or didn't work hard. This reasoning allows us to hold the belief that rich people and poor people deserve whatever money they have. The point here is not that merit is a bad idea. The problem is that this logic often leads us to overlook the societal barriers—such as poor nutrition, neighborhood gang violence, poor access to libraries and computers, dysfunctional families, and lack of child care—that hinder efforts by poor people to develop themselves to the fullest (Coakley, 2009).

Finally, our popular team sports also send messages about the importance of obedience and teamwork. These qualities are indeed valuable for most workers. Although some employers may need employees who can break out of traditional molds and be creative, most simply need people who can work well with others and follow directions. Because sport tends to reinforce obedience and teamwork, it helps maintain our current economic system.

Wrap-Up

Sociology of physical activity takes us beyond common, everyday understandings of sport and exercise. It illuminates patterns of change and stability, identifies social problems, and urges modifications aimed at enhancing equality and human well-being. So far, this subdiscipline of kinesiology has emphasized sport, but interest is growing in exercise, fitness, and societal conceptions of the human body. Although researchers have analyzed relationships and trends in social life, they have yet to completely nail down mechanistic factors underlying these trends and relationships. This circumstance is not unusual in the social sciences, because it is difficult to bring a slice of social life under the strict laboratory control needed to demonstrate exact mechanisms of social actions on patterns of physical activity.

If you are interested in what you have learned here, you might enjoy looking through sport sociology textbooks, taking an introductory course in this subdiscipline, and exploring its major journals. You might also wish to review research reports released by major research centers, such as the Institute for Diversity and Ethics in Sport at Central Florida University, the Centre for Sport Policy Studies at the University of Toronto, the National Research Institute for College Recreational Sports and Wellness at Ohio State University, and the Center for the Study of Sport in Society at Northeastern University. Sociology of physical activity provides information that kinesiologists can use to increase our understanding both of our own experiences (as participants, spectators, and professionals) and of the experiences of others. It also helps us think more clearly about the changes we would like to make in physical activity programs, as well as the things that we would like to keep the same.

Finally, sociology of physical activity invites an ongoing, critical engagement with the changing social conditions in which kinesiologists do their work. This intentional engagement leads to a question that is always relevant for kinesiologists: What do I need to know about the social, political, and cultural context for physical activity so that I do not mistakenly reinforce existing social inequalities?

Attributions of Failure: Lack of Merit or Lack of Money?

Think about the things that coaches typically tell players when a team loses a game. You can either recall your own direct experiences or talk with other people about theirs. How often did coaches tell players that they didn't work hard enough and that if they had practiced harder or given more effort during the game, then they would have won? How often did coaches tell players that they obviously tried hard but lost to a team that was just more talented? What other reasons did coaches offer for losing? Did any (or all) of these reasons lead players to feel that they really didn't deserve to win?

MORE INFORMATION ON SOCIOLOGY OF PHYSICAL ACTIVITY

Organizations

International Sociology of Sport Association
National Association for Kinesiology in Higher Education
North American Society for the Sociology of Sport
SHAPE America

Journals

International Review for the Sociology of Sport
Journal of Sport and Social Issues
Sociology of Sport Journal

REVIEW QUESTIONS

1. List and discuss the goals of sociological study of physical activity.
2. Describe the expanding research directions in sociology of physical activity from 1970 to the present.
3. List and discuss the six research methods commonly used in sociology of physical activity.
4. Describe the ties between participation in physical activity and power relationships based on gender, race and ethnicity, and socioeconomic status.
5. Describe the ties between leadership in physical activity programs and power relationships based on gender, race and ethnicity, and socioeconomic status.
6. Describe the ties between physical activity expressiveness and power relationships based on gender, race and ethnicity, and socioeconomic status.
7. Identify the contrasting beliefs underlying debate over the issue of mascots depicting American Indian figures. Which argument would a sociologist of physical activity be most likely to advance? Why?

Go online to complete the web study guide activities assigned by your instructor.

7

Motor Behavior

Katherine T. Thomas and Jerry R. Thomas

CHAPTER OBJECTIVES

In this chapter, we will

- explain what a motor behavior researcher does;
- present the goals of motor behavior, including motor learning, motor control, and motor development;
- explain the research process used by scholars in motor behavior; and
- present some of the principles of motor learning, motor control, and motor development.

Suppose that your roommate is starting a community service requirement for a campus club. The group decides to coach a youth basketball team in an afterschool program. As you and your roommate discuss how to help the young athletes learn the skill of shooting the ball, your roommate recommends using BEEF, an acronym that represents what the body should do during a free throw. However, you wonder if there is a better way to help these new players learn shooting skills? What are the instructions, cues, or practice actions that most improve shooting skill in young players?

This chapter introduces the subdiscipline of motor behavior and establishes its relationship to other scholarly subdisciplines in the discipline of kinesiology. The chapter will help you begin to understand why motor behavior holds interest for performers who want to improve their movement *expertise,* for scientists who contribute to *scholarly study,* and for professional practitioners who seek to help others develop their motor skills.

Why Use Motor Behavior?

Have you ever wondered why you enjoy certain physical activities and not others? Often, being skilled at an activity helps us enjoy it. If you pitch a great fastball, run a quick 5K, boast a killer volleyball spike, or hit a terrific backhand in tennis, then you are more likely to enjoy these activities. Someone who possesses less skill in these specific

Goals of Motor Behavior

- To understand how motor skills are learned, how processes such as feedback and practice improve learning and performance of motor skills, and how response selection and response execution become more efficient and effective
- To understand how motor skills are controlled, how the mechanisms in response selection and response execution control the body's movement, and how environmental and individual factors affect the mechanisms of response selection and response execution
- To understand how the learning and control of motor skills change across the life span, how motor learning and control improve during childhood and adolescence, and how motor learning and control deteriorate with aging

activities may find them frustrating and less enjoyable. Of course, there are notable exceptions; for example, some of us spend many hours and a great deal of money demonstrating that we have little skill at golf when compared with elite golfers—or simply par!

Think about a skill you like and do well—for example, riding a bike. How did you learn to ride a two-wheel bicycle? What types of practice experiences worked best for you? How did you learn to coordinate the movements of your feet in order to pedal, your hands in order to steer, and both while balancing? Do you remember early times when you moved slowly and wobbled, could not turn quickly, needed help starting, and often fell rather than stopping in a controlled manner? In what ways did your brain and nervous system develop and adjust so that you could improve your control and coordination?

Practice is an important factor in learning skills across the life span. Most people improve motor skills yet never understand how the nervous system adapts, how it develops or controls movement, or how best to use practice in order to improve performance. The study of motor behavior focuses on how skills are learned and controlled and how movement changes from birth through the end of life.

People generally believe that practice improves performance; after all, we see evidence of this notion at home, in school, and in the community. For example, parents urge children to practice playing the piano, athletes ask coaches for help in practicing sport skills, and students perform practice exercises to learn keyboard typing. Clearly, however, some forms of practice are superior to others. Thus, important questions about practice include *what, how,* and *how much*. Another important question is, "What can be changed with practice?" For example, is an expert better because of practice, or was she born with special talent? Research in motor behavior tells us that parents and teachers are partly right when they say, "Practice makes perfect." More precisely, they might have said, "*Correct* practice makes perfect"; as research tells us, the effectiveness of practice is influenced by many factors.

In the chapter-opening scenario, is the roommate's idea the best approach for learning the set shot? The BEEF cue directs the learner's attention to internally focused feedback, or what the performer does. In this case, that means focusing on the key words represented by the acronym: balance, eyes, elbows, and follow-through. A better approach for young players might be to use externally focused feedback: hold the ball like a waiter holds a tray, aim for a spot just above the rim, shoot the ball like it is going over a volleyball net, and make the ball spin backward (Perreault & French, 2015). Whereas the BEEF approach focuses on the player's body, the externally focused feedback deals with the ball. The consensus of motor behavior research indicates that externally focused feedback produces better long-term performance in elementary-age children. This sort of understanding is particularly important in sport, where we want to see permanent and positive changes—for example, changes that carry over from practice sessions to game play. Some of the motor behavior research on practice is summarized later in this chapter.

Teachers and coaches should be general experts on practice, because much of their training in kinesiology focuses on how to organize and conduct practice. However, we all may encounter situations that call for us to apply knowledge about practice and skill acquisition. Consider these examples:

- Your tennis partner asks you to watch her serve and advise her about how to avoid so many faults.
- You volunteer for a youth football league, where the topic of discussion is how best to group kids for fair competition.
- You drive to school every day and realize that on many days you do not remember a single event during your drive; you also realize that countless other people are driving in the same scary way.
- You work part-time at an assisted living center, where you see that many residents have difficulty standing up from a sitting position, balancing, and grasping small objects. You wonder why this is so and whether they can regain these skills.
- You and your dad are watching a professional baseball game and wonder how the first baseman caught the line drive and threw perfectly to second base for the double play. How would you explain some players' ability to attain such a high level of expertise?

You have probably encountered similar situations involving movement learning and control that you couldn't immediately explain. Although most of us can talk on a superficial level about motor skill performance—"Did you see how long he was off the ground? He floated through space to dunk that ball!"—few people really understand how a person controls, learns, and develops motor skill over time. The study of motor behavior provides answers to many of these questions and helps us understand the process of the development, control, and learning of motor skills (Thomas, 2006; Ulrich & Reeve, 2005). Understanding this process will allow you to better answer such questions and plan more effective practices.

College students may see themselves merely as students in a class rather than as everyday scholars engaging in more systematic scholarly study in kinesiology. Remember, however, that you have already accumulated a large body of experiential and embodied knowledge (see chapter 3), which you can use to learn about motor behavior. For example, you may recognize some of the theories and research methods used in the study of motor behavior from an introductory, developmental, or experimental psychology course. The study of motor behavior began as a branch of psychology that used movement or physical activity to understand cognition. The subdiscipline of motor behavior also incorporates information from biology and zoology because physical activity is also affected by factors such as heredity, aging, and growth (e.g., Thomas & Thomas, 2008). Researchers also apply principles and laws from physics to the study of humans in motion. Thus, what you know about physics, biology (or zoology), and psychology will help you understand motor behavior.

In addition, you have some applied experience with motor behavior. Remember your physical education teachers or coaches asking you to practice passing a volleyball, swinging a golf club, or dribbling a soccer ball repeatedly? Perhaps they gave you feedback about your practice, such as the following: "Bend a little more at the knees." "Follow through with a smooth stroke." "Keep the ball as close to your feet as possible." In doing so, they were implementing principles of motor behavior—specifically, that correct practice and appropriate feedback improve performance. Of course, some of us can recall adults who provided less helpful information, such as "practice harder" or "hit the ball." The study of motor behavior guides us in providing better situations for learning and practice, as well as understanding why some cues and feedback work better than others.

The subdiscipline of motor behavior is part of the behavioral sphere within the study of physical activity, along with psychology of sport and exercise. As people practice physical activity, they *experience* the essence of motor behavior by trying to control and learn movements. Scholars of motor behavior *study* how motor skills are learned, controlled, and developed in order to assist people as they practice and experience physical activity. The knowledge generated by motor behavior research is put into practice by professionals, such as coaches and teachers, who try to improve individuals' physical skills. Other examples include gerontologists who help improve the motor skills of people who are

Motor Learning Theory in Practice

Think about a time when you were injured and could not perform a skill in the way that you usually do. For example, perhaps you sprained an ankle and had to use crutches to walk. In such a situation—when you can use only one leg or need to minimize sideward forces that create pain—a simple sequence such as trying to go from standing to sitting and back to standing becomes a different kind of challenge. Over time, you probably got better and more confident at moving in the "new" way. In other words, practice improved your performance.

Once you healed, however, you could go back to walking, sitting, and standing in the way you had done before the injury. This reversion is possible because the sit-to-stand skill was so well learned that we refer to it as "overlearned." Other skills that you did not practice while injured may also have returned quickly to their former level once you were able to use the sprained ankle. These skills were learned and retained. You were even able to adapt some skills, such as walking and stair climbing, during the recovery period (e.g., marking time—leading with the same foot going up and down the stairs to reduce stress on the injured ankle and foot). This constitutes a form of transfer.

Thus, we see the motor learning concepts of practice, transfer, retention, and learning in our everyday lives. Unfortunately, injured people sometimes develop dysfunctional movement strategies and adaptations in strength and flexibility that are difficult to retrain. The resulting challenge of relearning functional skills can be addressed by kinesiology professionals who use motor behavior knowledge in therapeutic exercise careers.

In school settings, for well-learned skills, teachers may begin by having students use a skill in a game or gamelike situation. In such cases, instruction, demonstration, and feedback are less likely to include information and skill correction and more likely to focus on strategy.

elderly, physical therapists who help injured patients with rehabilitation, and athletic trainers who work to prevent injuries and rehabilitate injured players.

Kinesiology is the multifaceted study of human movement, and motor behavior is one of its spheres. Each of the subdisciplines contributes to our understanding of human movement; some are more closely related than others based on the methods or applications. Because motor behavior research must document technique and performance variables, motor behavior is also closely associated with biomechanics from the biophysical sphere of study. Motor behavior research also uses some of the same biomechanical research methods and equipment. The knowledge from motor behavior is applied with biomechanical knowledge in the treatment of disabilities and injuries and to the observed changes in movement across the lifespan. Movement skills change predictably during infancy and in old age; biomechanics and motor behavior research quantify these changes as demonstrated later in this chapter. Motor behavior produces knowledge that is unique and foundational in some applications. For example, the pedagogical areas of physical education and adapted physical education (for individuals with disability) should also be based on what we know about how people learn and control movements. These and other physical activity professions use knowledge about motor behavior to design instructional programs and evaluate motor skill. For example, a physical education teacher may use motor behavior information to plan a series of instructional classes for 6th graders to improve soccer dribbling techniques.

What Do Motor Behaviorists Do?

Scholars who study motor behavior are most often employed at universities, where they teach and do research and service, but some scholars work in research facilities unassociated with universities. Motor behavior research can be conducted in a university laboratory, in a clinical setting (e.g., a hospital), or in an industrial or military setting. Here are a few examples: A military researcher might investigate which training methods best improve fighter pilots' reactions to electronic threat signals. A motor behavior researcher working in industry might look for the optimal method for training workers to assemble a product. Medical and educational researchers often study motor learning, motor control, and motor development. They might investigate, for instance, how the nervous system changes with age in terms of controlling movement, what the best methods are for teaching rehabilitation protocols to patients in physical therapy, how the game performance of youth soccer players can be improved, or what causes movement deterioration in Parkinson's disease.

The duties of motor behaviorists at universities typically include research, teaching, and service. Their research may focus on learning, control, or

Contrasting Motor Control and Motor Learning

Do you remember learning how to drive a car? What was one task that was difficult for you to master (e.g., braking smoothly, signaling, parallel-parking)? Now think about the last time you drove somewhere familiar—for example, to school today. Sometimes we cannot remember anything about a familiar drive. Consider how your skill has changed as a result of learning to drive and then continuing to practice. When you were learning to drive, you had to think about every movement—moving your foot from the accelerator to the brake, using the turn signal, turning the steering wheel. Later, you thought only about the traffic and where you were going. Now, on an often-traveled route, you monitor the traffic with intermittent attentiveness.

The study of motor learning helps us understand how we learn skills such as driving a car so well that the skill becomes automatic. What other skills have you learned so well that you do not think about them either during or after the performance? In contrast, are there situations in which you do have to think about a well-learned skill? For example, when traffic chokes the road or a storm occurs, you must pay more attention to driving. When task demands change, we sometimes move from automatic control of movements to conscious control.

Thus, you can see that motor control and motor learning are related. Motor control is essential for every movement—from poorly skilled to well skilled. Motor learning, on the other hand, is responsible for the shift from poorly skilled to highly skilled movement. One way to judge when a movement is well learned or highly skilled is to recognize that the movement is automatic, or executed without conscious control. Because we may have less practice at driving in storms and heavy traffic, that type of driving is not well learned and therefore not automatic.

development. Their teaching duties may include courses in motor behavior or related courses such as biomechanics or sport psychology, research methods, measurement and evaluation, and pedagogy and youth sport. Service for a university faculty member may include evaluating motor disorders, managing a program for individuals with motor disorders, and conducting workshops or clinics about motor disorders. Whatever the setting, most motor behavior researchers also write grant applications and perform professional service, such as reviewing manuscripts for scholarly journals.

The subdiscipline of motor behavior produces knowledge of how people achieve motor skills across the life span. The subdiscipline involves three areas: motor learning, motor control, and motor development (a developmental view of motor learning and motor control). Motor learning research deals with the acquisition of skilled movements as a result of practice, whereas work in motor control seeks to understand the neural, physical, and behavioral aspects of movement (Schmidt & Lee, 2011). Both are often evaluated across the life span, resulting in a developmental view of how motor skills change.

The study of motor learning and the study of motor control are not distinctly different areas, but they tend to ask different questions. The goals of studying motor learning include understanding the influence of feedback, practice, and individual differences, especially as they relate to the retention and transfer of motor skill. The goals of studying motor control, on the other hand, include understanding how to coordinate the muscles and joints during movement, how to control a sequence of movements, and how to use environmental information to plan and adjust movements. In addition, scholars are often interested in exploring how motor learning and motor control vary across age groups, from children to senior citizens.

When you think of motor behavior, you may think of sport skills. However, consider the many other types of movements that people use in their daily activities:

- Babies learning to use a fork and spoon
- Dentists learning to control a drill while looking in a mirror
- Surgeons controlling a scalpel and microsurgeons using a laser while viewing a magnified video image of the brain
- Children learning to roller-skate
- Students learning to keyboard
- Teenagers learning to drive
- Dancers performing carefully choreographed movements
- Pilots learning to control an airplane
- Young children learning to control a crayon when coloring

All of these activities, and many others, involve motor behavior and thus hold interest for researchers and practitioners. Thus, the goals of motor behavior study are important not only in sport but also in the overall field of physical activity. As we saw in part I, understanding the learning, control, and development of movements plays an essential role in our culture and society. For example, parents' first clue that their baby is developing normally may be a reflex or motor milestone, such as grasping a finger rubbed across the palm (the Palmar grasp reflex) or taking the first steps around 1 year of age. Individuals have probably been asking questions about motor behavior since the beginning of time; scientists have been seeking the answers to those questions for more than 100 years.

History of Motor Behavior

Research has a long history in all three areas of motor behavior—motor learning, motor control, and motor development—and the focus of research in each area has changed dramatically (for a historical review, see Thomas [2006] or Ulrich and Reeve [2005]). For example, in the late 1800s and early 1900s, researchers were primarily interested in studying motor skills as a means to understand the mind; in other words, motor skill was viewed as a way to examine cognition (Abernethy & Sparrow, 1992). Adams (1987) identified five themes from this early work that have persisted through the years: knowledge of results, distribution of practice, transfer of training, retention, and individual differences. These five themes provide the foundation for the study of motor learning. Although these themes persist, researchers are now focusing on the motor skills themselves rather than simply studying them in order to understand cognition.

Similarly, motor control research can be traced to research in the late 1800s on the "springlike" qualities of muscle (Blix, 1892–1895). Other early research included Sherrington's (1906) seminal work on neural control, which is still useful in explaining how the nervous system controls muscles during movement. In both cases, the purpose was to understand not motor behavior but biology.

Great interest was taken in motor behavior research during the World War II era from 1939 to 1945 (Thomas, 1997) because the military needed

Kinesiology Colleagues

Photo courtesy of Randall Williams

Randall Williams

Randall Williams is completing his third year in the doctoral physical therapy program at the Fort Worth Health Science Center. His program of study began with a bachelor's degree in kinesiology and a minor in nutrition at the University of Houston. After volunteering as a subject in a research study and being inspired by his faculty mentors, he felt motivated to earn a master's degree in kinesiology, which he did at the University of North Texas. Now, by applying the subdisciplines of kinesiology in advanced study in physical therapy, he has found his passion—to help patients affected by neurological deficits achieve their life goals.

to select and effectively train pilots (Adams, 1987). Thus, motor skills were viewed as a necessary component in military efforts, but not because understanding motor skills was an important area of study in itself.

Beginning in the 1960s, and with increasing momentum in the early 1970s, the study of motor behavior evolved as a scholarly subdiscipline of kinesiology. With this change, the scholars doing motor behavior research were no longer neurophysiologists or psychologists; they were specialists in physical activity. Franklin Henry's paper on **memory drum theory** (Henry & Rogers, 1960) may have been the first major theoretical or landmark study (Ulrich & Reeve 2005) from the discipline of kinesiology (called *physical education* at the time). Henry's theory stated that reaction time was slower for complex movements because those movements took more planning time. In this view, for example, movements with several segments—involving moving from one position to a second position and then to a third position—required a longer reaction time than did single-segment movements, because the brain required more time to specify the needed information. The current work on motor programs (representations in the brain of plans for movement) evolved from Henry's memory drum theory.

Current research in motor control and motor learning usually focuses on understanding how the neuromuscular system controls and repeats movements. One purpose of this research is to understand and develop treatments for conditions such as Parkinson's disease and spinal cord injuries; another is to improve performance in sport and physical activity. In some cases, technological advances combine with motor control to allow patients to use complex prostheses and to allow quadriplegics to operate robotic arms with their brains. This cutting-edge technology is based on motor control theory and research.

KEY POINT

The subdiscipline of motor behavior continues evolving in response to research, technological developments, and theoretical models. The study of motor learning and the study of motor control are distinct yet both seek to understand human movement.

A developmental approach to motor learning and motor control (also known as *motor development*) originated in developmental psychology and child development. The study of motor development grew from "baby biographies" (many written before 1900) describing changes that occur in the reflexes and movements of infants. The early work used twins to establish the role of environment and heredity in shaping behavior (Bayley, 1935; Dennis, 1938; Dennis & Dennis, 1940; Galton, 1876; Gesell, 1928; McGraw, 1935, 1939). As with motor control and motor learning, the initial research in motor development took place not because scholars were interested in motor skills per se but because they were using the study of motor skills to understand other areas of interest.

In the 1940s and 1950s, developmental psychologists lost interest in the developmental aspects of motor skill. Study in this area might have ended if not for three motor development scholars—Ruth Glassow (Slone, 1984), Larry Rarick (Park, Seefeldt, Malina & Broadhead, 1996), and Anna Espenschade (1960). Their emphasis, which differed from that of the developmental psychologists, focused on how children acquire skills—for example, how fundamental movement patterns are formed and how motor performance is affected by growth. These three scholars maintained the focus on the developmental nature of motor learning and motor control in their research through the 1950s and 1960s.

Just as research in motor learning and motor control increased around 1970, developmental research addressing the questions of motor learning and motor control also became popular (Clark & Whitall, 1989; Thomas & Thomas, 1989). Motor development was considered part of the subdiscipline of motor behavior because it studied the same topics, in this case developmentally. However, two research themes in motor development continued from the years before 1970: the influence of growth and maturation on motor performance and the developmental patterns of fundamental movements. Growth influences the performance of motor skills in children over time; however, at the cessation of growth (e.g., maturity), growth is no longer a factor. Because growth cannot explain all improvement in motor performance during childhood, developmental scientists have turned to motor control, motor learning, and biomechanics for more information. Thus, we now see three lines of motor development research: motor learning in children; motor control in children; and the influence of growth on motor learning, motor control, and performance.

Research Methods for Motor Behavior

As you may be aware, research begins with a question stated as a hypothesis. The question helps the scientist decide who will participate in the research project, what type of task the person will do, and what the researcher will measure in order to answer the question. This process is used in most kinesiology research.

It is a complex endeavor to develop understanding of how movements are learned, how they are controlled, and how they change across the life span.

Professional Issues in Kinesiology

Expertise: When to Specialize in a Sport?

Research on motor expertise often compares novices with experts in order to determine how they differ in various sports. The study of motor expertise is an appropriate recent addition to the study of motor development because age and experience are associated with increases in expertise. The study of sport experts has brought forth new thinking about the age at which a young person should begin to specialize in one sport (Côté, Lidor, & Hackfort, 2009) and what benefits may result from engaging in more varied experiences during childhood and adolescence. Data on elite performers and sport dropouts suggest that participating in more than one sport per year is beneficial up to 15 years of age. At about that age, it is appropriate to begin specializing in a sport; the exception occurs in sports where peak performance comes before physical maturation, such as women's gymnastics. Thus, at about 15 years of age, athletes in most sports decide to continue recreationally (e.g., just for fun) or specialize with a goal of elite performance.

Early specialization has been influenced by two notions—first, that it is associated with expertise, and second, that 10,000 hours of practice are necessary. However, the work done by Côte and colleagues (2009) presents compelling evidence that for most athletes sports specialization should not be encouraged until after 12 years of age and as late as 15 years of age. Furthermore, a more accurate target for accumulated practice time to reach expert levels of performance would be 3,000 to 4,000 hours as opposed to 10,000 hours.

Citation style: APA

Côte, J., Lidor, R., & Hackfort, D. (2009). ISSP position stand: To sample or to specialize? Seven postulates about youth sport activities that lead to continued participation and elite performance. *International Journal of Sport and Exercise Psychology, 9*, 7-17.

In fact, in order to answer one question it is usually necessary to perform many experiments. Motor behavior researchers have concentrated on techniques for measuring movement speed and accuracy. Motor control and motor learning researchers use technology similar to that used by researchers in biomechanics. During the past 50 years, technology (e.g., computing, high-speed imaging, **electromyography [EMG]**) has been advanced and adapted to control the testing situation and precisely record and analyze movements for motor behavior studies. Moreover, noninvasive electronic measurement technologies have allowed the use of real-world movements instead of simple movement tasks invented in the laboratory just for research purposes. Motor behavior courses often include laboratory experiences so that students like you can repeat the classic experiments and theories that you study. This type of hands-on learning has demonstrated benefits to learning (Hofstein & Lunetta, 1982).

Types of Studies

There are many ways to answer any particular research question. Motor behavior research frequently uses three experimental designs or techniques: between-group, within-group, and descriptive. The between-group design compares two or more groups exposed to different treatments (interventions) but tests them using the same task. For example, a researcher could use this design to answer the question, "Does practicing a simple movement increase its speed?" This research design compares two groups (randomly formed) on the same task (movement speed). The treatment in this example is practice: If two groups practice differing amounts, will their performance on the task be different? In the second design, the within-group design, all participants are exposed to two or more different treatments and are tested on the same task.

Let's consider an example that could be conducted with either design. Suppose, for example, that you want to study whether a participant's **reaction time** (how quickly the movement begins after a signal) varies with the size of the target. You could tell the participant that after you say "go," she is to move a stylus as rapidly as possible to a target 30 centimeters (12 inches) away; the target is a circle either 2 or 4 centimeters (0.8 or 1.6 inches) in diameter (see figure 7.1). The question is whether the time between hearing the signal and beginning the movement (i.e., the reaction time) changes depending on the target size. In a between-group design, one group could move to the 2-centimeter target and the second group to the 4-centimeter target.

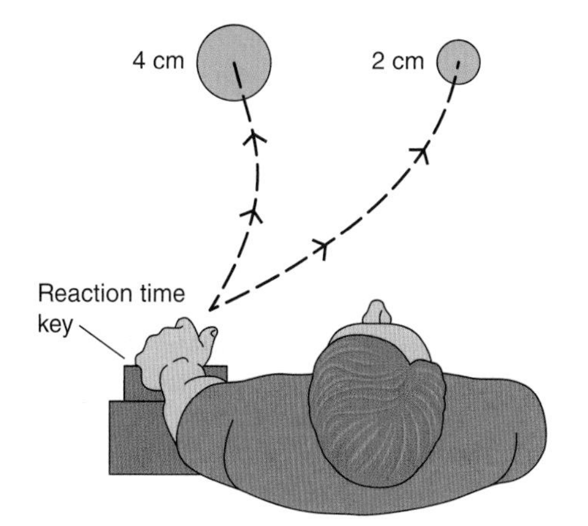

FIGURE 7.1 A reaction-time study might measure how fast a movement occurs after a signal and whether the movement is affected by the size of the target object.

In a single-group (within-group) design, half the participants could move to the 2-centimeter target first and then to the 4-centimeter target, and the other half could move to the 4-centimeter target first and then to the 2-centimeter target. In deciding which design to use, an investigator should consider whether changing target sizes interferes with performance. If so, then a between-group design is preferred; if not, then the investigator will use a within-group design because he or she will not have to test as many participants.

The third type of experimental design used in motor behavior research is descriptive research. Here, the investigator measures or observes participants performing a task. Sometimes the investigator observes the same participants several times in order to trace changes. For example, an investigator might measure reaction time in the same children when they are 4, 6, and 8 years of age in a longitudinal design. In other cases, the research compares the performance of different groups in a cross-sectional design; for instance, 4-, 6-, and 8-year-old children might be tested for reaction time. Researchers often use this cross-sectional research technique to describe age differences or differences between experts and novices. Descriptive research differs from research using the first two techniques because in this approach the participants receive no treatment.

Studying the Early Stages of Learning

Motor learning research has often used **novel learning tasks** to provide certainty that no participant has tried the task before. This approach eliminates the advantage that some participants might have because they have practiced the task before the experiment. Novel learning tasks tend to be simple—so simple, in fact, that in order to make the task challenging the participant is sometimes asked to wear a blindfold (see figure 7.2). Because the task is novel, participants are unlikely to have done it before; because it is simple, participants are able to master it in a short time; and because vision is occluded, learners must rely on what they remember about the movement (e.g., speed, distance, beginning and ending points). Vision is the dominant sense, and when vision is present (and not occluded), the other types of information about movement must be controlled.

You might question whether the results of these experiments can be applied to real-world, complex movement situations. This is a good point! These simple tasks do allow us to study improvement and have helped us understand a great deal about how movements are learned. However, using simple tasks in research limits what the researcher can learn. One reason is that this approach studies the outcome of the movement (the product) rather than the nature of the movement (the process) itself (Christina, 1989). Novel learning experiments are not helpful for researchers interested in physical activity or sport tasks in which performers have had thousands of trials (e.g., keyboard typing, baseball batting).

FIGURE 7.2 Linear positioning task with blindfolded participant.

KEY POINT

Before the early 1980s, motor behavior research often used simple, novel tasks to study early skill learning; such research helped us understand how beginners learn new motor skills.

Studying Expert Performers

To address the limitations of studying novel learning tasks, some scholars have focused their studies on experts. These investigators ask, "What do expert performers do during practice and competition?" One way to answer the question is to compare expert performers with novice performers and evaluate how they differ in terms of perception and knowledge—particularly, decision making, skill, and game performance variables. In these studies, expertise is established according to a criterion such as national team membership or ranking in the sport.

The knowledge and skills of a sport are often unique to that sport (Thomas & Thomas, 1994); for example, knowing about and being skilled in tennis does not help one play soccer. Knowledge does provide an early benefit in sport; for example, the game procedures of tennis and badminton are similar and so a tennis player playing badminton for the first time would have some understanding of the game goal and order of operations. However, the techniques of these two sports may interfere with play in the other sport. Evidence exists for transfer of pattern recognition, a cognitive skill, in sport experts (Abernethy, Baker & Côté, 2005). In another study, video recordings of badminton players were used to determine how age and expertise influence players' ability to predict where a birdie struck by an opponent would land. Participants looked at a video of a badminton player hitting a shot. The video had been altered (e.g., by erasing the head, arm, or racket or by zeroing in on certain motions), and in some cases information was erased that was needed for accurately predicting where the birdie would land. As you might guess, expert players could make better, faster predictions with less information than could novice players. Experts also looked at different body parts than novices did and therefore based their decisions on different sources of information. Their advantage seemed to derive from playing experience rather than age.

You can see how this finding might help a coach, teacher, or player improve badminton play. It also helps us understand something about skill learning because researchers have found similar patterns in regard to many sport and motor skills. This type of experiment would be nearly impossible to do in a motor learning class because the students might not represent the range of expertise necessary (novice to expert).

Measuring Movements

The tasks used in motor behavior research provide a number of ways to measure movements and their outcomes. For example, suppose that senior citizens perform a movement that involves reaching 30 centimeters (12 inches) and then grasping and lifting containers of different sizes and weights. We could simply count the number of times that each senior citizen successfully lifts a container (in this task, reaching and grasping are prerequisites to lifting) and use the number of successful lifts as an outcome measurement. By doing so, we might find, for instance, that the number of successful lifts decreased as the objects got smaller or heavier. Although that would be an interesting finding, our long-term goal would probably be to understand why or how the size or weight of the object influences success. Therefore, we would probably need to examine the process of the movement.

To examine process, we could use high-speed video of the movement, taken with two cameras, and evaluate how movement differs for 55-, 65-, and 75-year-old participants. From such videos, researchers can measure **kinematics** (location, velocity, acceleration—see chapter 9 on biomechanics). Other measures include muscle activation (EMG) of the arm and hand during the movement, the pinching force between the forefinger and thumb during the grasp, error in the movement, speed of the movement, reaction time before the movement begins, and accuracy of the movement. These process measures might help us understand how the reach, grasp, or lift changed for the various weights and sizes of objects.

Using our hypotheses (educated guesses), we could then begin to answer our original question. For example, we might hypothesize that the reaching and grasping phase would be the same for objects of the same size, even when weight varied, but that the lift would be different when the weight changed. Do you think reach, grasp, or lift would be different as we changed size but kept the weight the same? If we predict that reach or grasp changes, then it is critical to measure the process.

Characteristics of Movement Tasks

In addition to deciding what to measure, motor behavior scientists must consider the characteristics of the movements they study (Gentile, 1972). For example, some movements are more continuous (e.g., performing a gymnastics routine, riding a bicycle), whereas others are more discrete, involving a short period with a distinct beginning and end (e.g., striking a ball with a bat). Furthermore, some movements are more open in character, whereas others are more closed. One example of an open movement is hitting a thrown ball; the environmental characteristics change from trial to trial because the batter must respond to the speed and location of the oncoming ball. In contrast, a closed movement takes place in a more consistent environment, in which the performer tries to do the same thing each time (e.g., as in archery or bowling).

Researchers and teachers must identify the characteristics (open or closed, discrete or continuous) of the tasks that they study for two reasons. First, the results of two studies may differ if the task characteristics differ. Second, the characteristics of the task used in the study must match the questions asked. For example, if you want to know why something happens in bowling, you should not use an open, continuous task to answer the question. Teachers must also know how to draw distinctions between task characteristics because skills of one type may require a different instructional strategy than skills of another type.

Expertise: More Than Good Stats

Suppose you want to research the talent of the players on a women's softball team. You have the conventional data about each player, such as batting average, RBIs, home runs, and fielding percentage. You suspect, however, that these data do not always reflect the contributions that players make to the team. You also have video recordings of five games played by the team, taken with a wide-angle camera that shows the complete field. What kind of data could you get from the video recordings to answer the following questions?

- How many times did each player hit the ball hard while at bat?
- How many times did each player make a special effort that doesn't show up in the scorebook (e.g., backed up a throw, saved a bad throw, moved up a runner)?

Measuring Learning and Transfer

The idea of learning—as determined by retention and transfer—is related to the requisite number of hours of practice in the previous discussion about experts. That is, a motor skill can be examined at any point, from the first attempt to well beyond mastery. The goal of most practice is learning, and learning involves long-term change in performance; therefore, in motor behavior, learning is usually defined as performance in retention and transfer trials scheduled after practice sessions. Retention is measured by performing the task after a time without practice to determine recall. Transfer can be tested immediately to determine how practice on one task or in one context affects performance on a different but related task or in a different context. One example that holds meaning for most students relates to the three most threatening words in college: "comprehensive final exam." At the end of the semester, how much do you remember when taking a comprehensive final examination? The purpose of the examination is to determine whether you learned what was taught in the course. For those who have really learned something, a comprehensive final examination is not a problem. As you will see in your motor learning class, retention is also used to measure learning in motor skills.

Learning can also be measured in terms of transfer to other skills, in which one must do a slightly different version of the task. Experts are better at transferring information or skills than are novices; experts can also perform well after periods without practice (referred to as *retention intervals*). Many coaches feel frustrated when a team learns a play during practice but has trouble executing it in competition. This situation illustrates the lack of transfer; that is, it suggests that the team did not really learn the play since it could not execute when faced with new defensive pressures. The transfer in the practice-to-game example is a change of context by the addition of defense, speed, and other factors that were not present during practice. One example of a task that is well learned based on both retention and transfer is riding a bike. You can ride successfully after a winter off (retention), and you can negotiate new routes and terrain (transfer).

KEY POINT

To study motor skill acquisition, researchers must also study how well skills are retained and how they transfer to other, similar situations.

Motor learning research tells us the reason for retention and transfer. Once you learned to ride a bike, you continued to practice until the skill was automatic and "overlearned." Getting students to this point is the goal of comprehensive final examinations. By now you should also understand that performance in retention and transfer are the goals of most genuine efforts at learning. Therefore, the variables that affect retention and transfer are important to motor learning. Confusion can arise from the fact that everything we see in ourselves is our performance, but *some* of what we see is also learning. Performance can be influenced by variables that do not influence learning. Still, performance variables can hold interest for motor learning students and scholars, as you will see later in this chapter.

Overview of Knowledge in Motor Behavior

Now let us look at what motor behavior researchers have learned from years of study. Our goal is not to cover all of motor behavior but to present six major principles that are well supported in the motor behavior literature. These principles are based on research using the measurement and methods outlined in the previous section; they also represent evidence-based application of the field in many kinesiology careers. These principles come from the goals of motor behavior presented earlier in the chapter. In motor control, research and theory about development and learning were formed around a concept of how the brain and central nervous system work to create movements. We briefly describe that concept before presenting the principles.

To help scientists conceptualize the brain as the master controller in planning, organizing, selecting, and controlling movements, a model called *information processing* was adopted. This model represents the motor behavior system as a computer. In this view, general commands are sent from the brain (the central processing unit, or CPU) through the spinal cord (the wiring), which probably reduces the complexity of the information into relatively simple commands, and on to the muscles or muscle groups (the printer or screen). (If your brain were really a computer, then when you accidentally touched a hot burner on a stove, a message would pop up asking "Are you sure?" before you could pull your hand away!) From this perspective, the goal of motor behavior is to explain response selection (how the skill to be used is selected) and response execution (how the selected skill is performed).

Motor Learning

The two principles we have selected to represent motor learning hold that (1) correct practice improves performance and supports learning and (2) augmented feedback enhances practice and thereby learning. These principles were selected for several reasons. First, the variables they emphasize—practice and feedback—are treated at length in numerous motor learning textbooks (e.g., Fischman, 2007). Furthermore, these are likely the two most widely studied variables in motor learning, and scholars agree (Schmidt & Lee, 2011) that they are the two most important variables in motor learning. Indeed, two of our favorite historical adages for motor learning are "practice is a necessary but not sufficient condition for learning" and "practice the results of which are known makes perfect." Both of these evidence-based truths indicate that practice and feedback are independently important variables, and the relationship between them is also important.

The adage "You have to crawl before you can walk" is a good way to think of skill acquisition—an orderly process that sometimes leads to falls!

KEY POINT

Motor learning is an internal state that is relatively permanent; practice is required in order for it to occur, and it is difficult to observe and measure. Thus, transfer and retention tests are used to measure learning.

Consider why these variables are important for you. Assuming that transfer occurs from one learning situation to another—for example, from learning a motor skill to learning the content in anatomy—understanding how to practice for learning could be helpful. As you move toward a career, even if you are not a teacher, you will no doubt be called on to help someone learn something, and understanding how to do so will be to your advantage. For example, you might need to lead an exercise class or teach an employee how to do his or her job, or you might want to help your child as he or she learns a new skill. Physical therapists and occupational therapists also teach clients to perform tasks. Thus, understanding how we learn motor skills is critical in many fields.

The skill acquisition process constitutes an orderly progression. The learner begins by making many large errors while trying to understand the task. The cognitive demands are great in the early stages of learning; in fact, the task may be more cognitive than motor (Adams, 1987; Fitts & Posner, 1967; Gentile, 1972). With practice, the errors become more consistent; that is, rather than making a different error on each trial, performers make the same errors repeatedly. At this point, the demands are less cognitive and more motor oriented, and the errors are smaller and less frequent. Thus, response execution is improving. Once the learner can execute the skill with fewer and smaller errors and no longer has to think about the skill while performing it, then the skill is considered learned or automatic. Now, instead of thinking about what each body part is doing (response execution), the performer can think about strategy or about the opponent when deciding which response to select and execute.

Recall that the study of motor learning is an effort to explain and predict conditions that will make skill acquisition easier or faster and make learning more permanent. Such conditions include individual differences between learners, such as speed of movement and coordination. Task differences are also important conditions in skill acquisition, because tasks may be either more open (e.g., batting) or more closed (e.g., bowling). Learning may also be affected by environmental conditions, such as practice, **feedback (intrinsic and extrinsic)**, and transfer. Another way of thinking about the changes in movement during learning is that the portion that is automatic increases; that is, more of the movement becomes preprogrammed (Albers, Thomas, & Thomas, 2005). At the same time, less of the movement remains under sensory control; that is sensory information exerts less influence on how the movement is performed. More about this idea is presented in the section on motor control. The idea of automatic or learned skill is especially important in the study of experts because automatic performance allows the expert to focus on the strategy and not the execution of the skill.

At this point, you might wonder how we know when something is learned. Clearly, that is a challenge for motor learning research, theory, and application.

Whereas learning involves the relatively permanent acquisition of a skill, performance involves the degree to which someone can demonstrate that skill at any given time. In other words, performance is the current observable behavior—a snapshot of what the learner is doing right now. Sometimes performance reflects learning, as when a player can demonstrate a newly acquired skill. At other times, it does not. For instance, most of us have had the experience of turning in an examination and then remembering an answer that we knew but were unable to put on the paper during the exam. We would argue that we had learned the material but just could not produce it for the examination! In such a case, we are saying that performance does not represent learning. Thus, a single measure of performance may not indicate much about learning without other performance measures over time.

One way to distinguish between performance and learning variables is to remember that performance variables exert a temporary effect whereas learning variables exert a relatively permanent effect. Knowing the difference between these effects is a critical part of motor learning. For instance, you might have trouble typing a term paper after working late into the night and getting very tired. Yet the next day, after getting some rest, you might be able to type rapidly while making few errors. In this case, fatigue depressed your performance, but you had learned the keyboarding skills through practice and could demonstrate this learning once you were rested.

KEY POINT

Because motor performance can be influenced by variables such as fatigue, the best measure of motor learning involves tests of retention and transfer.

Because performance does not always reflect learning, researchers prefer to measure learning with retention and transfer tests (Christina, 1992; Magill & Hall, 1990). In sport, retention refers to how much of a skill a performer can demonstrate after a period of no practice. One example of retention is being able to type after a vacation during which you didn't type; another example is being able to swim at the beginning of the summer after having learned to swim during the previous summer but not swimming during the winter. Transfer tests, on the other hand, require that you use the information in a different way from how you originally practiced the skill—for example, using a principle from physics to solve a biomechanical problem. Transfer can be manifested in various ways—for example, transferring skills learned in practice to a game, or using experience in throwing a baseball when throwing a javelin for the first time.

Practice

Practice, of course, involves repetition, but there is more to understanding its importance than mere repetition. More practice is usually associated with better performance and is required for learning. However, it must be organized differently depending on whether the objective is performance (short-term) or learning (long-term). Has anyone (e.g., teacher, coach) ever said to you, "Oh, come on, it's easy," when you were struggling to learn a new skill? Hearing that remark can be maddening, because performing the skill is not easy for you at that stage of learning. Practice that leads to learning takes hard work with many errors, and good teachers and coaches understand this principle of motor learning. In your motor learning class, you will study practice in two phases, which comprise before-practice variables and during-practice variables.

Before Practice

- Goal setting (Locke & Latham, 1985)
- Instructions
- Demonstrations or modeling (McCullagh, 1993; Weiss & Klint, 1987)
- Mental practice

During Practice

- Scheduling of practice
- Context of practice (Chamberlin & Lee, 1993; Schmidt & Lee, 2014)

Once you have completed a motor learning course, you will understand the prepractice variables and how to schedule and organize practice.

Some of these variables are presented here in the sidebar titled Example of Practice Organization. For instance, constant practice involves doing the same thing over and over, whereas variable practice involves *not* doing the same thing on consecutive practice trials. The first step in practice planning is to consider the learner and the skill; the next step is to plan the appropriate type of practice.

Feedback—Knowledge of Results and Performance

As you may have experienced, feedback is an integral part of the practice regimen that leads to learning. In fact, one cannot learn a skill correctly without feedback. It guides the learner toward performing the task correctly and reinforces correct performance. Feedback can be intrinsic or extrinsic. Intrinsic feedback consists of information about performance that you obtain for yourself as a result of a movement. Extrinsic feedback (also called *augmented feedback*) consists of information provided by an outside source, such as a referee, judge, teacher, coach, or video recording.

KEY POINT

Knowledge of performance is about the process of movement and is important for beginners. Knowledge of results is about the outcome of a movement.

Besides distinguishing between intrinsic and extrinsic feedback, motor behaviorists also categorize extrinsic feedback as either knowledge of results (KR) or knowledge of performance (KP). While KR and KP are terms used very specifically in experiments, the distinction is likely arbitrary outside laboratory experiments. The role of extrinsic feedback in learning is to provide accurate information that the performer cannot provide for herself. Experiments on KR and KP may hide outcome information from the performer in order to understand underlying factors or negate experience; in sport and physical activity, outcomes are rarely hidden from the performer. As its name implies, knowledge of results consists of information about the result of a movement—for example, "You missed the target." The athlete can see the ball missing the target, and therefore KR from a coach or teacher is not necessary or helpful. KP focuses on the performance and issues like angle, speed or force, acceleration or deceleration, starting or ending location, and so forth. Early in learning motor skills, KP is critical and is provided by a coach or teacher; thus, it is extrinsic. As skill develops, the performer may detect performance information and begin to use that information to correct errors and to repeat correct movements. Figure 7.3 presents examples of intrinsic and extrinsic feedback. For instance, a good typist who is told that he keyed in "Hybe" instead of "June"

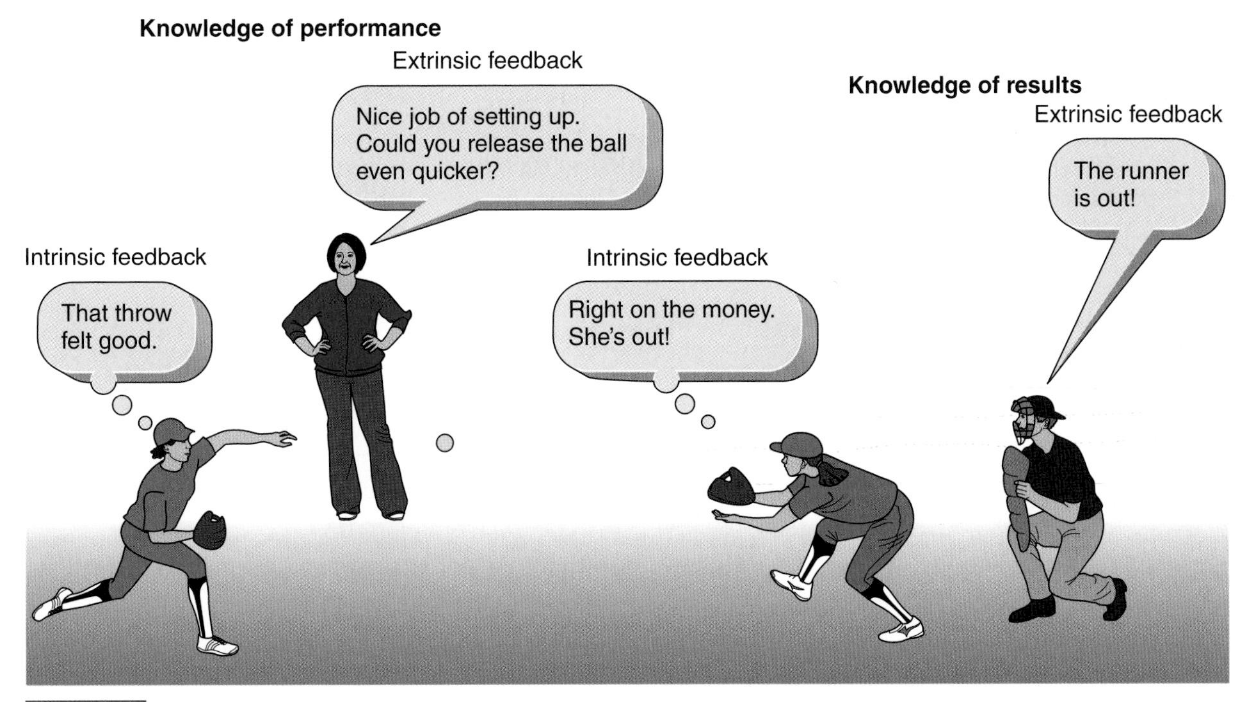

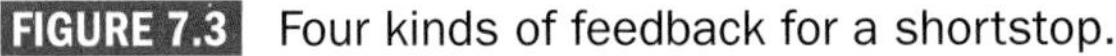
FIGURE 7.3 Four kinds of feedback for a shortstop.

Example of Practice Organization

Your sister has asked you to teach your 7-year-old nephew to swim. He remembers choking after inhaling water when he was younger and has avoided swimming lessons ever since. Fortunately, he looks up to you and is willing to give it another try. Unfortunately, you have not taught swimming before. Therefore, you face an important challenge, and you realize that in order to succeed you need to do some planning.

At this point, you know that practice and feedback are important to learning. We can consider planning practice in three parts: prepractice, practice, and postpractice. During practice, you also need to think about feedback. We know that our participant is a novice. What kind of skill is swimming? Good descriptors include *continuous* and *relatively closed*. As you plan the first practice session, consider what the learner is thinking, what you (as the teacher) are thinking, and what you should say. The process is demonstrated in the following table.

	The nephew thinks:	The teacher thinks:	The teacher says:
Prepractice (goal setting, simple instructions, modeling)	"I am afraid. This is going to be hard. I will probably fail."	"I should set a goal, demonstrate, and keep my instructions simple."	"It's okay to be afraid; you're smart to be afraid. The only way not to be afraid is to learn to swim. So today I want you to see how the water holds me up." (The teacher demonstrates a float.) "To float, you must put your face in the water. Once you get your face wet, then we will work until you keep your face in the water for 5 seconds."
Practice (constant practice, contextual practice, blocked practice, extrinsic feedback)		"So far, so good—he's not choking, and his face is in the water. Short practice trials with rest in between are working. I should show him what I want him to do. I can give him feedback when he comes up for air."	"That was great—you had your face in the water for 5 seconds. I'm proud of you. Next time, I want you to get your ears wet too, so put your face in far enough to get your ears wet. I know you can do it! Let me show you what I want." (The teacher demonstrates.)
Postpractice (verbal rehearsal, goal setting)	"That wasn't so bad. I can put my face in the water for 5 seconds, and I didn't choke. I could feel the water in my ears."	"Next, I need to increase the difficulty, but I also need to let him rest so that he doesn't get fatigued and make a mistake."	"You're making progress. Now we'll practice walking in the water, and then we'll try putting your face in again. Can you tell me what you've accomplished so far?"
	The nephew thinks:	The teacher thinks:	
	"I put my face and ears in the water."		"Good. You know, it took a lot of practice before I learned to swim. If we keep working, you'll learn to swim just like I did."

A few days later, the practice continues, as demonstrated in the following table.

	The nephew thinks:	The teacher thinks:	The teacher says:
Prepractice (goal setting, demonstration, verbal rehearsal)	"So far, this is easy. I'm pretty good at this. I can put my face in the water, walk in the water, and float."	"He puts his face in the water, walks in the water, and can do each of the parts (float, kick, and arms). I think he's ready to begin to swim. So, I want to provide verbal cues that he can rehearse and use for the progressive part of practice."	"Today you're ready for something new. You know how to count to three, and there are only three things you need to remember and do to swim: (1) Put your face in the water and float, (2) kick your feet, and (3) move your arms. Let me show you while you count for me." (The teacher demonstrates.) "What do 1, 2, and 3 stand for? Now it's your turn."
Practice (blocked practice, extrinsic feedback after several trials)	"Okay, 1 is face in to float, 2 is kick, and 3 is arms."	"He did it! Now we need to repeat this several times with just a little rest between. Should I change the practice? Let's see—this is a closed continuous skill, so constant practice is okay. Maybe we should practice in a different pool or at least a different place in this pool next time."	"Good job. You did all three. Now do it again." (Repeat several times.) "You're going about 10 feet [3 meters] before you stand up. That's good. You're working very hard. Let's see if you can go a bit farther. From where you are to where I'm standing is about 12 feet [3.5 meters]. Try to get to me. I'm sure you can do it!"
Postpractice (mental and verbal rehearsal, extrinsic feedback)	"I'm totally awesome. 1, face in and float; 2, kick; 3, arms."	"He's swimming. We've made a lot of progress, but I'm worried that the progress will slow down. I need to keep him motivated."	"You reached the goal—good work! Now say the three steps for me. Good. Can you think of the way it feels as you say the steps? Think about this and say the steps before next time."

would know that his right hand was one key away from the correct initial position on the home row. A beginner typist, however, might need a cue (e.g., "Check your home position") to correct the error; this typist is using KP feedback.

The ultimate goal of feedback is to help performers detect and correct their own errors. The study of feedback has focused on KP and KR with regard to frequency (how often), precision (how detailed), modality (auditor, kinesthetic, or other source), and processing time. Feedback can also be either reinforcing (e.g., "That was a good effort") or negative (e.g., "You did not try very hard"), which are equally good.

KEY POINT

Extrinsic feedback contains information that the learner could not obtain alone, should be corrective and provided for about half of the trials, and must be followed by sufficient time to make corrections before trying again.

Frequency of feedback can be addressed in terms of the percentage of practice trials for which feedback is given (e.g., 50 percent or every other trial versus 100 percent or all trials). On the surface, more feedback might seem to be better, but performers can become dependent when feedback is too frequent. Constant feedback is also counterproductive if the goal is to detect and correct one's own errors, because it prevents one from learning the processes of detection and correction. Thus, knowledge of performance (KP) should be given more often at the beginning of learning and then gradually reduced (Salmoni, Schmidt, & Walter, 1984); this process is often referred to as *fading* the feedback.

In your study of motor behavior, you will learn about two other variables in motor learning besides KP and KR—precision and modality of feedback. You are also likely to learn about the amount of time that children need in order to use feedback.

In baseball, the batter watches the pitcher for critical information that may give a clue about the pitch. In doing so, the batter ignores the crowd's yelling and tries to gather only relevant advance information about the pitch and where to hit it. Figure 7.4 provides an example of how long the batter has to make decisions about swinging the bat after the pitcher releases the ball. The figure demonstrates that a batter has more decision time (150 milliseconds versus 130 milliseconds) if he waits longer and speeds up the swing (140 milliseconds versus 160 milliseconds). The example also demonstrates how critical advance information is to successful performance—in this case, whether the pitcher is likely to throw a fastball or a breaking ball, as well as the likely pitch location. Thus, you may hear a baseball commentator say, "Three balls, one strike; he'll be looking for a fastball high and inside" (i.e., a certain pitch in a specific location). The batter is trying to reduce the decision to a go or no-go situation—he'll swing if the pitch is what he expects—rather than having to consider all possible options. The batter can do this because in this situation he can decide not to swing even if the pitch is a strike (because another strike will not make an out). If the count were three balls and two strikes, then the batter would be less likely to look for a specific pitch.

Motor Control

The first principle of motor control holds that the brain uses the central nervous system to initiate and control the muscles that make the desired movements. The second principle of motor control holds that one goal of most movements is to rely on the decision-making centers in the brain as little as possible once the movement is initiated. These two principles were developed on the basis of two theories and many research studies. One of these theories focuses on what are called *motor programs* and the other on *dynamical systems*. Motor programs (Schmidt, 1975) provide a theoretical explanation of how we produce and control movements. We can compare a motor program to a computer program that does math problems. First, you select the program to use; this is response selection. The program can add, subtract, multiply, and divide. If you input the relevant numbers and indicate which math operations to perform, the program outputs the answer; this is response execution. A motor program operates in the same way. First, you select the program (response selection) and indicate what it should do (operations). Then the program specifies how to do the skill and sends signals through the spinal cord to the muscles that perform the movement (response execution).

At a minimum, a motor program must do the following five things (Schmidt & Lee, 2014):

- Specify the muscles involved in the action.
- Select the order of muscle involvement.
- Determine the force of muscle contractions.
- Specify the relative timing and sequences of contractions.
- Determine the duration of contractions.

KEY POINT

A motor program is a proposed memory mechanism that allows us to control movements. As motor programs are developed, they become more automatic, thus allowing us to concentrate on using movement in performance situations.

If we had to remember how exactly to create every single movement we make, our memory would be overloaded. Thus, motor programs explain why we do not have a central storage problem for the great diversity of movement skills we use. Instead of storing in memory each movement that we have ever done, we store groups of movements with similar characteristics. These groups, called ***schemata***, serve as the foundation of motor program theory (Schmidt, 1975).

Some researchers, however, do not agree with the idea of motor programs drawn from schema theory, and they have proposed another theory. This approach, referred to as ***dynamical systems*** theory (Haken, Kelso, & Bunz, 1985; Kelso, 1995; Fischman, 2007), suggests a more direct and less cognitive link between motor action and information picked up by the perceptual system. This direct link is called a ***coordinated structure***, one characteristic of which is automated movement that relies on very little decision making or central control in the brain.

KEY POINT

Exciting and controversial results are likely to be produced by research that contrasts predictions and key elements of the motor program view of motor behavior with those of the dynamical systems view.

You may be wondering which theory—motor program or dynamical systems—more accurately represents the process of motor control. One decisive test for theories of motor control is to examine how well they explain motor learning and the development of motor expertise (Abernethy, Thomas, & Thomas, 1993). You will have a chance to examine and compare these theories in your motor learning class—and perhaps help advance them if you choose to pursue graduate study in motor control.

The study of motor control addresses five areas (Rosenbaum, 1991):

1. Degrees of freedom—coordination
2. Motor equivalency
3. Serial order of movements—coarticulation
4. Perceptual integration during movement
5. Skill acquisition

The brain initiates the planning of movements, and the nervous system then sends signals through the spinal cord to the muscles, which in turn make the movements. What is not known is how the brain represents the information to be sent to the muscles. The study of motor control is an effort to understand what the brain, nervous system, and muscles are doing to direct movements. Just as skill acquisition—including the notions of learning and improvement—is the focus of the study of motor learning, it is critical in the study of motor control as well. Indeed, skill acquisition accentuates the relationship between motor learning and motor control.

Once again, you may be wondering why understanding motor control is important. The answer is clear for professionals in physical education, physical therapy, or athletic training: Anyone who wants to learn or teach a skill uses the principles of motor control by trying to use the simplest movement possible at the early stages of learning. Moreover, some of what you will learn about motor control is less intuitive and thus often misunderstood by those who are untrained.

For example, consider the baseball batting example from figure 7.4. Here, motor control research tells us that the most rapid adjustments are made when we are not thinking about the flight of the pitch or the swing of the bat. That is, we turn control of the skill over to a more automatic part of the system called ambient vision. This process can be explained equally well by the two theories of motor control. However, the idea of automatic responses is not well understood by many sport enthusiasts. For instance, when you hear coaches or parents say, "Watch the ball all the way to the

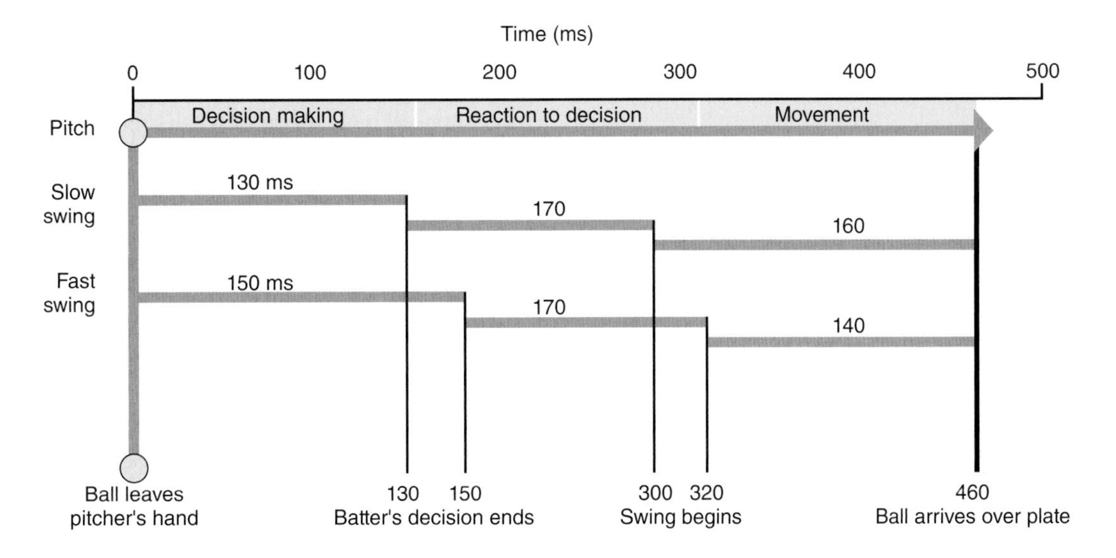

FIGURE 7.4 Time line showing the critical events in hitting a pitched baseball. The movement time is 160 milliseconds for the slow swing and 140 milliseconds for the fast swing.

Reprinted, by permission, from R.A. Schmidt, 2007, *Motor learning and performance: A situation-based learning approach,* 4th ed. (Champaign, IL: Human Kinetics), 153.

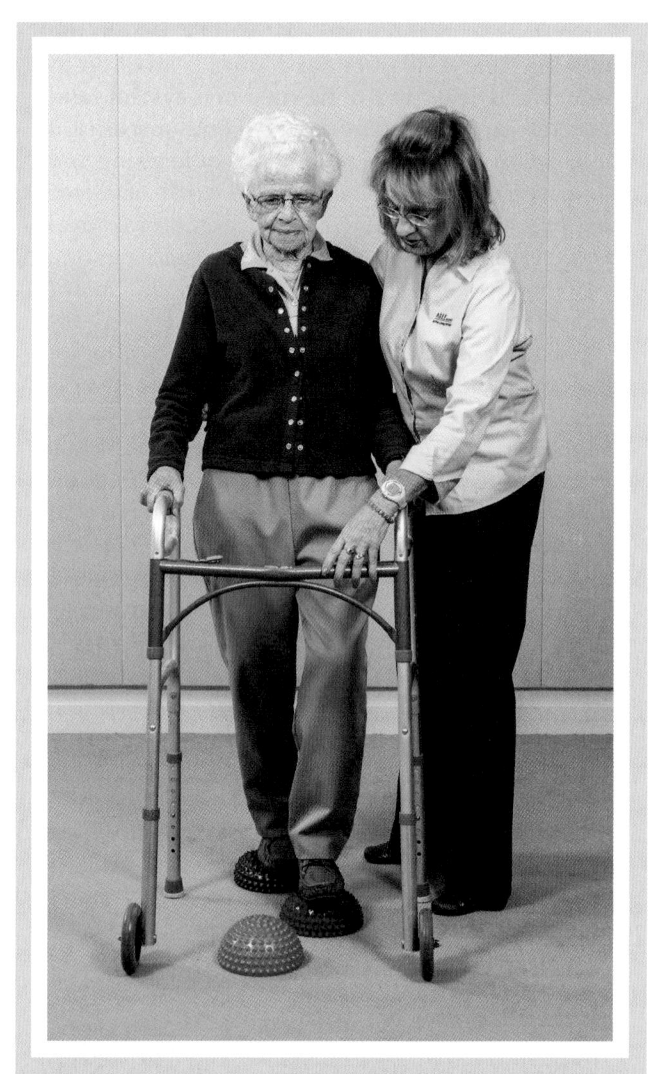

Understanding the theories of motor control enables this therapist to help her elderly client relearn balance skills.

bat," you know that they do not understand eye movements for tracking high-speed objects or how best to use the sensory system (ambient vision) in order to make rapid and accurate adjustments.

We hear comments alluding to the shift from central (decision-making) control of movements to peripheral (or sensory) control when athletes say things like, "It just felt right." The most rapid and accurate corrections are made when most of the movement is pre-programmed and sensory information is integrated only when necessary at the very end of the movement. In the study of motor control, this issue is a perceptual-motor integration problem: How do we best use sensory information to control movements? Rapid adjustments, like those of the baseball bat swing, might be explained by positing either that the information is not processed, which eliminates decision-making time (e.g., reaction time), or that there is no competition between responses (Wyble & Rosenbaum, 2016). The idea is that the demands are reduced to the simplest cases possible, thereby increasing the speed of adjustments. More choices or complexity decreases speed.

Motor control scientists try to explain what they observe, but they do not always agree on the explanation! We continue working to understand motor control because the range of motor skill performance—from normal to expert—depends on the shift from learning (central control, practice, and feedback) to more automatic movements programmed ahead of time and controlled peripherally.

Developmental Motor Learning and Control

The two principles of motor development (developmental learning and control) hold that (1) children are not miniature adults and (2) children are more alike than different (Thomas & Thomas, 2008). After watching a baby and a child perform a motor task, you can see clear differences and similarities related to age. Infants demonstrate more random movements, have a smaller repertoire of movements, and tend to do the same movement repeatedly. In contrast, children have a larger repertoire of motor skills, perform more voluntary skills, and exhibit greater skill than infants (Seefeldt & Haubenstricker, 1982) but less than adults. Thus, children's movements are not scaled-down versions of adult movements. One explanation for the relative inefficiency of children's movements as compared with those of adults may be their lack of practice. In addition, adults probably select and plan movements—which is a cognitive process—better than children do.

KEY POINT

To study physical activity, you must understand the physiology, biomechanics, and motor behavior underlying the development of movement in order to address problems such as how children gain control of movement skills.

Studying children is valuable because they typically change in an orderly fashion such that the order of change is the same across children even though the rate varies. Hence the second principle—that children are fundamentally alike—refers to gender, culture, and developmental age. When we understand how motor skills develop and what factors influence development, we can make predictions. Those predictions are important to pediatricians, teachers, and parents.

Virtually all the questions in motor learning and motor control can be studied as part of motor development by examining the same questions across age. In addition, some unique problems arise in studying infants, children, and persons who are elderly. For instance, children differ from adults in several ways, including physical growth, information processing, experience, and neurological development (Thomas, Gallagher, & Thomas, 2001).

Although some topics are unique to developmental motor learning and control, considerable overlap occurs with the previous two goals of motor behavior—motor learning and motor control (Ulrich, 2007). In some university programs, motor development is part of the course on motor learning and control, whereas in other universities the classes are separate. Motor development also includes several topics that are not part of motor learning—for example, physical growth and youth sport.

Developmental Changes in the Mechanics of Movement

The mechanics of a movement differ at different ages, in part because the skill is being executed by bodies of different sizes and proportions. For example, a baby's head accounts for one-quarter of her height, whereas an adult's head accounts for only one-eighth of her height. The legs also change dramatically during childhood. At birth, the legs typically account for less than one-third of body length; in adulthood, leg length often makes up more than half of body length. These physical characteristics influence balance and locomotion.

Another factor in motor behavior is growth. Females grow until about 13 or 14 years of age, and males grow until 18 to 20 years; in contrast, of course, growth is not a factor in motor learning studies with adults. Children grow in three physical dimensions: overall size, proportions (e.g., leg length, shoulder breadth, chest depth), and body composition (e.g., muscle increase). Growth influences motor performance, partly because children must contend with the changes in their bodies—an issue that adults do not have to deal with. Adult–child differences can also help explain child–child differences. For example, size and strength have a positive relationship in that larger children are generally stronger; therefore, as children grow, the increase in size produces increases in strength. However, strength also increases due to neuromuscular efficiency. Because physical growth lies beyond our control, we must understand how it influences performance in order to accommodate the challenges that it brings.

Life Span Development

In addition to addressing the effects of growth on motor performance, the study of developmental motor learning and control also deals with the effects of aging on motor performance. As a result, this area of study is sometimes referred to as *life span development* (Spirduso & MacRae, 1990; Stelmach & Nahom, 1992). In the course of the life span, changes in strength and motor coordination are most rapid at the extremes of the age continuum. Growth is less important after adolescence, because the changes in physical parameters are less dramatic in adults and elderly people than in infants and young children.

Cognitive processes also change during childhood, both because of the developing nervous system and because of experience. Children use fewer cognitive processes and use them less effectively than adults do. For example, when asked to remember a movement, a 5-year-old child might "put on his thinking cap"—not a very effective strategy for remembering. A 7-year-old might do the movement repeatedly, whereas a 12-year-old could repeat the movement in a series composed of several movements. Adults and older children know that they must do something to learn and remember, whereas younger children may not recognize the importance of deliberate practice or may select an ineffective practice regimen. One reason for the growing interest in cognitive processing during motor skill acquisition in children lies in the fact that the strategies used by adults can be taught to children in order to improve their performance. In other words, this research can enhance teaching.

Experience

As noted earlier, experience is a factor in children's performance. Of course, in situations where experience is helpful, those who are elderly may have an advantage, whereas many children have a disadvantage due to lack of experience. However, we can enhance experiences for children, and we can do so in two ways—quality and quantity. Research shows that children with experience can outperform adults with less experience, which means that practice and experience can help children (Chi & Koeske, 1983). This finding carries clear implications for pedagogy and curriculum development. In other words, curriculum and instruction need to provide experience in order to enhance sport performance in children. Furthermore, we can improve quality of experience by using what we know about information processing to help children get more out of practice. The key is that children need practice and, for it to be effective, it must help them retain information, skill, and decision-making ability that they would normally lose. We can enable this type of experience by providing children with adult learning strategies.

Changing Neuromuscular Systems

Children differ from adults in terms of neurology, or what we might characterize as neural hardware (analogous to a computer with its disk drives and monitor). In addition, as adults age, their neuromuscular hardware changes; more specifically, changes such as **sarcopenia** (loss of skeletal muscle), slowing of the nervous system, and reduced physical activity can all affect both health and performance in older adults. You may have observed older adults (generally, 65 or older) moving like toddlers with their arms extended and taking slow, wide steps. This movement pattern may result in part from physical inactivity due to injury, illness, or fear of falling. As a result, older adults who continue to perform tasks, whether related to exercise or work, tend to exhibit less decline than do those who stop performing as they age (Spirduso & MacRae, 1990). Moreover, attaining, maintaining, or regaining a physically active lifestyle predicts how long an older adult is likely to remain independent (Rikli, 2005; Rikli & Jones, 2012). Thus, in aging, we see a similar pattern to that of childhood—namely, a relationship between physical growth factors (e.g., muscle), practice, and performance.

Task performance by elderly persons can be negatively affected by deterioration in the nervous system. As people age, they lose neurons in the brain and motor neurons in muscles. These changes result in slowness and variability in movement control. On the other hand, older people benefit from experience and can use it to compensate for the loss of speed, strength, and control. Understanding such changes in the central nervous system is important because we have little opportunity to accelerate or decelerate them; if we live long enough, we are likely to face some decrement in the central nervous system.

KEY POINT

One key developmental question regarding motor behavior addresses how the brain and nervous system adjust to increases in cognitive function, body size, and strength during childhood and to decreases in those variables as people age.

Motor control issues are also important in developmental learning and control. Although the critical issues here are similar to those discussed in the previous section on motor control (for a review, see Clark & Phillips, 1991), there is one additional question: How does growth become integrated into the motor control system? As previously mentioned, children can change very rapidly, particularly at puberty (Malina, 1984; Malina, Bouchard, & Bar-Or, 2004). How does the motor control system account for changes in size, proportion, and mass? Consider a boy who played baseball from March until July at 12 years of age. By the next March, he is 13 years old, he has grown 4 inches (10 centimeters) in height, his arms have grown in length, and he weighs more. Moreover, he has not practiced batting or throwing since the preceding season, yet he can still bat and throw successfully. How does his nervous system compensate for his physical changes in order to continue producing coordinated movements?

At the other end of the age continuum, how does a system of motor control that has functioned for many years using one set of parameters account for loss in cognitive function, body mass, flexibility, and strength? No good explanation has been found for these important issues in the developmental aspects of motor learning and control. Some progress, however, is being made (for reviews, see Thelen, Ulrich, & Jensen, 1990; Thomas et al., 2001).

Growth and Gender in the Development of Overhand Throwing

We have all seen examples of people of different ages and genders who show markedly different physical abilities. Can we document such differences? Why do they occur? To explore these questions, let's look at what researchers have learned about overhand throwing. First, we need to examine physical growth and development (see figure 7.5 for a summary of growth in terms of stature and weight and figure 7.6 for proportion). Then we will see how these factors affect overhand throwing skills.

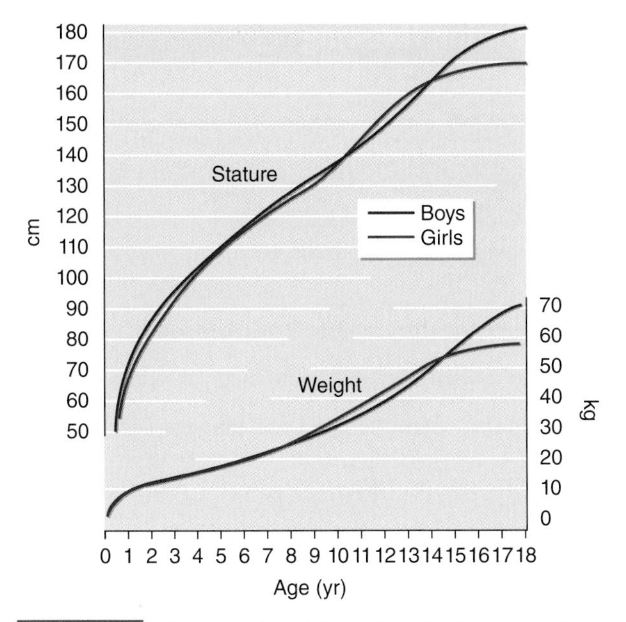

FIGURE 7.5 Average height and weight curves for American boys and girls.

Adapted from R.M. Malina, 1984, Physical growth and maturation. In *Motor development during childhood and adolescence,* edited by J.R. Thomas (Minneapolis, MN: Burgess), 7. By permission of J.R. Thomas.

Physical growth is rapid during infancy, constant during childhood, and rapid during the growth spurt associated with puberty (Malina, 1984; Malina et al., 2004). Children's limbs grow more rapidly than the torso and head, and the increase in proportion of total height attributed to leg length explains some of the improvement in running and jumping performance during childhood. In addition, the shoulders and hips grow wider, and males' shoulders become much broader on average than females' shoulders, which gives males an advantage in certain activities, including throwing.

Growth is orderly and marked by increases in length, then breadth, and then circumferences (figure 7.6 shows change in proportion from birth though adulthood). As a result, children in upper elementary school often look awkward because their legs and arms appear to be long and skinny. After most of the final length of a limb has been attained, the limb gains thickness as bone circumference and muscles (and fat!) increase, thus making the body look more proportional. In other words, children fill out; in fact, men's shoulders and chests usually grow (bones and muscle) until about 30 years of age, whereas most women stop growing by their late teens.

The changes produced by physical growth present three kinds of challenges: mechanical, adaptive, and absolute. The mechanics of movement change because of the different body proportions; the individual must adapt to a rapidly changing body. These changes are especially problematic in seasonal sports. Consider a wrestler who experiences a rapid growth spurt from the end of one season to the beginning of the next. His center of gravity has changed, which may influence his balance and the location of optimal points for exerting maximal force. Performance may also be influenced by absolute changes in size. Females gain fat at puberty, which adversely influences performance in most physical activities; in contrast, males gain muscle, which exerts a positive influence on performance. In these ways, physical growth during childhood interacts with all of the other factors that are devel-

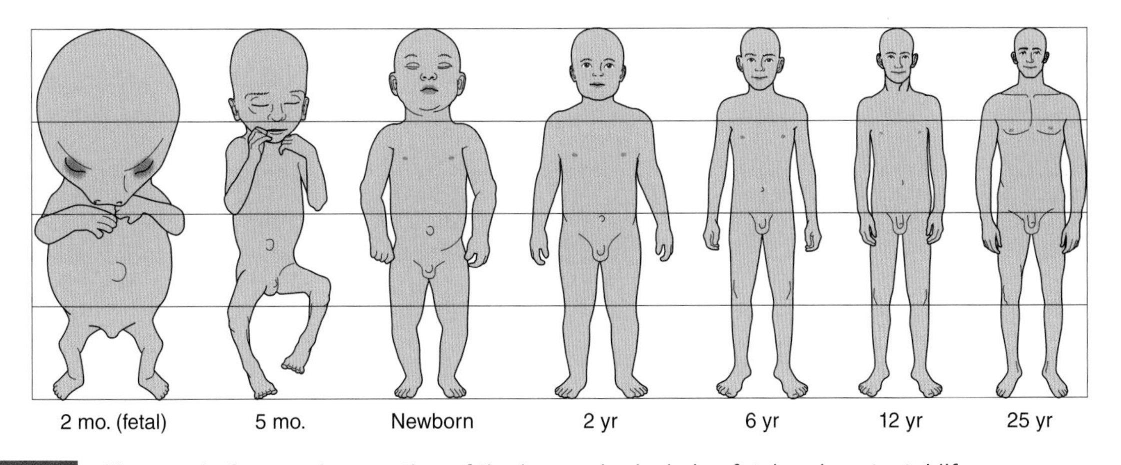
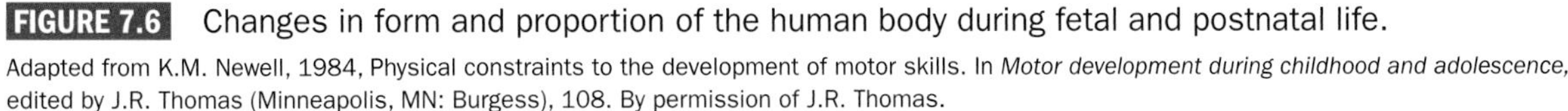

FIGURE 7.6 Changes in form and proportion of the human body during fetal and postnatal life.

Adapted from K.M. Newell, 1984, Physical constraints to the development of motor skills. In *Motor development during childhood and adolescence,* edited by J.R. Thomas (Minneapolis, MN: Burgess), 108. By permission of J.R. Thomas.

Development of Body Actions During Throwing

Level	Body actions
Preparatory arm backswing	
Level 1	The child's hand moves directly forward to release from where the child first grasps the ball.
Level 2	The forearm moves backward beside the head and then forward in ball release.
Level 3	The ball moves backward with a circular and upward backswing with elbow flexion.
Level 4	The ball moves backward with a downward swing below the waist and then up behind the head with elbow flexion.
Arm action	
Level 1	The upper arm moves backward and forward at an angle to the body. The forearm is linked to the upper arm movement and does not lag behind it.
Level 2	The upper arm is aligned with the shoulder and moves with it until near release, when the upper arm leads the shoulder. The forearm lags behind the upper arm (i.e., it appears to stay at the same point after the shoulder and upper arm have started forward).
Level 3	The upper arm lags behind the shoulder as the forward movement of the arm begins. The forearm lags even farther behind the upper arm and shoulder, reaching its final point of lag as the body is front facing.
Trunk action	
Level 1	Little or no trunk action occurs; the throw is accomplished by just the forward and backward action of the arm.
Level 2	The trunk rotation is in one block; the spine and hips move together in one rotating action.
Level 3	The hips lead in forward rotation followed by the spine (differentiated rotation).
Foot action	
Level 1	Feet do not move during throwing action.
Level 2	Step occurs but with the foot on the same side as the throwing hand.
Level 3	Step occurs but with the foot on the opposite side of the throwing hand.
Level 4	Same as level 3 except that the step is longer (about one-half of standing height).

Reprinted from M.A. Roberton, 1984, Changing motor patterns during childhood. In *Motor development during childhood and adolescence*, edited by J.R. Thomas (Minneapolis, MN: Burgess Publishing), 75. By permission of J.R. Thomas.

Student Learning Demonstrates Teaching Skill

How well did you do in planning swimming practice for your nephew? Compare the information presented in the sidebar titled Example of Practice Organization (earlier in this chapter) to the sections on practice, feedback, and transfer. How can you judge whether practice sessions influence the learning of a skill? If, in the example in the sidebar, your nephew can do what you presented in the first lesson the next time you meet, then you have evidence of retention. Remember, however, that retention is just one measure of learning. What characteristics of practice might disguise whether skills are being learned?

oping (e.g., motor programs) and therefore must be considered in both instruction and research.

One good way to examine how structural and functional change influences skill across childhood and adolescence is to consider overhand throwing. Figure 7.7 shows the changes in throwing for distance that occur across childhood and adolescence for girls and boys. The developmental nature of the overhand throw across childhood and early adolescence is detailed in the sidebar titled Development of Body Actions During Throwing. This description is separated into various body actions performed during the throw (e.g., arm, trunk, foot).

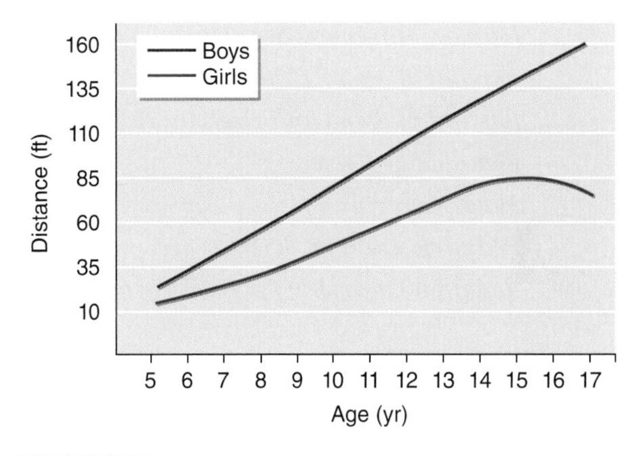

FIGURE 7.7 Age differences in throwing for distance.
Data from Espenschade 1960.

Given that prepubescent boys and girls are structurally similar, differences as large as those observed in throwing are unusual. Thomas and French (1985) noted that gender differences in throwing at 5 years of age were three times as large as those for any of the other 20 tasks they reviewed. Although it seems likely that these differences are accounted for largely by practice, encouragement, and opportunities, throwing may be one of the few motor skills in which biological factors play a significant role in gender-specific performance differences before puberty. Overall, however, we see that although we are unique, we (males and females) are more alike than different—both before and after puberty!

Wrap-Up

Although the study of motor behavior has produced important knowledge for human behavior since the late 1800s, it began to evolve as a significant subdiscipline in the field of kinesiology in the 1960s.

Knowledge developed through motor behavior has become increasingly important in all aspects of society. Although we often identify the main issues as learning and control of sport skills, our society depends on human movement in many ways—for example, babies learning to use spoons, surgeons using scalpels, pilots controlling airplanes, children learning to use pencils and pens, dentists using drills, and elderly persons working to remain independent. The study of motor behavior is aimed at understanding the development, learning, and control of these and other motor skills so that people can use them more effectively.

Knowledge of motor behavior is essential in several professions, including physical therapy, occupational therapy, physical education, coaching, and working with children in community organizations (e.g., YMCA, YWCA, Boys & Girls Clubs). Your school probably offers at least one undergraduate course in motor learning, motor control, and motor development.

MORE INFORMATION ON THE STUDY OF MOTOR BEHAVIOR

Organizations

American Geriatrics Society
Canadian Society for Psychomotor Learning and Sport Psychology
International Society of Motor Control
North American Society for the Psychology of Sport and Physical Activity
Society for the Neural Control of Movement
Society for Neuroscience

Journals

Human Movement Science
Journal of Motor Behavior
Journal of Sport and Exercise Psychology
Kinesiology Review
Motor Control
Pediatric Exercise Science
Research Quarterly for Exercise and Sport

REVIEW QUESTIONS

1. How does the study of motor behavior differ from the psychology of sport?
2. Explain the differences between motor learning and motor control in the field of motor behavior.
3. Why is the change in motor learning and motor control across the life span important?
4. Think about the practice issues discussed in this chapter, such as feedback, retention, transfer, goal setting, and scheduling. Choose a sport with which you are familiar and discuss how practice characteristics would influence your planning if you were a coach. Pick a specific age group or performance level, such as high school, college, or professional coaching.
5. Provide an example in which more difficult practice conditions result in better retention and transfer. Why does that happen? How could you plan practices to promote these benefits?
6. Discuss when it might be best to provide either knowledge of performance or knowledge of results to a person learning a motor skill.

Go online to complete the web study guide activities assigned by your instructor.

8

Sport and Exercise Psychology

Robin S. Vealey

CHAPTER OBJECTIVES

In this chapter, we will

- discuss what scholars and professionals do in sport and exercise psychology;
- describe how sport and exercise psychology evolved within the field of kinesiology;
- help you understand how professionals in sport and exercise psychology engage in research and practice;
- provide an overview of what research tells us about personality, motivation, arousal and anxiety, and group processes in sport and exercise settings; and
- explain imagery, attentional focus, and mindfulness as examples of how mental skills training is used in sport and exercise.

Suppose that you are an elite swimmer poised on the starting block for the 50-meter freestyle race at the Olympic trials. In the other lanes are the top swimmers in the country who, like you, have trained for years to hone their bodies into finely tuned physical machines—all in preparation for this moment. What are you thinking? Should you even *be* thinking? Where is your attention focused? Do you have negative thoughts and fears about not performing well, or do you feel a confident excitement that you are ready to meet the challenge? Do you feel a relaxed inner calm or a feverish emotional intensity? You have prepared yourself physically for this moment by swimming hours and hours for miles and miles. However, have you prepared yourself mentally?

Zumapress/Icon Sportswire

The psychologist William James wrote that "the greatest discovery . . . is that human beings, by changing the inner attitudes of their minds, can change the outer aspects of their lives." How true that is! Our participation in physical activity is greatly influenced by our "inner attitudes," or how we *think about* what we're doing. This mental aspect of physical activity provides the focus of study in sport and exercise psychology. Specifically, sport and exercise psychology focuses on the mental processes of humans as they engage in physical activity.

If you're the Olympic swimmer in the opening scenario, systematic training with a sport psychology professional has prepared you to think productively and focus optimally at this crucial moment. Your mental practice has prepared you to focus your attention on productive, energizing thoughts instead of worries and distractions. You have also programmed your mind and body to respond automatically at your preferred level of emotional intensity. Thus, you have mastered the mental focus and the physical readiness that you need for your best performance, and now your body and mind are one, waiting to explode from the platform for the performance of your life!

Of course, most of us will not compete in the Olympic Games, but the mental aspects of physical activity will challenge us repeatedly as we go through our lives. Consider the following situations that you may encounter:

- You take up mountain biking and are amazed at how the activity enhances your self-esteem and confidence along with your physical health.
- You begin an exercise program to enhance your personal fitness, but you lack motivation and quit the program after 2 months.
- You enjoy your weekly tennis league but feel very nervous when playing doubles because you don't want to let your partner down.
- As the coach of a local youth soccer team, you are deciding on the best ways to provide feedback and structure practices for children.
- You know that you need to attend your weekly shoulder rehabilitation session, but you lack motivation because you know some of the exercises and stretches will hurt.

Goals of Sport and Exercise Psychology

- To understand the social–psychological factors that influence people's behavior and performance in physical activity
- To understand the psychological effects derived from participation in physical activity
- To enhance sport and exercise experiences for those who participate in physical activity

All of these examples illustrate the critical and fascinating link between mind and body that is inherent in physical activity.

KEY POINT

Physical activity always involves the mind as well as the body. The mental aspect of physical activity is studied by sport and exercise psychology professionals.

Why Use Sport and Exercise Psychology?

Sport and exercise psychology involves the study of human thought, emotion, and behavior in physical activity. You can think of sport and exercise psychology as the study of the ABCs of physical activity. In this conception, *A* represents the term *affect,* which means emotion. For example, the emotions of anxiety and anger are studied to see how they influence athletes' performances. Emotional moods are influenced by exercise, which has been shown to decrease depression and enhance feelings of well-being.

B stands for *behavior.* It is fascinating to examine why people behave differently from one another when engaging in physical activity. That is, why are some people so committed to and persistent in their exercise routines? Why do some athletes train exhaustively to achieve excellence, whereas other talented athletes don't work as hard and thus don't reach their potential?

Finally, *C* represents *cognitions,* which means thought processes. Sport and exercise psychology examines how individuals' cognitions influence and are influenced by their participation in physical activity. For example, understanding why athletes choke under pressure requires that we understand their thinking processes. Similarly, exercise psychology professionals are interested in how physical activity can help individuals develop self-esteem and confidence or better ways to think about themselves. Remember these ABCs to keep in mind that sport and exercise psychology is the study of affect (emotions), behavior (actions), and cognitions (thoughts) related to physical activity participation.

This chapter discusses the psychology of physical activity in relation to the two main areas of kinesiology: sport and exercise. Sport psychology and exercise psychology have developed into two distinct areas of kinesiology. Sport psychology focuses more on the psychological aspects of competitive sport participation. Exercise psychology is devoted to the psychological aspects of fitness, exercise, health, and wellness. Because the two areas are closely related and share a great deal of theoretical content, this chapter covers both.

KEY POINT

Sport psychology and exercise psychology, as subdisciplines of kinesiology, focus on the study of human thought, emotion, and behavior in physical activity.

Although sport psychology and exercise psychology constitute areas of study in kinesiology, they also have close ties to psychology. The discipline of psychology is the science that deals with human thoughts, feelings, and behaviors; in turn, sport and exercise psychology deals with the thoughts, feelings, and behaviors of people involved in sport and exercise. For example, researchers attempting to understand why people begin, persist in, or discontinue participation in fitness programs often apply motivational theories developed in the discipline of psychology.

What Do Sport and Exercise Psychology Professionals Do?

Like all kinesiology professionals, sport and exercise psychology professionals attempt to enhance people's experiences in physical activity. Moreover, as in any subdiscipline of kinesiology, sport and exercise psychology professionals work in many career areas.

Exploring the ABCs of Physical Activity

Consider the following questions related to sport and exercise psychology:

1. Recalling that we've defined physical activity as intentional and purposeful movement (chapter 1), think of several forms of physical activity in which you participate. What is your motivation for performing each one?
2. Has participation in physical activity influenced your character? If so, what was the activity, and how did it affect you? If not, why is it that participation has not influenced your character?
3. Have you ever been involved in or watched any form of violence in sport? If so, what happened and why? What do you think caused the violence to occur?
4. Why do you think so few Americans participate in habitual exercise when it has been clearly shown to enhance health and longevity?

Kinesiology professors of sport and exercise psychology at universities fulfill the multiple roles of researcher, teacher, and service provider. For instance, a kinesiology professor specializing in exercise psychology may develop theory and conduct research on the psychological benefits of aerobic exercise, teach courses in exercise psychology, develop a health promotion program at the university, or conduct workshops for the community on motivation and stress management. A kinesiology professor specializing in sport psychology, on the other hand, might conduct research on stress in sport and athlete burnout, teach courses in sport psychology, consult with athletes from the university and the surrounding community, and offer regional workshops on coaching effectiveness.

Some kinesiologists trained in sport and exercise psychology focus more exclusively on practitioner or service-provider roles. For example, a sport psychology specialist might work with athletes on performance enhancement, personal development, and lifestyle management in a university athletic department, professional sport team, or private sport training academy. Similarly, an exercise psychology specialist might focus on worksite health promotion for a large corporation, develop fitness and wellness programs for a community recreation department, or coordinate physical activity programs for all age groups through a YMCA. Some professionals in sport and exercise psychology become consultants for sports medicine or physical therapy clinics to help injured people through the psychological aspects of injury rehabilitation.

It is important to differentiate between exercise or sport psychology professionals who are trained as physical activity specialists and those who are trained as clinical or counseling psychologists. Physical activity specialists focus on education or the teaching of skills to enhance either performance or personal fulfillment in individuals who are involved in sport or exercise. Clinical and counseling psychologists, on the other hand, are licensed to provide psychotherapy and consultation for individuals with clinical conditions such as depression, phobias, and anorexia nervosa. Although some clinical and counseling psychologists provide services for athletes and exercisers, the main focus of their practice differs from that of sport and exercise psychology as a subdiscipline of the physical activity field.

KEY POINT

Career opportunities in sport and exercise psychology include positions as university professors, performance enhancement specialists, fitness and health promotion specialists, and sports medicine consultants.

History of Sport and Exercise Psychology

Throughout the history of physical education and kinesiology, people have noted the connection between physical activity and mental health. However, sport and exercise psychology was not widely recognized as a subdiscipline of kinesiology until the 1960s.

Beginnings of Sport and Exercise Psychology

The relationship between exercise and psychological well-being was recognized as early as the 4th century BCE (Buckworth, Dishman, O'Connor,

Kinesiology Colleagues

Photo courtesy of Dede's Photography

Lindsey Hamilton

Lindsey Hamilton is a mental conditioning coach for IMG Academy in Brandenburg, Florida. After being an athlete in college, she acted on her interest in psychology by taking a social psychology research position at Stanford University. Then, following a stint with a tech company in the automotive industry, she found herself pursuing a master's degree in sport psychology at the University of Utah, where she got hooked on a career in applied sport psychology. Now, Hamilton works as a mental conditioning coach with soccer and basketball teams at IMG, as well as a variety of youth, collegiate, and professional athletes. She wakes up every day eager to challenge herself and those around her to find the best version of themselves by cultivating and practicing mental skills.

& Tomporowski, 2013). Around the turn of the 20th century, more physical educators began to write about the psychological benefits of physical activity. Meanwhile, researchers began assessing how the presence of others exerted social influence on motor performance, a research area that later became known as social facilitation. The first person to examine these social influences was Norman Triplett, who in 1898 studied the effects of the presence of other people on bicycling performance. The effects of exercise on depression were reported a few years later (Franz & Hamilton, 1905).

Griffith Era in Sport Psychology

The true beginning of sport psychology in North America came in the work of Coleman Griffith, who, as a professor at the University of Illinois, engaged in the first systematic examination of the psychological aspects of sport. Griffith established the Athletic Research Laboratory at Illinois, published numerous research articles, and wrote two classic books—*Psychology of Coaching* (1926) and *Psychology and Athletics* (1928). Griffith also interviewed sport celebrities such as Red Grange and Knute Rockne about the mental aspects of their sports. In 1938, Griffith was hired by Philip Wrigley as a sport psychology consultant for the Chicago Cubs. During this work, Griffith developed psychological profiles for specific players, including Dizzy Dean, and researched methods of building confidence and increasing motivation.

Unfortunately, Griffith may be thought of as a prophet without disciples because no one continued his significant work. Although sporadic publications about psychological aspects of sport and exercise appeared after Griffith's time, the subdiscipline largely lay dormant until a reawakening in the 1960s.

KEY POINT

Coleman Griffith began systematic research in sport psychology in the 1920s. However, because his early work was not extended, the area was not recognized as an academic subdiscipline of kinesiology until the 1960s.

1960s and 1970s

The 1960s saw an upsurge of interest in the social–psychological factors associated with physical activity. The research of this period focused primarily on personality traits related to sport participation and on social facilitation or audience effects on motor performance. One pioneer of the period, Bruce Ogilvie, a clinical psychologist at San Jose

State University, initiated early work in examining personality in athletes and applied psychological interventions with athletes. Another pioneer, Dorothy Harris of Pennsylvania State University, initiated a systematic research focus on women in sport.

In the midst of this work, three organizations became active in the field. The first meeting of the International Society of Sport Psychology was held in 1965, the first conference of the North American Society for the Psychology of Sport and Physical Activity was held in 1967, and the Canadian Society for Psychomotor Learning and Sport Psychology was founded in 1969. All three provided important forums for sharing knowledge in sport psychology.

In the 1970s sport psychology was formally recognized as an established scientific subdiscipline within kinesiology. Specifically, systematic research programs were established at leading universities; graduate study became available; and, in 1979, the *Journal of Sport Psychology* began publication. Much of the research done during this time was experimental work conducted in laboratory settings that involved testing theory from the parent discipline of psychology. For instance, Rainer Martens, a leading scholar during this time, pioneered the systematic study of competitive anxiety in sport.

Modern Sport and Exercise Psychology

The decade of the 1980s was marked by three trends. First, exercise psychology separated from sport psychology into a distinct subdiscipline within kinesiology. Previously, pockets of research had been done on the psychological components of exercise, but the 1980s brought the development of systematic research programs and graduate program offerings in exercise psychology. In particular, William P. Morgan was a major force in promoting the field of exercise psychology and served as the first president of Division 47, Exercise and Sport Psychology, of the American Psychological Association. Daniel Landers also served as a leader in this area by pioneering psychophysiological research in the field. Thanks to such work, the *Journal of Sport Psychology* became the *Journal of Sport and Exercise Psychology* in 1988.

The second trend of the 1980s consisted of explosive growth in field research in sport and exercise psychology. Much of the research of the 1960s and 1970s had been done in laboratory settings, seemingly to gain academic credibility for the young field. However, researchers now realized the need to study sport and exercise in the field in order to gain a better understanding of psychological processes as they operate in competition and in various forms of physical activity.

KEY POINT

The 1980s saw the emergence of exercise psychology, the growth of field research, and an explosion of applied mental training with athletes.

Logically following this move to field research, the third trend of the 1980s consisted of increased interest in applied sport psychology, or mental training with athletes. The expansion of the subdiscipline to include applied interests was indicated by the establishment of two new applied journals, *The Sport Psychologist (TSP)* in 1987 and the *Journal of Applied Sport Psychology (JASP)* in 1989, as well as a new organization, the Association of Applied Sport Psychology (AASP) in 1986. Influential professionals during this period included John Silva, founding president of AASP and first editor of *JASP;* Dan Gould, founding editor of *TSP;* and Terry Orlick, who pioneered applied sport psychology through his mental training books and consulting work with Canadian Olympic athletes.

During the 1990s, sport psychology and exercise psychology grew as separate subdisciplines through the accumulation of research and applied programs. In 1991, AASP established certification criteria in the form of minimum professional training standards for consultants in sport and exercise psychology. Ethical standards and guidelines for consulting in sport and exercise settings were approved, and the United States Olympic Committee created a registry of certified professionals eligible to consult with Olympic teams and athletes.

In the first two decades of the 2000s, the knowledge base of the field grew rapidly with an increase in published research, thus providing a sound foundation for sport psychology practice. These two decades have brought a large increase in qualitative research, which typically involves interview data, as opposed to quantitative research, which presents findings using numbers or statistics. Sport and exercise psychology practice has also expanded greatly, as both consulting services (e.g., Mental Training, Inc.) and applied materials (e.g., books, videos, websites) have proliferated. Online resources include Peak Performance Sports, Virtual Brands, and Sports Psychology Today by Mental Edge Athletics. In addition, two new journals have been established to serve practitioners (e.g., coaches, consultants, athletes): the *Journal of Sport*

Psychology in Action (2010) and *Case Studies in Sport and Exercise Psychology* (2016).

Research Methods for Sport and Exercise Psychology

By now, you understand the nature and historical development of sport and exercise psychology as a scientific area of study. How is science conducted in this subdiscipline? What methods do researchers use to ask important questions about the psychological aspects of physical activity participation? Researchers use six methods to systematically assess thoughts, feelings, and behaviors in sport and exercise psychology.

KEY POINT

Research methods in sport and exercise psychology include questionnaires, interviews, observation, physiological measures, biochemical measures, and content analysis.

› **Questionnaires.** Questionnaires are widely used in sport and exercise psychology. Some questionnaires take the form of survey instruments assessing demographic variables such as age, sex, and socioeconomic status; they may also assess general information such as the type, frequency, and duration of exercise engaged in during the past week. Most questionnaires, however, take the form of **psychological inventories**, which are standardized measures of specific forms of thoughts, feelings, or behaviors (Ostrow, 2002). For example, psychological inventories are used to measure the amount of anxiety, motivation, and confidence an individual feels about exercising or competing in sport. Questionnaires can be administered to participants using paper-and-pencil methods, but online programs (e.g., Qualtrics) are used more frequently to provide easier and quicker access to research participants via the Internet and to facilitate data analysis.

Consider for a moment the difficulty of accurately measuring the thoughts, feelings, and behaviors of people. Assessing individuals' levels of self-esteem is much different from measuring the amount of oxygen that they expend during a fitness test or how much weight they lose in the preseason. For this reason, psychological inventories must meet rigoroubptxs standards for uniformity of procedures in their development, administration, and scoring so that researchers can gain a valid and reliable assessment of behavior. Therefore, psychological inventories used by researchers do not simply list questions that have been thrown together; rather, they are carefully constructed and tested assessment tools that have met specific standards set by experts in the subdiscipline.

› **Interviews.** Interviews are used in sport and exercise psychology when the research question being pursued requires in-depth understanding of individuals' beliefs, experiences, or values. For example, interviews may be useful if one is attempting to understand why children drop out of youth sport programs, because interviews allow children to explain things in their own words rather than merely respond to questionnaires. As with any other scientific method, however, interviews must be structured to be systematic, and researchers must be trained in their use in order for them to be effective and valid.

› **Observations.** A third method used in sport and exercise psychology is observation of behavior. For example, researchers often observe the behavior of coaches during practice or competition to assess the frequency of various types of feedback and communication that they provide to athletes. Behavioral observation is also used in research examining the motivation of children to engage in physical activity or vigorous play activities. Typically, observation studies use video replay, along with some type of behavior checklist or coding system, to ensure that observation of the behavior is occurring within a particular set of parameters. Two observers are typically used to code behavior, and their results are checked against each other to ensure a consistent, reliable assessment.

› **Physiological measures.** Physiological measures of physical, mental, and emotional responses are sometimes referred to as *biofeedback*. For example, blood pressure and heart rate may be measured to assess the effects of psychological stressors on individuals. Exercise psychology professionals use measures such as these to study the effects of exercise on stress reactivity and existing anxiety levels. Sport psychology professionals might measure the amount of tension in muscles to assess how well athletes can learn to relax physically through mental training. Researchers can also assess levels of attention or relaxation by measuring brain waves. Brain-imaging technologies, such as fMRI and PET scans, can be used to view how the brain responds when athletes visualize performing their skills. For example, expert athletes activate

Have You Got Passion for Physical Activity?

Complete the following questionnaire by marking a response for each item as you think about your favorite physical activity.

Scoring your responses: totally agree = 7, agree = 6, sort of agree = 5, unsure = 4, sort of disagree = 3, don't really agree = 2, don't agree at all = 1.

	Totally agree	Agree	Sort of agree	Unsure	Sort of disagree	Don't really agree	Don't agree at all
1. I am completely taken with my activity.							
2. My activity reflects the qualities that I like about myself.	7						
3. My activity allows me to live memorable experiences.	7						
4. My activity is in harmony with other activities in my life.			5				
5. My activity is a passion that I manage to control.				4			
6. I cannot live without my activity.						2	
7. My mood depends on my being able to do my activity.	7						
8. I have difficulty imagining my life without my activity.			5				
9. I am emotionally dependent on my activity.				4			
10. I have an almost obsessive feeling for my activity.				4			

Total your scores for items 1 through 5 to get your positive passion score. This score represents how much you love to engage in your activity based on your own free will to choose to participate. A high score indicates that you gain great joy from the activity, that you don't feel you have to do it, and that it doesn't control your life. Next, total your scores for items 6 through 10 to get your obsessive passion score. This score represents how much you feel that you must engage in your activity, as if your activity controls you and you *have* to do it (e.g., for others, for approval, for yourself). Are you passionate about your physical activity? How much of that passion is positive, and how much is obsessive? You want your positive passion to outweigh your obsessive passion so that you feel in control of your life and gain joy from participation—not pressure to perform. Consider ways to lessen the obsessive passion you feel and increase your positive passion for your physical activity. Remind yourself what you love about the activity and focus on the joy of that movement. Participate on your own terms in pursuit of personal goals that are challenging and motivational, yet not obsessively controlling.

Adapted from Vallerand et al. 2003.

areas in their brains much differently than beginner athletes, and this difference indicates that elite sport performance is hardwired in a specific way in the brain.

› **Biochemical measures.** Biochemical measures are used less frequently in sport and exercise psychology, but brief mention here will be helpful. This type of assessment involves drawing and analyzing blood or urine for chemicals from the body that represent responses to stressors or emotions. Examples include epinephrine and cortisol, which are released by the adrenal gland in response to certain types of stressors.

› **Content analysis.** Researchers use content analysis to analyze written material from various sources, such as government documents, newspapers, magazines, and even video programming. For example, a researcher could analyze several popular video shows to assess the levels of physical activity being modeled to viewers. Written or dictated physical activity logs are also useful in exercise research because they provide detailed accounting of all or selected types of physical activity performed within a given period.

Thus, researchers in sport and exercise psychology have various methods available to them. Often, researchers use two or more methods in a single research study. For example, a study examining exercise adherence in an adult fitness program could use observation to assess participation rates, intensity of exercise, and instructor behavior; questionnaires to measure participants' self-confidence and perceived benefits in relation to exercising; and physiological measures of participants' resting heart rate and lung volume. All of these methods are designed to be used in a systematic manner to ensure that accurate and consistent measures are obtained. This systematic approach is the mark of science—and it is very different from the casual observations that we all make in everyday social interactions.

All scientific methods in sport and exercise psychology have limitations. For instance, questionnaires typically provide a systematic measure of certain phenomena, but they lack the depth and richness of interviews. However, it is harder to establish a consistent, systematic approach when using interviews than when using questionnaires. Given this mix of considerations, scientists in sport and exercise psychology look at the menu of research methods available and select the ones best suited to the research study at hand. This menu also includes other choices for researchers, such as whether to conduct the study in a laboratory or field setting and what type of participants to use in the research.

Overview of Knowledge in Sport and Exercise Psychology

This section takes a brief look at a few selected topics in sport and exercise psychology to give you a glimpse of the knowledge that researchers and practitioners have produced. The topics include personality, motivation, arousal and anxiety, social and group processes, and mental skills training in physical activity.

Personality

One of the most popular issues in sport and exercise psychology involves the relationship between personality and physical activity participation. We typically think of **personality** as the unique blend of characteristics that makes an individual both different from and similar to others. Our personalities determine our thoughts, feelings, and behaviors (i.e., the ABCs) in response to our environment.

Personality and Sport

Despite popular opinion, no distinguishable athletic personality has been shown to exist. That is, no consistent research findings show that athletes possess a general personality type distinct from the personalities of nonathletes. Also, no research has shown consistent personality differences between athletic subgroups (e.g., team athletes versus individual-sport athletes, contact-sport athletes versus noncontact-sport athletes).

Research has, however, identified several differences in personality characteristics between successful and unsuccessful athletes (e.g., Krane & Williams, 2015). These differences are not based on innate, deeply ingrained personality traits but rather involve one's level of motivation and one's effectiveness at thinking and responding to sport challenges. Specifically, as compared with less successful athletes, athletes who are successful are

› more self-confident,
› better able to cope well with stress and distractions,
› better able to control emotions and remain appropriately activated,
› better at attentional focusing and refocusing,
› better able to view anxiety as beneficial, and
› more highly determined and committed to excellence in their sport.

Olympic and world-champion athletes have defined *mental toughness* as the natural or developed psychological edge that enables them to cope with competitive demands and remain determined, focused, confident, and in control under pressure (Jones, Hanton, & Connaughton, 2007). These athletes identify the following as critical personality responses that represent mental toughness: loving the pressure of competition, adapting to and coping with distractions and sudden changes, channeling anxiety, being unfazed by mistakes in the process, being acutely aware of any inappropriate thoughts and feelings and changing them immediately to perform optimally when needed, using failure to drive oneself, learning from failure, and knowing how to rationally handle success—an impressive list of qualities that we all would like to have as part of a mentally tough personality!

KEY POINT

No uniform set of traits exists for an athletic personality, but successful athletes possess more positive self-perceptions, have stronger motivation, possess more adaptive perfectionism, and use more productive cognitive coping strategies than do less successful athletes.

Although most sport personality research has focused on the influence of personality on sport behavior, research has also examined the converse—that is, the effect of sport participation on personality development and change. It is commonly believed in U.S. society that organized sport builds character and that sport participation may help an individual develop socially valued personality attributes. Research shows, however, that competition *reduces* prosocial behaviors such as helping and sharing and that this effect is magnified by losing. Sport participation has also been shown to increase rivalrous, antisocial behavior and aggression and has been linked to lower levels of moral reasoning (Vealey & Chase, 2016).

Nevertheless, the character-related effects of sport do have a positive side. Research in a variety of field settings has demonstrated that children's moral development and prosocial behaviors (e.g., cooperation, acceptance, sharing) can be enhanced in sport settings when adult leaders structure situations to foster these positive behaviors (Hellison, 2003; Shields & Bredemeier, 2009). Interventions with children have succeeded in building character when naturally occurring conflicts arise and are

Professional Issues in Kinesiology

Expertise: Early Specialization in Youth Sport

The past two decades have seen a trend of pushing children to choose one sport and train in it year round in order to pursue elite status (termed *specialization*). To many, this seems to be a logical path because the young athlete receives more time to develop specific skills in the chosen sport. However, research indicates that early, exclusive specialization is not needed in most sports for most kids. More specifically, other than gymnastics, figure skating, and diving, most sports are designated as *late specialization*, which means that kids don't need to specialize in them until age 14 to age 16 (Balyi, Way & Higgs, 2013).

For example, a study of more than 700 professional baseball players found that their average age of specializing in baseball was 15, and 52 percent did not specialize until age 17 (Ginsburg et al., 2014). They did start playing baseball early (at an average age of 6 years), and this type of early engagement is important for priming the brain for eye–hand coordination, ball tracking, throwing, and striking. However, early engagement is not early specialization, and these athletes did not abandon other sports and activities just to play baseball. Moreover, early diversification (engaging in multiple sports and activities) helps kids develop better overall motor skills even if they later choose to specialize, whereas early specialization has been linked to burnout, dropping out, and injuries (Vealey & Chase, 2016).

Citation style: APA

Balyi, I., Way, R., & Higgs, C. (2013). *Long-term athlete development.* Champaign, IL: Human Kinetics.

Ginsburg, R.D., Smith, S.R., Danforth, N., Ceranoglu, T.A., Durant, S.A., Kamin, H., Babcock, R., Robin, L., & Masek, B. (2014). Patterns of specialization in professional baseball players. *Journal of Clinical Sport Psychology,* 8, 261-275.

discussed with children in order to enhance their reasoning and values related to both sport and life events. Indeed, we have seen a proliferation of positive youth development programs that use sport and physical activity to build important psychosocial characteristics in children (e.g., emotional regulation, self-worth, interpersonal competence). For example, a program called The First Tee has been successful in developing life skills in youth through the game of golf (Weiss, Bolter, & Kipp, 2016). The program trains instructors to use specific teaching strategies that are mastery driven, personally empowering for students, and focused on lifelong learning. The moral of the story is this: Sport doesn't build character—people do!

Personality and Exercise

As in sport, researchers have found no particular exercise personality or set of personality characteristics to predict exercise adherence; in other words, exercisers cannot be differentiated from nonexercisers based on an overall personality type. Researchers have, however, found three personality characteristics that are strong predictors of exercise behavior. First, individuals who are more confident in their physical abilities tend to exercise more than those who are less physically confident. A second important predictor is self-motivation; specifically, as you might expect, self-motivated individuals are more likely to begin and continue exercise programs, whereas less motivated individuals are more likely to drop out or never start at all. Third, researchers have identified an activity trait—that is, a tendency to be busy and energetic and to prefer fast-paced living—that is linked to people's choices to engage in exercise (Rhodes & Pfaelli, 2012).

Echoing the idea that sport builds character, exercise or fitness training has also been popularly associated with positive personality changes and mental health (Landers & Arent, 2007). The personality characteristic that researchers have examined most frequently in this area is self-esteem—our perception of personal worthiness and the emotions associated with that perception. In short, we can think of self-esteem as how much we like ourselves. Research has generally confirmed that

Carefully designed organized sport and individual fitness training can build character and lead to positive personality changes.

fitness training improves self-esteem in children, adolescents, and adults and that exercise positively influences perceptions of physical capabilities, or self-confidence. The research indicates that these changes in self-esteem and self-confidence may result from perceived, rather than actual, changes in physical fitness. Exercise training also improves body image, or how an individual perceives and feels about his or her own body. In addition, many aspects of intellectual performance have been related to physical activity through indications that cognitive functions respond positively to increased levels of physical activity.

Many people also associate exercise with positive changes in mood and anxiety. Most individuals say that they "feel better" or "feel good" after vigorous exercise, a fact that emphasizes the important link between physical activity and psychological well-being. In addition, research documents that anxiety and tension decline following acute physical activity. The greatest reductions in anxiety occur in exercise programs that continue for more than 15 weeks. Much research has been conducted to determine whether exercise or fitness reduces people's susceptibility to stress, and the generally accepted conclusion is that aerobically fit individuals demonstrate a reduced psychosocial stress response. One tentative explanation for this finding posits that exercise either acts as a coping strategy that reduces the physiological response to stress or serves as a sort of inoculator to foster a more effective response to psychosocial stress (Landers & Arent, 2007).

KEY POINT

Sport has not been found to build socially valued attributes, or character, but exercise has been shown to produce several benefits, including enhanced self-concept and psychological well-being and decreased anxiety and depression.

Prolonged physical activity is also associated with decreases in depression and reduction of depressive symptoms in individuals who are clinically depressed at the outset of exercise treatment. Explanations for these changes range from the distraction hypothesis, which maintains that exercise distracts attention from stress, to other explanations focused on physiological and biochemical changes in the body after exercise.

Motivation

Sport and exercise psychology professionals also seek to understand what motivates people to engage in physical activity. **Motivation** involves a complex set of internal and external forces that influence individuals to behave in certain ways. The behaviors most typically associated with motivation are choice, effort, and persistence. That is, we assume that people are motivated when they choose to join a fitness program, work intensely during the program, and continue to adhere to the training program when their lives become busy. Thus, motivation directs and energizes our behavior in sport and exercise. Unfortunately, motivation cannot be directly given to someone. Rather, in order to enhance motivation, we need to understand and tap into four important motivational fuels: enjoyment, competence, autonomy, and relatedness.

Enjoyment

Do you spend a lot of time doing things that you don't enjoy? One hopes that your answer is no! Obviously, athletes and exercisers gain motivation when they enjoy what they are doing, be it kayaking on a beautiful lake or competing in a rousing pickup game of basketball. Such activities provide us with the positive subjective experiences that constitute our major reasons for making physical activity part of our lives. The number one reason that children begin taking part in a competitive sport is to have fun, and research with elite athletes indicates that they engaged in enjoyable play experiences as children more than they did in grueling physical training (Vealey & Chase, 2016).

The fuel of enjoyment becomes even more motivational when it turns into **flow**—an optimal mental state characterized by total absorption in a task (Csikszentmihalyi, 1990). Exercisers may speak, for example, of a "runner's high," which is an elusive state in which physical activity feels effortless and immensely enjoyable. Athletes may train to achieve the ultimate high of "being in the zone" and enjoying the ultimate thrill of sport characterized by peak performance. Flow occurs when people experience a balance between their skills and the challenge that they perceive in the situation. Thus, as kinesiology professionals, we must take care to match the challenge to people's skills when designing exercise programs, sport training programs, and youth sport competitions.

Revisit Your Flow Experience

Think about a time when you experienced flow. Describe the activity that you were doing and what the experience felt like. What made it different from other experiences? How long did it last, and why do you think it went away when it did?

Competence

Motivation is also fueled by competence. The innate (inborn) human quest for competence begins in infancy and continues throughout our lives. Our need for competence motivates us to try new skills and activities. When we succeed in such attempts, we feel worthy and proud, which fuels our continued mastery attempts in those activities.

Thus, one important strategy for motivating people to engage in physical activity is to help them develop competence in their movements. For example, athletes who play ice hockey probably had effective instructors or coaches who taught them how to skate and handle a puck. Adults who begin exercise programs also need progressive instruction so that they feel competent in their physical activity training. Research shows that motivation is enhanced more when we define success and competence in personal and controllable ways rather than defining them in relation to others. We can do so by helping athletes and exercisers set personal performance goals based on individual improvement rather than comparing their progress with that of others. Another motivational strategy is to praise young athletes not for performance outcomes but for effort. Although they should be recognized for their accomplishments, excessive praise of outcomes (e.g., winning a heat in swimming) often leads to pressure and less motivation to work hard (Dweck, 2006). It's best to comment instead on their controllable effort and the persistent practice that enabled them to achieve the positive outcome (e.g., "Your hard work on your flip turns really paid off in that race—congratulations!").

The need for competence, and its importance as a fuel for motivation, can sometimes lead to incompetence avoidance. Athletes who focus on avoiding incompetence often experience fear of failure, which is really a fear of shame and embarrassment related to not being good enough. Another motivational orientation related to incompetence avoidance is perfectionism, in which an individual tends to strive for and focus on flawless performance. Some aspects of perfectionism are helpful, such as having high standards and strong organizational skills. However, maladaptive perfectionists are overly concerned with and unable to accept mistakes, which leads them to experience excessive anxiety, reduced confidence, and burnout.

Autonomy

Another basic motivational fuel is autonomy, also called *self-determination*, which consists of the feeling that one possesses control over oneself and one's actions. Think about how your motivation might wane when you have to sit through a boring required training class, as opposed to when you have a free afternoon to write up your own personalized training program. However, don't we sometimes need external motivators beyond what is inside of us? The answer, of course, is yes.

Motivation can be defined as either intrinsic or extrinsic. **Intrinsic motivation** involves doing an activity for the pleasure and satisfaction derived from engaging in that activity. For instance, kids play soccer and ice-skate and swim because they enjoy the feelings they experience when doing these activities. The reward is the activity itself. However, not all aspects of sport or exercise participation are intrinsically interesting to people. Often, in fact, extrinsic motivation is needed in order to fuel the repetitive practice required to build skill or the physical fitness training required for developing stamina and strength. **Extrinsic motivation** involves doing an activity in order to achieve outcomes or rewards beyond the activity itself, such as trophies, scholarships, approval from others, or enhanced muscle tone and body image.

Extrinsic motivation isn't always driven externally by the demands of others. Sometimes, it is autonomous, as when people choose to go on training runs or swim laps during a vacation. They do these activities to gain some reward outside of the activity, but they make the personal choice to do it and they own the behavior. It is important for us to develop autonomous extrinsic motivation, because activities are rarely driven solely by intrinsic motivation. For example, athletes need self-determined extrinsic motivation (along with intrinsic motivation) to commit to intensive training and invest the required time in practice.

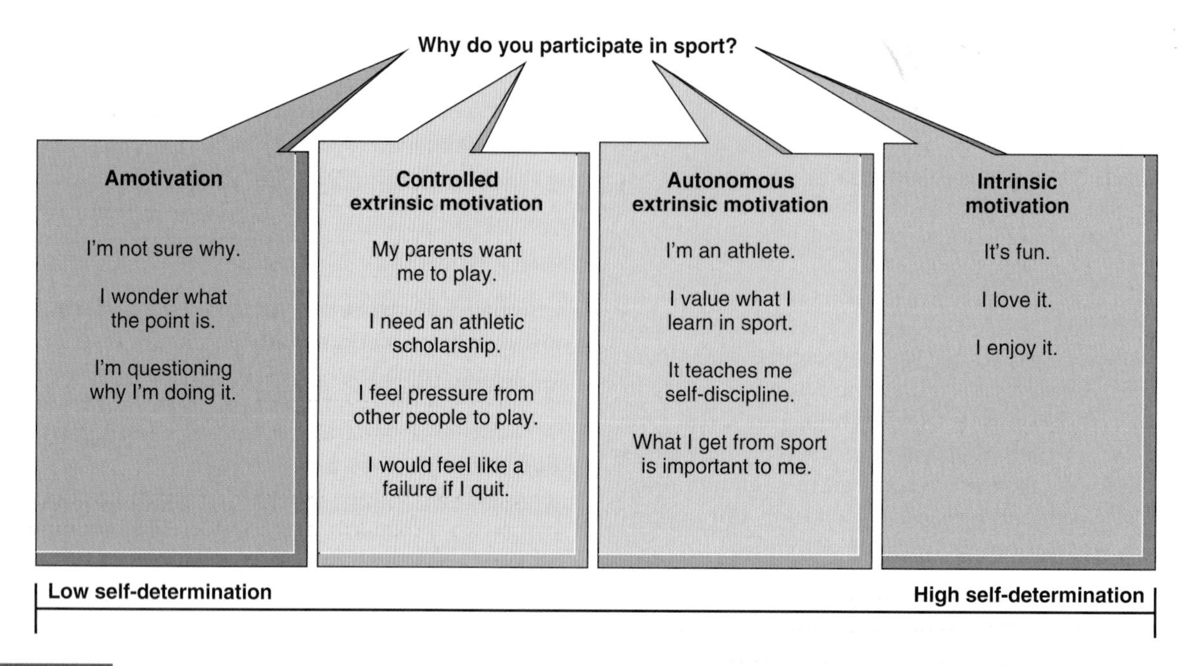

FIGURE 8.1 The continuum of motivation.

Reprinted, by permission, from R.S. Vealey and M.A. Chase, 2016, *Best practice for youth sport,* (Champaign, IL: Human Kinetics), 129.

One useful way to think about motivation is to situate its different forms across a continuum of self-determination (see figure 8.1). Intrinsic motivation anchors the high end of the self-determination continuum. Autonomous extrinsic motivation is also high in self-determination, because athletes internalize the extrinsic reasons for playing as valuable and as part of their identity. Controlled extrinsic motivation is the least self-determined form of motivation and occurs when athletes need to follow demands of others or obtain external rewards. Of course, amotivation is simply the lack of any type of motivation.

Different types of motivation often operate at the same time and work together to enhance behavior. For example, a person might love to bicycle (intrinsic motivation) because she values the strength and fitness that it provides (autonomous extrinsic motivation); she might also begin a challenging new training program with a coach in order to qualify for an upcoming race (controlled extrinsic motivation). There are times when we need controlled motivation, but it works bests when it is accompanied by intrinsic and autonomous extrinsic motivation. Self-determined forms of motivation have been linked to greater attendance and participation in physical activity, lower intentions to drop out of sport, and lower levels of burnout.

Relatedness

The final fuel source for motivation is relatedness, or our need to be connected to those around us and to experience a sense of belonging. In this regard, people who exercise as part of a group are more likely to be adherent than are those who exercise alone (Lox & Ginis, 2014). For instance, positive connections with friends and teammates lead to greater commitment and motivation in youth sport. Therefore, instead of just assuming that social connections are automatically made in exercise groups and sport teams, coaches and exercise leaders should thoughtfully and intentionally plan activities and strategies to provide the important motivational fuel of relatedness for all participants.

Arousal and Anxiety

To review, motivation involves intensity of behavior and the urge to experience enjoyment, competence, autonomy, and relatedness. However, the urge to be competent sometimes creates anxiety for individuals as they participate in sport or exercise. Has your performance ever been disrupted because you were tense or nervous? In contrast, can you remember performing "in the zone?" These two scenarios—gut-wrenching anxiety and the exhilaration of peak performance—illustrate two ways in which individuals experience arousal in physical activity.

What Fuels Your Motivation?

What is your favorite physical activity? Why do you do it? What fuels your motivation to participate in this activity? Consider how much your participation is fueled by enjoyment, competition or stimulation, desire to demonstrate your competence and feel good about that, desire to feel autonomous or in control of yourself, or desire to be part of a group and be accepted by others. What is most important to you? Consider how you might increase your motivation and commitment to your activity by creating the right fuel mixture for yourself every day.

Arousal is a state of bodily energy or physical and mental readiness. It results from the ways in which athletes' minds and bodies respond to competition, or the ways in which people involved in exercise respond to physical activity. Arousal falls along a continuum ranging from very low (as in deep sleep) to extremely high (the most "wired" you've ever been). When an individual experiences a high level of arousal and feels nervous or worried, we refer to this negative emotional state as **anxiety**. In other words, we can think of arousal simply as the experience of generic energy; when arousal becomes too high and people interpret it negatively (e.g., "I feel scared," "I feel nervous"), it is then labeled as anxiety. In contrast, high levels of arousal that are labeled positively (e.g., "I feel ecstatic!") are often described as joy or happiness.

Research shows that highly successful elite athletes do not have less anxiety than less successful athletes, but they do perceive their feelings of nervousness and anxiety as positive and helpful (Hanton & Jones, 1999). That is, they know that these feelings represent the increasing readiness of their bodies to perform, so they accept the nervousness, anxiety, and high arousal as normal and helpful. Try this yourself! The next time you feel nervous, instead of labeling it as negative and focusing on how bad it feels, tell yourself that your body is ready and that when the time comes to perform you will use that nervousness as positive energy or arousal to help you succeed. It works!

People often assume that anxiety is the same thing as stress, but stress serves the important purpose of stimulating growth. **Stress** may be defined simply as a demand placed on a person. In this formulation, we stress athletes all the time through weight training and exhaustive physical repetitions of their sport skills. In fact, careful progression of physical and mental stress builds athletes' stress tolerance so that they can withstand the stress of competitive performance. Thus, stress is not the same thing as anxiety, though it often creates anxiety. The type of anxiety studied most widely in sport psychology is based on the threat of failure or evaluation, as when athletes compete in public. As mentioned earlier, anxiety matters in exercise psychology because research has demonstrated that physical activity decreases feelings of anxiety, thus contributing to mental and physical health. Research has also demonstrated that exercise behavior can be influenced by social physique anxiety, or apprehension about one's body and appearance (Lox, Ginis, & Petruzzello, 2014). For instance, when people feel less apprehensive about their appearance, they are more likely to participate in activities that require them to reveal more of their bodies (e.g., gymnastics, swimming).

Overall, stress may induce high levels of arousal and even anxiety, but it often exerts a positive influence on performance. Think back to times when you performed well in either sport or academics because the stress motivated you to prepare or train to succeed in that situation. Stress becomes a problem only when individuals do not allow adequate recovery time from life or competitive demands. Chronic stress without adequate recovery burns out the energy reserves of athletes and others engaged in physical activity.

When individuals' arousal levels are in their optimal zones, the experience of flow occurs. Thus, when we help physically active individuals identify the special recipe of feeling states that leads to their optimal arousal zones and to flow, we help them enhance their performance and overall experience of physical activity.

Many sophisticated models are available to explain how arousal and anxiety influence performance in sport. One general rule holds that people perform better at moderate levels of arousal. This principle is referred to as the inverted-U model of arousal, in which athletes' performance increases as their arousal levels increase up to an optimal or moderate point, after which athletes go "over the top" and performance decreases as their arousal levels become too high. As shown in figure 8.2, there is an optimal zone of arousal for performance—not too low and not too high.

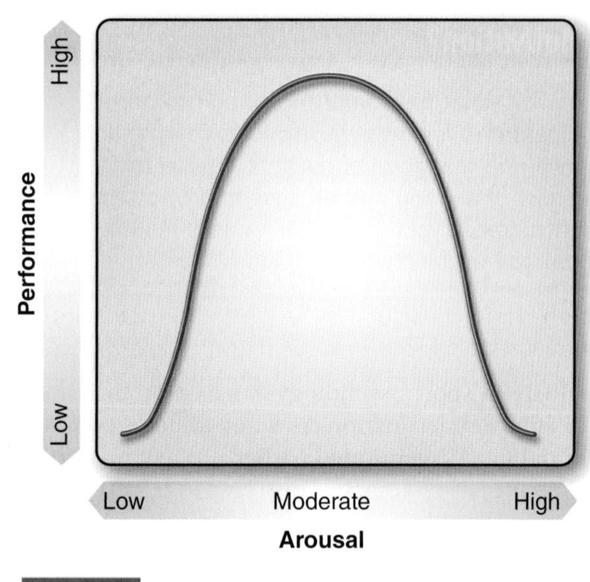

FIGURE 8.2 Inverted-U model of arousal.

The optimal intensity of arousal is unique for each person. Preferred arousal levels can be affected by individual factors such as personality characteristics, coping ability, and skill level. In addition, an athlete's optimal intensity of arousal differs according to the situational demands of different activities. For instance, a weightlifter can be at a very high arousal level and perform well, whereas a golfer's optimal level for performance is somewhat lower. Arousal influences performance through its effect on muscular tension and the performer's ability to properly focus attention. The muscles of athletes performing in sport often become tense and lose their capacity to make the smooth, coordinated movements needed to perform optimally. In this situation, the athlete's attention may turn inward toward worry and thoughts of inadequacy or narrow to the point where the athlete misses important competitive cues that are crucial to optimal performance.

Sport psychology experts help athletes learn to "find their numbers" and then develop focus plans in which they engage in individualized thoughts and activities (e.g., psyching up or calming down) to create the levels of arousal that work best for them in their particular sport activities. It takes some experimentation, but athletes can learn how to find their optimal numbers, how to monitor themselves, and how to adjust their numbers when needed. Think of this process as giving athletes the ability to optimize the challenge for themselves (i.e., optimize the stress) in order to create their desired levels of arousal and put themselves in position to avoid the negative effects of extreme anxiety and experience flow.

Social or Group Processes

Social or group processes can influence individuals' behaviors in various ways. This section addresses the effects of the presence of others and of group membership.

Presence of Others

How does your behavior when you are alone differ from your behavior when you are with other people? Why do some people prefer to exercise at home but others like to exercise with others in a public gym? Why do some athletes perform better in front of a big crowd? Does home advantage really exist? Since the turn of the 20th century, researchers have been fascinated with such questions—that is, with the effects of an audience on human performance.

Generally, the presence of other people increases our arousal, which in turn may either hurt or help our performance. Generally, spectators have a negative effect on someone who is learning a skill but a positive effect on someone who is very skilled. Think about a new golfer who is still learning the game and must tee off in front of a large group on the first hole. The presence of spectators increases her arousal, and because her skills are not well learned, this arousal causes her to hit a bad shot. In contrast, consider a professional tennis player who makes the finals at Wimbledon for the first time. The huge crowd at Centre Court is likely to inspire and elevate her performance.

Although this explanation may sound simple, the ways in which others affect our behavior are in fact complex. This effect is caused not by the mere presence of others but by performers' perceptions that others are evaluating them. We know from the previous section of this chapter that every individual has a different optimal arousal zone; therefore, the presence of others can influence performance differently based on individual responses to changes in arousal. People should avoid situations of excessive evaluation or analysis when learning sport skills; the pressure of such situations can hinder the learning process by adversely affecting beginners' arousal levels and quality of attention.

The presence of others applies a bit differently to exercise, because the people present are usually socially supportive workout partners or simply other individuals concurrently exercising on their own (e.g., others working out at a public facility). Family social support is a strong predictor of exercise maintenance, and individuals who exercise with their spouses have higher rates of exercise adherence than do those who exercise alone (Lox

& Ginis, 2014). Research has also demonstrated that the physical activity levels of children are related to the modeling of such activity, as well as shared activities, social support, and encouragement within families. Several studies have demonstrated that active children perceive more parental encouragement than do inactive children. At the other end of the age spectrum, the social isolation of many elderly people is problematic because social incentives to engage in physical activity decrease with age. Community-based programs can help provide the needed social support for habitual physical activity participation at all ages.

KEY POINT

Family support and modeling are highly predictive of the physical activity participation of children, whereas elderly people tend to become less physically active because of social isolation.

Group Membership

The area of group dynamics focuses both on how being part of a group influences behavior and, conversely, on how certain psychosocial factors influence collective group behavior. Are people more likely to adhere to exercise programs if they participate in groups rather than exercise individually? Why do some teams have better chemistry than others? And why do teams with better chemistry often perform better than more talented teams with poorer social and psychological cohesiveness?

A group performs better, and its members feel more satisfied, when the group is cohesive. **Cohesion** consists of the tendency for groups to stick together and remain united in pursuing goals (Eys, Burke, Carron, & Dennis, 2006). Thus, coaches and exercise leaders should strive to develop and nurture cohesion in their teams or groups, and they can do so in several ways. First, cohesion can be facilitated by emphasizing uniqueness or a positive identity related to group membership. For example, group identification and cohesion can be nourished through the use of nicknames such as the "Sweat Net" or the "Lunch Bunch." Athletes may also demonstrate solidarity and commitment to the group through actions such as shaving their heads or wearing team jackets. Cohesion also increases when individual members of teams understand and accept their roles within the group. Most people prefer to exercise with another person or in a

Professional Issues in Kinesiology

Expertise: Training the Quiet Eye

Everyone knows that you need to focus your visual attention on the target when performing a closed, accuracy skill such as a golf putt or basketball free throw. Research has shown, however, that some people are much more skilled than others at doing so and that many performers are unaware that they are not focusing optimally. This optimal visual focus is called the quiet eye, which means that performers lock onto the relevant stimulus (e.g., golf ball, front of basketball rim) prior to execution and hold this gaze for several hundred milliseconds before, during, and after the movement (Vickers, 2009).

Training one's quiet eye involves using visual tracking equipment, which was once used only for research purposes. In this training, the performer wears a computer-connected helmet with cameras focused on his or her eyes and on the visual field. This arrangement allows the athlete to perform repetitions of the task and then review his or her eye movements by watching the computer recording. Most performers are amazed to find that their eyes moved off of the target in random ways, and of course these random visual miscues affected the motor programs provided to their brains to guide performance. Quiet-eye training has been shown to improve performance in basketball shooting, golf, sharpshooting, and surgery. It has also improved throwing and catching in children with developmental coordination disorder (Miles, Wood, Vine, Vickers, & Wilson, 2015).

Citation style: APA

Miles, C.A., Wood, G., Vine, S.J., Vickers, J.N., & Wilson, M.R. (2015). Quiet eye training facilitates visuomotor coordination in children with developmental coordination disorder. *Research on Developmental Disabilities, 40*, 31-41.

Vickers, J.N. (2009). Advances in coupling perception and action: The quiet eye as a bidirectional link between gaze, attention, and action. *Progressive Brain Research, 174*, 279-288.

group—an arrangement that enhances enjoyment through affiliation and social support and strengthens commitment to the program. In fact, cohesion and friendship within a group constitute one of the strongest predictors of exercise adherence.

We also know that cohesion is bred by success. Thus, early successes are crucial in the development of group dynamics; indeed, when groups experience early success, cohesion develops accordingly. Even so, cohesion is a dynamic quality that is always changing within a group; for example, conflicts that require resolution may arise, but the group may still be cohesive. On the other hand, the quest for cohesion can go too far, resulting in extreme cohesion that involves conformity and elitism. Although a group needs a strong, positive shared identity, it should also celebrate and respect diversity among its members.

Group dynamics research also examines how group membership influences individual performance. For instance, the term **social loafing** refers to a decrease in individual performance within groups (Heuze & Brunel, 2003). This decline occurs because individuals believe that their performance is not identifiable and that other group members will pick up the slack. Social loafing is not a conscious process—people do not decide to loaf; rather, it is a psychological tendency when people perform in a group. Fortunately, it is easily reduced if we increase the identifiability of individuals by monitoring their performances. Common ways of doing so in sport include using video analysis to assess individual performance and compiling individual statistics that break out a single athlete's performance from the total performance of the team. Social loafing can also be monitored in exercise groups if performance totals (e.g., number of sit-ups performed, amount of weight lifted, distance run) are recorded so that progress can be checked. Identifying and monitoring individual efforts eliminate social loafing.

KEY POINT

Cohesion and group membership facilitate physical activity performance, but social loafing may occur unless individuals are monitored and their inputs are viewed as important to overall performance.

Mental Skills Training in Physical Activity

Mental skills training uses intervention techniques such as goal setting, relaxation training, imagery, and self-talk in conjunction with physical training to enhance important mental skills such as confidence, focus, and arousal management. The importance of mental skills has been documented by research. For example, a study of professional baseball pitchers found that their coping skills, confidence, and motivation were more important

Helping People Get Started With Exercise

- Avoid the "all-out or nothing" or "no pain, no gain" mentality. Advocate adding light activity (e.g., walking) that can enhance physical and mental health.
- Counsel individuals to choose activities that they enjoy or that give them a welcome mental health break (it shouldn't feel grueling or excessive).
- Start small, with easy exercise goals, to build confidence and momentum.
- Use the shaping technique by celebrating small achievements along the path to reaching a goal (i.e., use progressive, graduated exercise goals).
- Encourage individuals to schedule physical activity as an essential part of each day and to put it in their planners as an important appointment with themselves.
- Use immediate, informational feedback; provide encouragement; and help people regulate their confidence by practicing productive self-talk such as "I can do this," "it's a process," and "one step at a time."
- Encourage individuals to focus not on outcomes such as weight loss but on process goals such as moving every day. The physical payoff will come in time; for now, focus on moving daily and feeling better.

than their physical skills in predicting their success in baseball (Smith, Schutz, Smoll, & Ptacek, 1995). In addition, these mental skills were predictive of all players' career survival in professional baseball 2 and 3 years later (Smith & Christensen, 1995).

Mental skills training has been used to increase exercise adherence, improve sport performance, develop important life skills (among young people participating in physical activity), aid in rehabilitation from injury and disease, and enhance career transition and retirement from sport. For example, the intervention checklist shown in the sidebar titled Helping People Get Started With Exercise demonstrates how to set goals and monitor behavior in order to enhance one's commitment to physical activity. The rest of this section of the chapter gives you an idea of how mental skills training works. Specifically, we focus on the mental training tool of imagery, the important mental skill of attentional focus, and the emerging mental approach of mindfulness.

KEY POINT

Intervention techniques in sport and exercise psychology can increase exercise adherence, enhance sport performance, develop life skills, aid in injury rehabilitation, and ease career transitions and retirement from sport.

Imagery

The mental programming technique of **imagery** uses one's senses to create or re-create an experience in the mind (Vealey & Forlenza, 2015). Seeing is believing, as many athletes will tell you. For instance, Jack Nicklaus, perhaps the greatest golfer of all time, says that 50 percent of the key to hitting a good golf shot involves your mental picture of the swing. Swimmer Michael Phelps, the most decorated Olympian of all time, with 28 medals (23 gold), once said, "Nothing is impossible. With so many people saying it couldn't be done, all it takes is imagination, and that's something I learned and something that helped me."

Images guide our beliefs, and beliefs guide our performances. Indeed, much of our behavior is based on deep beliefs about ourselves that sometimes are not even accurate. To understand the power of imagery and self-belief, complete the sidebar activity titled Believe and Achieve.

Hundreds of studies have reported that when imagery is used systematically, it enhances sport and exercise performance; therefore, although imagery cannot take the place of physical practice, it is better than no practice at all. Many elite athletes use imagery, and they often cite it as an important mental factor in their success. In addition, novice athletes can use imagery to create positive patterns for successful performance, and exercisers can use

Believe and Achieve

Our self-images often serve as self-fulfilling prophecies. That is, we are what we dwell on, or what we think about ourselves. In the activity space provided here, use the left-hand column to identify at least one positive and one negative personal image or belief that you have about yourself—for example, that you're good at math, that you're not good at conversation, or that you're good at golf. In the right-hand column, describe how that image or belief affects your behavior—for example, that you freeze on exams, avoid meeting new people, or enjoy golf.

Next, take time to reflect on the items you listed and how they affect your life every day. Consider how your life is enriched by positive beliefs or images about yourself and constrained by negative ones. Commit to using imagery daily in order to visualize yourself in more positive ways and enhance your personal belief system.

Personal Image or Belief About Myself	Impact of That Belief or Image on My Behavior
______________________	______________________
______________________	______________________
______________________	______________________
______________________	______________________
______________________	______________________
______________________	______________________

imagery to visualize their muscles firing and becoming stronger when training for fitness. Exercisers can also use imagery to dissociate, or direct attention away, from repetitive exercise activity by visualizing motivational or pleasant things. For instance, many people use personal audio players to help them create pleasant mental images while running.

But how does imagery work? How in the world could imagining an experience improve your performance? Incredibly, when you vividly imagine performing a well-learned skill (e.g., throwing a ball), your brain fires in ways that are very similar to what it does when you actually perform the skills. Thus, imagery works by providing a mental blueprint for perfect responses; that is, a blueprint in your brain contains the physical and mental responses that represent perfect performance. These responses are modified and refined both when you physically practice them and when you engage in imagery. Therefore, imagery or mental practice works because it trains your brain in a way that resembles physically practicing the skill. This effect has been demonstrated with many sport skills (Vealey & Forlenza, 2015).

A great deal of research has been conducted in sport and exercise psychology to help us understand the most effective ways to practice imagery in order to achieve the greatest effects. Here are some tips about how to use imagery (Vealey & Forlenza, 2015).

› Use as many senses as possible: visual, auditory, tactile, olfactory, kinesthetic. In particular, the kinesthetic sense of feeling your body move in skilled physical activities is important for optimizing your imagery.

› Practice imagery using both internal and external perspectives. Internal perspective involves seeing the image through your own eyes, whereas external perspective involves watching from outside of your body, as if watching a video or yourself performing. Both perspectives can be used to enhance performance, but more recent work shows that an internal perspective is needed in order to truly mimic the brain activity of actually performing the skill.

› Load your images with productive behavioral, physical, and mental responses. Don't just imagine the environment and what you see; imagery is more productive when you imagine how you will respond to your environment. For instance, don't just see the track, the spectators, and your performance. Instead, imagine running smoothly and easily in perfect rhythm, feeling loose, fluid, strong, and fit. Use the energy of the crowd to increase your intensity. See the difference? Imagine the responses that make your performance successful.

› You don't have to be lying down to use imagery. It is helpful and more influential on performance to practice imagery while in your specific performance posture and, if applicable, while holding an implement that you use in your activity (e.g., hockey stick, golf club). Skaters, for instance, should stand and move somewhat as they listen to their music and imagine performing their routines perfectly.

› Use simple verbal triggers to prime the image that you want. For instance, thinking "dollar bill" in a sand bunker provides a golfer with the image of slicing her wedge through the sand behind and under the ball to splash the sand (and the ball) onto the green. Sprinters may think of their legs in the blocks as coiled springs ready to explode at the starting gun.

› Practice imagery daily. Elite athletes create and follow systematic mental training plans that include imagery practice. Don't be discouraged if your imagery skills are not as sharp as you'd like. Elite athletes will tell you that it takes time and practice to perfect your imagery skills (just as it takes time and practice to perfect your physical skills).

Attentional Focus

Maintaining attentional control and focus may be the most important mental skill for any type of physical activity. Here's how one elite golfer described the focus that he developed by training with a sport psychology professional: "I was thinking about exactly what I needed to do, staying in the present, thinking about what I was doing now and not about negative past experiences or the outcome" (Neil, Hanton, & Mellalieu, 2013, p. 123).

In sport, performance depends on the cues that athletes process from themselves and from the social and physical environment. In exercise, we can enhance our experience and our performance by learning how and when to focus on the body *(association)* and how and when to distract ourselves *(dissociation)*, perhaps with music or visual imagery. Attention, or focus, is trainable and crucial. Although athletes often think that distractions hinder their performances, the problem actually resides in how they choose to focus when distractions and obstacles arise. The elite golfer quoted in the preceding paragraph trained himself to choose to focus productively on the process of playing—not dwelling on the past or focusing on the future but thinking about one shot at a time.

Attentional focus is affected by one's arousal levels. Moderate arousal levels are better for sport performance than underarousal or overarousal. One explanation for this inverted-U model relates to attention. Easterbrook's (1959) famous model of attention and arousal (the "cue utilization" model) is shown in figure 8.3. The model shows that attention varies from very wide to very narrow depending on one's level of arousal. The attentional funnel in the middle shows that attention is too broad at low levels of arousal; as a result, irrelevant cues may be accepted that interfere with focused performance. Attention is predicted to be optimal at moderate levels of arousal, because once arousal becomes very high, attention becomes very narrow, in which case athletes fail to process important peripheral cues. Think, for example, of a quarterback who is so nervous that his narrow focus of attention interferes with his ability to read the defense and widely scan the field for open receivers. One way to optimize attentional focus, then, is to regulate one's arousal level to avoid the overarousal that impairs performance by causing attention to become too narrow.

Increases in arousal brought on by stress and pressure can cause another problem with attentional focus. Specifically, arousal and anxiety can make an individual more self-focused, which in turn can lead to the dreaded mental phenomenon known as "choking." As former tennis champion and current announcer John McEnroe has stated, "When it comes to choking, the bottom line is that everyone does it. The question is: how are you going to handle it? Part of being a champion is being able to cope with it better than everyone else" (Goffi, 1984, pp. 61-62). None of us are immune to this phenomenon; we are all faced with achievement situations in which we desperately want to perform well, only to have our strong need to do so unwittingly derail us. **Choking** is the sudden or progressive deterioration of performance below the typical and expected level of expertise for a person performing under pressure, wherein the person is seemingly incapable of regaining control over his or her performance. The problem is so common that the web is full of examples of the biggest and most famous chokes in sport history. Think about a time when you choked under pressure, and consider why it happened. What were you thinking, and how did that lead you to choke?

The increased arousal and anxiety that come with pressure situations can cause two problems with self-focus. First, people sometimes focus on their perceived inadequacies or fear that they won't be able to perform well. This inward focus then distracts them from the external focus that they need in order to complete the task with full concentration. The second, and larger, culprit of self-focused attention that leads to choking involves the attempt to consciously control performance. When people learn skills, they use conscious processing, which means that they think through exactly how to do the steps needed in order to perform the skill. We also use conscious processing when we try to fix a problem in our technique, such as working at the driving range to modify our golf swing. However, when we use conscious processing, high-level performance is not possible. To perform physical skills at a high level, athletes engage in automatic processing, during which they are not paying conscious attention to how they perform.

You can think of this mode as a form of autopilot. When athletes experience flow or peak

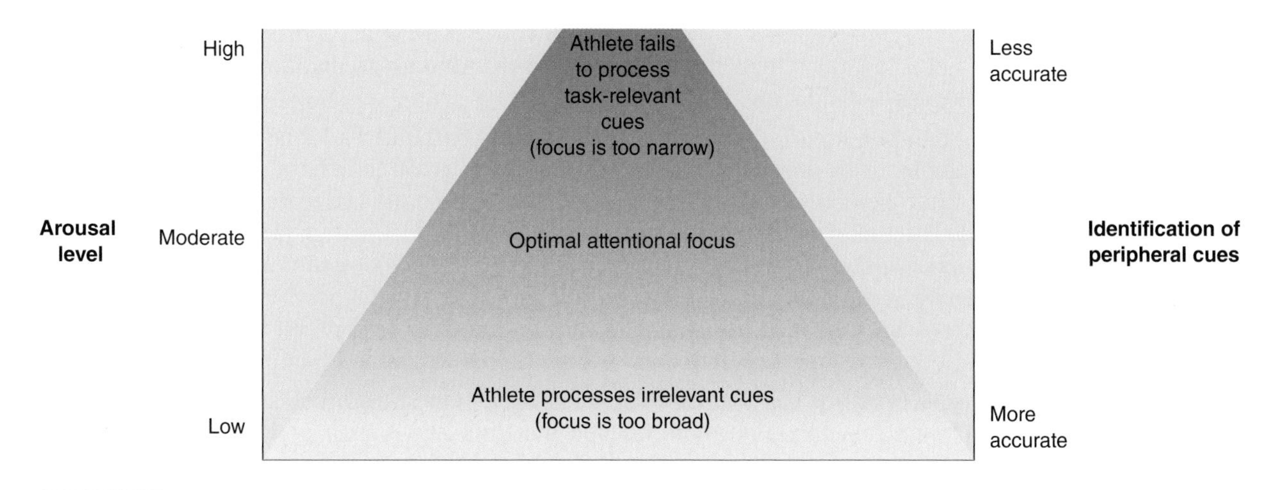

FIGURE 8.3 Relationship between attention and arousal–anxiety.

Reprinted from R.S. Vealey, 2005, *Coaching for the inner edge* (Morgantown, WV: Fitness Information Technology). By permission of R.S. Vealey.

performance, they typically struggle to describe what they were thinking or how they did what they did. This is the case because they were operating fully on autopilot, which is the performance zone that you want to be in so that your unconscious is controlling your performance.

Choking, then, occurs when athletes want so much to perform well that they pay attention to what they're doing—that is, they engage in conscious processing—which disrupts their autopilot and automatic performance response. And boy, do they pay! Anyone who has choked knows how devastating it is to try so hard (*too* hard, in fact) and yet fail. See the sidebar about choking for several tips on how to train to reduce choking and how to respond effectively in situations where choking may occur.

Mindfulness

Mindfulness is increasingly being used by people engaged in sport and exercise to enhance their performance and mental health. Consider the following quote by poet Apollinaire: "Now and then it's good to pause in our pursuit of happiness and just *be happy*." This is the point of **mindfulness**, which involves maintaining focus on the present moment in an open, nonjudgmental way. The intent of mindfulness is to engage fully in a present activity without being caught up in worries about the future or regrets about the past.

Practicing mindfulness has been shown to improve physical health in terms of stress release, improved sleep, lower blood pressure, and reduced chronic pain. It is also effective in treating depression, anxiety, substance abuse, and other mental health disorders. In addition, it is increasingly used as part of mental training in sport and has been shown to enhance athletes' performance and well-being (Baltzell, 2016). For example, mindfulness training is a major part of the mental preparation taught by sport psychology professionals at the U.S. Olympic Training Center.

Mindfulness is often taught through the practice of meditation. You can think of meditation as the art of focusing 100 percent of your attention on one thing. To give meditation a try, find a quiet location,

Choking: What Can We Do About It?

According to Sian Beilock (2010), who has studied choking in sport and other situations, we can reduce the tendency to choke, but we cannot eliminate it. It is a very difficult mental state to overcome, because we must release conscious control of performance. Here are some suggestions to reduce choking and to respond effectively when it happens.

- *Get used to it.* Practice under many types of pressure and get used to performing when others are watching and judging you.
- *Just do it.* The Nike slogan works here, as coaches recognize when they try to "ice" shooters or placekickers prior to their performances in high-pressure situations. Taking more time can lead to overthinking and thus overanalysis and controlled processing.
- *Distract yourself.* If someone tries to "ice" you, or if you're playing a sport with lulls in performance that allow too much thinking, use a preplanned strategy to distract yourself from becoming too focused on how you're performing.
- *Direct you attention out, not in.* Research on college baseball batters found that they performed best when focusing on where they wanted the ball to go off their bats, as opposed to thinking about what their hands were doing or what they wanted the bat to do (Castaneda & Gray, 2007). Similarly, elite golfers performed better when they focused on the desired flight of their shots, as opposed to thinking about their wrists or the position of the clubface (Bell & Hardy, 2009). More generally, many athletes pick an external focal point to look at (prior to batting or shooting or serving, for example) so that they can focus their attention on a target or on the outcome that they seek (rather than on *how* to make it happen).
- *Use holistic trigger words to cue your performance.* Research shows that skilled athletes perform better when they focus on more holistic thoughts about their performance (e.g., smooth, strong, sweep) rather than partial descriptions of how to move (e.g., quick hands, drive knee, arch back) (Mullen & Hardy, 2010).

put yourself in a comfortable yet stable position, and close your eyes. Take several deep breaths and attempt to breathe from your belly (not your chest). Try to maintain a passive attitude, without evaluation or concern about whether you're doing it right. When you have distracting thoughts, simply notice and acknowledge them, and then let them go. Try to focus all of your attention on your breathing for 60 seconds. When you catch your mind wandering (because it will), accept that as normal, let those thoughts go, and gently refocus your attention on your breathing. Don't worry if you struggle; it often takes people a lot of practice to meditate for 1 minute. Try this exercise throughout the day to restore your mind to the present moment and to create a sense of peace and calm focus.

Mindfulness and attentional focus are key skills for any performer in any physical activity situation, and these skills are trained through the use of mental techniques such as imagery and meditation. Other mental skills that are important for success in physical activity pursuits include confidence, energy management, self-awareness, and productive thinking. If you are interested in learning more about mental skills training, you can find many books and articles, including testimonials from athletes who have used these methods, in the scholarly and popular literature on sport and exercise psychology.

Even youth sport athletes are susceptible to burnout. To help prevent burnout, children should be exposed to many different sports and activities.

iStockphoto/Stuart Hannagan

Burnout

To cap off the chapter, this section focuses on a topic that is typically of great interest in sport and exercise psychology. Burnout has become more important because of the pressure we feel in modern society to perform, excel, and juggle many responsibilities at the same time.

"I don't know, I just feel burned out." How many times have you heard this statement from someone? What is burnout? Is it the same thing as stress? Is it real? Or is it a lack of mental toughness?

Burnout is real, and here's what it looks like. First, it involves feelings of mental, emotional, and physical exhaustion. Second, this exhaustion leads to negative moods and feelings (e.g., depression, despair) and a negative change in responses to other people (e.g., cynicism, aloofness, lack of empathy). Third, people experiencing burnout feel a lack of accomplishment even if they are quite accomplished, thus decreasing their performance level and feelings of self-esteem. Fourth, burnout causes people to become disillusioned with their involvement in an activity, such as sport, their career, or an exercise program. Burnout is a complex condition that occurs when certain personality characteristics of people interact with life stressors.

To help prevent burnout, coaches and trainers should encourage challenge and variety in training. For instance, people in fitness programs should cross-train in different types of activities—for example, biking, hiking, rowing, swimming, and perhaps even an activity such as orienteering or disc golf. Coaches should also build variety into sport training to break the monotony of repetitive training. In addition, they should help individuals distinguish between overload and overtraining. As the social rewards for sport success and fit bodies have become more glamorous, many people have come to think that more is better in training. In reality, people should focus on the quality of training and know when to push themselves and when to rest and adequately recover.

Furthermore, most people, especially children, should avoid intense specialization in one activity to the exclusion of all others. We enhance our mental health when we possess multifaceted identities or self-concepts with several areas of interest and expertise. Finally, burnout is not a sign of weakness or a simple response to stress. Thus, athletes and those involved in fitness training programs should watch for early signs of staleness and overtraining and attempt to stay committed in a positive, passionate way rather than feeling the pressure of *having* to engage in these activities.

Wrap-Up

As a young science, the field of sport and exercise psychology has only begun to scratch the surface of understanding the thoughts, feelings, and behaviors related to participation in physical activity. However, the knowledge base that has been developed over the last four decades is impressive, and researchers continue to study personality, motivation, arousal and stress, group processes, and mental skills training in physical activity. Kinesiologists are committed to extending and applying this knowledge in order to enhance participation in physical activity for all people.

If you are interested in what you've learned so far about sport and exercise psychology, you might want to take an introductory course in the subdiscipline, browse through journals in the field, and check out the websites of sport psychology organizations listed at the end of this chapter. You might also want to talk with a professor about career opportunities in either sport psychology or exercise psychology. We cannot all be Olympic athletes, but we can all engage in meaningful physical activity in order to derive personal fulfillment and optimal health and well-being.

MORE INFORMATION ON SPORT AND EXERCISE PSYCHOLOGY

Organizations

American Psychological Association

Association for Applied Sport Psychology

International Society of Sport Psychology

North American Society for the Psychology of Sport and Physical Activity

Journals

Case Studies in Sport and Exercise Psychology

International Journal of Sport and Exercise Psychology

Journal of Applied Sport Psychology

Journal of Sport and Exercise Psychology

Journal of Sport Psychology in Action

Sport, Exercise, and Performance Psychology

The Sport Psychologist

REVIEW QUESTIONS

1. What are the ABCs that are studied by kinesiologists in sport and exercise psychology? Identify questions that kinesiologists in this area might study based on these ABCs.
2. What was significant about Coleman Griffith's early work in sport psychology? Why did the subdiscipline not emerge again until the 1960s?
3. Identify the six methods used in sport and exercise psychology and provide one example of how each is used in the subdiscipline.
4. Does sport build character? Why or why not? Does exercise participation improve mental health? If so, how?
5. Explain the four sources, or fuels, for motivation to perform physical activity. Give specific examples of how these fuels can be put into practice by leaders in both exercise and sport contexts.
6. Explain cohesion in groups and discuss how it can facilitate sport performance as well as exercise adherence. What are some ways in which cohesion can be developed and nurtured in groups?
7. Although a distinctive athlete personality has not been found in terms of stable traits, what mental qualities have been shown to separate successful athletes from less successful ones?
8. Why do athletes choke? What are some ways in which sport psychology professionals can work with athletes to help them avoid the choking experience?
9. What is imagery, and how can it be used to make the physical activity experience more positive for people?
10. What is mindfulness, and how does the practice of meditation enhance it?

Go online to complete the web study guide activities assigned by your instructor.

9

Biomechanics of Physical Activity

Kathy Simpson

CHAPTER OBJECTIVES

In this chapter, we will

- describe what biomechanics is and what it encompasses;
- explain how biomechanics is useful to you and in careers in kinesiology and related areas;
- address what biomechanists and related specialists do;
- explain how biomechanics emerged within the field of physical activity; and
- introduce biomechanical concepts and the processes by which biomechanists and professionals in physical activity and related fields answer questions of interest in professional settings.

Aisha is a physically active, middle-aged female with mild degenerative knee osteoarthritis (OA; wearing away of knee cartilage) that is now causing stiffness, pain during certain movements, and decreased mobility. She wants to remain active at moderate to high-intensity levels. Therefore, she wants to know what she can do to prevent further cartilage degeneration and reduce the pain while staying active and fit.

To help her, imagine that you are serving in one of the following professional roles:

- Sport or personal coach, fitness instructor, or teacher
- Physical or occupational therapist, athletic trainer, or other rehabilitation specialist
- Medical personnel (e.g., orthopedist, nurse practitioner)
- Scientist in the research and development department at a company that manufactures knee braces
- Journalist writing to readers similar to Aisha
- Another relevant role of interest to you

Think of at least five pieces of information you would want to know that would help you answer Aisha's question. How might you obtain this knowledge?

In your role as a kinesiology professional in the situation just described, which of the following did you believe important to answering Alisha's question: movement technique; potential anatomical problems (e.g., knock-knee); injury, training, and physical activity history; weight; current footwear; other information? If you were to observe or measure Aisha's movements, how would you know if they posed a concern? How would you know whether information obtained from other sources was valid?

Kinesiology professionals use biomechanics to answer questions such as Aisha's by incorporating assessments of her movements and other information with research-based biomechanical knowledge of the factors affecting knee osteoarthritis. We will consider Aisha's situation throughout this chapter. Your opinions may shift as you learn about biomechanics and, later, knee OA.

Why Use Biomechanics of Physical Activity?

Biomechanics applies mechanical principles to understand how forces affect an organism's physical body (i.e., its structure) and the organism's ability to carry out its functions (i.e., pursue its purposes or goals) (Atwater, 1980; Hatze, 1974; Winter, 1985). Mechanics is the branch of physics

Goals of Biomechanics

- To understand how the basic laws of mechanical physics and engineering affect and shape the structure and function of the human body
- To apply this understanding to (a) improve the outcomes of our movements (i.e., performance effectiveness) or (b) increase or maintain the safety and health of our tissues

that documents motion (kinematics) and the causes of motion (kinetics). Within that branch, the field of biomechanics encompasses study of any living organism, be it bacteria, fungi, plant, animal, or human. Biomechanics can be applied to any part of an organism or to the entire organism. Because we focus here on biomechanics of physical activity—a subdiscipline of kinesiology—we use the term *body* to refer either to the entire human body or to any part of it (e.g., heart, muscle, leg).

Let's explore this definition of biomechanics and why biomechanics is useful. When Aisha exercises, one function that her heart must accomplish is pumping blood throughout her body. Meanwhile, her skeletal muscles function to move her body. For example, functions of her tibia (lower leg bone) include holding up her body and serving as a structure for muscles to pull against in order to create movement such as running. The tibia's structure consists of its anatomy, including its compositional elements (e.g., bone cells, minerals, proteins, water, marrow, connective tissue) and the way in which those components are put together and shaped. The tibia is shaped as a slightly bowed rod, with dense, compact bone on the outer part for strength and porous bone inside, which is arranged to help keep the marrow in place and reduce bone weight. This combination of structural elements allows the tibia to be lightweight enough to move rapidly yet also resist considerable force when muscles pull on it or when a performer hits the ground hard when landing.

Our structures help us to function—for example, to breath, run, text, and eat. Conversely, our movement (i.e., our functioning) affects our structure. For example, if we engage in heavy physical activity for months, then our tendons get thicker and stronger.

Forces are needed in order to accomplish any task involving movement; they also help us maintain the health of our body structures. A force consists of a push or a pull on an object or body. To kick a ball, for instance, you must apply force to the ball with your foot. For a hip replacement patient to stand up, she must be able to create muscle forces to push herself upward. Forces also are used by microscopic elements (e.g., proteins), cells, muscles, bones, connective tissues, and organs to accomplish their functions. For example, heart muscle creates force to pump blood. If we engage in sufficient physical activity—whereby forces are applied to bones, connective tissues, and muscles—then they become stronger and better able to carry out their functions. The converse is also true; that is, disuse

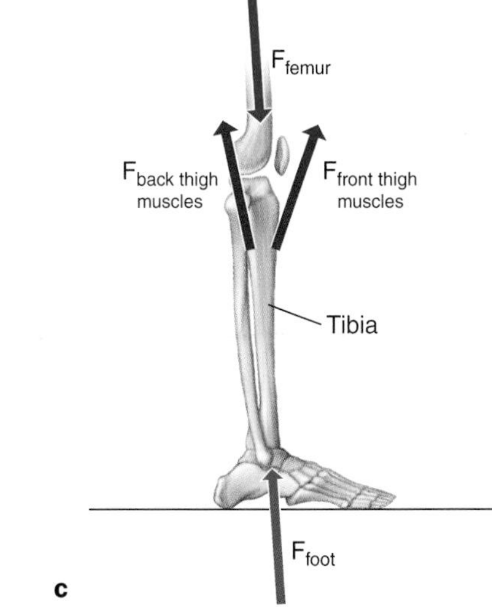

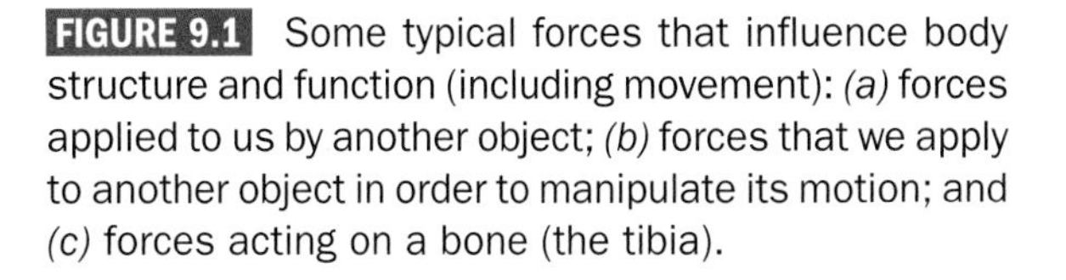

FIGURE 9.1 Some typical forces that influence body structure and function (including movement): *(a)* forces applied to us by another object; *(b)* forces that we apply to another object in order to manipulate its motion; and *(c)* forces acting on a bone (the tibia).

weakens body tissues. Excessive or repetitive forces can injure tissues, as in the brain cell damage that can be caused by a head impact.

The forces commonly involved in physical activity include those that you likely know about, such as air resistance, gravity, and friction (see figure 9.1). You have also experienced forces applied by another object (e.g., the ground) or person (e.g., collision with an opponent), as well as forces that you applied to another object (e.g., when lifting an object). Finally, you experience forces generated inside of your body (e.g., muscle forces pulling on bone, bones pushing or rubbing against one another at a joint).

Mechanical laws of physics and engineering are principles that explain how force is generated or manipulated or how force affects a system's structure or function (e.g., produces movement). Newton's law of acceleration tells us that a system's acceleration is higher when more total force acts on the system or when the system contains less mass. Thus, at the start of a sprint race, when high acceleration is needed, Aisha will want to generate as much muscle force as possible and carry only minimal excess fat mass. Another principle holds that if the amount of force applied to a structure exceeds a certain limit, then permanent damage occurs. This principle explains how certain injuries occur in bone, muscle, and connective tissue—for example, mild and moderate ankle sprains.

KEY POINT

Force is needed to (a) produce and control our movements and (b) maintain or improve the health of our tissues. Excessive or repetitive force, however, can injure tissues.

Biomechanics as a subdiscipline of kinesiology is often called *human movement* biomechanics. This is only one of several specialization areas in the field of biomechanics. For others, see the following list (Humphrey, 2003); note that human movement biomechanics can include other specialization areas (denoted by asterisks in the following list):

- Occupational biomechanics*
- Sport (dance) biomechanics*
- Orthopedic biomechanics*
- Forensic biomechanics
- Comparative biomechanics*
- Clinical biomechanics*
- Dental biomechanics*
- Continuum biomechanics
- Nanobiomechanics*

Goals of Human Movement Biomechanics

Human movement biomechanics pursues two main goals: discovery and application. The first goal is to discover the fundamental, universal theories that explain how forces affect our movements and our bodies' structures and functions. Here are some examples of fundamental theoretical questions:

- In making maximal-effort arm movements (e.g., throwing, striking, projecting an object as far and fast as possible), why do humans often generate muscle force in a sequential pattern, starting from the trunk muscles, moving outward down the arm, and ending with the hand muscles—like a whip?
- How are the structure and flexibility of connective tissues affected by mechanical loading factors, such as the speed and duration of stretching?
- What trunk movements apply the most and least force to spinal tissues?

The second goal is to use theory and biomechanical research to find answers to real-life questions and determine whether the theories work in real-life conditions or new theories are needed. Here are some examples of practical research questions asked by biomechanists in kinesiology:

- What weight-training exercises and loads are best for improving my client's sport performance?
- How do I know when it is safe for a student who had a moderately serious knee-ligament injury to return to physical education or dance class or to competitive sport?
- Should my patient, whose arm trembles when eating, use a heavier spoon to dampen some of the trembling effects and produce smoother hand movements?
- What can my client and I do to prevent her from sustaining another back injury at work?
- How can I convince coaches to purchase our company's protective gear when they assume that the gear will hinder athletes' movements?

If you become a kinesiology professional, biomechanical knowledge will help you answer questions about improving performance effectiveness, increasing the health of tissues, preventing and treating injuries, selecting or modifying assistive devices or joint or tissue replacements, and modifying sport equipment or rules in order to ensure safety.

The performance of any person can be improved in almost any situation that involves physical activity—for example, repetitive work tasks, activities of daily living, exercise, occupational or sport training, sport performance, music playing, dance, and even lying in bed. If we understand how living organisms can best exploit the mechanical laws that govern how motion is controlled, then we can intelligently select the best movement techniques for a given performer to use in completing a certain movement task. For example, a coach can determine whether a new technique may truly enhance performance or is merely a fad that offers little benefit. An occupational therapist can apply biomechanics in order to determine how to help a client with a neuromuscular disorder adapt his or her movements or use an assistive device (e.g., modified spoon) to perform daily movements.

We can also improve the health and safety of an individual's tissues. For instance, rehabilitation and training exercises are most effective if we possess theoretical and practical knowledge of how tissues are affected by conditions related to the application of force (e.g., speed of movement, amount of force). This knowledge also helps us understand how to reduce the risk of movement-related tissue injuries that can occur during work, daily life, and other physical activity. Consider this injury prevention question: Does stretching before we engage in strenuous physical activity reduce the risk of injury? See the sidebar on stretching for an answer.

Performance and tissue health can also be affected by the equipment and flooring that we use during human movement. Of course, the equipment can vary widely, ranging from bats to assistive devices (e.g., wheelchair, brace) to footwear and headgear. Biomechanical researchers can establish optimal characteristics for such equipment, including size, weight, racket shape, and characteristics of prosthetic limbs. In terms of flooring, we can use spring-loaded floors and mats in gyms and dance studios to reduce impact forces on the body during landings.

What Do I Need to Know to Use Biomechanics Professionally?

Most kinesiology degree-program specializations require at least one biomechanics course. In it, you will learn to apply the mechanical principles for manipulating movement, as well as other principles that can affect tissue health. You will also learn how to use practical methods and tools to answer human movement questions related to assessing or improving performance effectiveness or tissue health, including injury prevention (National Association for Sport and Physical Education, 2003). Because it takes considerable time and money to obtain most biomechanical measurements (see the section titled Research Methods for Biomechanics later in this chapter), most biomechanical assessments and diagnoses made by kinesiology professionals involve qualitative judgments based on observation (Knudson, 2013).

Practitioners and biomechanists often apply biomechanics in conjunction with knowledge from other disciplines and subdisciplines of kinesiology

Does Stretching Before Activity Prevent Injury?

You may have seen people get ready for physical activity by stretching and doing nothing else. They hold a stretch for a bit, then go on to the next static stretch. This type of stretching does not reduce one's risk of muscular or connective injury during subsequent activity; in fact, if a warm-up consists only of static stretches, it may result in injury.

A systematic review of well-done research, addressing many types of flexibility exercise protocols, showed mixed results depending on several factors, such as type of stretch and sequence of movements (Behm, Blazevich, Kay, & McHugh, 2016). So, what to do? The authors deduced that performing an appropriate warm-up prior to dynamic stretching—that is, stretching with slow movements, thus essentially blending stretching and warm-up—might reduce injury risk in muscle connective tissue and increase joint motion without reducing muscle force during the subsequent physical activity. In your own experience, has this warm-up strategy helped you prevent muscle pulls or connective-tissue tears?

(Knudson, 2013). For instance, anatomy, exercise physiology, and motor behavior are crucial to understanding how our body structures produce movement. A kinesiology professional recognizes that our movements are also affected by psychological, sociological, and cultural factors. Consider the question of why female athletes have a higher rate of anterior cruciate ligament (ACL) knee injury than do equally skilled male athletes (Hughes, 2014). Does this difference relate to sex differences in anatomy (answer: possibly) that cause females to land in a more "knock-kneed" position (yes, but not necessarily due to anatomy)? Or does it relate to effects of female hormone levels on ligaments (some contribution), or to gender differences in coaching (limited evidence), or to other cultural influences that affect how certain females perform movements or generate muscle force (limited evidence)? We would better understand the causes of ACL injury, and thus, how to prevent it, if this question were pursued not just in biomechanics but in an interdisciplinary fashion involving researchers from many subdisciplines of kinesiology.

KEY POINT

To make the best use of biomechanics in the kinesiology professions, we need to integrate work from the many kinesiology subdisciplines that help explain our health and physical activity behaviors.

What Do Biomechanists Do?

Career opportunities in biomechanics include such positions as researcher, clinical biomechanist, sport performance specialist, ergonomist, forensic biomechanist, and university professor. Related positions include certified orthotist and certified prosthetist.

Biomechanics researchers work in biomechanics laboratories, where they perform experiments that address problems of interest to various industries or assist with product development. For example, to design better footwear, a biomechanist working at a footwear corporation would collaborate with design engineers to understand the interaction between people's anatomy, the way they move, and the forces that act on them. A clinical biomechanist in a biomechanics laboratory at a hospital might work with physicians and therapists to understand how best to help patients regain normal walking patterns with medical treatment. The clinical biomechanist might also be expected to collaborate on research with physicians.

A performance enhancement biomechanist might work with collegiate athletes, elite athletes, or professional teams and their coaching staffs in order to improve athletes' performance. For example, a biomechanics company or sports medicine clinic could operate a facility in which biomechanists analyze athletes' techniques to assess performance effectiveness or detect injury-related errors. Several academic biomechanists have founded their own companies, which offer access to equipment, software, testing, and consultation in forensic biomechanics. This type of biomechanist may be hired, for example, by national-level athletic teams and dance companies to work with their performers as a consultant.

Biomechanists may also work in occupational settings as ergonomists or human factors engineers. Some ergonomists work in research and development departments as part of a team of people who design equipment such as gardening tools, commercial airplanes, factory equipment, and office furniture. Others may work for specialized ergonomics-focused corporations that perform job-site analyses. These analyses involve evaluating how and why employees perform their work tasks. The ergonomist generates data and then recommends appropriate modifications in tasks, equipment, employee training, or incentives in order to encourage employees to modify their behaviors in ways that improve their safety or efficiency. Meister (1999, p. 21) views human factors (in the North American terminology) and ergonomics (in the European terminology) as one field that covers "everything relating the human to technology."

Forensic biomechanists use biomechanical knowledge and principles to answer questions related to civil and criminal lawsuits (Schneck, 2005). For example, a forensic biomechanist might be hired as an expert witness in a civil lawsuit to testify about whether a biomechanical basis exists for an alleged work-related injury. Suppose that an employee develops a back injury and sues the employer. The employee alleges that the injury was caused by having to lift excessively heavy boxes of materials. The forensic biomechanist would try to determine whether the back-muscle forces required when performing the lifting task correctly exceed the maximums allowed by government standards and are high enough to injure this employee. The biomechanist would also have to determine whether it is likely that other factors caused or contributed to the injury, such as failing to use safe lifting techniques taught to workers.

Clinical biomechanists who work in medical settings, such as research hospitals that include

Kinesiology Colleagues

Photo courtesy of Kathy J. Simpson and Marika Walker

Marika Walker

Marika Walker is a PhD candidate at the University of Georgia who became interested and involved in biomechanics through an unusual route. In high school, after visiting the headquarters and research labs of a major sports apparel company, she thought, "Being a researcher in a sport company like this one—that's perfect. I would get to blend my love of sport, science, technology, engineering, and math in this type of work." Although her interests were defined early, her career path changed over time. She conducted research as an engineering intern at Nike and, as a graduate researcher, at the Textile Protection and Comfort Center at North Carolina State University. However, after talking with multiple mentors, she decided that in order to be a good researcher in the sporting goods and apparel industry, she also needed a strong background in human movement biomechanics.

biomechanics laboratories, perform biomechanical analyses of patients or research participants so that physical therapists and physicians can determine treatment. For example, several Shriners hospitals have a biomechanical gait analysis laboratory. In this setting, the surgeon, physical therapist, and biomechanist work together to assess whether a child with a disorder (e.g., cerebral palsy) requires treatment. If the child does receive treatment (e.g., surgery, physical therapy), they then work together to determine whether the treatment sufficiently improved the child's ability to walk. The biomechanist's responsibility is to perform and interpret the gait analysis and report the findings to the treatment team.

Although prosthetists and orthotists are not biomechanists, they use many biomechanical concepts and methods. Prosthetists and other professionals (e.g., physical therapists, podiatrists) sometimes also become certified orthotists. A prosthetist's main focus is to help a client obtain and use a replacement body part, such as an artificial hand. Orthotists, on the other hand, assess a client's body structure and function to understand what may be preventing a body part from completing its tasks effectively. For instance, injury or pain can result from anatomical deviations, which in children may result from abnormal bone growth, improper muscle functioning, neural defect, or some combination of these factors. The orthotist then fits the patient with an orthosis, a limb-supporting device such as a molded plastic brace that holds a child's foot in the proper position. Prosthetists and orthotists also reassess their clients' movements, watch for potential problems (e.g., injury, skin issues, ineffective movements), and make adjustments as needed.

Many biomechanists work in college or university departments of kinesiology, engineering, medicine, or other disciplines. A professor in a kinesiology department teaches biomechanics to students with interests in a variety of fields related to physical activity, rehabilitation, or medicine. The professor also conducts research in an area of biomechanics. Some biomechanists may also be clinicians and integrate patient care with their research and teaching.

History of Biomechanics

Throughout history—both for practical purposes such as survival and to satisfy our innate need to know and understand—we humans have been applying natural laws (often unknowingly) to investigate the structure and function of living organisms. The field of biomechanics, however, is relatively young.

Does a Career in Biomechanics Interest You?

To determine whether you're interested in a career in biomechanics, answer the following questions:

- Do you like observing and analyzing how people move?
- Do you find yourself trying to figure out a better way to perform a task?
- Do you enjoy trying to solve puzzles?
- Do you like the idea of applying the biology, physics, and math that you've learned to movement situations of interest?
- Do you enjoy using your mathematical skills?
- Are you interested in investigating the internal structures of the body, such as tendons, to determine how they act and are affected by factors such as exercise?

If you answered yes to some of these questions, consult the website of the International Society of Biomechanics for information about the field of biomechanics, job opportunities, graduate programs, and biomechanists available for contact. You can also talk with biomechanics professors to help you identify programs aligned with your interests.

Antecedents

Beginning thousands of years ago, antecedents of biomechanics were influenced by the science, math, and medical contributions of Middle Eastern, African, Near Eastern, Asian, and Greco-Roman civilizations. A second wave of antecedents began with the scientific Renaissance in Europe. One example can be found in Giovanni Alfonso Borelli (1608–1679), who is considered by some to be the founder of biomechanics. He understood correctly how muscles act and use leverage to move the body and how to calculate forces acting on or produced by organs (Pope, 2005). You may be more familiar, however, with the name René Descartes (1596–1650). Although he was a mathematician, he strongly hoped to understand the human body in order to cure disease and slow down aging. He found no cures for disease or aging, but he did conclude that the physical body was simply a machine that could maintain itself without assistance from a soul or essential being.

These and other antecedents of biomechanics established the philosophy that our bodies function simply as machines, a notion that still influences biomechanical thinking today. For instance, we talk about how the kneecap acts like a pulley enabling the front thigh muscles to extend the knee or how we might improve the "leverage" of the outside hip muscles in order to help patients after hip replacement surgery. At the same time, we also know now that movement is influenced by more than peoples' structures.

Beginnings

By the late 1800s and early 1900s, rapid progress in both knowledge and technology had made the time right for kinesiological sciences, including biomechanics, to bloom. Concurrently, the Efficiency Movement (late 1890s to early 1930s), popular in industrial countries, was thought to reduce waste and improve the quality of life (and increase profit for business owners). Frederick Winslow Taylor, an American engineer and a leader of the movement, scientifically studied the motions and time requirements of work tasks and generated principles for increasing workers' efficiency in order to improve productivity—for example, using certain movement techniques to reduce muscle fatigue (Taylor, 1911). Taylor-style efficiency analyses may have been some of the first assessments of human movement based on quantitative movement data. Taylorism also laid the foundation for the field of ergonomics or human factors engineering, and efficiency analyses began to be applied to sport movements in the 1920s.

In Russia, Taylor's ideas influenced Vsevolod Meyerhold, who was perhaps the first to use the term *biomechanics*, although its meaning then differed somewhat from that of today. Meyerhold was an innovative but controversial theater director who viewed movement as the most powerful mode of theatrical expression (Cash, 2015). He developed his version of biomechanics between 1913 and 1922 as a method of acting and theater production that integrated "laws of biomechanics" (from mechanical physics) with other components such

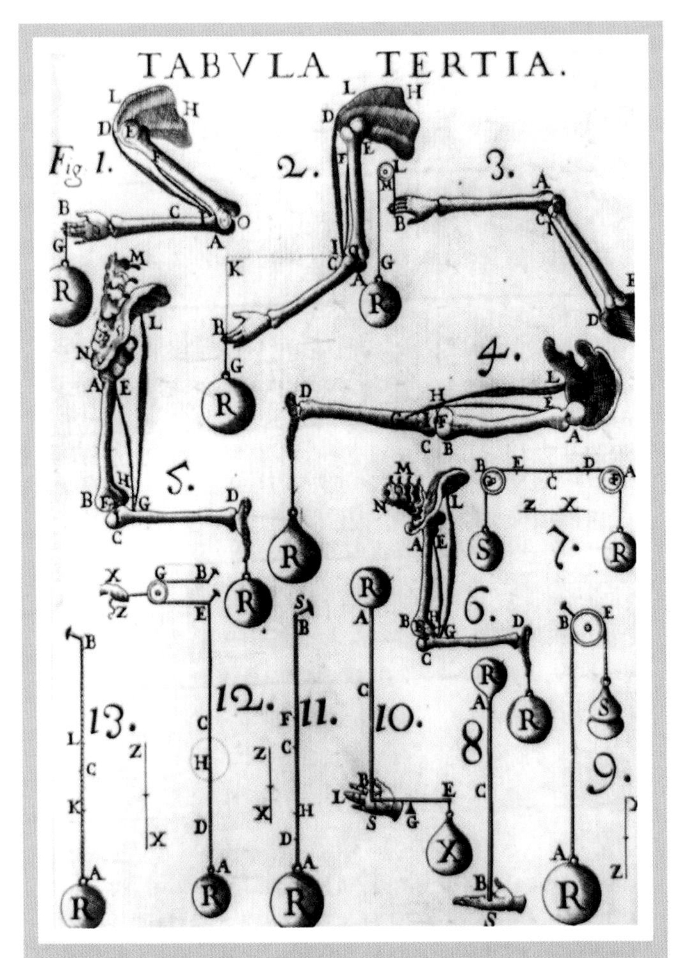

This drawing by Giovanni Alfonso Borelli compares the movement of human appendages and pulley systems using principles of mechanics and statics.

Courtesy of the Library of Congress, LC-USZ62-95253

KEY POINT

Human movement biomechanics was initially called *kinesiology* (in the late 1800s and early 1900s). The term *biomechanics* later became the accepted name for the entire field of biomechanics with its current definition.

as movement efficiency, balance, coordination, and rhythm (Pitches, 2004).

The development of biomechanics and other kinesiology subdisciplines in physical education, sport, and dance applications began in the very late 1800s and early 1900s. In the United States, two Swedish men (Posse, 1890; Skarstrom, 1909) interested in gymnastics were among the first to apply the term *kinesiology* (initially used in North America instead of *biomechanics*) to the analysis of muscles and movements in an education setting (Atwater, 1980). Posse's and Skarstrom's definition of *kinesiology* as the mechanical analysis of muscles and movements is still used today in some allied health and medical fields.

Mechanical analyses of basic movements were not emphasized as part of the formal training of dance and physical educators until 1923. At that time, Ruth Glassow and her students at the University of Wisconsin began three decades of work to classify activities into categories such as locomotion, throwing, striking, and balance. They also quantified movements (e.g., joint motions) using Glassow's pioneering film-analysis technology and applied fundamental principles of mechanics to understand the skills in each category (Atwater, 1980). Glassow used this knowledge to develop learning objectives for teaching physical education long before this practice became common in education (Sloan, 1987).

Many research leaders in university kinesiology programs in the 1930s and 1940s came from areas outside of kinesiology (e.g., medicine). Those who researched questions that were best answered by biomechanics, such as Glassow, had a profound effect on human movement biomechanics. They were the first modern scientists to answer such questions, which required them to develop the needed research methodology and tools. They also educated the first specialized biomechanists and were among the first professors to teach kinesiology preprofessionals how to use biomechanical principles.

Era of Contemporary Biomechanics

In the 1960s, biomechanics began to get established as a scholarly area and a recognized subdiscipline of university kinesiology programs. For one thing, newly created scientific societies and journals for biomechanics produced the first generation of an international community of biomechanics researchers. In addition, biomechanics courses and graduate-level university programs (Atwater, 1980) created pathways in the field of kinesiology for learning biomechanics or becoming a biomechanist. Since then, the world has seen a rapid expansion in the number and scope of national and international professional organizations and university programs in biomechanics (see the International Society of Biomechanics' website). Moreover, the many areas of biomechanics are now represented in these organizations and can be studied not only in kinesiology but also in other disciplines.

Ruth Glassow was one of several early biomechanists who profoundly affected the field. Her groundbreaking work led to an important foundation of biomechanical understanding of human movements.

University of Wisconsin—Madison Archives

KEY POINT

The era of contemporary biomechanics began in the late 1960s. Many biomechanists from kinesiology contributed to the development of biomechanics as a new area of scholarly study.

Research Methods for Biomechanics

What tools do biomechanists use to accomplish the goals of the field? For example, how do they perform gait analyses, research the cushioning properties of newly designed shoes, analyze the performance of Olympic athletes, and perform research to answer fundamental questions about how we move? In the pursuit of answers to such questions, biomechanics research has benefited greatly from technological advances.

What tools could you use to measure movements made or forces generated by a performer? Examples that may come to mind include timer and camera devices and apps for measuring time and motion; barbells and free weights for measuring lifting force; and tape measures for measuring distances or lengths. In addition, a variety of sports equipment (e.g., bats, basketballs, footballs, golf clubs) can now be equipped with sensors and software to indicate the equipment's speed and perhaps its accuracy. With video and still images and simple phone apps, you can also measure the body's speed and direction of travel and the distance that a limb traveled, quantify body movements and positions such as knee-joint angles during running, and assess trunk flexibility. Biomechanical researchers have used many tools, ranging from the ones listed here to more sophisticated systems. Figure 9.2 shows parts of instrumentation systems typically used in a biomechanics laboratory for research, clinical, ergonomic, performance, or other types of analysis.

Motion Measurement Devices

Did you know that animation methods used in video-game and movie animation began with biomechanical technology? The use of cameras and other motion-detection technology is called *motion capture* (or *mocap* for short). Digital high-speed cameras are used to trace the motion of reflective markers placed at selected points on a human body in order to reconstruct the motion of the various body segments. MEMS sensors, like those in a smartphone and in some sports equipment, combine miniature mechanical sensors (e.g., gyroscopes) with microcircuitry on a tiny chip. MEMS sensors used for mocap detect some combination of the following factors: acceleration, gravity, magnetic fields, and gyroscopic data. When one MEMS unit is attached to each body segment of interest (e.g., head, arm), the movements of those segments can be calculated.

For research purposes, mocap output can be used to understand movements or to provide feedback to performers about their movements in many activities, such as sport, dance, music, and physical rehabilitation. This technology can also be used to help practitioners analyze movements observed in occupational and clinical settings in order to improve people's functioning or prevent injury. In addition, mocap achieved through modified laser technology is used in some commercial games (e.g., Kinect, Move, Wii).

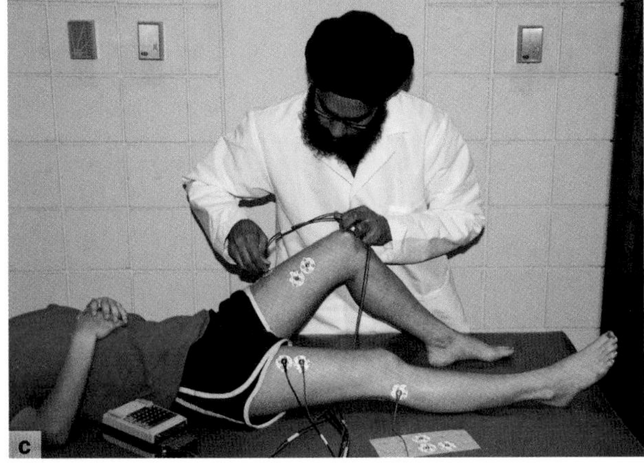

FIGURE 9.2 Biomechanical instrumentation. Components for motion measurement include *(a)* a digital, high-speed camera that tracks the location of reflective markers identical to those used in video game and movie animations; *(b)* microelectromechanical systems (MEMS) sensors that track body movements via miniature gyroscopes, accelerometers, and magnetic field detectors; *(c)* electrodes used in electromyography to measure electrical activation of muscle; and *(d)* force platforms used to measure ground reaction forces.

Photos *a*, *c*, and *d* © Kathy J. Simpson; photo *b* courtesy of XSENS.

Force Measurement Devices

When we need to know what is affecting a performer's movement or contributing to injury, we can measure quantities related to force. To measure the force placed on a joint, ligament, or object, tiny force-measuring devices (force transducers) can be surgically attached to or inserted into living tissues, artificial joints, or nonliving tissues. Transducers can also be used to measure how much force a patient can exert against a strength-measuring device. For instance, force platforms can be used to measure ground reaction force (GRF) when a performer applies force to the ground; this approach can help diagnose postural and balance disorders. GRF values are also used in calculating internal force loading on bones and cartilage during weight-bearing movements by means of various biomechanical models and laws of mechanics.

Pressure is determined by the amount of force applied to a given amount of surface area. Areas of high pressure applied to human tissue can cause

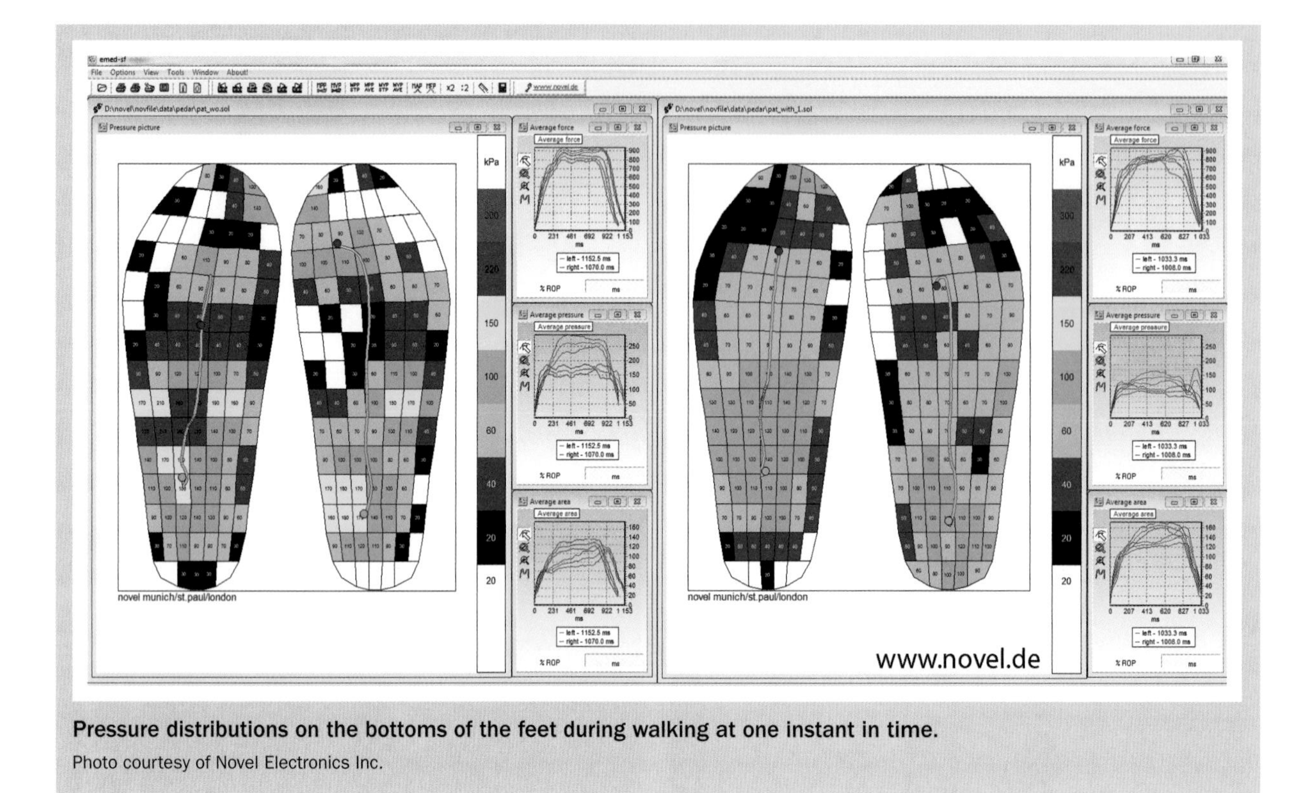

Pressure distributions on the bottoms of the feet during walking at one instant in time.

Photo courtesy of Novel Electronics Inc.

health problems. For example, in a person with diabetes who cannot sense foot pain, too much pressure repeatedly placed on one area of the foot can cause the skin in that area to deteriorate and become infected. Therefore, identifying high-pressure areas on the feet of a patient with diabetes helps the clinician tailor a shoe insert to spread the forces more evenly and thus prevent further tissue damage (Bus et al., 2016). Pressure devices can also be placed on wheelchair seats to test seat cushions in order to minimize pressure on patients who have little sensation in the lower body.

Because of the difficulty of directly measuring muscle forces, motion and force platform data are used to calculate estimates of muscle forces using biomechanical models. Another method of estimating muscle forces involves measuring the electrical activity of the membranes of the muscle cells (fibers) when nerve cells stimulate them to contract. This method, called electromyography (EMG), uses electrodes placed on or in muscles to record their electrical activity. The amount of electrical activity generated is a measure of muscle activation and can be converted into an estimation of muscle force when calculated using a mathematical model of human anatomy and muscle biomechanical properties.

EMG also is used for other purposes, such as identifying muscles that are active during a particular movement so that we know which muscles to target for training. Understanding muscle activation (timing and intensity) has allowed researchers to build a wearable device that stimulates the appropriate muscles with electricity at the right time, thus enabling a person with paralysis to walk. Electromyography can also be used by neurologists to assess whether muscles and nerves are functioning incorrectly.

Computer Simulations

All of the instrumentation described in the previous paragraphs measures one quantity or another. However, many internal biomechanical quantities (e.g., joint angles, muscle forces) cannot be directly measured; therefore, they must be derived through mathematical computations. Often, these techniques also require mathematical models of the body's anatomy and other structures, such as muscle attachment locations and bone lengths. Another approach is to use computer simulations to create virtual humans or other systems. Simulations can help us understand basic mechanics at the cellular, molecular, and whole-body levels. They also allow us to try out hypothetical scenarios.

How Do You Measure Up?

What instrumentation would you like to use to help answer a biomechanical question that you have? Returning to the chapter-opening scenario, what instrumentation could be used to assess the movements of Aisha's leg with OA and her healthy leg?

For example, simulated motion of the knee joint can be used to model the anatomy and forces involved in various structures (e.g., ligament, tendon, cartilage, bone, muscle), thus helping us to understand typical motion in the knee joint. We can then change the model's knee joint alignment to understand how poor knee alignment contributes to degeneration of knees with OA. For instance, a sport biomechanist could change the aerial technique of a virtual gymnast and watch the resulting performance to assess safety before having real gymnasts try the technique.

Overview of Knowledge in Biomechanics

Much of biomechanics consists of learning why and how to apply the mechanical laws of physics and engineering in order to answer questions about movement or tissue health. Consider the following: When going to your next class, you probably walk; you don't amble on your hands and feet like a 4-footed animal. Why not? Contemplate this: While all of her competitors (upper leg amputation classification) at the 2006 Paralympics wore their leg prosthesis to perform the high jump, Hai Yuan Zhang did not; she hopped on her intact leg to build the momentum she needed, and she won the event. Does this mean that all competitors in this classification should use this movement technique to jump higher? Could hopping on one leg increase the risk of injury compared to running? These situations are examples of two major questions to be explored: Why do we move in certain ways and not others? Is there such a thing as the perfect movement technique?

Moving Masterfully

"Why do people move the way they do?" "Is there a best technique for performing a given movement?" These are two of the biggest questions in human movement biomechanics, and practitioners must consider them carefully, regardless of whether they are working with clients or designing new equipment or assistive devices.

Movement technique involves using precise body positions and movements in a prescribed manner with proper timing. For example, for a squat exercise, you might position your feet shoulder-width apart and then bend at your hips, knees, and ankles without arching your back. Even before thinking about technique, however, we must always remember that how and why we move as we do is shaped by specific characteristics in the following five categories: task, performer goal, performer characteristics, environment, and laws of nature (including biomechanics). In addition, these factors interact with each other. Let's now consider an example.

The *task* of getting to school by riding a bike shapes how you move because it reduces the possible movements to those that can be done while riding a bike. The *performer's goal* in this situation is to get to the classroom before class starts. *Performer characteristics* can include anything related to the performer: body and mind, emotions, motivation, skill level, strength, and culture. For example, the biker's legs can move only in certain directions because bone and muscle structures constrain possible movements. Another factor is the

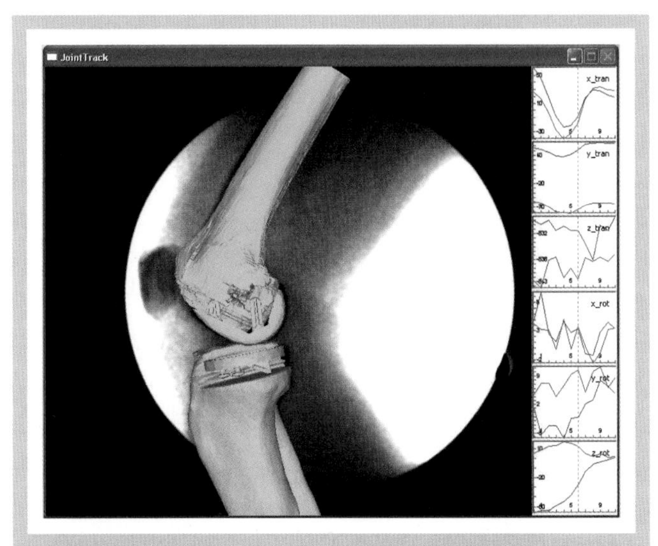

Modeling the movements of a knee joint implanted with a partial knee replacement.

© Yang-Chieh Fu

environment, which in this case includes weather (e.g., wind), the bike itself, the terrain, and traffic conditions. Relevant *laws of nature* include physical principles related to air resistance, gravity, and friction, as well as biomechanical principles of how tissues respond to forces applied to them. *Interaction* of factors might occur, for example, between rough-textured baggy clothing (environment) and air resistance (law of nature); by increasing the air resistance, the clothing would require the performer to generate more pedaling force to achieve the performance goal. Biomechanists are often most interested in the effects of interaction between mechanical laws of nature and the other factors.

Next, contemplate the notion of a "correct" or "best technique" in relation to the goals of the task and the performer. In competitive movements in which technique is judged (e.g., dance, gymnastics), movement technique is everything. However, what about other competitive activities? Ask yourself and others whether performing a given movement (e.g., high jump, running, soccer playing) with the "right" technique is important for performance success and why (or why not). How do we know whether the current technique or a new one provided by a coach is the "right" one? Referring back to the chapter-opening scenario, how is running technique relevant to Aisha's situation? Now, consider whether there is a "right" technique for movements with other purposes, such as work tasks (e.g., lifting objects or people) and daily activities (e.g., standing, sitting). Also consider characteristics that might affect "best technique"—for example, the performer characteristics of age, gender, skill level, and medical condition.

For competitive activities, we often assume that the correct technique is that of the world's best performers. After all, they must be doing it right in order to be the best—right? What if a movement technique works well only for, say, very tall individuals? Sport biomechanists compare the biomechanics of elite performers to lesser-skilled athletes in order to identify potential biomechanical factors related to high-level performance.

For factory work tasks, the employer's goals of high productivity and minimal production costs may influence what is considered "correct" technique. Potentially, if workers are paid by the number of tasks completed, then an employee's goal may be to complete the maximum number of tasks. In this case, the employee may use techniques that are the fastest but not necessarily the safest. In such a case, a responsible employer might use an ergonomist (occupational biomechanist) to determine what techniques will accomplish production goals while maintaining worker safety. Even then, however, it is not always possible to determine a "best" movement technique, particularly for tasks that are unusual or involve awkward work spaces.

Lastly, for activities of daily living, what technique is "correct" for individuals with movement disorders? The goals of a client with weak leg muscles may be to be stand up or sit down safely during daily life without assistance and without "looking foolish." Such a client will demonstrate a different pattern of movements than those who possess adequate muscle function. However, though we often assume that the movement techniques displayed by "healthy" individuals set the standard to achieve, they may or may not be appropriate for a given client. Even so, they may be useful in assessing a client's movement.

Here are the author's deductions about "best" movement technique. Do you agree?

1. Because how and why we move as we do are influenced by the characteristics of the five main categories discussed earlier and their interactions . . .
   - Most humans display relatively comparable basic movement techniques for some fundamental movements (e.g., running, hopping) due to similarities among us for some characteristics of the five main categories (e.g., basic skeletal structures, being subject to laws of nature such as gravity).
   - Variations among humans of those characteristics also means that the "best" technique for a given person is not necessarily the best for another. For instance, baseball pitchers all throw somewhat differently from each other, due in part to different shoulder joint and overall body structure. Similarly, a runner with shorter legs takes quicker, shorter steps to keep up with a longer-legged runner.
2. A "successful" movement technique is one that, regardless of how the movement is performed, enables the performer to accomplish movement goals (i.e., performance effectiveness) and health and safety goals at the highest level that the person can or wants to achieve.

- Better choices for movement technique are available when the performer takes advantage of biomechanical principles that best enable the performer to generate and control forces and movements. Given Newton's law of acceleration, for example, when shooting a basketball, a player whose arm muscles are weak and unable to produce enough force with one hand to accelerate the ball sufficiently can create more total force by shooting with two hands.
- The "best" movement technique may be optimal for achieving some goals but not others. This dilemma can be seen in both factory work and sport, where performance goals may be achieved at the cost of safety goals.

Each individual has a unique set of potentially conflicting goals and interactions among characteristics of the five categories. Therefore, it can be difficult for biomechanists to determine appropriate movement techniques that are common to most performers and challenging for professionals.

Balancing Performance Effectiveness and Safety

Let's explore how biomechanical principles and analysis can help an individual achieve both performance and tissue safety goals. Specifically, we return to Aisha's scenario from the chapter opening. Her performance goal is to win her age group during a national 10K running road race; her tissue safety goals are to minimize progression of her knee OA and remain injury free. Watch for conflicts between these goals due to force effects on running performance and knee joint tissues.

Performance Effectiveness

To ensure that Aisha can run at the fastest pace that she can maintain, we can look at the biomechanics of her entire body during various intervals of time within a single running cycle. Here, we only examine one small interval—the braking phase, or roughly the first half of the time period when her foot is in contact with the ground. First, we describe some of the motions and body positioning. A runner's foot typically strikes the ground slightly in front of the body, and then the person's body rotates over the foot. During the braking phase, the runner loses a small amount of speed and forward momentum; of course, she wants to minimize this loss of momentum.

Second, because forces affect Aisha's motion, let's identify the forces acting on her during this phase (see figure 9.3). There is of course the force of gravity on her body weight. We also consider whether air resistance and wind forces are acting (assume that they are negligible). Next, we search our biomechanical toolbox for principles to help us identify forces that could be generated between the foot and the ground. According to Newton's third law (action–reaction), for every force applied by the foot to the ground, the ground will apply an equal and opposite-acting force back to the foot. Because Aisha's foot is in front of her body at contact, her momentum causes her foot to push forward against the ground (action force); in reaction, the ground applies a backward force (ground reaction force) to her body, thus opposing her motion and slowing her down. Simultaneously, she creates another set of action–reaction forces in the vertical direction by pushing down against the ground; the reaction force here is upward.

Our third step is to determine the biomechanical principles that could help Aisha minimize her loss of momentum. Here, we want principles that explain how forces affect momentum. We identify the principle stating that the amount of momentum gained or lost by a system depends not only on the amount of force but also on the length of time for which the force is applied. Therefore, the higher the backward GRF acting on Aisha, and the longer it acts on her, the more momentum she loses.

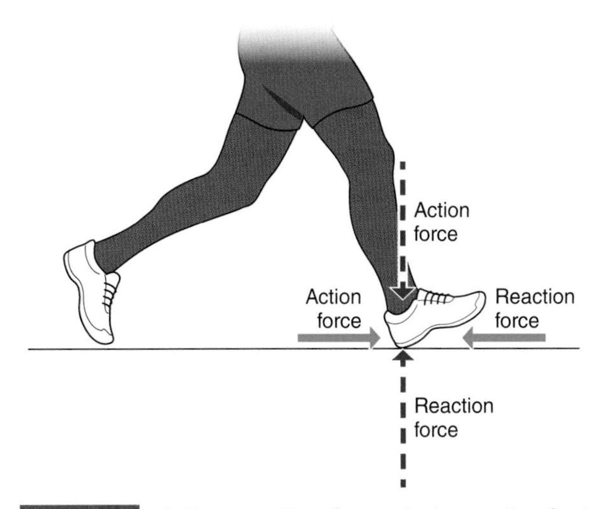

FIGURE 9.3 Action–reaction forces between the foot and ground during the braking phase. The ground reaction forces acting on Aisha affect her body's vertical and horizontal momentum.

Risk Factors for Knee Osteoarthritis

Osteoarthritis causation exemplifies how forces can affect tissue structure and health. OA has no cure and can become severe, causing excruciating pain and disability, due primarily to continued wear of the joint cartilage and development of bone abnormality (National Institute of Arthritis and Musculoskeletal and Skin Diseases, 2015) (figure 9.4).

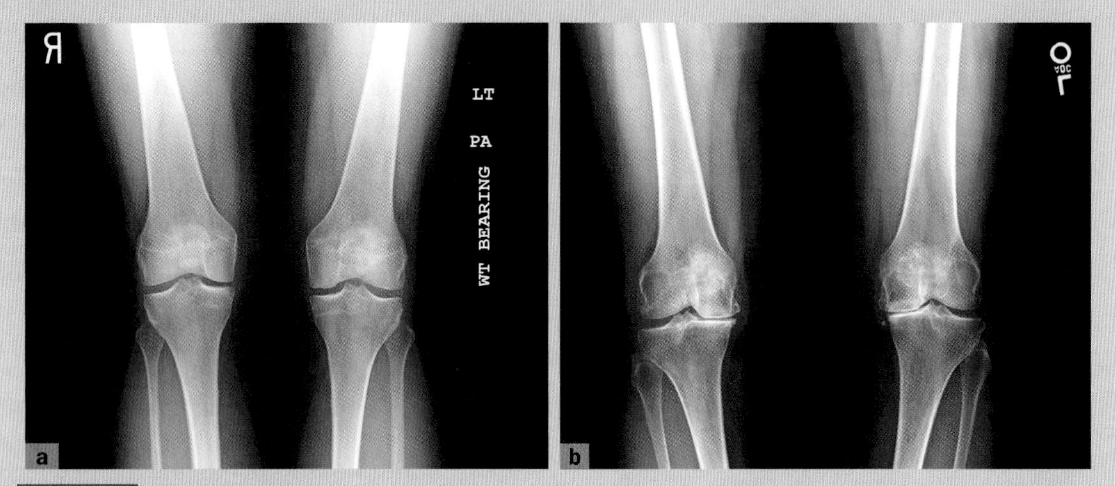

FIGURE 9.4 Radiographs of *(a)* a healthy knee and *(b)* an osteoarthritic (OA) knee. The space between the OA bones is narrowed because the cartilage (which isn't visible on a radiograph) is worn down, particularly on one side of the knee joint.

What might be your highest risk factors? For reasons that are poorly understood, previous knee joint injury is the predominant factor causing young-adult knee OA, which tends to occur 12 to 14 years after the injury (Roos, 2005; Vannini et al., 2016). Otherwise, obesity is the highest risk factor for adults in developed countries because it causes a continuing cycle of damage. Specifically, obesity creates abnormal biochemical processes and high tissue loading due to high body weight, which leads to atypical structural changes in bone and weaker, damaged cartilage (Silverwood et al., 2015). Another risk factor is the length of time spent during one's life performing high-impact, quick-stop activities, which apply high loads to knee cartilage ranging from two to four times body weight during level walking to as much as twenty times body weight during landings in certain sports (Vannini et al., 2016).

Fourth, we seek to determine how to manipulate the crucial quantities, and this is where we consider Aisha's movement technique. Specifically, how can we reduce the amount of backward-directed GRF or the time spent in the braking phase? If Aisha is placing her foot too far in front of her, or landing on her heel instead of on her forefoot, then she is likely creating too much backward-directed GRF that will act for too long. This is also where we find the first potential conflict between performance and tissue safety: If Aisha changed her footstrike to a forefoot landing in order to shorten the time acted on it by GRF, she might also increase the loading force on her calf muscles or foot connective tissues that could eventually cause tissue injury.

Tissue Health and Safety

In our ongoing example involving Aisha and her knee, the goals for tissue health and safety are to prevent injury and slow down the OA progression. Here again, we examine the braking phase of running, but our interest now focuses on the biomechanics of the tibia. We also limit our analysis here to the vertical direction, because it involves the highest bone and knee-joint forces that could damage tissue. First, for simplicity, we identify only some of the vertical forces acting on the tibia. The downward force that the femur applies to the tibia's articular surface is created, in part, by Aisha's downward momentum at touchdown, which is a function of her mass and the speed at which she

Professional Issues in Kinesiology

Brawny Bones and Overload

According to Klentrou (2016), who reviewed factors related to building bone strength, about one in every three women and one in every five men will suffer an osteoporosis-related bone fracture. Bone strength is the critical property of bone health, and it is built during childhood through the early adult years. During this period, appropriate mechanical loading is essential to building bone strength. Bone is stimulated to become stronger when it experiences an "overload," in which bone load is greater and applied more rapidly than is typical for the individual. Therefore, physical activity is crucial. Because muscle and ground reaction forces are primary sources for bone loading, bone strength can be improved by enhancing muscle strength and performing movements (e.g., jumping) that generate high forces and are performed rapidly. In contrast, insufficient physical activity reduces mechanical loading and is related to bone health deficits. At the same time, bone is also weakened by excessive overload. Researchers are still investigating the optimal loading conditions for building bone without inflicting damage during the critical age period.

Citation style: APA

Klentrou, P. (2016). Influence of exercise and training on critical stages of bone growth and development. *Pediatric Exercise Science, 28*, 178–186. doi:10.1123/pes.2015-0265

is moving. The rest of the femur force is created as a reaction to the tibia being pulled and pushed upward against the femur by knee joint muscles and vertical GRF transmitted to the tibia.

Next, we seek mechanical principles that explain how tissue health and safety are compromised. We know that we can help keep bone, connective tissue, and cartilage healthy by applying mechanical load (force) in optimal amounts for an optimal number of repetitions. However, another principle holds that high total force combined with too many repetitions (i.e., number of steps) over a long period of time can weaken these tissues, which means that less force is required to create permanent damage when loaded. Thus, high volumes of running in training coupled with high tibial forces could result in overuse injuries in the tibia (e.g., stress fracture) or connective tissue (e.g., tears), or, for Aisha, progression of knee OA.

Now, we analyze how we might reduce the forces and the number of times that they are applied to Aisha's tissues. In terms of movement technique, running more slowly could decrease all of the vertical forces that affect tibial loading, but that is unrealistic in this case because she wants to compete. Another option would be to strike the ground using the heel rather than the forefoot, which would create 14 percent less knee-joint force (Rooney & Derrick, 2013), perhaps due to a reduction in the vertical GRF and muscle forces transmitted to the tibia. Here again, we see the conflict between performance and safety, because a forefoot strike might be better for performance but might also increase the risk of stress fracture or OA progression. In terms of environmental considerations, Aisha could run on relatively soft, stable surfaces to reduce vertical GRF. She should also evaluate her training schedule to ensure that her training volume is no higher than needed and thus avoid excessive repetitions of tibial loading.

Wrap-Up

The physical laws of nature shape our movements and our body tissues. Biomechanists and kinesiology professionals can apply mechanical principles and laws to enhance performance effectiveness; maintain or increase the health of tissues and prevent injury; and select or modify assistive devices, joint or tissue replacements, equipment, sport rules, and prescribed techniques. When we understand how mechanical principles influence our body functioning, movement, and structure, we can apply that knowledge to work, leisure activities, sport, exercise, dance, daily tasks, rehabilitation—indeed, to any action involving movement or forces acting on or within the body.

MORE INFORMATION ON BIOMECHANICS OF PHYSICAL ACTIVITY

Organizations

Gait and Clinical Movement Analysis Society
International Society of Biomechanics
International Society of Biomechanics in Sports
International Society of Electrophysiology and Kinesiology
International Society for Posture and Gait Research

Journals

Applied Bionics and Biomechanics
Clinical Biomechanics
Gait & Posture
Human Movement Science
International Biomechanics
International Journal of Experimental and Computational Biomechanics
Journal of Applied Biomaterials and Biomechanics
Journal of Applied Biomechanics
Journal of Biomechanical Engineering
Journal of Biomechanics
Journal of Dance Medicine & Science
Journal of Electromyography and Kinesiology
Journal of Prosthetics and Orthotics
Sports Biomechanics

REVIEW QUESTIONS

1. What is similar and what is different between the meanings of the terms *biomechanics*, *human movement biomechanics*, and *kinesiology*?
2. What is meant by *structure* and by *function*? Give two new examples for each term.
3. What are the goals of the study of biomechanics?
4. What are the primary application areas of biomechanics in kinesiology?
5. In what types of settings do biomechanists typically work?
6. What major instruments are used in biomechanics research?

Go online to complete the web study guide activities assigned by your instructor.

10

Physiology of Physical Activity

Jennifer L. Caputo

CHAPTER OBJECTIVES

In this chapter, we will

- cover the key features of the subdiscipline of physiology of physical activity and the employment opportunities available to exercise science professionals,
- explain how physiology of physical activity fits within the discipline of kinesiology,
- review the history and development of physiology of physical activity as a subdiscipline,
- identify the research methods used by kinesiologists working in exercise physiology, and
- examine how the body responds to physical activity and how these changes relate to physical performance and health.

Does using a standing desk at work improve health? This is the question asked by staff members in the human resources department of an insurance company. In an effort to decrease the company's health care costs, the staff are looking into purchasing desks that allow employees to stand while reading and reviewing policies during the work day.

A coach who is designing a conditioning program wonders when to schedule cardiovascular and strength training sessions for her athletes. She has read research reporting that strength training benefits may be reduced when cardiovascular and strength training are conducted concurrently, whether in the same session or close in time to each other.

Many people now use wearable technology to measure their physical activity (e.g., monitor intensity), estimate their caloric expenditure, and even monitor their sleep patterns. While shopping at a local fitness store, a customer reads the boxes of several fitness monitors and wonders about the accuracy of the measurements and the marketing claims.

As these scenarios suggest, individuals who work in the kinesiology subdiscipline known as physiology of physical activity have the opportunity to help people in a variety of settings and situations. Exercise physiologists generate key knowledge in kinesiology that can enable evidence-based recommendations for such situations.

The questions raised in the chapter opening scenario represent some of the areas of specialization available in physiology of physical activity. If you are curious about how your body and your health status are altered by either physical activity or a lack thereof, then this subdiscipline will be of great interest to you. With knowledge of this field, you can help athletes achieve peak performance; help people participate safely in physical activity; conduct research about preventing and treating disease by means of physical activity and exercise; and even help maximize the capacity of people who work in extreme conditions of heat, cold, altitude, underwater settings, or even the microgravity of space.

Students of physiology of physical activity seek to understand both how the body responds in order to meet the immediate demands of being physically active and how it adapts to repeated bouts of exercise. For instance, when you chase down a shot on the tennis court or go for a jog, your heart rate and breathing rate increase; in addition, your blood pressure increases, and you may begin to sweat. These changes are your body's way of meeting the increased demand for oxygen, energy, and temperature regulation. If you were physically active across several weeks, you might notice changes in your body weight and muscle mass. These physiological responses to physical activity and exercise are just a few of the topics that intrigue students of physiology of physical activity.

Often referred to as **exercise physiology**, this is a subdiscipline of the biophysical sphere, along with biomechanics of physical activity (discussed in chapter 9). However, whereas biomechanists apply principles of physics and engineering to the mechanical causes of physical activity, exercise physiologists apply principles of biology and chemistry to understand the acute and chronic responses to physical activity.

Why Use Physiology of Physical Activity?

From chapter 1, we know that exercise is a form of physical activity used to improve or regain performance, health, or physical appearance. Exercise physiology primarily focuses on the exercise components of physical activity, including training, detraining, and how participating in physical activity can improve health and decrease the risk of all-

Goals of Exercise Physiology

- To understand how to enhance physical performance
- To understand how to improve physical function in altered environments, such as those characterized by high temperature or high altitude
- To understand how physical activity and exercise improve health and fitness
- To understand how exercise can be used in treating and preventing disease and alleviating symptoms of disease
- To understand adaptations in physiology and pathophysiology in response to physical activity

cause mortality. The following sections summarize some of the benefits derived from understanding the subdiscipline of exercise physiology.

Enhancing Sport Performance and Training

Applying physiological techniques in order to understand and improve human exercise performance has been a major goal of exercise physiology since its inception. The principles of physiology of physical activity form the foundation for developing conditioning programs for various athletes. This area of exercise physiology, often called *sport physiology,* involves applying "the concepts of exercise physiology to enhancing sport performance and optimally training the athlete" (Kenney, Wilmore, & Costill, 2015, p. 3). Think about a track coach who needs to develop running programs for sprinters, middle-distance runners, and distance runners. The variety of speeds and distances covered by these athletes poses special challenges for the coach, who must optimize the training program for each group of runners.

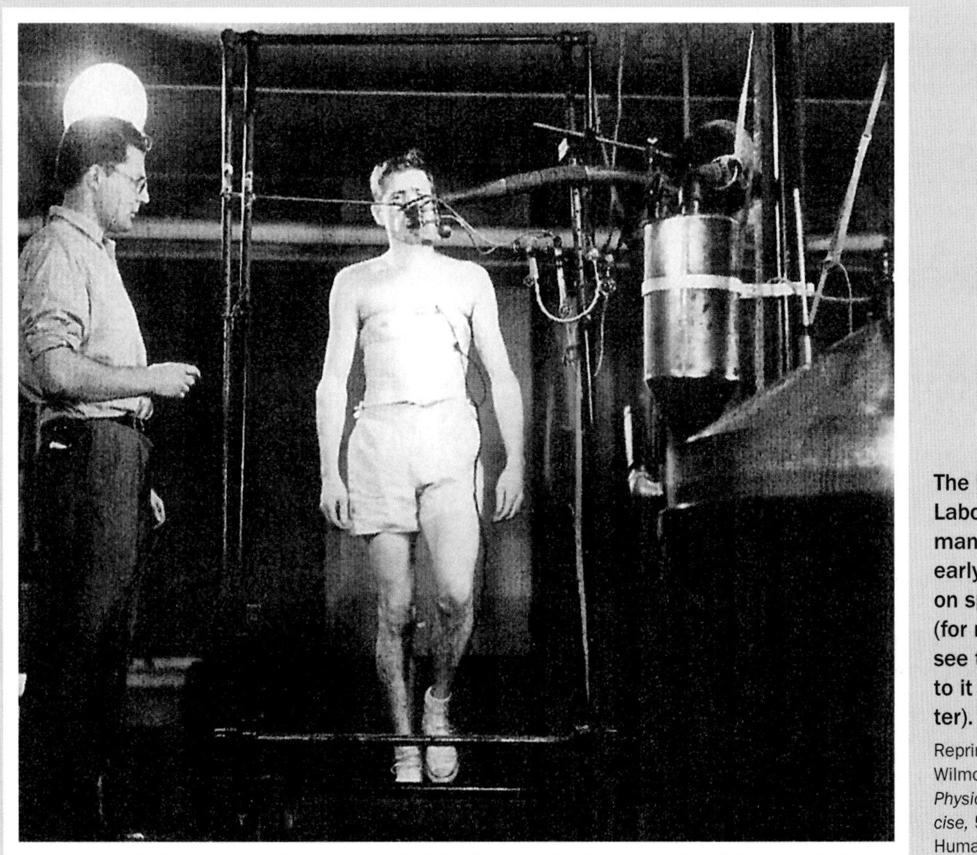

The Harvard Fatigue Laboratory conducted many studies in the early years of research on sport performance (for more about the lab, see the sidebar devoted to it later in this chapter).

Reprinted from W.L. Kenney, J. Wilmore, and D. Costill, 2012, *Physiology of sport and exercise,* 5th ed. (Champaign, IL: Human Kinetics), 7.

Some of the earliest studies conducted on physiological responses to strenuous exercise and training came from the Harvard Fatigue Laboratory (Dill, Talbott, & Edwards, 1930; Margaria, Edwards, & Dill, 1933). This research was influential given the dramatic changes in physiological responses seen in vigorous exercise and sport. Research by sport physiologists includes topics such as fluid and food recommendations before, during, and after exercise; the role of genetics in exercise and physical activity performance; sex differences in brain injuries during athletic events; and the use of wearable technology to monitor performance. Sport physiologists also use information and techniques from other disciplines and subdisciplines to study ways of enhancing performance. For example, they use information gleaned from nutrition research in working with sport nutritionists to manipulate athletes' diets in order to enhance carbohydrate storage and improve performance.

Sport physiologists are also concerned with the effects of the environment on sport performance. For instance, data from sport physiology studies have been used to develop guidelines to avoid heat illness in sport and prevent health problems brought on by prolonged immersion of scuba divers in deep water. In another example, before the 1968 Summer Olympics in Mexico City—which sits 7,218 feet (2,200 meters) above sea level—leading exercise physiologists conducted numerous studies on the acute effects of, and acclimatization to, high altitude. Sport governing bodies used the knowledge gained from these studies to better prepare their athletes for the competition.

Improving Physical Fitness

Studying exercise physiology also helps us understand the physiological determinants of physical fitness and the ways in which training programs improve fitness. In the 1950s, interest in improving physical fitness was sparked by two events in the United States—President Dwight D. Eisenhower's heart attack and the publication of a study reporting that American children were less fit than European children (Berryman, 1995). These events combined with concerns about young men's fitness for military service to spark interest in and development of exercise physiology research on physical fitness.

Research conducted in exercise physiology and other areas of kinesiology over the past 50 years has resulted in recommendations for optimal intensity, frequency, and duration of training programs in order to develop various components of physical fitness. Just as no distinguishing athletic personality has been discovered, early kinesiology research concluded that there is no uniform quality of athletic ability; instead, fitness for physical activities is specific to identifiable components. Individuals who work in fitness centers need to understand this research, as well as findings about how to adapt fitness programs to make them safe and effective for many kinds of clients—young and old, sedentary and trained, and those with special conditions (e.g., pregnancy, diabetes). For instance, pediatric exercise physiologists specialize in studying and working with children and adolescents, and other exercise scientists focus on gerontology (the study of aging). In addition, the influence of heredity on physical fitness components and trainability has attracted considerable research interest among exercise physiologists. Much of this interest has been stimulated by research on identical twins conducted by Claude Bouchard and colleagues (1986).

Promoting Health and Treating Disease With Physical Activity

Exercise physiology serves as a foundation for understanding why physical activity and exercise help reduce the risk of disease and help treat some forms of disease. In 1996, the Surgeon General of the United States emphasized that "significant health benefits could be obtained by including a moderate amount of physical activity on most, if not all, days of the week" (U.S. Department of Health and Human Services [USDHHS], 1996). For instance, many researchers have examined the benefits of physical activity and physical fitness in

Strength and Conditioning Specialists

When working with athletes in the weight room, strength coaches need to understand how to alter conditioning programs in order to meet the needs of athletes from many sport teams. Strength coaches learn how to manage the amount of weight lifted, the number of repetitions and sets performed, and the length of rest periods assigned in different ways in order to produce appropriate responses in athletes as diverse as sprinters, basketball players, softball players, and football players.

preventing and even reversing coronary heart disease (Lawler, Filion, & Eisenberg, 2011; Thompson et al., 2003). Researchers have also documented the role of physical activity in preventing and treating other diseases, such as osteoporosis (Vuori, 2001) and non-insulin-dependent diabetes (Conners, Morgan, Fuller, & Caputo, 2014). Researchers are now developing exercise tests to screen for obstructive sleep apnea and examining the role of exercise in decreasing the risk of dementia and Alzheimer's disease.

Some exercise physiologists also focus on the big picture—how physical fitness can prevent disease in large populations. In these cases, exercise physiologists might collaborate with specialists in epidemiology in conducting two types of studies: longitudinal (comparing the same group of people over a period of time) and cross-sectional (comparing different groups of people at the same time).

Physiologists who study the role of physical activity in disease management and rehabilitation are known as *clinical exercise physiologists*. Students who train in this area become familiar with basic concepts in several medical specialties, such as cardiology and pulmonary medicine, and work under the direction of a clinician. For example, students interested in cardiac rehabilitation need to learn the forms of cardiovascular disease, how they are diagnosed, and the appropriate medical and pharmacological treatments. They also gain knowledge of the cardiovascular system and its responses to exercise. In addition, they learn to read an electrocardiogram (ECG), which traces the heart's electrical activity. Cardiac rehabilitation specialists learn how to modify exercise programs for people who have high blood pressure, have had a heart attack (myocardial infarction), or have had open heart surgery.

Understanding Physiological Changes From Physical Activity

Early work in exercise physiology focused on the ways in which exercise affects the functioning of organs and body systems. More specifically, exercise physiologists have examined how the cardiovascular, respiratory, muscular, and endocrine systems respond functionally and adapt structurally to different types of physical activity. Recently, research has focused on the responses of the reproductive, skeletal, and immune systems to acute and chronic exercise.

One branch of exercise physiology is closely linked to **biochemistry**—that is, the chemistry of living things. For example, research techniques developed in biochemistry (and physical chemistry) have been used by investigators to examine the fuel sources used by skeletal muscles during exercise. In another example, one of the most important advances in exercise physiology resulted from the use of muscle biopsy needles by Bergstrom and Hultman (Hultman, 1967) to sample muscle tissue and examine muscle glycogen (stored carbohydrate) concentration during exercise. Biochemical techniques have been used to develop our understanding of lactate production and use of energy stores during exercise.

In addition, research in molecular biology, one of the newest subdisciplines of biology, has greatly enhanced our understanding of how cells function. Biological techniques are used by molecular scientists to determine how genes regulate protein synthesis. Exercise physiologists use techniques of molecular biology to study how muscle protein synthesis is turned on and off by changing levels of muscle activity (Booth, 1989). This molecular knowledge is important in understanding how muscles increase and decrease in size in response to changes in physical activity; it also contributes to understanding of how muscles are damaged and how they recover from injury.

What Do Exercise Physiologists Do?

Scholars who study exercise physiology typically work as university faculty members, although employment opportunities also exist in clinics, hospitals, and research centers. Researchers and professors usually hold a doctoral degree in exercise physiology. University faculty members typically teach courses in exercise physiology and conduct research on topics such as the effects of conditioning programs on sport performance and the effects of physical activity on reducing the risk of chronic disease. To support their research, faculty members often write research grant proposals to federal agencies (e.g., National Institutes of Health) and foundations (e.g., Robert Wood Johnson Foundation). Exercise physiology scholars also work in corporate and government laboratories. For example, research funded by the U.S. military and by the National Aeronautics and Space Administration (NASA) is conducted in university, government, and medical laboratories.

Graduates who hold bachelor's or master's degrees from kinesiology programs may also have careers in clinical exercise physiology, often with additional training and certifications. These individuals provide clinical exercise testing and prescription in affiliation with medical clinics or

Kinesiology Colleagues

Photo courtesy of Jennifer L. Caputo

Sandy Stevens

Walking poses a challenge for thousands of people living with disability. Some find a source of help in Sandy Stevens, a faculty member in exercise science in the Health and Human Performance Department at Middle Tennessee State University, who puts her knowledge of kinesiology to work every day in helping people once again place one foot in front of the other. Stevens trains people with spinal cord injuries to walk using underwater treadmills; over time, her participants often develop the strength and skills to begin gait training on land. The program, which receives financial support from the National Institutes of Health, draws participants from around the world. Stevens and her kinesiology colleagues apply their knowledge of how to train the human body to the unique characteristics of an aquatic environment in order to make a difference in the lives of people living with disability—one step at a time.

hospitals to help patients with cardiac and pulmonary rehabilitation. Kinesiology graduates can also work as exercise instructors or personal trainers in commercial and corporate fitness centers or as strength and conditioning coaches. Current knowledge of exercise physiology is essential in providing the most effective training for physical activity.

History of Physiology of Physical Activity

Physiology of physical activity evolved from physiology in the 18th century after Antoine Lavoisier and Pierre de Laplace developed the methodology to measure oxygen consumption and carbon dioxide production during respiration in animals (USDHHS, 1996). Across the following century, these techniques were further developed and applied to studying how humans responded to physical activity and daily tasks (e.g., lifting loads).

Early Beginnings

Two of the early contributors to our understanding of physiology of physical activity were August Krogh of the University of Copenhagen and A.V. Hill of University College, London. Krogh developed one of the first cycle ergometers (i.e., exercise bikes), which he used to study physiological responses to exercise. He received the Nobel Prize in Physiology in 1920 for his research on the regulation of microcirculation (Åstrand, 1991). In 1921, Hill received the Nobel Prize for his work on energy metabolism. In a classic paper, Hill and Lupton (1923) presented many of the basic concepts of exercise physiology related to oxygen consumption, lactate production, and oxygen debt (i.e., excess oxygen consumption following exercise).

Our early understanding of exercise physiology was also shaped by research conducted at the Karpovich and Cureton laboratories established in physical education departments in 1927. Dr. Peter V. Karpovich established an exercise physiology laboratory at Springfield College in Massachusetts (Kroll, 1982). Karpovich became well known for his research on the effects of ergogenic aids on physical performance and served as one of the founders of the American College of Sports Medicine (ACSM). The second laboratory, established in 1944 by Thomas K. Cureton, Jr., was located at the University of Illinois. This lab became well known for research on physical fitness, and many leading investigators of the physiology of exercise were trained there.

Significant Events Since 1950

One of the most important events that stimulated research in physiology of physical activity after

1950 occurred in England. In 1953, Jeremy Morris and colleagues published a study on coronary heart disease and physical activity in which they found that drivers of double-decker buses in London had a significantly higher disease risk than did the conductors on those buses (Morris, Heady, Raffle, Roberts, & Parks, 1953). This study stimulated interest in epidemiological research on physical activity, physical fitness, and chronic disease (Paffenbarger, 1994) that continues to this day. Morris' research also stimulated studies of the effects of fitness and endurance training on risk factors (e.g., serum cholesterol, blood pressure) for coronary heart disease.

Based on the results of many of these studies, U.S. government agencies released two official statements in the 1990s regarding the role of physical activity in preventing chronic diseases. The first statement followed the NIH Consensus Development Conference on Physical Activity and Cardiovascular Health held in December 1995. Among other things, the consensus development panel concluded that "physical *inactivity* is a major risk factor for cardiovascular disease" and that "moderate levels of regular physical activity confer significant health benefits" (NIH Consensus Development Panel, 1996, p. 245; italics added). The other statement came in the form of the Surgeon General's first report on physical activity and health, published by the U.S. Department of Health and Human Services in 1996. The report concluded that regular physical activity can not only reduce the risk of heart disease but also reduce the risk of diabetes, hypertension (high blood pressure), and colon cancer and help control body weight (USDHHS, 1996).

KEY POINT

Both the National Institutes of Health and the Surgeon General concluded that moderate physical activity is beneficial in reducing the risk of chronic diseases (e.g., heart disease, diabetes, hypertension, and colon cancer).

Research Methods for Physiology of Physical Activity

Kinesiologists working in the subdiscipline of exercise physiology use many methods to measure and determine how the human body responds and adapts to physical activity. This work is conducted both within and outside of the laboratory.

Laboratory Work

Much of the equipment used to monitor and evaluate physiological responses is located in exercise physiology laboratories. Working in a laboratory space is beneficial because it provides greater opportunity to control factors that can affect responses to exercise, such as variations in temperature and humidity. In order to enable comparison of work conducted in different laboratories, researchers are encouraged to use standard protocols and techniques.

Many techniques are available for assessing the health-related components of fitness and the traits associated with lower risk of hypokinetic disease. For example, cardiovascular fitness can be assessed either through direct measurement or through estimation of oxygen consumption,

Harvard Fatigue Laboratory

The Harvard Fatigue Laboratory, founded in 1927, was the brainchild of L.J. Henderson, a physical chemist. Directed by David Bruce Dill, the laboratory included a room containing a treadmill borrowed from the Carnegie Nutrition Laboratory, as well as a large gasometer, a room for basal metabolism studies, an animal room, and a climatic room (Dill, 1967). Research undertaken by members of the laboratory included environmental studies conducted at high altitude, in the desert, and in steel mills.

Although the laboratory existed only until 1946, it profoundly affected research on exercise physiology in the United States and Europe during the second half of the 20th century. Many young investigators received their formative training at the laboratory as postdoctoral or doctoral students, including Ancel Keys, R.E. Johnson, Sid Robinson, and Steve Horvath. The many international scientists who spent time at the laboratory included Lucien Brouha of Belgium; Rodolfo Margaria of Italy; and E.H. Christensen, Erling Asmussen, and Marius Nielsen of Denmark (Dill, 1967). All of these investigators went on to establish their own laboratories and were responsible for training many of the leading investigators in physiology of physical activity in the following decades.

which requires measuring the volume of oxygen and carbon dioxide in expired (exhaled) air. Early investigators collected expired air in Douglas bags and analyzed the gas concentrations using chemical analyzers. They then used a spirometer or volume meter to measure how much gas was in the bag. Today, investigators measure gas concentrations with electronic analyzers and gas volumes with flow meters. When interfaced with computers, these devices provide nearly instantaneous and continuous information about oxygen uptake as people rest or exercise.

By measuring oxygen uptake, researchers can obtain information about how the muscles use oxygen and how much energy is expended during physical activity. The most widely accepted method of measuring cardiovascular fitness is to determine **maximal oxygen uptake, or $\dot{V}O_2max$**. This can be done by having an individual exercise while the intensity is increased progressively (i.e., using a graded exercise test) until the person can no longer maintain the required exercise intensity. Physiologists also use submaximal exercise tests to estimate cardiovascular fitness. These tests do not require individuals to exercise at maximal level; instead, prediction equations using heart rate responses to submaximal exercise are used to estimate maximal oxygen uptake.

Ergometers are used during exercise tests to measure the external work performed by muscles. Common ergometers used in exercise science include motorized treadmills and leg and arm cycles. These devices enable physiologists to monitor factors such as the performer's heart rate, blood pressure, breathing rate, oxygen consumption, and carbon dioxide production. Physiologists may also draw blood samples to monitor changes in lactic acid and glucose.

Because body weight is supported during cycling, cycle ergometers are useful for comparing people of different body weights. Leg cycle ergometers also enable researchers to obtain blood pressure measurements and blood samples during the exercise bout because they leave the arms stationary. The major disadvantages of the cycle ergometer are that oxygen uptake is generally lower and fatigue occurs earlier because the rider uses only the leg muscles. Motorized treadmills allow people to be tested while they walk or run at different speeds; in addition, the slope (grade) of the treadmill can be increased in order to increase exercise intensity at a constant speed.

For studies of moderate to vigorous exercise, treadmills are preferable to cycle ergometers because running is more strenuous than cycling for most people. When using treadmills, however, researchers must account for body-weight differences. When walking or running at the same speed and grade, heavier people work harder than lighter people do because these activities require performers to lift their body weight with every step. Treadmills also make it more difficult to measure blood pressure and sample blood, especially when the performer is running, because the arms are moving.

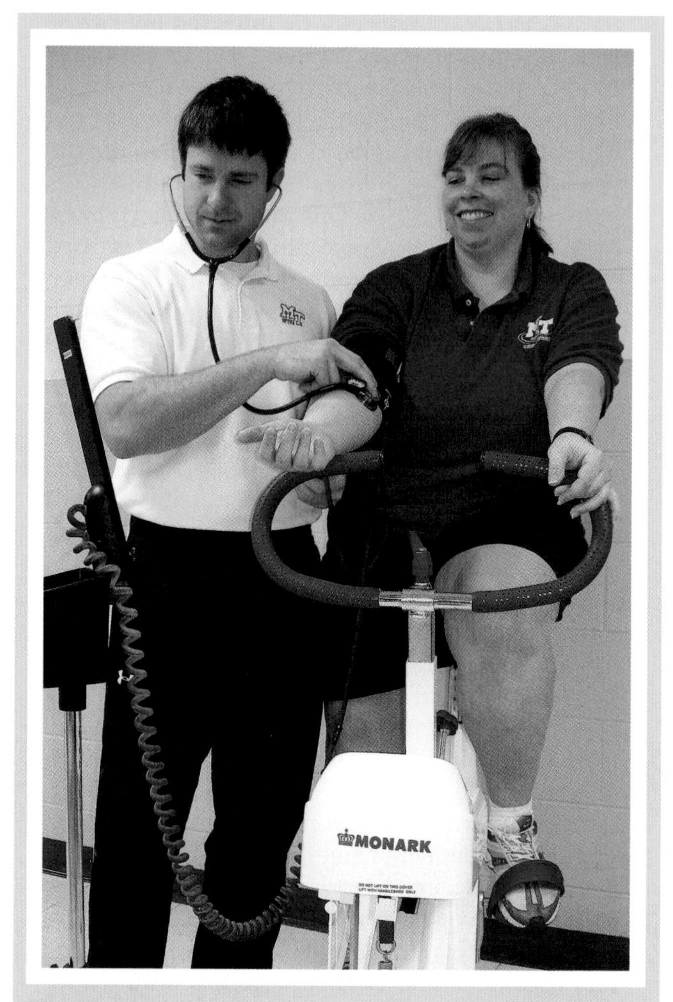

Blood pressure is easily monitored as a person rides a leg cycle ergometer.

To study upper body exercise, exercise physiologists can use arm ergometers. For instance, the swimming flume, developed in Sweden, has been used to study physiological responses to swimming, and rowing ergometers are useful in studying rowers.

Another measure commonly assessed by kinesiologists is body composition. This assessment is used to determine an individual's percentage of body fat or ratio of lean tissue to adipose tissue.

The traditional gold standard for determining body composition in humans involves hydrostatic weighing, also known as **underwater weighing**, in which, as the name indicates, the individual is weighed while submerged in water. This technique makes use of Archimedes' principle, which states that the weight of the water displaced by an object is equal to the volume of the submerged object (density = mass ÷ volume). Once body density has been determined, equations are used to estimate the percentage of body fat. Other techniques for estimating body density include measuring total body water using isotopes (e.g., deuterium), measuring the thickness of subcutaneous fat with skinfold calipers, determining tissue impedance through bioelectrical impedance analysis (BIA), and measuring body fat with dual-energy X-ray absorptiometry (DEXA).

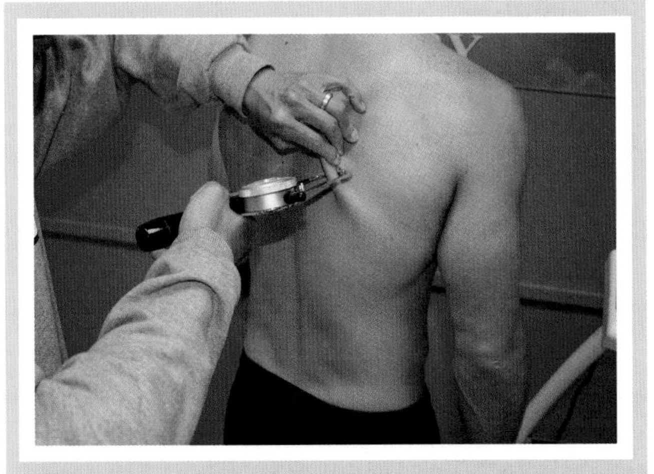

Body fat can be assessed through use of skinfold measurements.

Biochemical Methods

Exercise physiologists use biochemical methods to examine changes at the tissue and cellular levels during and following physical activity. These more invasive techniques include blood sampling and muscle biopsy. Blood samples are obtained from either venipuncture of superficial arm veins or finger pricks. Monitoring changes in the concentration of blood constituents can help physiologists determine utilization of substrate (stored carbohydrate and fat), acid–base balance, dehydration, immune function, and endocrine responses. One commonly used biochemical technique in exercise physiology is to measure lactate from a blood sample taken with a finger prick during exercise as an indicator of the use of **anaerobic** (without-oxygen) energy-producing systems. Physiologists can also use muscle biopsies obtained before and after exercise to examine changes in stored carbohydrate and fat, lactate production, and enzyme activity; the chemical methods used to analyze tissues help determine muscle fiber types in the biopsy samples. Examination of muscle tissue samples under an electron microscope helps determine structural changes following different types of training, as well as structural damage after exercise.

Animal Models

The effects of physical activity on some organs cannot be studied easily in humans; examples include the brain, heart, and liver. In some instances, researchers use animals (e.g., rats) to examine both the functional and structural changes that occur in response to single and repeated bouts of exercise on animal treadmills and running wheels. One of the major advantages of using animals in physiological research is that both the animals and the environment can be more carefully controlled than when working with human volunteers. The use of animals with genetic abnormalities has also proven helpful for examining the effects of exercise on certain clinical disorders, such as obesity and hypertension. In another advantage, some experimental techniques that are not approved for use with humans can be used to study physiological responses in animals—for example, the injection of radioactive isotopes of iron to study changes that occur in tissue iron stores with training.

On the other hand, animal research is criticized for the fact that not all physiological and chemical mechanisms observed in animal species are identical to those observed in humans. This assertion does carry some truth. For example, growth hormone produced in humans is not identical to that of any other species. In addition, tissue changes due to training may also differ between species. Therefore, investigators must be careful to select the animal, usually a mammal, that most closely reflects the human responses to physical activity in which they are interested.

KEY POINT

Technological advances increase opportunities to conduct research outside the laboratory. Over the past century, however, research conducted both inside and outside the laboratory, using both humans and animals, has generated knowledge of exercise physiology.

Field Work

Whether working with individuals or conducting research, exercise physiologists often go beyond the laboratory. Working in the field can present difficulties with respect to monitoring physiological responses, controlling exercise intensity, and controlling environmental conditions. Therefore, when scientists work in the field, it is helpful to use tests that require minimal equipment and enable the screening of many people in a short time. There is always a trade-off between good experimental control in lab-based research and the ecological validity of field-based research.

Kinesiologists often conduct research studies in schools. Although it is not practical to bring a treadmill and a cycle ergometer to a school, alternative measures of cardiovascular fitness exist. For example, we can determine how quickly students can run or walk a mile marked off on the school field and then use their completion times to estimate maximal oxygen consumption. Another example is the PACER (Progressive Aerobic Cardiovascular Endurance Run), a multistage fitness test adapted from the 20-meter shuttle run, which can be performed in school gymnasiums. In addition to testing multiple students at once, these tests require minimal equipment and offer flexibility when space availability differs from one facility to another.

Alternative means also exist for measuring physiological responses while working in the field. For instance, whereas heart rate can be monitored in the laboratory with an ECG machine, it can also be tracked outside of the laboratory with a low-cost, battery-operated heart-rate monitor. Some heart-rate monitors allow information to be collected over several days before it is downloaded to a computer or uploaded over a wireless network. These monitors are also useful in assessing and providing information about exercise intensity in the field.

Physical activity can also be monitored in the field by means of motion sensors, which provide an alternative to potentially unreliable self-report measures. For instance, pedometers and accelerometers (worn at the hip, ankle, or arm) can be used to measure the quantity or intensity (or both) of physical activity and even to estimate caloric expenditure. These instruments have become increasingly popular due to their low cost, small size, and ease of use. Whereas pedometers generally measure only step count, accelerometers can detect the intensity of movement using an acceleration-versus-time curve. Some pedometers and accelerometers have memory capacity, which allows for monitoring of weekly activity and progress toward caloric expenditure goals. Many such monitors are now commercially available and are used by people in the community to monitor their own physical activity.

Overview of Knowledge in Physiology of Physical Activity

Over the past century, exercise physiologists have conducted extensive studies of responses to single and repeated bouts of physical activity. Much of this research has centered on three body systems: muscular, cardiovascular, and respiratory. Physiologists have also examined the influence of environmental factors such as temperature, diet, and altitude on physiological responses and performance. In this overview, we examine how physiological systems respond and adapt to physical activity; review factors that influence physiological responses; and then consider the relationships between fitness, activity, and health.

Skeletal Muscles

Movement is produced by the contraction of skeletal muscles. Each muscle is composed of many muscle cells, which are called *muscle fibers*. Inside a muscle cell, many myofibrils run the length of the fiber. These **myofibrils** contain the contractile elements that shorten to generate force and move your bones during physical activity. Human skeletal muscles contain three primary types of muscle fibers.

Muscle Fiber Types

Did you ever wonder why some people can run extremely fast for short distances (e.g., the straightaway on a track) whereas other people are much better at running long distances (e.g., a marathon)?

Improving Sport Performance or Fitness

What aspect of your sport performance or physical fitness would you like to improve? If you were to begin a training program to improve this aspect, which tool would you use to measure physiological changes related to your performance or your fitness level before training and after training? Why would this tool be helpful in measuring the changes from your training?

People in these two groups likely have varied fiber type distribution in their leg muscles. Muscle fiber types are classified according to the speed at which they contract: fast-twitch (FT) or slow-twitch (ST). Fast-twitch fibers are further subdivided according to how energy is generated. Fast-twitch fibers that use almost exclusively anaerobic energy systems are called *fast glycolytic* (FG) fibers, whereas fast-twitch fibers that use both **aerobic** (with oxygen) and anaerobic (also called *glycolytic*) energy systems are called *fast oxidative glycolytic* (FOG) fibers. Slow-twitch fibers primarily use aerobic energy systems.

During light- to moderate-intensity exercise, such as walking to class or jogging, slow-twitch fibers are recruited first. Because these fibers fatigue slowly, light to moderate physical activity can be sustained for prolonged periods. As the intensity of activity increases, FOG fibers are recruited next, and the FG fibers are recruited at the highest intensities (e.g., for an all-out sprint). Although the FG fibers produce much greater force and power, they fatigue rapidly. Thus, we can sustain the highest intensities of physical activity for only a short time (less than a minute).

Resistance Training

Muscular **strength** is the maximal amount of force exerted by a muscle or group of muscles. Muscular power, in contrast, is the product of the force and the velocity of shortening or lengthening. A muscle's ability to exert force repeatedly over a prolonged period is known as **muscular endurance**. Resistance or weight training programs designed to improve muscular strength may use exercises that are **isometric** (producing tension without changing muscle length), **isotonic** (producing changes in muscle length without changing resistance), or **isokinetic** (producing changes in muscle length with joint movement at a constant angular velocity). Isometric contraction occurs, for example, if you put the palms of your hands against a wall and

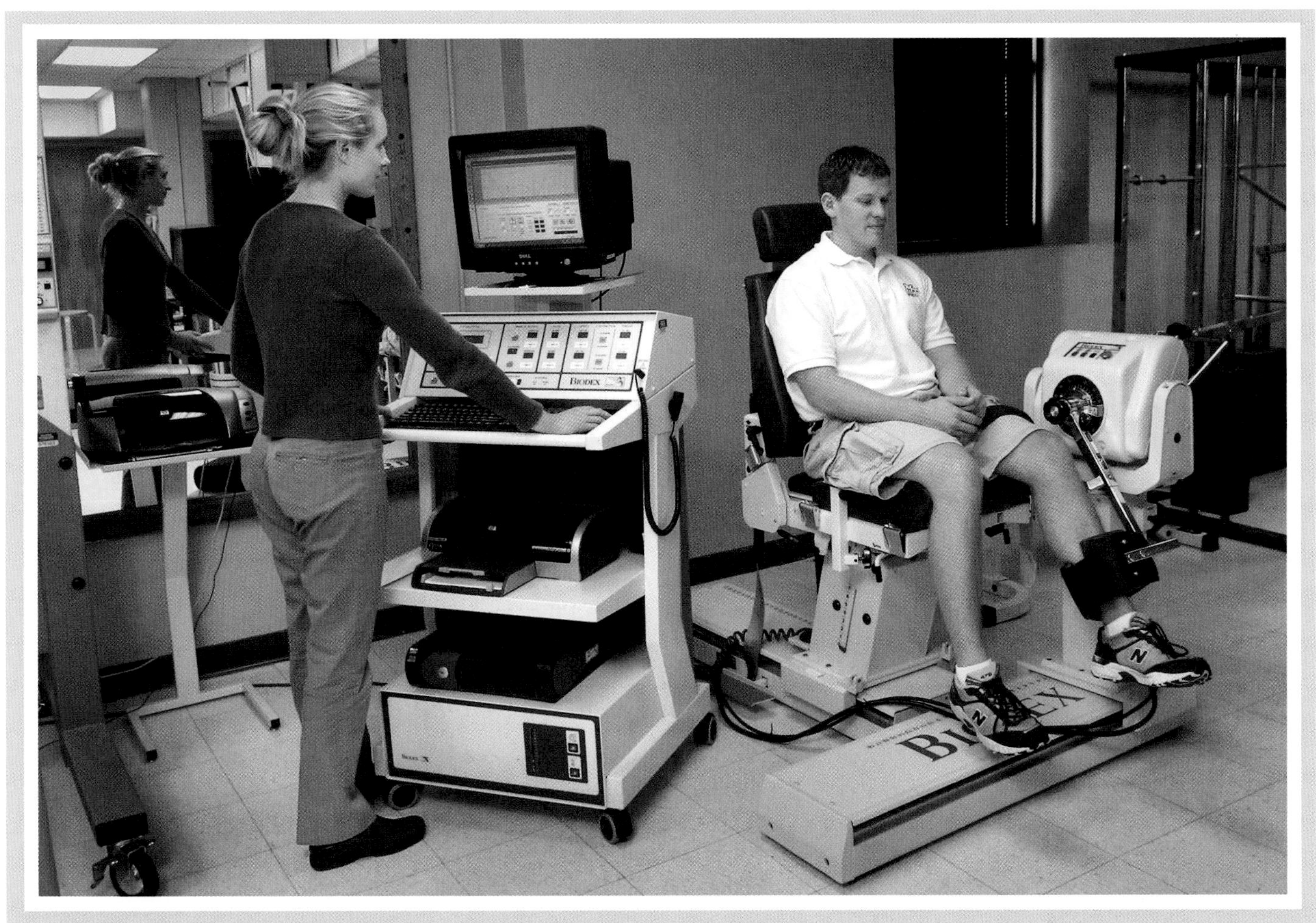

Isokinetic machines are often used in research, clinical testing, and rehabilitation to measure muscular strength.

Photo courtesy of Middle Tennessee State University

push as hard as you can. Isotonic contraction, on the other hand, occurs when you lift a free weight (e.g., barbell). An isokinetic **dynamometer**, such as a Biodex, can measure joint **torque** throughout the range of motion for angular velocities of joint rotation between 500 (**concentric**) degrees and −300 (**eccentric**) degrees per second.

Regardless of the type of exercise used in resistance training, the exerciser must overload the muscle group in order to increase muscular strength, power, or endurance. This is the training principle of progressive overload, which holds that stress must be gradually increased on a physiological system or tissue in order for improvements to be made. Likewise, when you stop overloading the system, training adaptation is lost. This is the training principle of reversibility.

KEY POINT

Increases in muscular strength, power, and endurance result from progressive overload in resistance training.

Training is also subject to the principle of specificity. This principle holds that if you want to use resistance training to improve your performance in a given sport, then you must select exercises that match, as closely as possible, the techniques and energy systems used in the chosen sport. Strength gains will be greatest across the range of movement and velocity of the resistance exercises used (Morrissey, Harman, & Johnson, 1995). For instance, in some physical activities, the ability to develop force rapidly is critical to performance. For example, a volleyball player must rapidly extend his or her legs in jumping to block a spike at the net. If your sport requires you to move at high speed, then higher-velocity isokinetic or isotonic training will be more beneficial than low-velocity or isometric training.

Increases in muscular strength are believed to result from two factors: muscle size increase and neural adaptation. Individual muscle fibers increase in size following resistance training (**hypertrophy**). Although evidence from studies of resistance training in cats suggests that muscle fibers split and increase (**hyperplasia**) (Gonyea, 1980), researchers have not observed increased numbers of fibers in other animal species (Gollnick, Timson, Moore, & Riedy, 1981; Timson, Bowlin, Dudenhoeffer, & George, 1985). Neural adaptations may include recruitment of additional muscle fibers (motor units), better synchronization of muscle fiber contraction, and reduction in neural inhibition. Neural adaptation are more prominent in the early stages of weight training programs prior to hypertrophy.

Cardiovascular System

Transporting oxygen throughout the body is a primary function of the cardiovascular system, which is composed of the heart and blood vessels. To increase the supply of oxygen to skeletal muscles, the cardiovascular system responds immediately to physical activity in several ways in order to increase the volume of blood pumped and the distribution of blood flow to the active muscles. In addition, chronic physical activity results in several important physiological adaptations in cardiac function that improve exercise endurance.

Cardiac Output

Cardiac output, defined as the amount of blood pumped out of the heart each minute, is a function of both heart rate (number of beats per minute) and **stroke volume** (amount of blood pumped per beat). Resting cardiac output is remarkably constant in adult humans at about 5 liters (or quarts) per minute. During physical activity, cardiac output increases as the muscles use more oxygen. Oxygen uptake (written as $\dot{V}O_2max$), or the amount of oxygen used by muscle tissues, increases in direct proportion to the intensity of exercise until maximal oxygen uptake ($\dot{V}O_2max$) is reached. The amount of oxygen delivered to the tissues depends on how much oxygen is in the blood and how much blood the heart is pumping.

Physical Activities and Resistance Training

The type of overload that you place on your muscles must match the requirements of the physical activity for which you are training. Think about the physical activities in which you participate most often. Which activities require muscular strength, and which require muscular endurance? Do you need to maintain low-intensity contractions for an extended time, or do you need to rapidly produce force for a short duration? Given the speed and duration of the movements required, what type of resistance training would be most beneficial in improving your performance in this activity?

As more oxygen is delivered to the active muscles, cardiac output increases. At lower exercise intensities, this increase in cardiac output results from increases in the amount of blood pumped per beat (stroke volume) and the number of beats per minute (heart rate). As your exercise intensity increases above 40 percent of $\dot{V}O_2$max, your stroke volume plateaus because you have reached the limit of how much blood your heart can pump per beat. Your heart rate, however, continues to increase along with oxygen uptake until you achieve your maximal heart rate (see figure 10.1). At this point, you have also reached your limit for oxygen uptake. Thus, the capacity of the heart to pump blood—the maximal cardiac output—appears to be one of the primary factors limiting maximal oxygen uptake and maximal exercise intensity. Because $\dot{V}O_2$max and maximal heart rate are reached simultaneously, fitness instructors can use submaximal exercise tests to estimate $\dot{V}O_2$max. This method is useful in exercise testing and prescription because it does not require the client to exercise to fatigue.

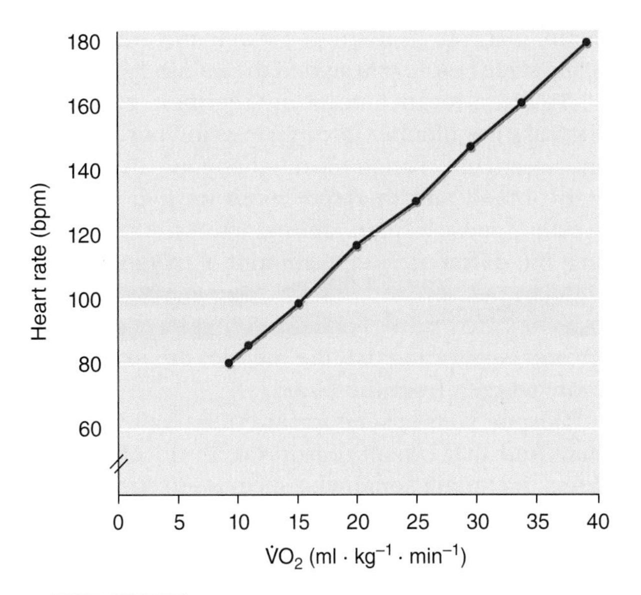

FIGURE 10.1 Relationship between heart rate and $\dot{V}O_2$max during a graded exercise test.

Blood Flow Distribution

When you are at rest, most of your blood flow (cardiac output) goes to your brain and internal organs (e.g., liver, kidneys); only 15 percent goes to your skeletal muscles. When you become physically active, your muscles need more blood to receive oxygen and nutrients and to remove waste products. At the onset of activity, your blood vessels constrict (vasoconstrict) in regions of your body that need less blood flow and dilate (vasodilate) in your active skeletal muscles and heart. During heavy exercise, two-thirds of your blood flow goes to your skeletal muscles through a shift in blood flow away from your kidneys and digestive organs.

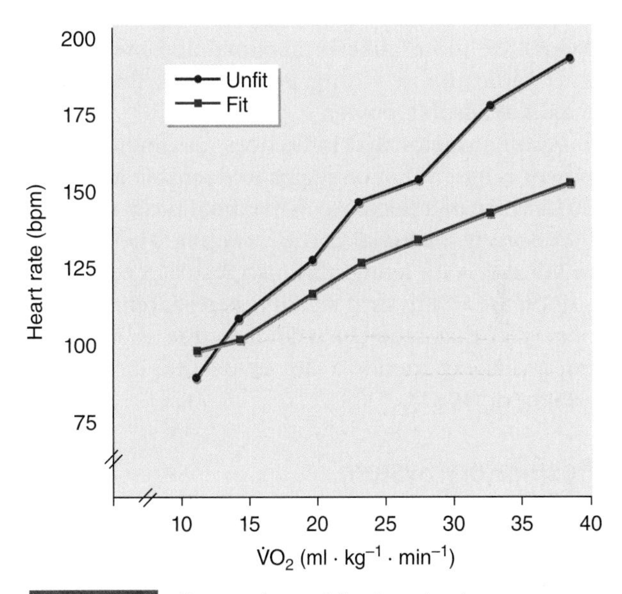

FIGURE 10.2 Comparison of the heart rate responses of trained and untrained people during a graded exercise test.

Cardiovascular Adaptations to Training

Your ability to exercise at moderate to heavy intensities for prolonged periods is referred to as your *aerobic* or *cardiovascular* endurance. You have learned that one of the best indicators of aerobic endurance is your $\dot{V}O_2$max, also known as maximal aerobic power. You can increase your maximal aerobic power through endurance training. Much of the improvement in your $\dot{V}O_2$max results from an increase in stroke volume as the cardiac muscle of your heart increases in size and contracts more forcefully in response to endurance training. As a result of training, you have increased stroke volume both at rest and during exercise, and your heart rate decreases. In other words, you are able to pump more blood with less work!

Of course, you will want to know whether your training regimen succeeded (see figure 10.2). To this end, exercise testing should be conducted both before and after an endurance training program. This approach establishes your baseline fitness level and allows you to determine the effectiveness of the training program. One way to check the efficacy of your training program is to conduct a graded exercise test, in which the intensity increases progressively. At each stage of the test, you should notice that your heart rate is lower than it was at that same stage before you started the training program.

Your endurance training regimen will also increase the number of capillaries in your skeletal muscles. This change means that you can receive more blood and therefore more oxygen in your working muscle fibers during activity. After training, the difference in the amount of oxygen in the arteries (which move blood from the heart) and veins (which move blood back to the heart) is greater, suggesting that the muscles are extracting more oxygen from the blood.

Despite your best efforts at fitness training, you may find that others remain far ahead of you in gains, including some who seem to do so without trying. This point brings us to a discussion of the other factors that affect maximal aerobic power. One is genetics. Studies of the $\dot{V}O_2$max of identical and fraternal twins have shown greater variation between fraternal twin pairs than between identical twin pairs (Bouchard et al., 1986). In other words, individuals with identical genes (e.g., identical twins) are more alike in maximal aerobic power, thus indicating a strong genetic contribution to maximal aerobic power.

Another factor that influences maximal aerobic power is age. $\dot{V}O_2$max begins to decrease after age 30 because of a decrease in maximal heart rate and therefore in maximal cardiac output. The decline in $\dot{V}O_2$max with aging may also result in part from a decrease in physical activity, as researchers have observed that athletes who maintain their training levels experience a slower decline in $\dot{V}O_2$max (Hagberg, 1987).

Respiratory System

As you now know, the cardiovascular system transports oxygen to skeletal muscles. However, where does that oxygen come from? The respiratory system regulates the exchange of gases, including oxygen, between the external environment (air) and the internal environment (inside the body). For this exchange to occur, air must move from the nasal cavity or mouth through the respiratory passages to the alveoli in the lungs. After the fresh air enters the alveoli, oxygen can diffuse into the blood in the pulmonary capillaries, and carbon dioxide can leave the blood and enter the lungs for exhalation into the environment. The process of moving air in and out of the lungs is known as **ventilation**. The amount of air exhaled per minute is known as the minute volume $\dot{V}O_2$max; it is the product of the amount of air exhaled per breath (**tidal volume**) and the number of breaths per minute (breathing rate). The minute volume of a person at rest is approximately 6 liters (6.3 quarts) per minute.

KEY POINT

During high-intensity exercise, a greater volume of air is moved in and out of the lungs, first by breathing deeper and then by breathing faster.

At the beginning of physical activity, you may notice that your breathing rate increases rapidly during the first minute until it reaches a plateau. Researchers believe that the stimulus to increase ventilation comes from sensory **receptors** in the moving limbs (e.g., muscle spindles, joint receptors) as well as the motor cortex (i.e., the part of the brain that stimulates muscles to contract). At low exercise intensities, the increase in minute volume results primarily from an increase in the amount of air moved with each breath; the number of breaths remains constant. At higher exercise intensities, however, increases occur in both tidal volume and respiratory frequency; in other words, you take bigger and more frequent breaths.

As your level of exercise intensity increases, your breathing also increases—steadily at first and then much more rapidly at higher intensities (see figure 10.3). The point at which your breathing begins to increase rapidly is known as the **ventilatory threshold**, which occurs at exercise intensities between 50 percent and 75 percent of $\dot{V}O_2$max.

Determining Your Resting Heart Rate

The average resting heart rate for adults ranges from 60 to 100 beats per minute, though it can be as low as 40 beats per minute in well-trained endurance athletes. You can determine your resting heart rate by counting your pulse for 15 seconds and multiplying the result by four. The best time to assess your resting heart rate is when you first wake up in the morning. On the basis of the amount and types of physical activities in which you participate daily, what might explain why your resting heart rate is above or below average?

After you have been training for a while, your maximal ventilation—the amount of air entering and leaving your lungs—increases and your ventilatory threshold occurs at a higher exercise intensity.

Variation in Temperature and the Response to Physical Activity

In addition to studying the three primary body systems affected by physical activity, exercise physiologists have studied the influence of environmental factors on our experience of sport and exercise. One such factor is temperature variation. Our bodies have complex and effective ways of dealing with temperature changes in the outside environment. Humans are able to regulate their internal (core) temperature so that it remains relatively constant over a wide range of environmental temperatures (Haymes & Wells, 1986). When you begin to exercise, muscular contractions produce heat. Blood flowing through these regions of the body warms up and distributes heat to other regions of your body. As your body temperature rises, your skin blood vessels begin to vasodilate and you begin to sweat. In comfortable ambient temperatures, your body core temperature reaches a plateau in 20 to 30 minutes (see figure 10.4). The higher the intensity of your physical activity, the higher your core temperature will be when it reaches a plateau.

Have you ever wondered why you sweat more profusely during physical activity in warm environments? When air temperature approaches or exceeds skin temperature, the body's cooling mechanism of bringing warm blood to the surface is less effective. In this case, the major avenue of heat loss becomes the evaporation of sweat; however, even this method is limited in hot, humid environments. This results in a higher body temperature and increased risk of heat illness (e.g., heat exhaustion, heatstroke). We tolerate physical activity in cool environments better because our bodies lose heat from blood close to the skin as well as from the evaporation of sweat.

In cold water, however, heat loss from the skin increases dramatically because water is an excellent conductor of heat. Submersion in cold water stimulates shivering and an increase in **metabolic rate** (the rate at which the body uses energy). Swimming in cold water also elevates the metabolic rate, which may lead to an earlier onset of fatigue. Similarly, individuals whose clothing becomes wet during physical activity in cold environments lose heat more rapidly, begin to shiver, and are at risk of **hypothermia** (below-normal body temperature).

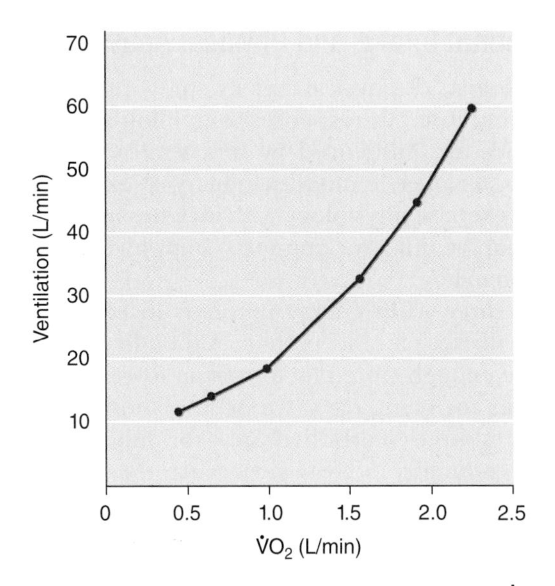

FIGURE 10.3 Relationship of ventilation to $\dot{V}O_2$max as exercise intensity increases progressively during a graded exercise test.

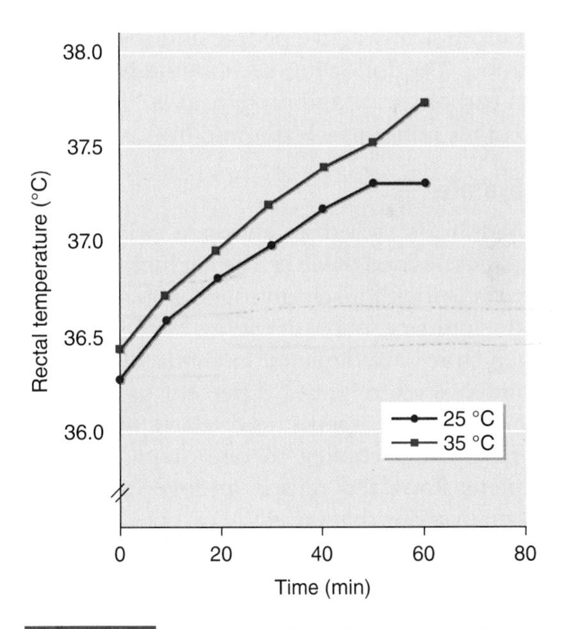

FIGURE 10.4 Increases in body core (rectal) temperature during exercise in comfortable ambient temperatures (77 degrees Fahrenheit, or 25 degrees Celsius) and warm ones (95 degrees Fahrenheit, or 35 degrees Celsius).

KEY POINT

To cool the body during exercise, blood is distributed to the skin as the skin blood vessels dilate and sweating increases.

Nutritional Intake and Physical Activity

Physiologists of physical activity study nutritional intake because our responses to training and performance are influenced by biochemical energy sources. As a result, interdisciplinary research that blends exercise physiology with dietetics improves personal health and supports high-level sport performance.

The three main energy nutrients in foods are carbohydrate, fat, and protein. Although most of us have enough stored fat to sustain low-intensity activities for many days, we need a shorter-term energy source—carbohydrate—for moderately heavy endurance activities. Because the amount of carbohydrate stored in the body is less than 1 pound (0.45 kilogram), daily carbohydrate intake is crucial to athletic performance. Although most athletes do not need to alter their daily consumption of protein, there are higher recommendations for grams of protein per kilogram of body mass (American College of Sports Medicine, 2004). Dietary fat intake recommendations do not differ between physically active people and the general population. The following sections highlight the roles of carbohydrate and protein, as well as those of two other nutrients—water and iron.

Carbohydrate

Carbohydrate is stored as glycogen primarily in skeletal muscles and the liver. During high-intensity physical activities, muscle glycogen stores serve as the body's primary source of energy. Normal muscle glycogen stores are depleted in about 90 minutes of continuous exercise at 75 percent of VO_2max. The amount of glycogen stored relates directly to the carbohydrate content of one's diet. Coaches and athletes know that muscle glycogen storage is increased by a diet that is high in carbohydrate (70 percent or more of the calories) consumed for 2 or 3 days. In light of this fact, some endurance athletes use a technique known as *carbohydrate loading* before competition. Carbohydrate loading is most effective in activities that last more than 2 hours (e.g., marathons); it offers no apparent advantage in activities that last less than 1 hour.

Maintaining carbohydrate levels during prolonged exercise is important because low blood glucose is associated with fatigue. Consumption of carbohydrate drinks containing 6 percent to 8 percent carbohydrate by weight during prolonged exercise has been shown to improve performance and delay the onset of fatigue (Coggan & Coyle, 1991). Drinks containing a higher concentration of carbohydrate (more than 10 percent by weight), however, delay gastric emptying and can lead to gastric distress (Davis, Burgess, Sientz, Bartoli, & Pate, 1988). Glycogen stores can also be depleted by intermittent physical activity and sport participation. For example, soccer players often deplete their muscle glycogen stores during the second half of a game. More specifically, Saltin (1973) showed that players with depleted glycogen stores covered less distance and spent more time walking and less time running during the second half of a game.

In addition, some athletes with low carbohydrate intake (less than 50 percent of total calories) may progressively deplete their glycogen stores through daily training. To maintain adequate glycogen stores, physically active individuals should consume a diet in which carbohydrate accounts for 55 percent to 60 percent of total calories. It is also helpful to eat foods high in carbohydrate immediately following exercise in order to increase the rate of muscle glycogen storage (Ivy, Katz, & Cutler, 1989).

Protein

The recommended dietary allowance (RDA) for protein is 0.8 gram per kilogram per day. Athletes who perform high-intensity work have elevated protein needs to support repairing and replacing damaged proteins, maintaining the function of metabolic pathways, supporting increases in lean tissue, and aiding the adaptations of bone, tendons, and ligaments to physical activity. For elite endurance athletes, the protein requirement ranges from 1.2 to 1.4 grams per kilogram per day (American College of Sports Medicine, 2004). For elite athletes adding muscle mass through high-intensity strength training, the requirement ranges from 1.2 to 1.7 grams per kilogram per day (American College of Sports Medicine, 2004).

Even though protein requirements are elevated during high-intensity endurance training and strength training, athletes do not automatically need to consume additional protein. In fact, in the United States, the dietary consumption of protein already exceeds the RDA at a mean of 1.5 grams per kilogram per day (Moshfegh, Goldman, & Cleveland, 2005). Therefore, an athlete should meet with a dietician before revising his or her diet. Of course, some athletes consume low levels of protein—for example, those who do not consume meat—and investigations are now being conducted to determine the most beneficial forms of protein and the optimal time to consume it (i.e., before or after working out).

Fluid Intake

Water makes up 55 percent to 60 percent of the human body. During physical activity, you lose some of this water through sweating. When you exercise on a warm day, even just walking to class, you will notice that you sweat even more. Because sweating can cause substantial fluid loss, fluid replacement is important. Sweat loss decreases body fluids both within and between cells; it also decreases **plasma volume** (the fluid portion of blood). When plasma volume decreases, less blood returns to the heart, which reduces the amount of blood in each heartbeat (stroke volume). To compensate for the reduced stroke volume, heart rate increases. If fluid losses are not replaced, the sweat rate decreases, which causes an increase in body temperature. This increase puts the individual at greater risk of developing a heat illness, especially heat exhaustion. Also, as plasma volume is lost, the blood becomes more viscous (thicker), which necessitates a greater force of contraction by the heart in order to move the blood through the blood vessels. This change leads to an increase in blood pressure.

KEY POINT

Inadequate fluid intake during and after physical activity results in elevated body temperature, greater risk of heat illness, and decrements in performance.

Failure to adequately replace fluids can decrease performance in some physical activities. For example, Armstrong, Costill, and Fink (1985) found that running velocities at distances ranging from 1,500 to 10,000 meters (about 1 to 6 miles) decreased by 3 percent to 7 percent following a 2 percent decrease in body weight. Dehydration is less likely to affect short, high-intensity activities such as sprinting. Fluid replacement is crucial during many types of physical activities, including hiking, bicycling, aerobics classes, and outdoor work. To ensure adequate fluid replacement during physical activity, drink fluids (e.g., water, carbohydrate-electrolyte drink) at regular intervals (every 15 to 20 minutes). Fluid replacement can be difficult to accomplish during play in certain sports, such as soccer and field hockey, where the rules do not allow time-outs

Energy gels are an easy way to ingest carbohydrate during physical activity.

during competition except for injury. Thus, players in these sports should drink additional fluid before the contest and during halftime in order to reduce the fluid deficit.

Following activity, continue to drink fluids even though you may not feel thirsty. Unfortunately, thirst is not an accurate indicator of the need for fluid. During activities lasting less than 1 hour, consuming carbohydrate-electrolyte drinks offers no advantage over plain water. In more prolonged activities, fluids containing a small amount of carbohydrate are more likely than water to be beneficial in enhancing fluid absorption and maintaining blood glucose.

Iron Intake

As you know, physical activity cannot take place without the transport of oxygen-carrying blood to the muscles. Oxygen is transported in the blood with the help of iron. Specifically, oxygen essentially rides on the iron atoms in the **hemoglobin** found in red blood cells. Each hemoglobin molecule contains four iron atoms that bond with four oxygen molecules. A person with low iron cannot adequately synthesize hemoglobin. As hemoglobin concentrations decrease, the amount of oxygen transported to the tissues decreases. Eventually, the person becomes anemic. **Anemia** is defined as a hemoglobin concentration below 12 grams per deciliter of blood in women and below 13 grams per deciliter in men. People with anemia experience reductions in $\dot{V}O_2$max and in endurance (Celsing, Blomstrand, Werner, Pihlstedt, & Ekblom, 1986).

KEY POINT

Low iron stores can lead to anemia, which reduces $\dot{V}O_2$max and exercise endurance.

Iron deficiency is one of the most common nutritional deficiencies in the United States, especially among adolescent girls and women. The most likely causes of iron depletion in physically active women are inadequate iron intake and excessive blood loss through the menses. Low iron intake in female athletes most commonly occurs in those who restrict their caloric intake and those who consume diets that are low in meat (Clarkson & Haymes, 1995). Heme iron found in meat, fish, and poultry is more highly absorbed from the gastrointestinal tract than is the non-heme iron found in other food sources. People can typically avoid iron deficiency by consuming iron-rich foods daily.

Physical Activity, Fitness, and Health

One need not exercise strenuously in order to gain health benefits from physical activity. To the contrary, a large body of scientific evidence (e.g., Pate et al., 1995) indicates conclusively that health benefits can be obtained by engaging in moderate-intensity physical activity on a regular basis. Moderate-intensity physical activities include brisk walking (3 to 4 miles per hour, or 4.8 to 6.4 kilometers per hour), playing doubles tennis, cycling less than 10 miles per hour (16 kilometers per hour), swimming at moderate speed, golfing (pulling a cart), climbing stairs, and mowing the lawn.

A 1996 report titled *Physical Activity and Health: A Report of the Surgeon General* recommended accumulating 30 minutes or more of moderate-intensity physical activity on most (preferably all) days of the week (USDHHS, 1996). The activity need not be continuous for 30 minutes; it may include intermittent activities as well, in blocks of time as small as 10 minutes. The goal should be to perform activities that use approximately 150 to 200 calories over a 30-minute period.

A 2008 report titled *Physical Activity Guidelines for Americans* affirmed the health benefits of physical activity and recommended accumulating 150 minutes per week of moderate-intensity physical activity or 75 minutes of vigorous-intensity physical activity such as jogging or cycling (Physical Activity Guidelines Advisory Committee, 2008). The recommended duration of activity can be accumulated in various ways. For example, the 150 weekly minutes can be met 30 minutes at a time on 5 days per week or 50 minutes at a time on 3 days per week. The recommendation for moderate-intensity physical activity is aimed at increasing the percentage of the adult population that engages in regular physical activity.

KEY POINT

The recommended amount of weekly physical activity for improved health and decreased disease risk is 150 minutes of moderate-intensity activity or 75 minutes of vigorous-intensity activity.

Greater volumes and intensities of exercise confer additional benefits in both health and fitness. To improve cardiovascular endurance, the American College of Sports Medicine (2014) recommends that adults take part in physical activity for 20 to 60 minutes per day on 3 to 5 days per week at moderate to vigorous intensities. Exercise intensities of 64 percent to 76 percent of maximal heart rate are labeled as moderate exercise, and exercise intensities of 77 percent to 93 percent of maximal heart rate are considered hard or vigorous exercise. Activities that are aerobic and can be maintained for prolonged periods are best for improving cardiovascular endurance. These activities include running, hiking, walking, swimming, cross-country skiing, bicycling, aerobic dancing, stair climbing, and rowing. Cardiovascular fitness can also be aided by participation in sports—such as soccer, field hockey, and tennis—that involve high-intensity, intermittent activities carried out over prolonged periods.

One of the most common reasons for stopping an exercise training program is injury (Pate et al., 1995). For this reason, exercise programs designed to improve fitness and health should begin with moderate-intensity activities and gradually increase the intensity and duration of the activities as fitness improves. For example, an exercise program for sedentary adults might begin with walking at a speed of 3 to 4 miles per hour (4.8 to 6.4 kilometers per hour), which is close to the moderate intensity of 64 percent of maximal heart rate, and gradually increase the duration from 20 to 60 minutes. People can begin by walking, progress to alternating between walking and jogging, and advance to jogging.

Effects of Age on Fitness

Although research has shown that maximal oxygen uptake ($\dot{V}O_2max$) declines with age at the rate of about 10 percent per decade, the decrease is smaller (5 percent per decade) in individuals who remain physically active (Hagberg, 1987). The decline in cardiovascular endurance in older individuals may result in part from a reduction in the intensity and duration of physical activity. This possibility is suggested by the fact that master athletes who maintain their training intensity experience little change in maximal oxygen uptake as they age.

$\dot{V}O_2max$, the primary indicator of cardiovascular endurance in adults, increases in absolute terms (liters per minute) with growth in children. Because body mass increases with growth, more oxygen is required to supply the active tissue. When $\dot{V}O_2max$ is expressed per kilogram of body mass, however, it remains relatively constant or decreases during growth (Bar-Or, 1983). Results of several studies of children suggest that $\dot{V}O_2max$ per kilogram does not increase with training in prepubescent children until the peak of the growth spurt occurs (Zwiren, 1989). This finding does not mean, however, that children's endurance does not improve before puberty. Instead, improvements in children's endurance with training may relate to improvements in other factors, such as running economy and anaerobic capacity (Rowland, 1989).

Physical Activity, Fitness, and Coronary Heart Disease

In general, higher levels of fitness are associated with improved health status. This generally held assumption is based on research conducted by exercise physiologists. For instance, Blair and colleagues (1989) found that fitter individuals have lower relative risk of developing or dying from cardiovascular disease or cancer than those who are less fit. The investigators based their determination of physical fitness on a cardiovascular endurance test. The risk of cardiovascular disease, as well as colon cancer, is also reduced by participation in physical activities, as distinct from one's level of fitness. Paffenbarger (1994) estimated that engaging in moderate-intensity physical activities added approximately 1.5 years to life. Moreover, using the number of calories expended weekly in walking, stair climbing, and leisure-time sport as an index of physical activity, his study showed that Harvard alumni had a 46 percent lower risk of dying from cardiovascular disease if they expended 2,500 calories or more per week in physical activities. Participants in moderate-intensity activities (e.g., brisk walking, swimming, cycling) had a 37 percent lower risk of cardiovascular disease than did those who did not participate in leisure-time sport.

KEY POINT

High levels of physical fitness lower the risk of developing and dying from cardiovascular disease.

Physical activity and fitness also play a role in reducing the risk of hypertension, or high blood pressure. Over time, endurance training lowers blood pressure in hypertensive individuals by about 10 mmHg (millimeters of mercury). In addition, lower-intensity exercise programs (40 percent to 70 percent of $\dot{V}O_2max$) are as effective in reducing blood pressure as are those of higher intensity (Fagard, 2001). Among people with normal blood pressure, those who are more physically active have lower blood pressures than do those who are less active.

Physical Activity and Weight Control

Participation in daily physical activity helps control body weight. Weight gain and weight loss are

Professional Issues in Kinesiology

Evidence-Based Practice: Weighing the Risks and Benefits of Physical Activity

The risk-to-benefit ratios of participation in some youth sports have recently come into question. Although we know that sport participation can improve one's physical and psychological health—and even academic performance in children (Strong et al., 2005)—there are also inherent risks of head injury associated with activities such as soccer, football, skiing, and baseball (Centers for Disease Control and Prevention, 2011). Concussions and the risk of repeated subconcussive hits have been the topic of a major motion picture, television specials, and research funded by the Department of Defense, the Centers for Disease Control, and the National Institutes of Health. Researchers are studying the long-term health risks of sport-related concussions and how repetitive concussions sustained in sport by youth and professional athletes may contribute to degenerative brain disease. Sport leagues are reviewing policies and rules in order to minimize the risk of head injuries, and parents want to know and understand the potential risks to their children.

Citation style: APA

Centers for Disease Control and Prevention. (2011). Nonfatal traumatic brain injuries related to sports and recreation activities among persons aged ≤19 years — United States, 2001–2009. *Morbidity and Mortality Weekly Report, 60*(39), 1337–1342. Retrieved from www.cdc.gov/mmwr/preview/mmwrhtml/mm6039a1.htm

Strong, W.B., Malina, R.M., Blimkie, C.J.R., Daniels, S.R., Dishman, R.K., Gutin, B.H, . . . Trudeau, F. (2005). Trudeau, F. (2005). Evidence-based physical activity for school-age youth. *Journal of Pediatrics, 146*, 732–737.

determined by the interplay between the consumption and the expenditure of calories. Put simply, excess caloric consumption results in weight gain. Exercise physiologists have added to our understanding of caloric expenditure by investigating the process by which the body expends calories and uses energy. The metabolic rate—the rate at which the body uses energy—increases in direct proportion to the intensity of activity. Lower-intensity activities, such as walking at 3 miles (4.8 kilometers) per hour, increase the metabolic rate threefold, to a caloric expenditure of about 4 calories per minute. Higher-intensity activities, such running at 7 miles (11 kilometers) per hour, increase the metabolic rate tenfold, expending 12.5 calories per minute. Total energy expenditure also depends on the duration of physical activity.

Moreover, energy expenditure does not return to resting levels immediately after an exercise bout. Rather, researchers have shown that the metabolic rate remains elevated during recovery from physical activity. Thus, the total energy expended due to a single bout of exercise is somewhat greater than the energy cost of the activity itself, especially following high-intensity, longer-duration activity.

KEY POINT

Energy expenditure during physical activity is directly proportional to the activity's intensity and duration.

Whereas some people seek to maintain body weight through exercise, others want to lose body weight and body fat. The recommended exercise dose for weight loss is approximately 150 to 250 minutes of weekly physical activity (Donnelly et al., 2009). When caloric intake exceeds caloric expenditure, the excess calories are converted to fat and stored in adipose tissue. Sedentary individuals are more likely to be overweight or obese, a condition that puts them at increased risk of developing coronary heart disease (Willett et al., 1995); they are also at higher risk of developing non-insulin-dependent diabetes mellitus (NIDDM) (Paffenbarger, 1994). Reducing body weight is effective in lowering serum triglycerides and reducing blood pressure in individuals with hypertension. Furthermore, participation in daily physical activity is beneficial not only in reducing the risk of NIDDM but also in helping individuals with this condition regulate their blood glucose levels.

Now that you have gained some basic knowledge about the physiological systems and some of their responses during physical activity, we will examine one of the most pressing problems facing physiologists of physical activity today: how to decrease obesity rates. Large-scale population studies have shown that obesity has increased dramatically in the United States over the past 25 years. These population-based investigations often use body mass index (BMI) as a surrogate measure of obesity. BMI is calculated using a person's body mass and height, as follows:

How Far Do You Need to Run to Lose 1 Pound?

An excess caloric expenditure of 3,500 calories is needed to lose 1 pound (0.45 kilogram). If a person burns 100 calories per mile when running (jogging), how many miles would this individual need to run to lose 1 pound? How many days per week do you run? How many miles do you usually run each time? Assuming that your diet is constant, calculate how long it will take you to lose 1 pound from running.

BMI = body mass in kilograms ÷ the square of height in meters

This practice is accepted because BMI assessment is easy and noninvasive and elevated values are indicative of obesity-related disease risk. A BMI of more than 30 is classified as obese, and the percentage of adults in the United States classified as such has increased steadily. In 2011–2012, for instance, the rate had reached 35 percent of adults in the United States (Ogden, Carroll, Kit, & Flegal, 2014). The problem of obesity is also apparent in younger segments of the U.S. population. For instance, the prevalence of obesity in U.S. children aged 2 to 5 years in 2011–2012 was about 8 percent (Ogden et al., 2014). The rates are even higher for children aged 6 to 11 years (17 percent) and for adolescents from 12 to 19 years of age (21 percent).

These statistics are alarming, especially given that the risk of developing many chronic diseases is greater in individuals who are obese. More specifically, obese men and women have a higher risk of developing heart disease, hypertension, and NIDDM; these diseases are also associated with physical inactivity. Impaired glucose tolerance, a preliminary stage in the development of NIDDM, was found in 25 percent of obese children and 21 percent of obese adolescents (Sinha et al., 2002). Overweight and obesity are also related to an elevated risk of some forms of cancer in adults, including endometrial (inner lining of the uterus) and breast cancer in women and colon cancer in both women and men.

People can lose body weight by reducing caloric intake, increasing physical activity, or a combination of the two. Exercise physiologists are particularly interested in studying the effects on body weight of increased physical activity and of programs that combine dietary components and exercise. Researchers have found that creating a caloric deficit through a combination of physical activity and diet management is as effective as dieting alone in reducing body weight. As this research suggests, collaboration between exercise physiologists and researchers in other fields, such as dietetics and exercise psychology, may help determine the best methods of improving body composition.

The challenge is how to get more people to participate regularly in physical activity. Only half of adults in the United States regularly participate in 150 minutes of moderate-intensity or 75 minutes of vigorous-intensity physical activity per week, and about one in four adults participates in no leisure-time physical activity (U.S. Department of Health and Human Services et al., 2015). Women are less likely than men to participate regularly in physical activity, and older adults are less likely to do so than young adults. Several strategies have been proposed for increasing daily physical activity, including increasing the amount of physical activity in physical education classes; providing more opportunities for children and adolescents to participate in physical activity before, during, and after school; building walking and bicycle paths that are separated from roadways; and increasing opportunities for adult participation in physical activity at worksites, in shopping malls, and in community facilities.

Wrap-Up

This chapter provides an overview of the knowledge base of physiology of physical activity. This subdiscipline of kinesiology centers on the acute and chronic changes that occur in the physiology of the human body in response to physical activity. Exercise scientists have studied and measured these changes in order to help people be healthier and improve their physical performance.

Exercise science professionals work in many arenas, including clinical, corporate, and commercial fitness programs and universities. In addition, research opportunities exist in the military and other government organizations. If you are interested in learning more about how the body responds and adapts to physical activity and exercise, you can begin by taking courses in exercise physiology, anatomy, and physiology. You are also encouraged to visit the web resources listed at the end of this chapter in order to learn more about the educational and certification opportunities available to you in exercise physiology and other spheres of kinesiology.

MORE INFORMATION ON PHYSIOLOGY OF PHYSICAL ACTIVITY

Organizations

American College of Sports Medicine
American Council on Exercise
American Society of Exercise Physiologists
Canadian Society for Exercise Physiology
Exercise and Sports Science Australia
National Strength and Conditioning Association

Journals

ACSM's Health & Fitness Journal
Applied Physiology, Nutrition, and Metabolism
Biology of Sport
British Journal of Sports Medicine
European Journal of Applied Physiology
Exercise and Sport Sciences Reviews
International Journal of Exercise Science
International Journal of Sports Physiology and Performance
Journal of Applied Physiology
Journal of Exercise Physiology Online
Journal of Physical Activity and Health
Journal of Science and Medicine in Sport
Journal of Sports Medicine and Physical Fitness
Journal of Sports Sciences
Journal of Strength and Conditioning Research
Measurement in Physical Education and Exercise Science
Medicine & Science in Sports & Exercise
Pediatric Exercise Science
Research Quarterly for Exercise and Sport
Sports Medicine

REVIEW QUESTIONS

1. How does physiology of physical activity fit within the discipline of kinesiology?
2. Describe the contributions made to physiology of physical activity by A.V. Hill, David Bruce Dill, and other key scholars.
3. Give examples of how $\dot{V}O_2max$ can be measured in the laboratory and estimated in the field.
4. Explain how knowledge of exercise physiology might be used to help each of the following: a college athlete, a cardiac patient, a person trying to lose weight.
5. Calculate your maximum heart rate, then measure your heart rate while you are performing each of three physical activities. Which of the activities meets the intensity guideline to help improve your cardiovascular endurance?
6. Get up and move at high intensity for a few minutes. List the physiological changes that you notice and explain why each occurred.
7. What health improvements result from being physically active on a regular basis?

Go online to complete the web study guide activities assigned by your instructor.

PART III
Practicing a Profession in Physical Activity

© Human Kinetics

For some people, experiencing physical activity and studying it are ends in themselves. Like some psychology or history majors, some kinesiology majors make their choice not because they plan to carve out a career in the field but simply because they are interested in acquiring as much knowledge about it as they can. Thus, the popularity of kinesiology as a liberal arts subject has increased over the past few decades.

For most kinesiology majors, however, experiencing and studying physical activity are also means to a specific end: preparing for a career in a physical activity profession. This preparation requires mastery of another type of knowledge beyond those described in parts I and II of this book. Professional experience in physical activity—the application of skills to help clients, students, athletes, or patients achieve specific goals—rests on knowledge obtained from professional experience in physical activity. If this sounds circular, remember that knowledge derived from professional practice is often incorporated into the discipline of kinesiology. When systematic knowledge—usually developed by careful observation of how various manipulations of physical activity affect clients—is included in kinesiology curriculums, it becomes part of the discipline. Because this knowledge of professional practice is part of the curriculum, you will be required to master it by demonstrating both your understanding of it and your ability to use it.

Those who work in the field of physical activity assume the responsibilities and obligations that belong to all professionals. Chapter 11 expands your understanding of what it means to be a professional, of the obligations and opportunities that belong to professionals, and of the steps that you can take now to enable your success as a professional. The rest of part III explores five professional career areas: health and fitness, therapeutic exercise, teaching physical education, coaching and sport instruction, and sport management.

To help you organize your thoughts about the types of work and worksites available to kinesiology majors, we have grouped the many careers according to general objectives and goals, working environments, and qualifications for professionals. Part III does not cover all careers open to kinesiology majors. Some graduates seek advanced degrees in kinesiology and devote their lives to studying, researching, and teaching about physical activity at an advanced level as a college professor or research scientist. In addition, advancements in science often mean that the future will hold careers that are currently not obvious.

If you're like some of your peers, you may be somewhat uncertain about which physical activity career you will pursue. This uncertainty is normal and shouldn't cause you any anxiety at this point. In fact, one of the objectives of this part of the book is to help you determine whether you are suited to a career in kinesiology and, if you decide that you are, to help you chart a course for your professional future. Thus, part III should move you closer to selecting a career. As you go through this process, keep in mind that this choice may be only the first of several career choices that you make over the course of your working years—two, three, and even four changes in career are likely for individuals currently entering the workforce. Nevertheless, you should make the most intelligent career choice possible, and this section will help you accomplish that goal.

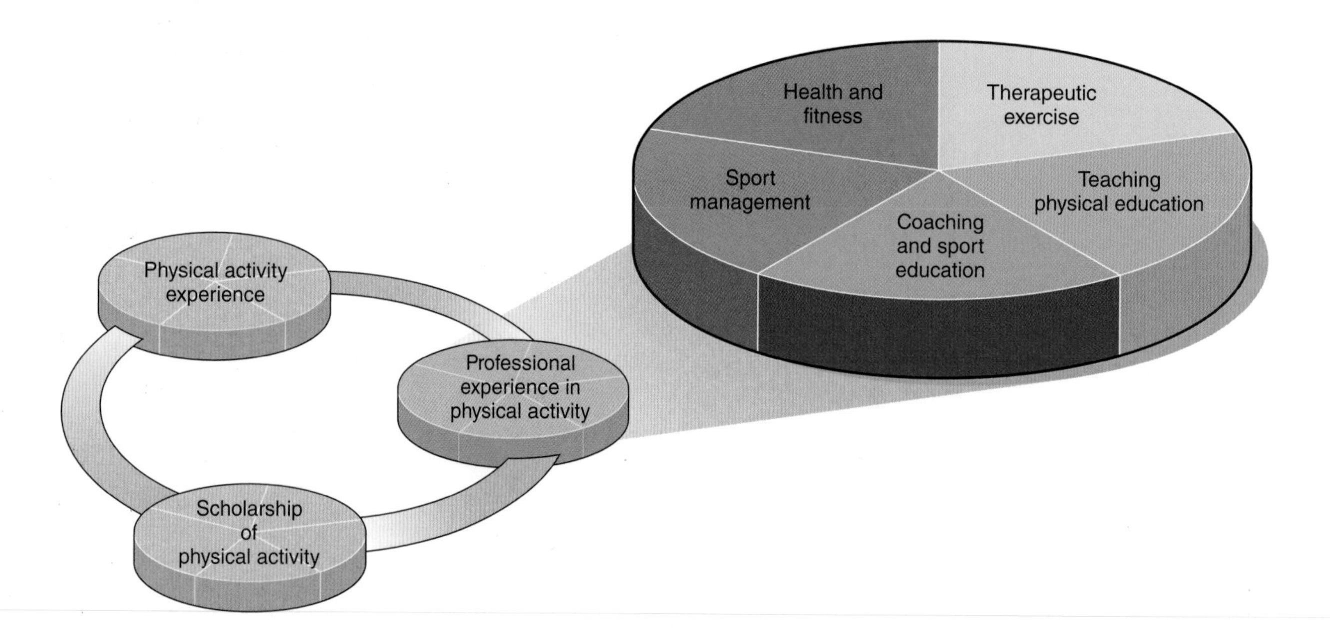

11

Becoming a Physical Activity Professional

Shirl J. Hoffman

CHAPTER OBJECTIVES

In this chapter, we will

- acquaint you with what it means to be a professional and, more specifically, a physical activity professional;
- explain the types of knowledge and skills essential for performing professional work;
- discuss what you need to do during your undergraduate years in order to gain entry into and succeed in your chosen professional field; and
- help you determine whether you are suited for a career in a physical activity profession.

Marilyn was an accomplished athlete, so friends and relatives assumed that she would pursue a career as a physical education teacher and coach. Marilyn wasn't so sure. She loved sport and physical activity but thought that a different type of work might present new and different challenges.

During the summer between her junior and senior years, Marilyn worked as a counselor at a camp for children with physical or intellectual disability. Going into the role, she wasn't sure she would like it, but she soon found herself amazed at the grit and determination possessed by the children. She felt inspired and noted the satisfaction felt by the nurses, therapists, and doctors, to whom the gratitude shown by their patients was worth more than money.

Thus, it came as no surprise to Marilyn's fellow camp counselors when she enrolled at a local college as an adapted physical education major. With its focus on delivering modified physical education programs for individuals with physical or mental challenges, the program meshed well with her interests and experience. After earning her degree, she secured a position as an adapted physical education teacher and found a rewarding career in assisting students with a variety of disabilities.

In order to improve her teaching and service to children, Marilyn also became active with colleagues in a state-level professional organization and in other disability-support professions. In addition, she went through the necessary steps to achieve professional recognition as a certified adapted physical educator.

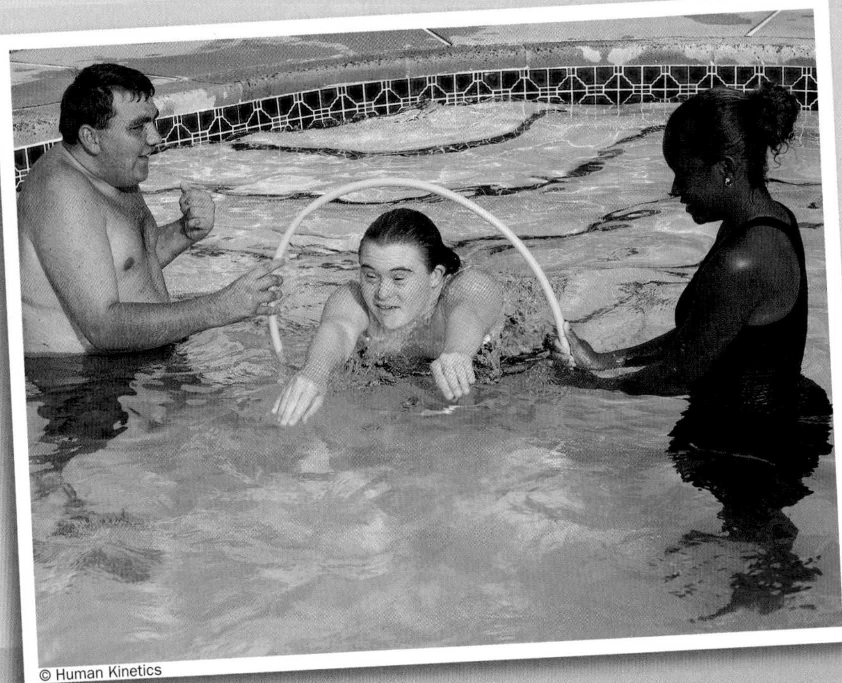

Committing to a career early in your college experience can be a scary proposition. It can be daunting to think about making a decision at this early stage that will form the center of a major aspect of your identity for the rest of your life. If you feel this fear, consider the fact that people often change careers more than once, though what may be described technically as a career change sometimes takes place within the same general field. In fact, according to some estimates, more than 5 million people in the United States between the ages of 44 and 70 are embarking on an "encore" (second or third) career at any point in time.

This chapter and those that follow are designed to help you make a career choice in kinesiology as early in your studies as possible. Granted, in all likelihood, your first career choice will not be your last. However, there are distinct advantages to choosing a career early. First, midstream career changes can disrupt your personal and family life, and they inevitably set back your progress toward an ultimate career goal. Second, students who identify a career track early in their college experience are much better able to relate their course work to their anticipated career. In addition, because they know the type of work they like and want to do, these students can gain experience in professional roles and settings and begin to develop a network of associates in the profession while still in college.

So, if you have already committed to a specific career in the field of kinesiology, you are on track. You're likely to enjoy your study of kinesiology a bit more and derive more from it than students who have yet to make that commitment. If you are still trying to decide on a career, this part of the text should help you narrow your career choices. In fact, even if you have already decided on a career, the chapters in this section should interest you as a budding kinesiologist who shares with his or her colleagues a special interest in physical activity.

For most students, getting a respectable job tops the list of reasons for going to college. Accordingly, you probably expect your college degree to help you find a well-paying job that is respected by the community, offers good working conditions, and provides a reasonable amount of time off for leisure pursuits. No doubt you also want your job to count for something, to contribute to society in some substantial way. In addition, many people want a job that is so enjoyable to them that they feel energized by their work and willingly invest more time than the job requires. Furthermore, the ideal job should be one that not just anybody can do; that is, it should require specialized skills that set you apart from others in the workforce. Ideally, it also puts you in control of most decisions about your work. Last, but certainly not least, you want a job that centers on physical activity!

Having the freedom to make decisions is one of the advantages of being a professional. However, with freedom comes responsibility, because the quality of your decisions determines how successful you will be in your chosen profession. Will you be a good decision maker? Will the decisions you make increase the health, well-being, and enjoyment of the students, clients, patients, or customers you serve? Or will your decisions result in wasted time and effort and perhaps jeopardize the safety of those who depend on you? Will your decisions be grounded in the knowledge and skills that you acquired in your kinesiology program? Or will you make decisions entirely by the "seat of your pants"? An undergraduate education in kinesiology teaches you to make correct decisions based on available knowledge.

If this description applies to the position you hope to secure following graduation, then you have already narrowed your choice down to the professions, but there are many professions. This chapter is intended to help you decide whether a career in a physical activity profession, specifically, is right for you. The chapter also explores the general curriculums adopted by most kinesiology departments to help you prepare for your chosen field.

As you know, not all workers are professionals. A profession is a particular line of work, and a professional is a particular type of worker. Teachers, doctors, lawyers, and physical therapists are professionals. Janitors, housepainters, plumbers, electricians, carpenters, computer technicians, athletes, and other tradespersons may describe their expertise as "professional," but the character of their work generally lacks characteristics associated with the professions. Knowing why this is the case will help you better understand the distinctive roles, responsibilities, educational requirements, and compensation in the type of work to which you aspire.

The kinesiology curriculum may prepare you for a specific career in the physical activity professions or it may serve as the first step to a career in a physical activity profession requiring postgraduate education. To be clear, when we speak of a physical activity profession, we mean a profession in which physical activity plays a central role—for example, teaching physical education, coaching, serving as a corporate fitness consultant or fitness director, or serving as an athletic trainer or conditioning specialist for a sport team. Most of the physical activity professions can be grouped into relatively distinct **spheres of professional practice** (see the figure in the part III opener). Keep in mind that these spheres represent general categories and that each is made up of several professional occupations.

The chapters that follow in this part of the book describe some of the most popular careers available in the physical activity professions to students who graduate from a kinesiology program. Before we go any further, however, you need to know exactly what we mean by the term *professional* since, as a physical activity professional, you will be expected to exhibit the general attributes and expectations—and share in the obligations—of those who work in professional fields, whether they are lawyers, counselors, clergypersons, or physicians.

What Is a Profession?

Often we use the term *professional* in an informal way to describe the general quality of a person's performance or the work that a person does. For instance, the work of a cabinetmaker may be described as "very professional," or a person who is an excellent plumber may be described as "a real pro." At other times, we use the term to describe individuals who play or teach a sport as a full-time source of income (e.g., professional baseball players, tennis and golf pros who teach at clubs) in order to distinguish them from people who play only part-time and receive no remuneration (i.e., amateurs). These usages, however, do not convey the strict, formal meaning of the term. Normally, we do not think of cabinetmakers or athletes as members of a profession in the same way that we think of lawyers, physicians, teachers, counselors, physical therapists, rehabilitation specialists, or clergypersons as members of a profession (Etzioni, 1969). Surely, members of a profession are known for doing their work well and for doing it full-time and for pay, but that is far from the whole story.

KEY POINT

Individuals who are preparing for physical activity professions should know what a profession is, what type of work professionals do, how one gains entry to and acceptance in a profession, and the obligations of professionals.

Professional types of occupations have been described in terms of a cluster of attributes (Greenwood, 1957; Jackson, 1970). Although some characterizations of a profession may seem a bit idealistic, they provide useful models for helping us understand the distinctions and responsibilities of this type of work. You will discover that most types of work are not easily classified as being either completely professional or completely nonprofessional. Although physicians may be located at the extreme professional pole and day laborers may be located at the extreme nonprofessional pole, most types of work probably fall at some point along the continuum depicted in figure 11.1. To underscore this point, some of the physical activity professions are termed *minor professions* or *semiprofessions* (Lawson, 1984). Workers in the jobs that are positioned closest to the professional pole of the continuum

- master complex skills that are grounded in and guided by systematic theory and research;
- perform services for others, known as *clients* or *patients*;
- are granted a monopoly by the community to supply certain services to its members;
- are guided by formal and informal ethical codes intended to preserve the health and well-being of their clients; and
- meet the expectations and standards prescribed by their professional subcultures.

Mastering Complex Skills Grounded in Theory

The hallmark of professional work is that it draws on a complex body of knowledge and theory developed through systematic research. This is not required for strictly manual work (e.g., day labor, some restaurant work, some lawn work). Because the workplace in general is becoming more technology oriented, advanced specialized knowledge is becoming more important, even in occupations once considered nonprofessional. For example, welding is generally considered to be a fabrication craft; however, because it increasingly involves complex technology and sophisticated evaluation of the structure of materials, it is being touted as a profession (Albright & Smith, 2006). This point deserves emphasis: Determining what is and what isn't a profession can pose a difficult challenge.

Professionals are sometimes referred to as **practitioners** (derived from *practice*), a term signifying that they are not merely gifted thinkers but also gifted *doers*. Professionals establish their reputa-

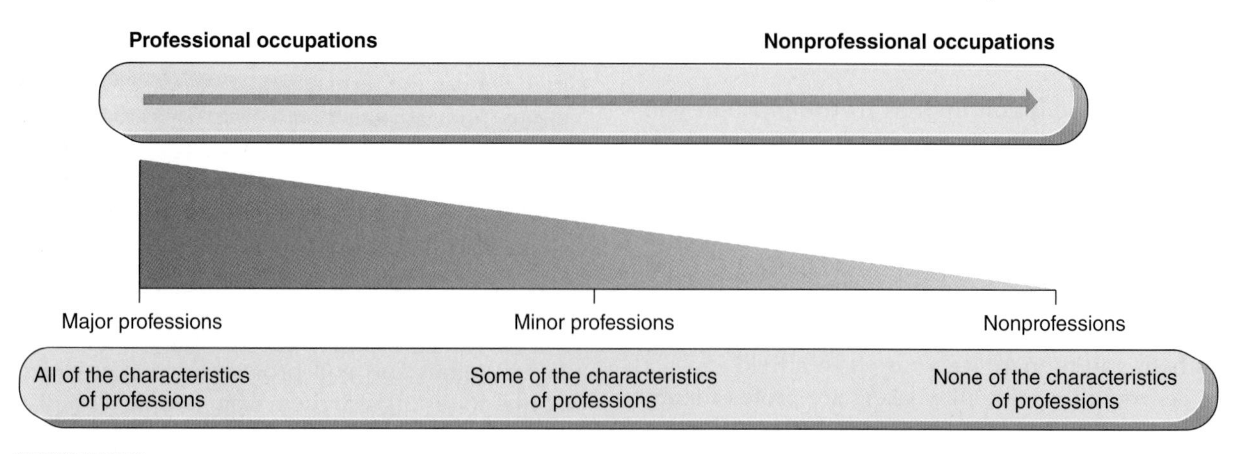

FIGURE 11.1 Occupations can be located along a continuum ranging from profession to nonprofession.

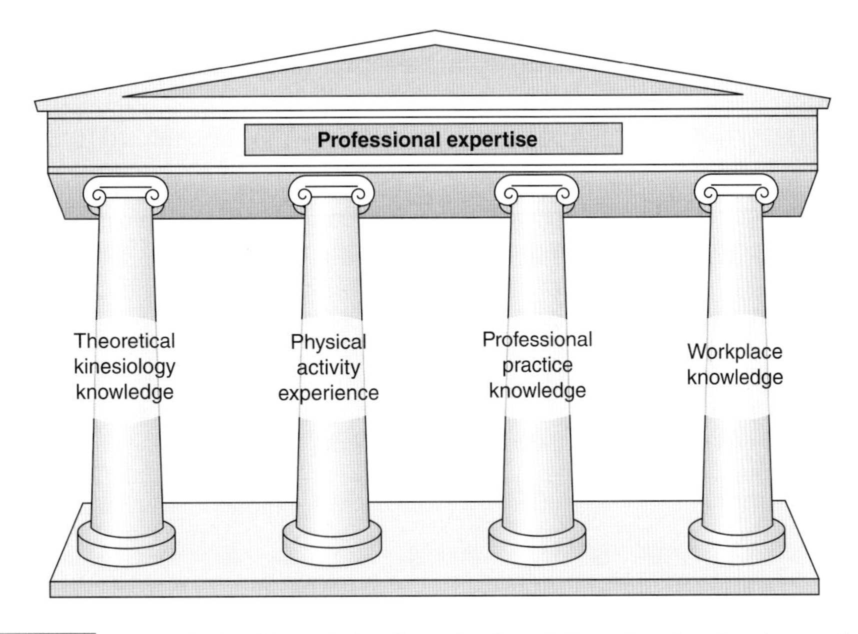

FIGURE 11.2 Many kinds of knowledge form the foundation of professional expertise.

tions by bringing about predetermined outcomes efficiently and effectively, usually on behalf of others. Doing so requires skill; in fact, the expert skill of professionals is what sets them apart from other types of workers. We usually associate the concept of skill with moving—for instance, the speed of a world-class sprinter or the agility of a pro basketball player. Surely, some professions do require deft motor skills—dentistry and surgery, for example—but motor skills are merely one category of skill. Here are three other categories of skills that are critically important in professional work: **cognitive skills,** which include analysis, deduction, diagnosis, prescription, and high-level reasoning; **perceptual skills**, which enable a person to identify and recognize existing problems or anticipate ones likely to occur; and **interpersonal skills**, such as listening, communicating, and motivating.

Development of these skills is essential if professionals are to make expert decisions regarding their professional practices. For example, biomechanics or ergonomics professionals who work for a corporation must have a keen eye for movement (perceptual skill), be able to analyze workers' movements in order to determine the reasons for injuries (cognitive skill), be able to devise strategies in order to correct the movements and reduce injuries (cognitive skill), and possess the ability to explain their findings to workers and supervisors (interpersonal skill). Athletic trainers must possess perceptual skills that enable them to spot hazards and potential sources of injuries, physical skills for conducting orthopedic tests, cognitive skills of analysis and deduction for determining correct emergency procedures, and communication skills for relating their findings to physicians or motivating athletes to adhere to rehabilitation schedules. Becoming a professional, then, involves recognizing the relevant factors in any situation; knowing the steps that must be taken to bring about a desired result based on research and theory in the field; possessing the skill required to achieve the result; and communicating a plan of action to a client, student, or patient.

How are these important skills acquired? In many cases, professional expertise is developed from experience in performing and watching others perform physical activity (see figure 11.2). This type of knowledge is discussed as physical activity experience in part I of this book. For example, although not all soccer, basketball, or gymnastics coaches were elite performers, most have spent time performing and watching the activity of their chosen sport and are familiar with and comfortable in the contexts in which it takes place. Later in this chapter, we talk more about the contribution of physical activity experience to your professional development.

The skills of a physical activity professional also derive from knowledge of theory about physical activity. As discussed in chapter 1, **kinesiology theory is embedded in concepts and principles**

A Simple Rule of Thumb

Here is a practical tip for professionals from someone who has succeeded in the business world. The sentiment he expresses applies to most professional settings (Satell, 2010):

> Over the years, I've come up with a simple litmus test that has been extremely useful in making personnel decisions and choosing partners: *Professionals solve more problems than they create*. Whenever I need to make a decision about entering or extending a relationship, I apply that standard. . . . Problems are rarely caused for want of competence but usually stem from ill-placed pride and a lack of common decency. Many feel that because they work for a big company, have a big title, and know a lot of things (especially acronyms), they are entitled to make it difficult for others. How else can they advertise their stature? True professionals look to solve problems. They want to help people, seek out "win-win" situations, and are willing to shoulder the burden when things go wrong. Their self-worth is defined by what they do for others, not what they can do to them and get away with.

Which one are you?

related to physical activity that are taught in the subdisciplines. Of course, not all kinesiology theory applies directly to every task that a professional performs, just as not all theoretical knowledge in medicine applies to every case that a physician treats. For example, learning how to calculate the theoretically optimal angle for the projection of a shot put in biomechanics class may not translate into useful prescriptions for a track coach, and learning the details of pulmonary function in exercise physiology class may not be directly applicable to an athletic trainer's work.

Nevertheless, because kinesiology theory focuses on broad principles of physical activity, it may exert a potent, albeit indirect, effect on the way professionals think and the skills they use to solve problems. Understanding the big picture can help professionals adapt knowledge to new situations, integrate knowledge of the principles of physical activity, and think flexibly. Contrast this, for example, with bricklaying and carpentry, which are truly respectable forms of work known as *trades* or *crafts*. These jobs certainly require the use of cognitive and perceptual skills, but the skills needed are almost always applied to a fairly stable set of problems and contexts. Because one brick wall is pretty much like another, even laypersons can buy books about home repair and, simply by reading, serve as their own bricklayers. In contrast, the work of a physical activity professional entails solving a wide variety of complex problems involving complex human beings. People cannot learn to do this simply by reading manuals.

For example, physical education teachers may observe errors in students' motor performance that they have never seen before. When faced with such a situation, they can systematically apply their knowledge of biomechanics, motor learning, and pedagogy—and use a diagnostic eye developed through practice—to devise practice experiences that enable students to correct errors and improve performance. Similarly, an athletic trainer confronted with a rare injury may rely on her in-depth knowledge of biomechanics, orthopedic testing, emergency medicine, and other kinesiology concepts to determine the best course of emergency action. Thus, mastering theoretical knowledge of kinesiology enables professionals to understand the fundamental bases of problems (e.g., physiological, social), ask intelligent questions about their work, and devise flexible strategies for solving problems—even those they have never before encountered.

KEY POINT

Professionals develop cognitive, perceptual, and motor skills—anchored in theoretical, workplace, and practice knowledge—that enable them to achieve predetermined outcomes efficiently and effectively.

Physical activity experience and theoretical knowledge, though crucial, are not sufficient in themselves. Professionals' skillful behavior is also guided—and their reputations determined—by their wealth of professional practice knowledge. This type of knowledge is not acquired so much by reading, nor can it be fully evaluated by written examinations. Rather, professional practice knowledge is what some have called "knowledge in action" (Schon, 1995). It can be seen, for example, in a physician's competence in diagnosing diseases and in the skill with which a lawyer assembles a convincing case for a court trial. In kinesiology, professional practice knowledge may be reflected in an athletic director's expertise at recruiting and retaining effective coaches, in a fitness leader's skill at prescribing effective exercises for a client who is recovering from a heart attack, and in an athletic trainer's skill at treating an on-field injury. In many cases, such highly competent professionals are unable to explain in any detailed fashion how they manage to bring about such effective results.

Finally, professionals rely on **workplace knowledge**, which is knowledge about performing the relatively mundane tasks associated with their jobs. Workplace knowledge requires little formal

What Does Being a Professional Really Mean?

The term *professional* is used in many ways. Consider whether it is used appropriately or inappropriately in each of the following examples.

- An athletic trainer shares information about one athlete's injury with another athlete. Her supervisor reprimands her for not conducting herself as a professional.
- An award is bestowed on a physical education professor for having provided 25 years of distinguished professional service to the university.
- A personal trainer is heard repeatedly giving clients misinformation about nutrition and exercise. The fitness director tells him that he is not behaving in a professional manner.
- An athletic director praises a laundry room employee for having the highest professional standards.
- A physical education teacher tells her colleague that he should join SHAPE America because it is the professional thing to do.

Widely varying definitions of terms such as *profession*, *professional*, and *professionalism* are used by people in a variety of fields that offer services to the public. The definition offered by the Professional Standards Councils, which works to improve standards of practice and protect consumers in Australia, is applicable to many professions:

> A profession is a disciplined group of individuals who adhere to ethical standards. This group positions itself as possessing special knowledge and skills in a widely recognized body of learning derived from research, education and training at a high level, and is recognized by the public as such. A profession is also prepared to apply this knowledge and exercise these skills in the interest of others. (Professional Standards Councils, n.d.)

What then is *professionalism*? According to the late Eliot Freidson, a sociologist who spent most of his life studying the topic, "Professionalism exists when an organized occupation gains the power to determine who is qualified to perform a defined set of tasks, to prevent all others from performing that work, and to control the criteria by which to evaluate performance" (Friedson, 2001, p. 12).

A more expansive definition has been offered by a representative of the international bar association (McGuigan, 2007) who believed that lawyers should reflect professionalism by

- dedicating themselves to serving clients before self,
- dedicating themselves to serving the public interest, improving the law, and improving the profession,
- devoting themselves to honesty, integrity, and good character, and
- maintaining competence in a specialized body of knowledge and skills, which are freely shared with other professionals.

What similarities and differences (either literal or implied) do you find in these definitions?

education or training. For example, cardiac rehabilitation technologists must know how to maintain the electronic equipment they use, how to schedule patients for examinations, and how to follow proper procedures for opening and closing the facility. Such tasks are not professional in their scope and do not require highly sophisticated knowledge, but they are important. For example, a cardiac rehabilitation technician who does not ensure that the equipment is kept in proper working condition—or a strength and conditioning coach who doesn't maintain a clean workout facility—may soon be unemployed, regardless of the quality of his or her professional skills.

Performing Services for Clients or Patients

Most professional work takes the form of services performed for others. **Service** involves giving assistance or advantage to others, usually in a spirit of helpfulness and concern. Individuals who are employed in service work such as nursing, social work, family counseling, public health, recreation, and kinesiology are increasingly referred to as members of the **helping professions** (Lawson, 1998b). The beneficiaries of a professional's services are known as *clients*. (In the allied medical fields, clients are patients; in teaching, they are students.) Most professionals like meeting and working with people, feel comfortable in social settings, and derive great enjoyment from helping people meet needs and accomplish important goals.

You always should keep in mind the following three important aspects of service:

› **Professionals are committed to helping others.** One often hears professionals describe their career as a calling, which suggests that their

Professionalism in the Workplace

Certain key values, attitudes, and behaviors are important in any type of professional work. The following list presents recommendations from a leader in the communication industry (Feignbaum, 2012).

Personal Appearance

At work, perception matters. Thus, a person with talent, skill, and ability may not be taken seriously if he or she comes to work ungroomed, disheveled, or dressed inappropriately. Moreover, in many health care professions, certain clothing and grooming are required, not only to ensure a respectful look but also to protect patients from disease and adverse reactions.

Emotional Baggage

Being professional includes checking your personal issues at the door when you come to work. Doing your job properly and professionally means not letting what happens at home affect your work performance. It also means not letting people push your emotional hot buttons.

Respect

Respect provides the foundation of workplace success with superiors, co-workers, employees, customers, and clients. Whatever your company culture may be, everyone responds to respect. Keep in mind also that your workplace conduct represents not just your organization—it also represents you.

Communication

Savvy professionals know how to communicate. Knowing who to keep informed, when to use discretion, and when to clearly disseminate information is part of what separates the good from the great. Listening to and really understanding people does not go unnoticed.

Integrity

Along with respect, integrity is one of the highest professional values. In fact, some professions (e.g., law, medicine, journalism) have established a professional code of ethics. Even when the industry doesn't mandate it, maintaining your own code of ethics helps you be an invaluable and respected professional.

primary motivation is to enhance the health, education, enjoyment, and general well-being of those they serve. In fact, concern for the well-being of clients often supersedes professionals' concern for their own comfort and satisfaction. For instance, you may have known physicians or lawyers who sacrificed lucrative practices in order to work with people who are poor and disadvantaged. The same spirit appears in a teacher who volunteers her time to give remedial instruction to a student on the weekend or a geriatric fitness leader who stays late after each session to give special attention to an elderly man who feels too embarrassed to exercise with others present. Obviously, becoming an effective professional need not entail such sacrifices, but wanting to help others is a prerequisite for such work.

› **Professionals render expert service.** Being a professional means more than simply being willing to help. Professionals know how to determine a client's needs and how to initiate and evaluate an appropriate course of action to meet those needs. As a result, professionals are granted a certain amount of autonomy to carry out their work. Their opinions are respected, and they usually have a great deal of freedom to make decisions about how they help clients. Some professionals have more autonomy than others. For example, a personal trainer in private practice may enjoy much more autonomy in decision making than a fitness instructor who works for a commercial gym. Similarly, a professional golf instructor who is self-employed enjoys more autonomy than a physical education teacher, whose actions may be limited by school board policies, the dictates of a school principal, and a curriculum guide.

› **Professionals should not encourage their clients to become totally dependent on their services.** It is inevitable that clients, patients, and students become somewhat dependent on authority figures in the early stages of treatment, but such dependence can be undesirable when it results in professionals' adopting a paternalistic attitude toward those whom they serve. Ideally, a professional and a client work collaboratively to achieve common goals. This is not to say that a client should have abject faith that a professional will keep his or her best interests in mind when making decisions. One of the most egregious violations of professional conduct is that of exploiting the vulnerability of a client by making decisions that are self-serving or that stretch far beyond the professional's area of expertise. Therefore, a coach who orders an injured player back into the game will quickly lose the respect of colleagues, as will a physical therapist who, not wanting to suffer financial loss, fails to refer a patient to a competitor even though the competitor is better equipped to meet the patient's specific needs.

KEY POINT

The touchstone of professional work is the delivery of expert services to improve the quality of life for others, always with priority given to the client's welfare.

Possessing a Monopoly on the Delivery of Services

Society values professionals because they provide needed services that are unavailable from any other occupational group. If you need a cavity filled, you go to a dentist. If you need spiritual guidance, you may go to a member of the clergy. If you need to rehabilitate an injured knee to play basketball again, you go to a physical therapist or athletic trainer. Governments and professional organizations control access to professions through requirements for education and licensure.

KEY POINT

Professionals are granted monopolies because only they possess the necessary knowledge and skills to meet particular needs of the community.

The physical activity professions came into existence and continue to flourish because society recognizes that people in such positions make a unique and valuable contribution to the health and well-being of all. In the case of kinesiology, no other discipline prepares people with the background knowledge and skills needed to meet such a diverse range of needs related to physical activity. In this sense, kinesiology and the physical activity professionals it prepares possess a monopoly on the delivery of physical activity services. That monopoly is protected by various certification and licensing arrangements. Teachers, for example, are usually required to possess a certificate granted by the state or province in which they operate. The purpose of this certificate is to ensure that only professionals who have met the minimum standards of educational preparation will be admitted to the monopoly allowed to practice the profession of teaching.

Ensuring High Standards and Ethical Practices

Professionals recognize that society has granted them authority to regulate their own conduct and expects them to do it. Violation of accepted professional standards by a single professional threatens all members of the profession by bringing into question the legitimacy of the monopoly they have been granted. Thus, professionals have a huge stake in maintaining or improving not only the quality of their own practices but also the quality of service delivered by all members of their profession. In this sense, although professionals may work at clinics, schools, hospitals, or other agencies, they view themselves primarily as representatives of the profession as a whole.

Because the knowledge base for the physical activity professions is expanding continually, exercise leaders, teachers, trainers, sport leaders, and other professionals can easily become outdated in their techniques. Professionals whose practices are not in accord with the latest information become a threat to all professionals in the field. For example, a clinical exercise physiologist who fails to keep abreast of her field is in danger of recommending activities to clients that recent research has shown to be inferior or even dangerous. Such a person may face accusations of unprofessional practice.

Keeping updated will require you to participate in professional organizations; regularly read professional books and articles; and attend conferences, workshops, and other meetings sponsored by professional societies. Unlike salespersons, business executives, or others in the commercial marketplace, who tend to protect information from their competitors, members of a profession relate to each other as colleagues and willingly share information in order to serve the best interests of their clients. If you want to be a top-notch professional, you must commit to educating yourself continually; in fact, it is not too early to join a professional organization while still in your undergraduate years. In addition, in order to hold a professional license or certification, a professional must usually obtain continuing education units (CEUs) through professional readings, conferences, or trainings to maintain currency and expertise.

Operating on a less than adequate knowledge base is not the only way in which professionals breach standards of conduct; they can also be sanctioned for unethical behavior in the workplace. Unethical conduct occurs when a professional's behavior at the worksite conflicts with generally accepted standards of practice in the field. To ensure that members clearly understand ethical expectations, professional organizations often publish a **code of ethical principles and standards**. The primary objective of such codes is to protect the rights of students, clients, and patients. A code typically instructs professionals to describe and advertise their competencies accurately, defines the nature of the relationship that should exist between clients and professionals and between subordinates and professionals, and clarifies matters involving financial transactions with clients.

The excerpt presented in the sidebar titled American Council on Exercise Code of Ethics is representative of most ethical guidelines. The code is intended to guide the practice of those working in the fitness industry who have been certified by the council. Every professional association has a code of ethics which, in most cases, is available on its website.

KEY POINT

Professionals have a stake in maintaining standards of conduct for all practitioners in their profession. To do so, professional organizations provide members with opportunities for continued education and publish guidelines specifying acceptable and unacceptable conduct.

In order to ensure that members understand the prevailing code of conduct and are competent to implement safe and effective practice, a profession may license its members. A license constitutes formal authority granted by law to practice a profession. Professions that require licenses can suspend a professional's license for various types of infractions, effectively terminating the professional's authority to practice.

Although the medical and legal professions have made effective use of licensing provisions to control the quality of professional services, relatively few physical activity professions are licensed. Two exceptions are physical therapy and physical education teaching. In addition, some U.S. states are now licensing athletic trainers, and that trend is spreading across the country. In Canada, some provinces are licensing kinesiologists to provide health-related physical activity prescriptions beyond any rehabilitation treatment provided by physiotherapy (physical therapy).

The lack of licensing for some physical activity professions doesn't mean that unethical or

American Council on Exercise Code of Ethics

As an ACE Certified Professional, I am guided by the American Council on Exercise's principles of professional conduct whether I am working with clients, the public or other health and fitness professionals.

I promise to:

- Provide safe and effective instruction.
- Provide equal and fair treatment to all clients.
- Stay up-to-date on the latest health and fitness research and understand its practical application.
- Maintain current CPR and AED certificates and knowledge of first-aid services.
- Comply with all applicable business, employment and intellectual property laws.
- Uphold and enhance public appreciation and trust for the health and fitness industry.
- Maintain the confidentiality of all client information.
- Refer clients to more qualified health or medical professionals when appropriate.
- Establish and maintain clear professional boundaries.

ACE Professional Practices and Disciplinary Procedures

The professional practices and disciplinary procedures of the American Council on Exercise (ACE) are intended to assist and inform certificants, candidates for certification and the public of the ACE Application and Certification Standards relative to professional conduct and disciplinary procedures. ACE may revoke or otherwise take action with regard to the application or certification of an individual in the case of:

- Ineligibility for certification.
- Irregularity in connection with any certification examination.
- Unauthorized possession, use, access, or distribution of certification examinations, score reports, trademarks, logos, written materials, answer sheets, certificates, certificant or applicant files, or other confidential or proprietary ACE documents or materials (registered or otherwise).
- Material misrepresentation or fraud in any statement to ACE or to the public, including but not limited to statements made to assist the applicant, certificant, or another to apply for, obtain, or retain certification.
- Any physical, mental, or emotional condition of either temporary or permanent nature, including, but not limited to, substance abuse, which impairs or has the potential to impair competent and objective professional performance.
- Negligent and/or intentional misconduct in professional work, including, but not limited to, physical or emotional abuse, disregard for safety, or the unauthorized release of confidential information.
- The timely conviction, plea of guilty, or plea of nolo contendere in connection with a felony or misdemeanor, which is directly related to public health and/or fitness instruction or education, which impairs competent and objective professional performance. These include, but are not limited to, rape, sexual abuse of a client, actual or threatened use of a weapon of violence, the prohibited sale, distribution, or possession with intent to distribute, of a controlled substance.
- Failure to meet the requirements for certification or recertification.

improper conduct by physical activity professionals goes unpunished. Usually, local entities administer appropriate sanctions at the local workplace, often in consultation with a professional organization. For example, a local school board may terminate a schoolteacher whose conduct violates ethical codes, and a university may fire a conditioning coach who administers illegal anabolic steroids to athletes. Similarly, although the American College of Sports Medicine (ACSM) cannot suspend a cardiac rehabilitation therapist whose conduct puts a client at risk, the clinic director or other supervisor at the worksite may decide to terminate the worker's employment after consulting ACSM guidelines.

Adhering to Standards of Professional Subculture

Although professionals work in a variety of different locales and contexts, they tend to hold high expectations for how members of the profession present themselves and relate to others. For example, professionals emphasize respect and politeness not only to those they serve but to colleagues as well. Professionals tend to be well organized in their work, are not clock watchers, and don't hesitate to work extra hours when required. They dress according to accepted occupational standards and attend to matters of personal grooming and hygiene. A sport program director who comes to a board meeting in dirty and disheveled clothes, a cardiac rehabilitation technician who forgets to wear her lab coat, a physical education instructor who teaches classes in street clothes—these individuals are not dressed professionally. Likewise, a physical therapist who never combs or brushes his hair, a personal trainer who smells of body odor—both are guilty of presenting themselves in an unprofessional manner. These general expectations apply to professionals in any field.

In addition, physical activity professionals often face a more specific set of expectations. For example, fitness leaders and physical education teachers may be expected to model a physically active life, to refrain from smoking, and to maintain a level of fitness appropriate for their age.

More specifically, particular physical activity professions are likely to require unique types of knowledge, employ different types of skills, operate according to certain standards and codes of conduct, require specialized language and terminology, require specific certification and continuing education procedures, and often involve certain types of dress and on-the-job conduct. When an individual learns to adapt to these profession-specific expectations, he or she is said to have been socialized into the profession. In other words, the person has learned to adhere to the roles and responsibilities of the workplace. The degree to which a professional has been socialized into the professional role is reflected in the ease, confidence, and efficiency with which he or she performs daily activities and communicates with clients and co-workers.

Are All Fitness Instructors Professionals?

Researchers who examined the fitness knowledge of a random sample of 115 health fitness "professionals" in the areas of nutrition, health screening, testing protocols, exercise prescription, and general training knowledge regarding special populations found that 70 percent of them of did not have a degree in a field related to exercise science (Malek, Nalbone, Berger, & Coburn, 2002). Those who did not have a degree scored 30 percent lower on the test than those who had a degree. Strong predictors of knowledge included a bachelor's degree in kinesiology or exercise science and certification by either ACSM or the National Strength and Conditioning Association (NSCA), as opposed to other certifications; in contrast, knowledge was not related to years of experience as a fitness instructor. The researchers argued that "these findings suggest that personal fitness trainers should have licensing requirements, such as a bachelor's degree in exercise science and certification by an organization whose criteria are extensive and widely accepted, before being allowed to practice their craft" (p. 19).

This study was conducted in 2002, and there is reason to believe that the preparation of fitness instructors has improved in recent years. However, the nature of the qualifications possessed by fitness workers continues to be an issue. Certification in fitness instruction is reportedly now offered by more than 100 organizations, many of which require little more than a short online experience or weekend workshop. The most respected certifications are still those awarded by ACSM and NSCA.

The sooner you become socialized into your profession, the more valuable you will be to an employer. For that reason, students in professional programs are increasingly seeking opportunities to volunteer at clinics, YMCAs, schools, and community agencies. Although usually not counted as course work, such extracurricular experiences can give you a head start in learning how to assume professional roles and provide you with valuable references in the field. Similar growth and leadership opportunities are available in student major clubs, athletics, student government organizations, and other on-campus activities.

How Do Our Values Shape Our Professional Conduct?

Hal Lawson (1998a) described two value orientations that may characterize professionals in any field—**mechanical, market-driven professionalism** and **social trustee, civic professionalism**. These orientations are ideal types, which is to say that most professionals don't fit completely into one category or the other, but they can serve as reference points that help you determine how you want to operate as a professional.

KEY POINT

Mechanical, market-driven professionals value the profession, profit, personal prestige, and status over the rights and needs of clients. Social trustee, civic professionals value clients and the social good more than themselves or their profession.

Mechanical, Market-Driven Professionalism

Mechanical, market-driven professionals tend to be so focused on their own professional aspirations that they lose sight of their clients' needs and desires. In many cases, the work of these professionals becomes its own justification, valued by them and their colleagues regardless of whether it provides any real benefits to society. The medical, nursing, and law professions are sometimes accused of this type of professionalism, but some of the physical activity professions can be guilty of it too.

Mechanical, market-driven professionals are likely to serve their clients in a fragmented, compartmentalized fashion. That is, these professionals view their professional contributions within the strict limits of their specialization and ignore the variety of relevant forces acting on the client, such as family disintegration, illness, or drug or alcohol dependency. Such professionals may dismiss these other problems as "somebody else's responsibility" and never go out of their way to see that proper referrals are made. They may treat clients humanely but with little feeling, believing instead that an objective, arm's-length attitude is critical to professional success.

Mechanical, market-driven professionals also are likely to devote much time to enhancing their status and competing with other professionals, even if it means sacrificing the quality of service they give to their clients. They pride themselves on their expertise, view themselves as superior to those they serve, and often encourage clients to develop a dependency on them and their services.

Social Trustee, Civic Professionalism

Social trustee, civic professionals hold to a much different set of values. They believe that the worth of a profession is measured by its effectiveness in promoting the social welfare, enhancing social and economic development and democracy, and ensuring social responsibility and social justice. The actions of a social trustee, civic professional are guided not by a fascination with technology and technique or by a concern for status and prestige but by a vision of the good and just society. They operate according to this rule: "Healthy people and a good society first, me and my profession second in service of this greater good" (Lawson, 1998a, p. 7). They recognize that clients live in a multiplicity of worlds (e.g., work, school, family, church), and they understand the ways in which worlds can interact to affect their clients' lives.

Because of this recognition, these professionals often work in teams with other specialists to achieve desirable goals. They make no pretense about being objective in their relationships with clients. Their professional practices reflect their personal values. They don't view clients as dependent on them; instead, they believe that professional–client interaction can be a mutual growth experience in which each benefits from the other's knowledge and skill.

How Does Your Personality Match Up With Your Career?

Career counselors often use occupational attitude inventories to determine the best matches between an individual's abilities and personality and possible career choices. Many of these tests are based on researcher John Holland's six vocational personalities and work environments.

- *Realistic personality:* Values practical and mechanical experiences such as operating machines or farming. Does not enjoy social interaction and would not enjoy teaching. Possible careers include firefighting, electrical work, or carpentry. Opportunities are limited in the field of kinesiology.
- *Investigative personality:* Skilled at understanding and analyzing science and math problems. Not usually leaders or salespersons; more comfortable in laboratory environments. Those who pursue careers in kinesiology as researchers in exercise physiology, biomechanics, sport psychology, or sport sociology might have investigative personalities.
- *Artistic personality:* Enjoys creative activities such as art, drama, crafts, dance, music, or creative writing. Best suited to work as writer, actor, artist, graphic designer, musician, or dancer.
- *Social personality:* Enjoys helping people; values friendliness and trustworthiness in others. Tends to excel in careers such as nursing, counseling, physical therapy, and medicine. Socially-oriented individuals often pursue careers in athletic training and sport medicine.
- *Enterprising personality:* Ambitious, energetic, and sociable. Tends to be a self-starter who enjoys being in charge. Less interested in analyzing or conducting scientific studies than in getting something practical accomplished. Often pursues a career in business or politics. Managers of fitness facilities, coaches, and school principals often have enterprising personalities.
- *Conventional personality:* Likes to work with numbers, machines, facts, and records. Prefers highly structured, orderly work that follows a set plan. Often is a bookkeeper, bank teller, or clerk. Most careers in kinesiology are not compatible with this type.

Which personality type do you think best describes you? What careers in kinesiology do you believe your personality best matches?

How Are Physical Activity Professionals Educated for the Workforce?

By now you are aware that professional work requires advanced knowledge that you can obtain only through formal education at the undergraduate and graduate levels. If you're like most college students, however, you may not always understand why certain courses are required or what the purposes are of various phases of the curriculum. You may be interested to learn that kinesiologists themselves are constantly engaged in a debate concerning the structure of the kinesiology curriculum. Although the curriculum at your institution may not be an exact replica of the curriculum described here, it is probably similar in its general features.

The kinesiology curriculum model incorporates five general types of academic experiences that are essential in preparing kinesiologists for their professional roles:

- Course work in liberal arts and sciences
- Course work in physical activity
- Course work in theoretical and applied theoretical knowledge in kinesiology
- Course work in professional practice knowledge and skills for particular professions
- Apprenticeship or internship experience at a worksite

The best professional preparation programs immerse students in all of these types academic experiences.

Course Work in Liberal Arts and Sciences

All college graduates, regardless of their career aspirations, are assumed to be educated people. Used in this sense, the word *educated* describes someone who has struggled with the great works of literature, learned something about the historical development of human civilization, developed critical skills essential for analyzing arguments logically, learned how to determine the truthfulness of ideas and identify the values embedded in them, developed sufficient knowledge about the arts to appreciate the richness that they can add to one's life, and developed insights into the problems and prospects of their own and other cultures. Accomplishing this goal requires one to take courses in science, philosophy, history, literature, sociology, mathematics, art, music, and the dramatic arts, among other subjects. These courses are usually part of the "general education" or **liberal studies** requirements in college and are taken during the first 2 years of undergraduate study. They form the foundation for your professional education, which usually comes later.

Liberal education derives its name from a sense of freedom: Only by immersing yourself in the arts and sciences can you liberate, or free, yourself from dependence on the thinking of others or the past. Liberally educated people develop a capacity for seeking the truth and making intelligent choices when they confront life's problems and opportunities. They also understand that nobody is ever completely educated—that everyone is always in the process of becoming educated. Thus, these people are aware of what they do not know, and they continually seek opportunities to expand their intellectual and cultural horizons by reading books and newspapers, attending concerts, patronizing the arts, and engaging in other lifelong educational activities.

Thus, the liberal education you receive in college is just the beginning. Whether this initial immersion in the liberal arts and sciences accomplishes its purpose will be reflected in your level of interest in continuing to educate yourself after graduation. What types of books will you read? What kinds of music will you listen to? How will you speak and write? Will your analysis of the claims of politicians and other public figures be logical and based in fact? Your approach to these and hundreds of other life experiences should bear witness to the fact that you are a liberally educated person. Of course, a liberal education should also have short-term effects. The most important effect is to provide you with the broad-based knowledge and intellectual skills essential for undertaking more advanced study in your professional curriculum. A liberal education frees you from depending on others to do your thinking for you. Thus, it provides the foundation on which all other educational experiences are constructed.

Course Work in Physical Activity

Most kinesiology departments require their majors to enroll in physical activity courses, either as part of their liberal education requirements or in conjunction with their professional programs. As noted in part I, you entered college with a wealth of experiential knowledge about physical activity and no doubt found it to be a vehicle for learning much about yourself, others, and the world around

So You Want to Become an Expert in Your Field?

According to a survey by Eagan and colleagues (2015), 60 percent of your classmates are determined to become authorities or experts in their chosen fields. Being an expert means that you have mastered an area of knowledge, a set of skills, or both to such a high level that you are recognized as an authority by your colleagues and clients. This is easier said than done. Here, in a nutshell, is some advice offered by scholars who have spent their careers studying expertise:

> Consistently and overwhelmingly, the evidence . . . [has shown] that experts are always made, not born. These conclusions are based on rigorous research that looked at exceptional performance using scientific methods that are verifiable and reproducible. . . . The development of genuine expertise requires struggle, sacrifice, and honest, often painful self-assessment. There are no shortcuts. It will take you at least a decade to achieve expertise, and you will need to invest that time wisely, by engaging in "deliberate" practice—practice that focuses on tasks beyond your current level of competence and comfort. (Ericcson, Prietula, & Cokely, 2007, p. 116)

Kinesiology Colleagues

Terry and Jan Todd

Courtesy of H.J. Lutcher Stark Center for Physical Culture and Sports at the University of Texas at Austin

Since they joined the kinesiology faculty at the University of Texas at Austin in 1983, Terry and Jan Todd have pursued the goal of preserving the history of physical culture. Both are former lifting champions who set national and world records. Now they have established what historians have labeled the single most important archive in the world on the topic. The archive contains an estimated 25,000 volumes and some 200,000 magazines and journals, along with pamphlets, clippings, photographs, and training equipment. Thanks to generous gifts from benefactors, the collection is housed in a beautiful, 27,000-square-foot structure on the Texas campus and is open to scholars and members of the public investigating historical issues related to physical culture. Named the H.J. Lutcher Stark Center for Physical Culture and Sports, the facility also houses a museum and serves as an excellent example of how kinesiology can be integrated with the liberal arts.

you. However, the physical activity experiences offered within the framework of a liberal or professional education should result in qualitatively richer experiences than the experiences you had before you came to college. Let's take a moment to consider how physical activity courses at the college level might affect you.

If you've already taken a physical activity course at your college, you may have discovered that the course was more intense and taught at a more advanced level than were your classes in high school. Moreover, everything we learn occurs in a context of the totality of our personal experiences. Thus, as your liberal education continues to expand your horizons, your approach to physical activity experiences will change as well. You should be a more intelligent analyst of your performances and a more persuasive critic of the performances of others. You should also be able to play your role more astutely in planning and preparing for your own physical activity experiences. Finally, you should be able to draw connections between your physical activity experiences and the world around you.

KEY POINT

College courses in physical activities are often taken in conjunction with liberal studies and form an essential part of professional preparation in kinesiology.

For instance, a course in martial arts should attune you to the nuances of pace, tempo, and rhythm; similarly, a course in fencing should foster an appreciation of beauty, force, and form in movement. Your course work in physical activity may also sensitize you to ethical and moral questions that you might not have noticed before you were exposed to a liberal education.

Physical activity courses are also required as part of your kinesiology major. This requirement may seem immediately reasonable to you if you plan a career in teaching, coaching, or fitness leadership. However, why does it exist for students in sport administration, pre-rehabilitation sciences, and other professions in which physical activity performance is not part of the professional responsibilities? If you're struggling with this question, think about the general model of physical activity described in chapter 1, which recognizes that knowledge gained through physical activity provides one of the three legs of the stool on which the discipline of kinesiology rests. Whatever career you enter, performance of physical activity will be at center stage, and you should therefore acquire a wealth of firsthand physical activity experiences—not only to help you identify with physical activity problems confronted by your clients, but also to broaden and enrich your comprehensive understanding of the phenomenon of physical activity. Many kinesiology professions also require good physical fitness in order to perform physical tasks without injury.

A liberal education expands your approach to physical activity by incorporating the totality of your personal experiences.

Edie Layland - Fotolia

Course Work in Theoretical Kinesiology

The chapters included in part II give you a general overview of the subdisciplines of kinesiology. Courses in these subdisciplines teach theoretical knowledge not only about sport and fitness but also about physical activity in general. Much of this knowledge is applicable to professional practice. For example, theoretical knowledge about how the body responds to exercise (drawn from the subdiscipline of physiology of physical activity) seems directly applicable to the work of a swimming coach or cardiac rehabilitation specialist. Some other theoretical knowledge—in sport history, for example, or biomechanics—may not seem immediately applicable to work; over time, however, it may profoundly affect how a professional thinks and thus acts.

Some professors would argue that students should not expect all theoretical knowledge in kinesiology to apply to specific careers, just as theoretical knowledge in literature, philosophy, biology, and physics should not be expected to directly equip students with occupational skills. These professors would claim that the value of theoretical knowledge lies in its power to sharpen students' perspectives on themselves and the world around them and to help them develop new, critical, and creative ways of thinking. Thus, some would say that a "discipline" of kinesiology (theoretical kinesiology) should not directly apply to professional work any more than should other liberal studies, such as poetry and medieval history.

This view is not shared by all kinesiology professors. To the contrary, some believe that kinesiology is better understood as a professional field and that its best justification to the world rests on the fact that it prepares people for jobs in the physical activity field. A detailed discussion of this debate lies beyond the scope of this textbook, so let us just say simply that it is probably not an either–or matter.

Figure 11.3 shows that theoretical kinesiology involves knowledge of physical activity with roots in the traditional parent disciplines in the humanities; the biological, behavioral, and sociological studies; and the arts. Theoretical knowledge in kinesiology builds on these foundational areas by relating each to the emerging body of knowledge about physical activity. The result is a stimulating body of knowledge for students who plan to enter the physical activity professions. At present, this body of knowledge is represented by isolated courses in exercise physiology, motor behavior, and so on. The important challenge for kinesiologists is to integrate this knowledge in the way that practitioners must do in order to make decisions in their work.

The evolution of theoretical knowledge of kinesiology has been an exciting development in the field of physical activity, but questions continue to surface about how much of this knowledge should be required of students embarking on careers in the professions. Should future athletic trainers, physical education teachers, and aquatic therapists be expected to master the entire body of knowledge of kinesiology? Similarly, sport management majors may well ask why they must take courses in exercise physiology or biomechanics. Students who plan to be conditioning coaches may ask why they should be expected to know about the history and philosophy of sport, especially when the concepts may seem not to apply to the professional problems that they will face in the workplace.

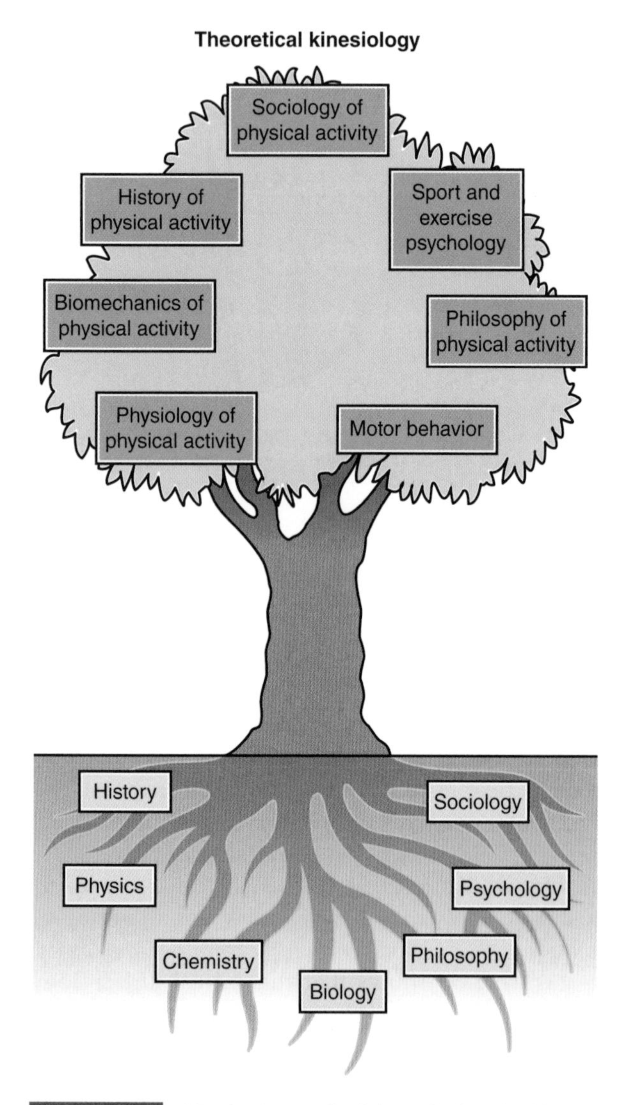

FIGURE 11.3 Much theoretical knowledge in kinesiology is derived from parent disciplines. What new subdisciplines do you think will grow in the kinesiology of the future?

In fact, a good case can be made for requiring all undergraduate kinesiology majors, regardless of their career aspirations, to learn the body of knowledge of kinesiology (Newell, 1990). For starters, it seems only logical to expect students who are earning a degree in a given field to master at least a general level knowledge about that field in its entirety. In addition, the wisdom of history, philosophy, and sociology of physical activity is relevant to kinesiology professionals when they face changes in fitness fads, inconsistent research results, politics, and other forces that affect their practice.

Another argument for requiring students to master theoretical kinesiology is to ensure that people who work in the field are united around a common core of physical activity theory—for example, the American Kinesiology Association core described in chapter 1. Without such a core, kinesiology might fragment into a number of completely different fields. Those who believe that all physical activity professionals should be required to study theoretical kinesiology also argue that doing so may benefit future professionals in ways that we do not yet fully understand. Just as a liberal education alters students' ways of approaching problems and opportunities in their lives, so an immersion in the theoretical knowledge of kinesiology may equip physical activity professionals with general concepts and ways of thinking about physical activity that will help them solve the myriad problems they will face in professional practice.

Most undergraduate programs in kinesiology require students to take core courses in kinesiology theory regardless of their career aspirations; these courses may vary from institution to institution. The American Kinesiology Association has recommended that all programs offer core course work in four areas (American Kinesiology Association, 2003):

- Physical activity in health, wellness, and quality of life
- Scientific foundations of physical activity

Professional Issues in Kinesiology

Evidence-Based Practice

Kinesiology professionals often need knowledge from numerous subdisciplines to help people be more physically active. For instance, most people know they should be physically active and exercise regularly, but, as with eating their vegetables and practicing other healthy habits, many find it difficult to follow through. Given this reality, most kinesiology professionals should, as part of their work, integrate research evidence from psychology and sociology in order to help motivate people to be physically active and sustain an active lifestyle. In other words, it's not enough to simply have the research documenting physical activity's benefits; the field and its professionals need to continue researching how to implement programs and motivate clients in the real world.[1] Evidence-based research on implementing programs in the messy and complex real world is called *translational research*. Thus, professionals must be committed to continuous learning in relation to all evidence related to their practice—not just the topics they find interesting or initially think are most important.

Citation style: AMA

1. Segar ML, Guerin E, Phillips E, Fortier M. From vital sign to vitality: Selling exercise so patients want to buy it. *Curr Sports Med Rep*. 2016;15(4):276-281.

- Cultural, historical, and philosophical dimensions of physical activity
- The practice of physical activity

Do the core requirements in your department fit this general scheme?

Course Work in Professional Practice Knowledge and Professional Skills

Your preparation as a professional wouldn't be complete if you didn't acquire the professional practice knowledge essential for success in your career. Whereas theories of kinesiology tend to focus on physical activity alone, mastering professional practices requires knowledge that will help you identify clients' needs and desires, analyze impediments to meeting these needs and desires, and implement actions appropriate to the context and other relevant factors. Professional practice knowledge is incorporated into **practice theories**, which are developed through applied research and professional experience. A practice theory may, for example, guide a professional in selecting the best way to perform cardiovascular tests on an elderly person at a nursing home with limited equipment, determining how to motivate a soccer player who is having academic difficulties and misses off-season training sessions, or selecting the types of balls and bats to use in order to speed the learning of children with autism.

Becoming Familiar With Your Academic Program

Do you know what core courses in your kinesiology department are required of all kinesiology or exercise science majors? Does the core requirement differ for different career tracks? Why are these courses required? Do some courses have prerequisites? Are students in your department required to enroll in physical activity classes? Why, or why not? Are you required to complete an internship (or student teaching) as part of your degree requirements? Again, why, or why not? How does a student in your program apply for an internship placement? Does the internship have prerequisites? Must you maintain a minimum grade point average in order to enroll in higher-level courses? If so, what is it?

One barometer of students' depth of engagement in their field of study is their familiarity with the requirements of the academic program in which they are enrolled. If you aren't familiar with the requirements of your program, consult the bulletin or calendar for your institution that lists courses required in your major.

Unlike kinesiology theory, practice theories and the skills of professional practice cannot be mastered simply through studying them; they are learned through action. Students acquire these theories and skills by participating in hands-on practical experiences centered on solving real-life professional problems. You can learn about administering a stress test by reading a book, but you learn *how* to administer a stress test only by following up your reading with practical experience. By committing errors and receiving instructive feedback from your professors, you will gradually develop and sharpen your command of professional practice knowledge.

KEY POINT

Disciplinary knowledge alone does not equip you to perform the tasks required in your chosen profession. You develop the necessary competencies through course work that provides you with practice in performing these skills.

These initial attempts at putting practice theory into practice will probably take place in relatively smaller classes on campus, in which your performance will be evaluated more intensely than it might have been in introductory courses in kinesiology theory. In these courses, your classmates or mannequins may serve as simulated clients; therefore, your initial attempts to perform professional tasks will occur in the safe environment of the classroom, where your mistakes will not lead to serious consequences. If you perform up to departmental standards in these courses, you will be permitted to enter the final stage of your professional education: field experiences, or internships.

Internships

An apprentice learns how to perform a job by working closely with an experienced worker. An apprentice usually assists a veteran for little or no pay in return for the opportunity to learn the skills required in order to enter the full-time workforce. The term *apprentice* is used most commonly in connection with crafts or trades (e.g., plumbing, electrical contracting, carpentry) in which apprenticeship is the only type of training offered for the job. In professional education programs, the apprenticeship phase is referred to as *field experience* or **internship**, and it constitutes the culminating experience of 4 years of study. Unlike apprenticeships for the crafts or trades, internships for professional careers are preceded by a rigorous course of study.

KEY POINT

For professionals, an internship constitutes the culminating educational experience, in which they apply knowledge and skills under the supervision of a trained professional. An internship is also an evaluative experience that tests your preparedness.

In the physical activity professions, an internship may be known by any of a variety of names. The term *internship* may be used in sport management and fitness instructor programs, whereas *student teaching* may be used in physical education teacher education programs, and *clinical training* may be used in athletic training programs. Regardless of what it is called, the experience serves two purposes:

- Teaching you how to apply the knowledge and skills you have learned in your professional program to a real-life situation
- Testing your level of preparedness to enter professional practice

For most students, this is the first opportunity to assume the role of a member of their chosen profession. The internship usually is the most challenging experience of the undergraduate years—and the most enjoyable. For these reasons, most students feel that it is the most valuable experience of their undergraduate education.

An internship differs from traditional course work in three ways. First, it occurs at a worksite rather than at the college or university. This arrangement means that your conduct is dictated by the subculture of the worksite rather than the subculture of your undergraduate department. Thus, you will be evaluated not only on the knowledge and skills you display but also on how well you conform to the rules of the particular workplace in which you are acting as an intern. For example, if you were engaged in an athletic training internship for a Division I athletic department, you would need to pay attention to the local rules that govern behavior in the training room. Are dress codes strictly enforced? Does the supervisor of the training room expect you to arrive 15 minutes before your scheduled starting time? What local terms or abbreviations are used to describe interventions, locations, or procedures? What safety measures are you expected to follow? What are the established emergency procedures? Does the supervisor want interns to show initiative, or does she prefer that you ask for advice or permission before taking action? You must consider these and other such issues in order to ensure success in the internship as well as in the profession.

A second way in which an internship differs from your regular course work is that your performance will be supervised on a day-to-day basis by an experienced professional who is probably not a member of the faculty in your department. A university supervisor may occasionally visit the internship site, but the primary overseer and evaluator of your work will be the professional to whom you are assigned. Faculty members choose these on-site supervisors, or preceptors, because they are considered to be good role models for future professionals. However, you should not be surprised if their opinions, philosophies, and methods of operation differ from those that you have learned in your professional course work. Exposure to a variety of philosophies and methods usually expands your knowledge of the profession.

The third way in which an internship differs from typical course work is that you will be serving real, not simulated, clients. Therefore, your actions will carry important consequences, which can be serious. Serving real clients also means that you must learn the social graces expected of a professional. You will need to demonstrate poise, courtesy, alertness, and initiative, even when clients are not easy to serve. You will also be expected to demonstrate a level of confidence that makes clients feel comfortable and encourages them to place their trust in you.

Internship requirements vary across institutions and professional specializations. If you are studying sport management, your internship may involve a semester of 20-hour weeks working in an athletic administration office, where you might be required (like most professionals) to perform a range of duties, including mundane tasks such as handling mailings in the ticket office or answering the phone in the information office. You will also be given measured amounts of responsibility in performing the types of tasks that will be required of you as a sport management professional. If you are undertaking a student teaching experience in physical education, you may be required to spend an entire semester of full days in which you assume major responsibility for teaching classes. If you are a physical education intern (i.e., a student teacher), you will be responsible for logging a preestablished number of hours specified by your department and the state education department.

An internship provides real-world experience, which can help you decide to continue your course of study.

Are You Suited for a Career in the Physical Activity Professions?

Now that you know something about the types of work that professionals do and the educational experiences required to prepare you for such work, it's time to ask yourself how well suited you are for a career in the physical activity professions.

Do My Attitudes, Values, and Goals Match Those of Professionals?

The first step is to consider how compatible you are with professional work in general, regardless of the particular field. Approach this task honestly and objectively. If you decide that you're not well suited to professional life, consider alternative careers.

In this step, take stock of your strengths, weaknesses, and potential in relation to the personal attributes and preferences likely to be required in most types of professional work. Nobody is in a better position to judge our attitudes, attributes, and preferences than we ourselves are, but sometimes our self-perceptions can be a bit clouded. One way to bring our own attitudes into sharper focus is to compare them with those of our friends and peers. To this end, you can find a convenient template for assessing your own attributes and preferences in the results of a survey, conducted by the Higher Education Research Institute at UCLA, of more than 140,000 students entering 4-year institutions in the United States in 2015 (Eagan et al., 2015).

Here are some results from the survey. If you are a typical undergraduate student, chances are that you have already identified a career track (only 14 percent said they have not) and enrolled in college in order to gain a general education and appreciation of ideas (72 percent) and "get a better job" (85 percent). This is a good start for future professionals, and the survey also included some encouraging signs that incoming college students are becoming more serious about their academic performance. For example, 22 percent of students in the survey reported spending 6 to 10 hours per week studying or doing homework in high school—a trend that has been on the rise since 2005. (Expect to do a lot more than this in college!) An upward trend was also seen in the percentage of first-year college students (70 percent) who said they frequently took notes during class as high school seniors, whereas a decrease was reported in the percentage of students (34 percent) who reported being frequently bored during class. In another good sign, 54 percent said there was a good chance that they would discuss course content outside of class.

The survey also raised some red flags. For instance, only 44 percent of students planned to communicate regularly with their professors, and only 54 percent planned to become involved in clubs or other campus activities. In contrast, such forms of engagement are viewed as very important by most college faculty members.

On a more encouraging note, 75 percent of first-year students planned to pursue graduate education. As a rule, advancement in a profession is linked closely to educational credentials, and higher-paying jobs usually include some administrative and supervisory responsibilities. Graduate degrees open doors to such career paths. In fact, because educational standards for professional practice usually increase over time, undergraduate students will increasingly discover that a master's degree has become the minimal educational qualification for entering many of the physical activity professions.

Another good sign appeared in the fact that a relatively large percentage (88 percent) of first-year students indicated that they had performed volunteer work in the previous year. It is troubling, however, that only 37 percent plan to continue volunteering or doing community service work in college. A student who is not interested in volunteer work may not be well suited for a career in the professions, given that a spirit of service is the lifeblood of professional work. On a more pragmatic note, employers and graduate school admissions officers usually view continuing commitment to volunteer work throughout college, even if only for 2 or 3 hours per week, as an indication that an applicant understands the serious service responsibilities that come with a career in the helping professions.

About one-third of students (approximately 34 percent) reported feeling overwhelmed by all of their obligations during their high school years, but professionals in the workforce sometimes feel this way, too. Thus, stamina and robust health are especially important personal attributes for physical activity professionals. Yet the survey results indicated that only 53 percent of respondents exercised or played a sport more than 6 hours per week. On a positive note, it is quite likely that the percentage of students who plan to engage in sport and exercise is much higher among the subset of students who plan to major in kinesiology.

After giving the matter some thought, you may have concluded that you really aren't cut out for a professional career. Perhaps you don't find working with people attractive, don't particularly like helping to solve other people's problems, or would rather work for a supervisor than take responsibility for acting on your own authority. Maybe the prospect of continually educating yourself to keep abreast of the professional literature or attending to the ethical ramifications of your work behavior is not part of the vision you have for your career. If so, it seems unlikely that a professional career will interest you, regardless of the area of specialization. We suggest that you continue to collect more information by reading the other chapters in this section of the book, talking with faculty members, and consulting with the career counseling center on your campus.

What if you are convinced that you would like a career in the physical activity professions but you recognize that you have a glaring deficiency (e.g.,

Consequences of Harmful and Underage College Drinking

Drinking is an indelible part of life at most colleges. To imagine, however, that drinking doesn't pose a problem for students is a bit naïve. Consider the facts:

- Each year, an estimated 1,825 college students between the ages of 18 and 21 die from alcohol-related unintentional injuries, including motor vehicle crashes.
- From 1998 to 2001, about 696,000 students between the ages of 18 and 24 were assaulted by another student who had been drinking.
- About 97,000 students between the ages of 18 and 24 report experiencing alcohol-related sexual assault or date rape.
- College students have a higher rate of binge drinking (about 40 percent) than their noncollege peers and are more likely to be arrested for driving under the influence of alcohol.
- One-fourth of college students say their academic performance has suffered because of drinking.
- Thousands of students are transported to emergency rooms each year for alcohol poisoning, which can lead to permanent brain damage or death.

How might a history of alcohol abuse, particularly one documented on social media, hinder your preparation for a profession or your application to a graduate or professional school? When do personal choices interfere with professional and social responsibilities?

Adapted from National Institute of Alcohol Abuse and Alcoholism 2014.

fear of speaking before large groups, lack of organizational skills)? If this is the case, consult with your faculty advisor about addressing the weakness by including appropriate academic experiences during your undergraduate years.

Am I Interested—Really Interested—in Physical Activity?

If what you have read about the physical activity professions excites you and matches your interests and attitudes, the second step is to consider how much you like physical activity. If you're not intrigued by performing, studying, and watching physical activity; if you aren't curious about the mysteries of body movement in all of its manifestations (biophysical, behavioral, and sociocultural); and if you're not prepared to apply your intellect and skills to exploring these mysteries, then you may be better off selecting another profession.

Do My Attitudes, Interests, and Talents Lend Themselves to a Specific Physical Activity Profession?

The third step in this personal evaluation is to consider whether your attitudes, interests, and talents match the specific characteristics of the occupation that you are considering. For example, if you envision a career in therapeutic exercise or physical therapy, consider whether you enjoy interacting with people who are sick or injured. Of course, most people—even medical students—require some time before they feel completely comfortable in medical environments, but you need to understand, fairly early on, your degree of comfort with the environments in which allied health professionals work.

If you like being around people who are vigorous, healthy, and active, then you might enjoy a career in physical fitness counseling, personal training, coaching, or athletic training. Or do you have a special place in your heart for people with disability, such as intellectual disability, autism, blindness, and deafness? Would you find it challenging to help such people maximize their physical and mental potential? If so, then a reasonable goal might be a career in adapted physical education. If, on the other hand, you like being the center of attention; organizing and speaking to large groups; planning, monitoring, and evaluating activities that others perform; and accepting responsibility for the actions of people younger than you, then perhaps you would like a career in teaching fitness, coaching, or physical education. Take a moment now to complete the sidebar about perceived strengths and weaknesses in order to assess your general strengths and interests that may have a bearing on your career choice.

What Are Your Perceived Strengths and Weaknesses?

Place a check mark next to each quality for which you judge yourself to be above average or in the top 10 percent for your age.

	Above average	Top 10 percent
Academic ability	_____	_____
Computer skills	_____	_____
Cooperativeness	_____	_____
Creativity	_____	_____
Drive to achieve	_____	_____
Emotional health	_____	_____
Leadership ability	_____	_____
Physical health	_____	_____
Public speaking ability	_____	_____
Self-confidence (social)	_____	_____
Self-understanding	_____	_____

Now, compare your results with the averages among first-year students:

Academic ability	74 percent	Leadership ability	65 percent
Computer skills	33 percent	Physical health	55 percent
Cooperativeness	72 percent	Public speaking ability	41 percent
Creativity	55 percent	Self-confidence (social)	47 percent
Drive to achieve	80 percent	Self-understanding	57 percent
Emotional health	51 percent		

Adapted, by permission, from K. Eagan et al., 2015, *The American freshman: National norms fall 2015.* (Los Angeles: Higher Education Research Institute, UCLA). Available: http://www.heri.ucla.edu/monographs/TheAmericanFreshman2015.pdf

Does working at a desk dressed in business attire and adhering to a 9-to-5 workday appeal to you? If so, you might be well suited to a career in sport management. Are you more attracted to fluid and informal working environments in which your work schedule might be more flexible, even though work might intrude on your weekends and evenings? If your answer is yes, then you might be suited to a more entrepreneurial professional position, such as personal trainer, professional sport instructor, or university professor. More complete descriptions of physical activity professions are provided in chapters 12 through 16 to help you assess which careers may suit you best.

If you become convinced that your needs, interests, and attitudes do not closely match the specific professional occupation you had in mind, explore other career possibilities in the physical activity professions. If you're unsure, consider taking advantage of opportunities to volunteer at clinics, commercial or governmental agencies, schools, or other physical activity settings. Volunteering gives you firsthand experience in various working environments and thus increases the probability that the career you choose will match your interests, talents, and preferences. If possible, coordinate these volunteer experiences with your faculty advisor.

Will My College or University Program Prepare Me Well?

The fourth step is to determine whether the program in which you are enrolled will prepare you well for your chosen career. If, for example, you are most interested in a career in teaching and coaching but your department offers only concentrations in exercise science, sport administration, and health promotion, then the department's curriculum would seem to be a weak match for your career objectives. The same would be true if the

What Should You Look For in a Kinesiology Department?

Deciding which college you will attend often revolves around a variety of practical issues, such as the school's location, tuition charges, and family traditions. However, the most important consideration is whether the school can offer you first-rate preparation for the career you wish to pursue. Here are some recommendations to help you decide whether the school and program in which you are currently enrolled are compatible with your career interests.

- Select a college or university that suits your general tastes. Are you afraid of getting lost in the shuffle of a large state university? If so, there are scores of smaller, liberal arts institutions that offer programs in kinesiology. These smaller schools usually offer the opportunity to establish close relationships with faculty.
- Examine the academic catalog. Does the kinesiology department offer ample course work to support your study? Does the program offer only a concentration in kinesiology, or a full-fledged degree program?
- Review the departmental website. Does the department seem alive? Are exciting things going on there?
- Look at the faculty directory of the department (on the website). If faculty members' curriculum vitae (resumes) are included, take time to review them. Does it appear that some faculty members share your particular interests? Do the faculty seem fully engaged in their profession? For example, are they active in academic and professional societies? Do they sponsor intriguing projects? Do they publish in academic and professional journals?
- Plan to visit the department. Set up an appointment with the department head or with faculty who work in an area that interests you. If possible, send them an e-mail to tell them something about yourself before meeting with them.
- If possible, talk with students during your visit. Generally, students can offer a unique perspective on the department that others cannot. Caution here: It is dangerous to accept one person's opinion; if a student has had a bad experience with a professor, that experience can taint his or her feelings about the department. It always is best to talk with a number of students.

Adapted, by permission, from S.J. Hoffman, 2011, Examining the big picture. In *Careers in sport, fitness, and exercise*, edited by American Kinesiology Association (Champaign, IL; Human Kinetics), 12-13.

program slants heavily toward the preparation of physical education teachers or professionals in the fitness industry but you are most interested in a career in sport management or exercise science. If you have doubts about the fit between your career aspirations and the course requirements of your program, talk with a faculty member or the department head.

Sometimes students change their career goals after enrolling in an undergraduate program. Although this would occur less often if students took the five-step approach recommended here, changing career goals during the undergraduate years is quite common. You should keep in mind that switching majors or concentrations within a department (e.g., from physical education to exercise therapy) can delay your graduation for a year or more. If you switch to an entirely new field, you may push your graduation even farther into the future. Making such a drastic change also requires you to get socialized into another department, become accustomed to new professors, and form new alliances with classmates.

Still, however inconvenient switching your major might be, it is better to endure delays and inconvenience than to continue training for a career in which you have little interest. On the other hand, if you are deep into your undergraduate program and find that your career interests have changed to another physical activity profession, the best action may be to complete your present program and then pursue a master's degree in your area of interest. Faculty members in your department will be ready and willing to advise you, even if doing so means referring you to other departments or institutions.

Asking questions about the relevance of your program to your life goals is risky business. You may not like how you answer your own questions!

But this vital exercise is one that students too often ignore. Remember that individuals who make effective career decisions early in their undergraduate years are more likely to enjoy their undergraduate studies and appreciate their relevance, particularly in their major courses.

How Committed Am I to Preparing to Be the Best Professional Possible?

At the first practice of the season, coaches often rhetorically ask their teams, "How bad do you want it?" They refer, of course, to the team's willingness to invest the time, energy, and hard work necessary for winning a championship. So, how badly do *you* want it? How much energy are you willing to invest in order to succeed in your chosen profession? This fifth and last step in the process may be the most important because it supersedes all of the others. You may have had little difficulty answering the questions up to this point, but if you aren't committed to success, then there's little point in continuing to pay tuition only to have your career plans end up on the scrap heap. Are you willing to commit to preparing yourself to be the best professional possible? Let's examine some of the ways in which this commitment might be manifested.

Excellence in Academic Work

Graduating with a superb academic record is regarded by most employers as a baseline indicator of your level of commitment to becoming an outstanding professional. Of course, some individuals are more academically gifted than others, but students can often overcome slight shortcomings in academic ability by means of effort. Few things make a more indelible impression on professors than a student's willingness to work hard to achieve academic success.

KEY POINT

Perhaps the best predictor of your success in the physical activity professions lies in your level of commitment to preparing yourself to be the most knowledgeable and skilled practitioner possible.

What types of behaviors suggest a willingness to work hard? Attending class regularly and on time, regularly using the library or its electronic resources, asking questions and contributing to class discussions, and reading unassigned journal articles and books about the topics that you are studying in class—all are signs that you take your academic work seriously. In addition, most professors view students who approach them after class with questions about the day's material as more committed than students who avoid them at all costs.

Early Identification With the Professional Field

Commitment can also be indicated by how early in an undergraduate program an individual identifies with his or her chosen career. Perhaps the most reliable indicator of commitment is the act of obtaining an undergraduate membership in a professional association, many of which are referred to in the chapters that follow. Professional associations usually admit preprofessionals at reduced rates and offer them reduced registration rates at conferences. Members receive the organization's publication and other information on a regular basis.

In addition, many kinesiology departments host student clubs that align or affiliate with professional organizations—for example, Phi Epsilon Kappa, a national professional honor society in health, kinesiology, recreation, and related fields. You can also indicate commitment by attending professional conferences. Professional associations usually hold annual regional and national meetings that feature lectures, workshops, and exhibits of equipment used in professional practice. Attending professional meetings can be expensive, but, then again, taking a spring break trip to the beach is also costly—and it does not provide professional connections. Remember, keep your eye on your long-term goals and be willing to sacrifice to attain them.

KEY POINT

To get a head start on a successful career, takes steps to identify early with your chosen profession. Specifically, join appropriate professional organizations, attend conferences, establish alliances with veteran professionals, and obtain professional certifications.

Students who identify with a career early tend to approach their studies with added excitement and vigor. Their orientation changes from that of a student who views courses as obstacles to be overcome in obtaining a degree to that of a professional who tries to gain from each course the knowledge and skills that will help him or her develop into the best qualified professional possible. When you identify with your chosen profession, you also seek advice from veteran professionals and observe them in practice; in this way, you begin establishing

Kinesiology Colleagues

Photo courtesy of Karli Champ

Dennis Johnson

By the time he was a junior in high school, Dennis Johnson knew that he wanted to be a physical education teacher and wrestling coach. After a stellar high school athletic career, he earned a wrestling scholarship to Marshall University, where two influential professors inspired him to give as much attention to his academic goals as he was giving to wrestling. Upon graduation, true to his plan, he became a high school teacher and coach. For 13 years, he experimented with using physical education and sport experiences to teach life skills and social responsibility to young people, especially those from disadvantaged backgrounds.

For the next chapter in his career, Johnson entered a doctoral program at the University of North Carolina at Greensboro. Following his graduation there, he joined the faculty at Wingate University. During the next 15 years, he taught courses in the physical education teacher education program, coached cross country, and, for 5 years, served as assistant dean in the school of education. Johnson has been heavily involved in the North Carolina Alliance for Athletics, Health, Physical Education, Recreation, Dance, and Sport Management and in the Leadership Academy of the National Wrestling Coaches Association. Reflecting on his career, Johnson offered the following perspective: "One never knows where one's career will lead; for me, it led to many different and exciting professional ventures."

a communication network with practicing professionals. You also establish the habit of reading the journals in your field. Thus, you learn the language of the field and begin to feel comfortable around experienced professionals while still in the preparation phase of your career.

KEY POINT

Becoming involved in activities both within and outside of your department is one indication of your commitment to developing a successful professional career.

Another way to identify early with a physical activity profession is to obtain certification in an area in which you plan to work. Certifications are available from such groups as the American College of Sports Medicine, the National Strength and Conditioning Association, the American Council on Exercise, and the Athletics and Fitness Association of America.

Engagement in College or University Life

In addition to academic performance, one of the best indicators of your level of commitment to becoming a successful professional can be found in your level of engagement in your department, school, and community. Being engaged means being connected to what is happening around you. Engaged students take responsibility for their academic experiences and their professional futures. They are curious, and they constantly test their personal limits by seeking new ways to become involved in life experiences. They are active, energetic leaders who view the department as their department and seek ways to participate in its operational life. They are likely to serve as leaders in their major clubs or in student government or as coordinators of charity events or other community activities outside of the academic environment.

Besides becoming engaged in social and organizational extracurricular activities, you should also become involved in learning. Go beyond doing well on course assignments and exams by asking yourself, "How can I take on even more responsibility for learning what I need to know in order to succeed?" Does a professor need students to assist on a research project? Does your institution sponsor undergraduate research assistantships? Does your department sponsor community-based projects on which you can work? Will a professor sponsor you for an independent study on a topic of interest to you? Don't be afraid to express your interest in such projects. Independent study not only reflects a high level of commitment on your part but also offers an excellent opportunity to develop the leadership skills and knowledge that are essential for success in the world beyond college.

Participating in Volunteer Services

We have seen that the driving force for professionals, especially those in the helping professions, is a desire to improve quality of life for others in the community. Although it is easy for preprofessionals to say that they want to serve others, nothing speaks louder than actions. As a result, if two students have approximately equal academic records, the advantage always goes to the one whose resume displays clear evidence of volunteer service to community agencies. These groups might include such programs as Meals on Wheels, Boys & Girls Clubs, local food banks, Boy Scouts and Girl Scouts, shelters for homeless persons, church-based programs, walkathons for various charities, and the American Red Cross. If a volunteer experience bears some relation to the profession for which you are preparing, then it offers the added benefit of socializing you into your future profession. Potential volunteer opportunities that you might want to explore include physical therapy and sports medicine clinics, YMCAs and YWCAs, youth sport organizations, schools, Special Olympics programs, and other agencies associated with the physical activity professions.

KEY POINT

Volunteering regularly for a community agency indicates that you share your profession's commitment to service and constitutes an investment in your future career in the physical activity professions.

Attending Graduate School

Evidence of commitment to a profession can also be reflected in plans to pursue advanced graduate work, and some 70 percent of entering college students say that they intend to do so. In fact, in some professions, such as physical therapy and athletic training, a master's degree is a minimal entry degree. In addition, permanent certification for teaching physical education requires a master's degree or its equivalent in many states. You may not be prepared to think about graduate school at this early point in your undergraduate career, or you may have decided to delay that decision until a few years after you have graduated. This may be a sensible approach for many students. Overall, however, making an early decision to continue with advanced graduate education in the physical activity professions is another indication of commitment.

Generally, master's programs in kinesiology offer advanced education both in a number of subdisciplines (e.g., exercise physiology, motor learning, sport history) and in professional areas (e.g., teaching physical education, fitness leadership, athletic training, sport management). In determining which institutions to consider for your own graduate work, you might find a faculty member at your institution to be an excellent source of information. You might also find posters or brochures advertising graduate study at various institutions displayed in your departmental office.

Master's programs typically require 30 to 36 credit hours of work, depending on the specialization. In most cases, full-time master's students are able to complete their degree requirements in 1 1/2 to 2 1/2 years. Programs that require a thesis involve slightly fewer hours of course work but include an in-depth research project that may take up to a year to complete. Some master's programs do not require a thesis but may include additional course work; some programs that do not require a thesis do require an internship. If you delay attending graduate school until after you have secured a professional position and plan to do your graduate work by attending classes in the evenings, you may need as long as 5 years to complete a master's degree. Doctoral education in kinesiology should be considered by those who envision a career as a professor at a college or university.

The Decision-Making Process

Now that you've read through the previous sections, how would you answer these questions:

1. Do my attitudes, values, and goals match those of professionals?
2. Am I interested—really interested—in physical activity?
3. Do my attitudes, interests, and talents lend themselves to a specific physical activity profession?
4. Will my college or university program prepare me well?
5. How committed am I to preparing to be the best professional possible?

Snapshot of Your Peers

Academic Information

Completed precalculus or trigonometry.	86 percent
Did remedial work in math.	16 percent
Reported being bored in class.	34 percent
Averaged a grade of A– or better in high school.	59 percent
Tutored another student.	61 percent
Asked teacher for advice after class.	85 percent
Spent less than 5 hours per week studying.	55 percent
Spent 16 hours or more per week studying.	12 percent

Reason for Enrolling in College

To get a better job	85 percent
To make more money	70 percent
To get training for a specific career	76 percent
To prepare for graduate or professional school	58 percent

Goals Considered Essential or Very Important

Raising a family	72 percent
Being well off financially	82 percent
Becoming an authority in one's field	60 percent
Developing a meaningful philosophy of life	46 percent

Personal Information

Attended a religious service frequently or occasionally.	70 percent
Socialized with someone of another racial or ethnic group frequently or occasionally.	96 percent

Beliefs and Opinions

Abortion should be legal.	64 percent
Racial discrimination is no longer a problem in U.S. society.	19 percent
Colleges have the right to ban extreme speakers from campus.	43 percent
Students from disadvantaged backgrounds should be given preferential consideration in admissions.	52 percent

Expectations

To socialize with someone from a different ethnic or racial background	71 percent
To get a job to help pay for college expenses	46 percent
To communicate regularly with professors	44 percent
To participate in volunteer or community service work	37 percent
To make at least a B average	68 percent

Adapted, by permission, from K. Eagan et al., 2015, *The American freshman: National norms fall 2015.* (Los Angeles: Higher Education Research Institute, UCLA). Available: http://www.heri.ucla.edu/monographs/TheAmericanFreshman2015.pdf

If you answered yes to all five questions, then you are well on your way to a successful career in a physical activity profession. The next five chapters give you an in-depth look at some of the professions from which you can choose.

If you hesitated on some questions or were unsure of your answers, then the next five chapters may help you make a decision. Use this as an opportunity to assess honestly and objectively whether you are suited to professional work, and more specifically to professional work in one of the physical activity professions.

Wrap-Up

Career decisions are among the most important choices that people ever make. Such decisions are not irrevocable; in fact, you are likely to change your career several times during your working years. Nevertheless, you will gain nothing by delaying your commitment to a career, and you have much to lose by doing so. Students who make an early career commitment put themselves in a better position to benefit from their undergraduate education than do students who make only tentative plans about the type of work that they will do when they graduate. If you have decided to attain a degree in kinesiology by default (e.g., nothing else seemed interesting) and life beyond graduation looms like a great void, then you should take steps to assess your compatibility with a career in the physical activity professions.

As we have explained, the process of career planning necessarily involves learning what it means to be a professional and learning about the types of work and work environments associated with each of the physical activity professions. It also involves making a realistic assessment of your level of excitement and commitment to kinesiology and the physical activity professions. If this chapter has achieved its goal, then you have begun this self-examination process. Now it is time to learn more about the physical activity professions by studying the next five chapters. These chapters, written by experts in the professions being described, provide specific information about what it is like to work in each area. When you have finished reading them, you should be in a good position to make important decisions about your career.

REVIEW QUESTIONS

1. List three ways in which professional work differs from nonprofessional work.
2. List two benefits of attending professional conferences and reading the professional literature.
3. List three differences that you might observe between a community sport program leader who adheres to a mechanical, market-driven professionalism and one who adheres to a social trustee, civic professionalism.
4. What is important about the liberal arts courses required of kinesiology majors?
5. Why is an internship an important experience in preparing kinesiology students for professional practice?
6. What are five questions that all kinesiology students should ask themselves before deciding to major in kinesiology?
7. If you were an employer, what evidence would you use to determine an applicant's general suitability for a professional position?
8. What would be the ideal work history of a kinesiology graduate who seeks to enter the athletic training field? The health fitness field? The sport management field?
9. If, during your second year of a kinesiology major, you decide to pursue a career as a physical education teacher after having entered as a sport management major, what is the best course of action to take?

Go online to complete the web study guide activities assigned by your instructor.

12

Careers in Health and Fitness

Warren D. Franke

The author acknowledges the contributions of Sandra Minor Bulmer and Jeremy Howell to this chapter.

CHAPTER OBJECTIVES

In this chapter, we will

- introduce you to a variety of professional opportunities in the sphere of health and fitness;
- explain the purpose and types of work done by health and fitness professionals;
- explore how the sphere of health and fitness is changing on the basis of public policies, societal trends, and research in kinesiology and public health;
- summarize the educational requirements and experiences necessary for becoming an active, competent professional in health and fitness; and
- help you determine whether one of these professions fits your skills, aptitudes, and professional interests.

A lawyer wants to join his co-workers in a 150-mile (240-kilometer) cycling event to raise money for a nonprofit organization. However, he hasn't biked more than 5 miles in 20 years. He wants to hire you as his personal trainer to guide him in safely improving his cardiovascular endurance and fitness for this event. How should you proceed?

You are talking with a new member of your facility. He tells you that he was relocated to your city by his employer and hopes to have as much success with you as with his previous trainer. He lost more than 100 pounds (45 kilograms) of weight while training at his old gym. Before he left his former facility, his personal trainer told him that he was "gaining about 8 pounds a month of muscle with his program," and "because some of his fat was being turned into muscle, he was probably losing more fat than the scale indicated." Is it physically possible for someone to gain 8 pounds of muscle monthly while also losing this much body fat? Can body fat be converted to muscle? Do these comments make sense to you? Are they scientifically possible? No, of course not. How do you tell this new client that his former trainer was not well educated and gave him incorrect information?

A health and fitness director realizes that her facility should expand its programs next year in order to better reach clients with disability. What additional staff will she need to hire? What should their qualifications be? What new equipment and facility renovations will be required? Will her budget allow for this expansion?

While touring a health and fitness facility, a working mother mentions that she wants to begin an exercise program because she understands how beneficial it will be, both for herself and as a model for her children. However, she has trouble fitting exercise into her busy lifestyle; she feels overwhelmed and constantly pressed for time. What strategies should the health and fitness professional use to help this busy mother find time to exercise?

Simone van den Berg/fotolia.com

These types of scenarios occur daily for health and fitness professionals. In each case, the professional must make decisions that serve the best interests of the participant and client and match the goals, budget, and objectives of the environment, program, or facility.

Health and fitness, which is one of the spheres of professional practice for kinesiology graduates (see the figure in the part III opener), includes many dynamic professions. These professions encompass such positions as group fitness instructor; health and fitness specialist; health and wellness coach; personal trainer; health and fitness director; and specialty positions such as clinical exercise physiologist, inclusive fitness trainer, strength and conditioning coach, and public health educator.

The scope of the professional work in this sphere is expanding to include a more multifaceted view of health and fitness. For decades, a health and fitness facility was viewed simply as a place to exercise; in other words, participants focused on improving their fitness. Today, however, many participants are motivated to join a health and fitness facility in order to improve their physical health. Exercise and other forms of physical activity are increasingly appreciated as important tools for reducing the risk for common noncommunicable diseases (e.g., cardiovascular diseases, diabetes) and contributing to a longer, healthier life span. As a result, the core goal of the health and fitness professions today is to improve both the physical functioning and the physical health of individuals and communities.

These professions have traditionally been practiced in worksite, clinical, commercial, and community settings. In recent years, the lines between these settings have blurred as each type of program or facility has expanded its offerings to reach more people with additional services. This blurring has created a variety of career opportunities for kinesiology graduates.

What Is the Health and Fitness Profession?

The 21st century is a great time for seeking a career in the health and fitness professions. Because many people in the United States and other knowledge-based economies are not nearly as active as they should be for good health (see chapter 2), the opportunities for well-trained physical activity professionals have never been better. In fact, the health and fitness settings in which physical activities take place—and the jobs that health and fitness professionals perform—have been expanding for years as more adults recognize the value of physical activity and are willing to pay for it, both for themselves and for their children.

Scientific research and popular culture have popularized the understanding that regular physical activity can lead to a wealth of positive changes in the human body. Exercise can increase lean body mass and improve heart and lung function, immune function, cognitive function, muscular strength, flexibility, bone density, balance, and mental health. Perhaps the most important research finding is that physical activity contributes substantially to lowering the risk of death due to heart disease, which is the leading cause of mortality in the United States.

Numerous government reports have extolled the indisputable virtues of physical activity for good health. In 1996, the U.S. Department of Health and Human Services (USDHHS) published a document titled *Physical Activity and Health: A Report of the Surgeon General.* For more than a century, the Surgeon General has used such reports to focus the nation's attention on the most important public health issues of the day, and they have led to innovative public health initiatives. This 1996 report differed from previous ones in an important way: It reflected a dramatic shift in focus from primarily communicable diseases to noncommunicable ones. For example, in 1910, 46 percent of deaths were due to infectious diseases and only 6 percent were due to heart disease. By 2010, infections accounted for only 3 percent of deaths and heart disease accounted for 24 percent (Jones, Podolsky, & Greene, 2012).

Cardiovascular diseases, of which heart disease is but one, hold the dubious distinction of being the leading cause of death in the United States for 99 of the past 100 years. Physical inactivity is a powerful contributor to the development of many cardiovascular diseases, and physical activity may be the most potent factor in improving health and longevity. With such concerns in mind, the 1996 Surgeon General's report concluded that Americans should engage in a minimum of 30 minutes of moderate activity on most days of the week in order to reduce their risk for poor health.

In 2008, the USDHHS published *Physical Activity Guidelines for Americans,* which described the department's updated recommendations for physical activity. This document reflected an increased understanding of how physical activity can best contribute to improved health and decreased mortality from many diseases. As a result, the new guidelines recommended that young children and teenagers engage in 60 minutes or more of physical activity per day. Adults were encouraged to engage in at least 150 minutes per week of moderate-intensity aerobic activity or 75 minutes per week of vigorous-intensity aerobic activity, or some combination of the two. The report also recommended performing muscle-strengthening exercises on 2 or more days per week.

Most recently, the USDHHS (2010) published *Healthy People 2020*, a document that laid out overall health objectives for the nation. The objectives were based on a set of 26 leading health indicators, one of which is physical activity. One objective of the report was to reduce the proportion of adults who engage in no leisure-time physical activity and increase the proportion of both children and adults who meet the 2008 *Guidelines* recommendations. Other objectives related to physical activity were to increase the proportion of the nation's schools that require daily physical education, provide regularly scheduled recess, and allow access to the school's physical activity spaces and facilities outside of normal school hours. Yet another objective was to increase the proportion of physician office visits that include counseling or education related to physical activity.

Table 12.1 provides a list of selected objectives related to physical activity, as well as baseline measures. The target goal for all of these objectives is a 10 percent improvement in 2020 as compared with 2010. The full list of physical activity objectives and other resources can be found at the Healthy People 2020 website (USDHHS, 2010).

All these evidence-based government documents argue that the public should not view physical activity as merely a recreational or physical fitness endeavor. Rather, physical activity should be seen as a powerful strategy for health promotion and disease prevention that is integral to our U.S. society and its public health policies. However, as the Healthy People 2020 measures indicate, it is not easy to find creative solutions for healthy living and make substantial progress toward meeting these goals. Changes in our understanding of exercise and fitness often find only slow acceptance by many people.

These changes also affect the responsibilities placed on health and fitness professionals. These professionals are still asked to assess and train clients for improved fitness, but they also need to be able to develop individual and group exercise programs that connect to other physical, intellectual, emotional, social, and spiritual dimensions. Similarly, better health and fitness strategies need to be put in place in community, commercial, corporate, clinical, and population settings. These changes—new evidence, new services, and expanded settings for health and fitness professionals—create diverse opportunities for kinesiology graduates in this sphere.

Settings for Health and Fitness

The health and fitness field continues to evolve with expanded job descriptions and a greater diversity of positions available to both entry-level and experienced professionals. Traditionally, health and fitness professionals have operated in four primary settings: worksite, commercial, clinical, and community. In earlier years, these settings differed in their missions, target markets, and programming objectives. Now, however, for the reasons outlined earlier, the various health and fitness settings have become more similar than dissimilar. Examples include commercial settings that offer worksite

Table 12.1 Selected Healthy People 2020 Physical Activity Objectives

Objective	Baseline (%)	Year baseline data acquired	2020 goal (%)
Reduce the proportion of adults who engage in no leisure-time physical activity.	36.2	2008	32.6
Increase the proportion of adults who engage in aerobic physical activity of at least moderate intensity for at least 150 minutes per week, vigorous intensity for at least 75 minutes per week, or an equivalent combination.	43.5	2008	47.9
Increase the proportion of adults who engage in aerobic physical activity of at least moderate intensity for more than 300 minutes per week, vigorous intensity for more than 150 minutes per week, or an equivalent combination.	28.4	2008	31.3
Increase the proportion of adults who perform muscle-strengthening activities on 2 or more days per week.	21.9	2008	24.1
Increase the proportion of adolescents who meet current federal physical activity guidelines for aerobic physical activity.	28.7	2011	31.6
Increase the proportion of adolescents who meet current federal physical activity guidelines for muscle-strengthening activity.	55.6	2011	61.2
Increase the proportion of physician office visits that include counseling or education related to physical activity.	13.0	2007	14.3

Adapted from Healthy People 2020, *Physical activity objectives.*

health promotion programs, medically oriented settings that operate for-profit fitness centers, and commercial swim facilities that offer rehabilitation pools. The lines have been further blurred by the expansion into a fifth setting: population-based promotion of physical activity in the field of public health. This new approach emphasizes promoting physical activity on larger scales by integrating the effects of social, environmental, government, and private organizations.

KEY POINT

Health and fitness professionals now operate in five overlapping types of settings—worksite, commercial, clinical, community, and public health-oriented—that focus either on increasing physical fitness or on reducing the risk of noncommunicable diseases, or both.

Each of the settings for health and fitness professionals pursues specific objectives, identifies a specific population to address, and offers a particular menu of programs. However, when seeking employment in one of these settings, look at each potential employer on its own merits and avoid putting too much emphasis on the distinctions between worksite, commercial, clinical, community, and population settings. As described in this chapter, these settings overlap considerably, and all are rapidly expanding the scope of the health and fitness programs they offer. According to *IHRSA's Profiles of Success* (IHRSA, 2015), more than 144 million people worldwide are members of facilities that IHRSA categorizes as either commercial, not-for-profit (e.g., JCCs and YMCAs; hospital-based clubs; residential, municipal, university, and military facilities), or miscellaneous (e.g., corporate fitness centers, resorts, spas, hotels, country clubs). Many exciting professional jobs are available in each setting. Your ideal job will be the one that best matches your goals as a professional. The following sections describe each type of health and fitness setting and provide an overview of the objectives, types of programs, and target markets for each setting. They also describe key changes and emerging trends.

Worksite Settings

One of the *Healthy People 2020* objectives (USDHHS, 2010) is to increase the proportion of employed adults who have access to and make use of employer-based exercise facilities and exercise programs. Worksite health and fitness programs (often referred to as *worksite health promotion programs*) have experienced considerable growth during the past several decades. In the early 1980s, fewer than 5 percent of employers offered worksite health and fitness programs (O'Donnell & Harris, 1994). That figure is increasing, though not rapidly enough in some sectors. For example, more than two-thirds of employers with 750 or more employees now offer programs or services to promote physical activity (Linnan et al., 2008). On the other hand, small and low-wage companies are less likely to offer exercise programs (Hannon et al., 2012).

Worksite health promotion programs are designed to improve employee health, but for most companies they also make good business sense. In 2007, the Community Task Force of the Centers for Disease Control and Prevention (CDC) reviewed more than 50 research studies and concluded that programs designed to improve a range of health behaviors at the population level produce substantial positive outcomes (Goetzel, Roemer, Liss-Levinson, & Samoly, 2008). Strong evidence also indicates that such programs provide economic value to the stockholders of these companies. In fact, companies with the best programs typically perform better in the stock market compared to companies that do not have such good programs (Fabius et al., 2013; Grossmeier et al., 2016).

This difference may result from any of a host of possible reasons. For one thing, healthier workers tend to be less absent and more productive (Merrill et al., 2013); they also tend to have lower health care costs. For more than 25 years, high-quality research has shown a strong association between modifiable health risk factors of employees and health care expenditures by their employers (Goetzel et al., 1998; Goetzel et al., 2012). Moreover, 10 risk factors have been consistently linked to 25 percent of all health care spending by employees and employers. For example, employees with self-reported diabetes or hypertension had health care expenses 32 percent greater than those who did not. Costs were 27 percent higher for those who were obese, 15 percent higher for tobacco users, and 15 percent higher for those with poor exercise habits (Goetzel et al., 2012). Clearly, then, it is in the best interests of most companies to invest in a high-quality worksite health promotion program; it is good for the employees *and* good for the business.

Early worksite programs focused primarily on physical fitness, nutrition, weight control, stress management, and smoking cessation and were available exclusively to employees of the sponsoring company. Today, although programs vary across companies, additional services have been added to create a wellness culture centered on the needs of the employee. Examples include "health education classes, access to local fitness facilities, company policies that promote healthy behaviors such as a tobacco-free campus policy, employee health insurance coverage for appropriate preventive screenings, a healthy work environment created through actions such as making healthy foods available and accessible through vending machines or cafeterias, and, finally, a work environment free of recognized health and safety threats with a means to identify and address new problems as they arise" (Centers for Disease Control and Prevention, 2016).

These strategies result from increased emphasis on taking a more holistic approach to employees' well-being (Niebuhr & Grossmeier, 2015). For example, work–life balance initiatives have become popular as methods to promote employee wellness in all dimensions of health. These initiatives include flexible work schedules, on-site child care, and leave-of-absence policies for caregiving and other life events. Other popular worksite programs include recreational activities, volunteer service opportunities, financial planning, and **employee assistance programs**. In an effort to reduce the medical costs of all individuals covered by the company health plan, some worksite programs now also offer health and fitness services to employees' family members and to company retirees.

KEY POINT

Employers have found that health and fitness programs are a good business practice. They can help reduce health care costs, increase productivity and morale, decrease absenteeism, and improve the organization's corporate image.

Many companies have also built elaborate health and fitness facilities as a way to recruit and retain highly qualified employees. Other companies have contracted with local health clubs to offer access to exercise facilities at a reduced rate. In either case, the trend is for companies to establish partnerships with outside organizations with expertise in health and fitness to manage their worksite health promotion facilities and programs. For instance, EXOS provides services for nearly 1 million people in more than 300 locations on six continents. The company provides a variety of contracted services, including facility management, on-site programming, and web-based programming for employees who work from home or in smaller satellite branch offices.

Perhaps the most innovative change in workplace health promotion has come about as many large corporations have become self-insured. This change has been driven by the relationship between personal health practices and soaring health care costs for both the employer and the

Worksite health and fitness programs, such as this yoga class, can lead to increased productivity.

employee. According to a large 2015 employer survey, the average annual health insurance premium for family coverage has been steadily increasing. In 2015, the figure was $17,545, which reflected a 61 percent increase over the 2005 figure of $10,880; about 30 percent of the premium was paid by the employer, and the remainder was paid by the employee (Claxton et al., 2015).

In response to these rising costs, many large companies have decided to operate their own health insurance plans rather than purchase coverage from an insurance company. Usually, the employer still pays a third party (such as an insurance company) to administer the plan that it has designed, but the employer can provide it at a much lower cost. However, while the company saves money on costs, it also takes on additional risk in becoming responsible for all claims that are made. In other words, an employee with a severe illness could cost the company a lot of money. Therefore, many companies choose to invest in disease prevention and health promotion for their employees.

Moreover, according to a 2010 research report by the National Institute for Health Care Reform, the interest of self-insured employers in health promotion and wellness programs has increased dramatically in recent years. Since this report was issued, this increase may be due in part to the passage of health care reform, such as the Patient Protection and Affordable Care Act of 2010, which provides support for businesses to expand wellness programs and services to employees. This support includes financial incentives in the form of reduced insurance premiums for employees who meet health goals. As a result, many employers are motivated to implement wellness programs in order to reduce either direct medical costs or indirect medical costs such as insurance claims. Moreover, employers increasingly view health promotion programs as an effective business strategy for improving the company's reputation and enhancing its ability to attract and retain the best employees (Tu & Mayrell, 2010).

For example, the grocery chain Safeway Inc. created a solution to providing its employees with health insurance. CEO Steven Burd described Safeway's Healthy Measures program as a completely voluntary insurance plan that 74 percent of the nonunion workforce have signed up for (Burd, 2009). As with most companies, Safeway employees pay a portion of their own health care costs through insurance premiums. However, in the Healthy Measures program, the amount of that premium is reduced based on the results of tests associated with risk factors such as tobacco use, obesity, hypertension, and high cholesterol.

In addition to encouraging reductions in disease-causing risk factors, the company provides programs to encourage healthy behaviors, such as a 24-hour information line and health advisor, wellness and lifestyle management programs, healthier food options, large employee discounts for healthier food, and state-of-the-art fitness facilities at the corporate office. The goal of such programs is to create a "win-win" environment—employees become healthier and, in so doing, help their employer keep health care costs as low as possible. In Safeway's case, from 2005 to 2012, costs rose at only 25 percent of the pace seen among similar employers (Bradley, 2013).

Commercial Settings

Commercial health and fitness facilities typically operate with the objective of generating a profit. The method by which a facility generates this profit places it in one of two distinct categories: **sales-based facilities** and **retention-based facilities**. These two business models are compared in the following sections. However, many health and fitness facilities do not fit neatly into one of these two categories. Instead, they use some practices from each in order to maximize their profit-making potential.

KEY POINT

Because commercial health and fitness facilities operate for profit, they compete for customers. This requirement encourages innovative programming and, ideally, employment of well-trained staff. These facilities are either sales-based, retention-based, or a mix of both.

Commercial facilities offer health and fitness programs similar to those offered in many workplaces. Popular programs include fitness assessments; health risk screening; individual or group fitness training sessions; spa services; and access to other professionals such as massage therapists, physical therapists, and registered dietitians. Another type of offering, which is increasingly popular, consists of specialized exercise programs for children, prenatal and postnatal women, older adults, cancer survivors, and individuals with chronic disease (e.g., Parkinson's disease, obesity, diabetes). Commercial facilities also cater to niche markets, such as women's fitness, sport- or performance-oriented training programs, personal training, dance, and yoga. The nature of a for-profit business encourages innovation as a way to compete for customers in the marketplace. Consequently, the commercial health and fitness business is dynamic and offers many opportunities for health and fitness professionals to become involved.

Sales-Based Facilities

As the name suggests, sales-based health and fitness facilities focus primarily on membership sales activities. These facilities rarely place limits on the number of memberships sold and often offer discounted rates as a way of attracting a large number of new customers. Historically, another characteristic of sales-based facilities has been the sale of long-term membership contracts. Within the industry, the key element of the philosophy under which sales-based facilities operate is often referred to as a "future service contract." To receive the best possible rate, a new member must make a long-term commitment rather than pay for membership on a monthly basis.

These contracts often work in the facility's favor; even in programs of very high quality, more than half of new exercisers are likely to drop out within 6 months of beginning the program. In these cases, the facility obtains the income from the long-term contract without having the member continue to use the facility. Consequently, sales-based facilities may be best suited for people who are already exercising but need a venue to do so, can exercise regularly at hours when the club is not full, and do not need extensive assistance from expert personnel. Nevertheless, many of these facilities occupy a useful niche in the marketplace by providing low-cost access to fitness equipment and indoor facilities.

Sales-based fitness facilities present two major barriers to improving the health and fitness of the general population: their marketing practices and their sales strategies. These facilities often use advertising that features slender, muscular young models. Although these models may be intended to show what could happen by joining the facility, they may deliver the message that exercise is only for those who are already fit and healthy. Such messages may have a discouraging effect on the very people who would most benefit from health and fitness programs—those who are at risk for chronic diseases such as hypertension, diabetes, and obesity; who are older; who have special health

What barriers does this fitness facility present to potential clients?
Andres Rodriguez/fotolia.com

concerns; or who feel self-conscious about exercising. This advertising is shortsighted given the fact that most Americans fall into these latter categories rather than in the "lean and fit" category. In other words, it would serve a facility's best interest to cater to these other population markets.

Another challenge lies in the fact that sales-based facilities may lure prospective members in through manipulative and deceptive sales strategies that offer substantial membership discounts in exchange for long-term commitments. As described earlier, customers who make a long-term purchase but whose needs are not met may stop using the facility. Even more concerning, this negative first experience with a health and fitness facility may discourage them from getting involved with other, more suitable facilities or programs.

Retention-Based Facilities

Selling memberships is an important aspect of any commercial health and fitness facility, but retention-based facilities have a financial stake in making a sincere effort to meet the long-term needs of current members. These commercial facilities hold a "voluntary dues" philosophy. Members pay dues voluntarily from month to month rather than paying one large membership fee up front. The philosophy is that the facility will provide such good services that members will stay engaged and not be interested in looking elsewhere. As a result, these facilities typically focus on delivering high-quality programs and services so that current members will continue their membership month in and month out and will be more likely to recommend the facility to others.

Because retention-based facilities may not invest in extensive advertising, referrals from current members are essential to achieving the goal of profitability. These facilities limit the number of memberships sold, usually on the basis of a ratio of members to available space. Staff members engage in sales activities only to replace members who have left the club. As a result, they focus primarily on meeting the needs of current members by providing high-quality facilities and programming. Because of the additional cost of offering high-quality programs and hiring qualified personnel, monthly dues in retention-based facilities tend to be higher than in sales-based facilities.

Clinical Settings

The **clinical setting** category includes hospitals, outpatient medical facilities, and sports medicine and physical therapy clinics. This setting, which has been a large growth area for health and fitness programs in the past decade, has its own professional organization—the Medical Fitness Association. Many administrators in clinical settings recognize the value of providing health and fitness programs beyond the immediate phase of rehabilitation. The objective of these programs is to help each patient manage medical conditions and keep the subscribers of their health insurance partners and programs healthy in order to avoid expensive medical procedures in the future. In other words, health and fitness facilities have been integrated into the community's local medical environment and are often viewed as extensions of the health care services provided in that environment. Many of these facilities also offer memberships to the public.

Examples of popular clinical programs include preventive screenings, physical therapy, cardiac rehabilitation, chronic disease management, water exercise therapy, childbirth and parenting education, weight management, and nutrition counseling. By design, fitness professionals who work in clinical settings have more contact, and work in partnership, with both medically based professionals and clients with diagnosed medical conditions. In fact, the American College of Sports Medicine recently pioneered the "Exercise is Medicine" concept. The overall goal is to reduce the progression of chronic diseases by integrating the activities of health care providers with community resources such as high-quality health and fitness facilities.

Current trends in the clinical setting are not unlike those in the worksite and commercial settings. Organizations in clinical settings are also building fitness facilities and offering a wide range of health and fitness programs delivered by qualified staff. In addition, these services and programs extend into other dimensions of health besides physical fitness. Much like worksite settings, some clinical settings have also developed partnerships with commercial facilities or agencies that provide expertise in the management of health and fitness facilities.

This combination of (1) commercial settings with expertise in facility and program management and (2) clinical settings with expertise in specialized health services has become a popular business model that presents qualified college graduates with new job opportunities. One company that has been partnering with hospitals to design and manage health and fitness facilities is MedFit Partners, based in Evanston, Illinois. This company's many hospital partnerships include the Wilfred R. Cameron Wellness Center of the Washington Hospital in Washington, Pennsylvania, and the Dedham Health and Athletic Club in Dedham, Massachusetts.

The latter facility has developed successful partnerships with several local medical providers, including a diabetes management partnership with the Joslin Clinic and a physician referral program for individuals with diagnosed medical conditions such as hypertension, obesity, and depression.

Community Setting

The **community setting** includes many types of health and fitness facilities. One of the largest segments of this setting consists of local branches of nonprofit organizations such as the YMCA, the YWCA, the JCC Association, and city parks and recreation departments. Community settings may also include health and fitness facilities and programs that operate in churches, universities, and apartment complexes. Health and fitness programs offered in community settings may be funded by these organizations, by government grants, or by other philanthropic donations. This funding sometimes reduces the cost of community health and fitness services in comparison with that of commercial settings.

Typically, health and fitness programs in community settings seek to fill a specific community need. Most target a specific community group and attempt to reach as many individuals in that group as their resources allow. Many facilities in community settings receive outside funding and are classified as nonprofit organizations for tax purposes. This classification enables them to offer health and fitness programming at low cost and to provide financial assistance to individuals who could not otherwise afford to participate. For example, more than 200 YMCAs across the United States offer a diabetes prevention program (YMCA of the USA, 2016) specifically targeted to at-risk individuals. This work is part of the National Diabetes Prevention Program offered by the Centers for Disease Control and Prevention. This low-cost program encourages patient referrals from health care providers.

The types of programs offered in community settings are as diverse as those offered in worksite, commercial, and clinical settings. They may include walking groups for seniors, community cardiac rehabilitation programs, aquatic exercise, group exercise, and yoga classes. Resources are typically more limited in the community setting, which leads to greater reliance on volunteers or minimally trained employees in order to meet many staffing needs. For example, the American Diabetes Association offers Project POWER, a program aimed at reducing diabetes and specifically designed to be offered by churches in African American communities. Some programs may be offered on a for-profit basis as a way of generating revenue to deliver other community service programs at little or no cost to participants. For example, many YMCAs sell health club memberships to the public as a method of generating revenue to offset the cost of nonprofit programs such as youth summer camps and programs for individuals with special needs.

Community-based nonprofit organizations are not the only entities that offer community health and fitness programs. In addition, an increasing number of companies and clinical entities have expanded their missions to include contributing to the health of surrounding communities; these groups do not typically operate as nonprofit entities. The motives behind these activities are varied and can include a desire to help, to improve public relations in the community, to offer volunteer opportunities for employees, to reduce tax exposure, and to improve long-term business strength.

Commercial Fitness with a Medical Mission

In March 2005, Augie Nieto, co-founder of Life Fitness, the world's largest commercial manufacturer of fitness equipment, received a diagnosis of amyotrophic lateral sclerosis (ALS, or Lou Gehrig's disease). ALS is a progressive disease that destroys the nerve cells controlling muscles, ultimately causing complete paralysis. Under the banner Augie's Quest, Augie has dedicated his time and energy to fundraising and research in order to find more effective treatments and, ultimately, a cure for ALS. The International Health, Racquet & Sportsclub Association (IHRSA), the trade association of high-quality commercial clubs, has assisted Augie by hosting an annual fundraising gala at its annual conference and trade show. Many IHRSA member clubs have also participated via their own fundraising programs. Since 2006, Augie's Quest has raised more than $45 million. Through this funding and a partnership with the ALS Therapy Development Institute, scientists are searching for the elusive treatment and cure. To learn more about Augie's Quest—and how a health and fitness community can come together in innovative ways—visit their website.

In one example, the International Health, Racquet & Sportsclub Association recognizes the importance of community service by presenting an annual award for outstanding community service to a member that has demonstrated a long-standing commitment to making a difference in the community. In 2016, the recipient was the Claremont Club of Claremont, California. Its community outreach activities include a Living Well After Cancer program, Project Walk (dedicated to improving quality of life for clients with spinal cord injuries), summer camp scholarships, and an adopt-a-family program during the Christmas season.

Population-Based Setting

A major shift has occurred in the diseases that plague the United States, as well as much of the world. Most of the major noncommunicable diseases (NCDs) occur as a consequence of lifestyle choices. For example, lung cancer is the leading cause of cancer death in the United States, and tobacco use is a major contributor to it. The development of the most lethal cardiovascular disease, ischemic heart disease, is markedly affected by tobacco use, dietary habits, physical activity habits, and obesity. Similarly, contributors to type 2 diabetes include dietary and physical activity habits, as well as excess body fat.

Encouraging lifestyle change is an important focus in most health and fitness settings, but the population-based setting does so in a different way. The other settings discussed here focus on working with people in relatively small groups, ranging from the one-on-one interaction of a personal trainer to the small groups of a fitness instructor and on to the larger groups of a facility or community setting. In the population-based setting, however, as the name suggests, the focus is placed on an entire population—for example, a city, a county, a state, or even a nation—with the goal of reducing the population's overall health burden. A population may also be defined as a group of people whose physical activity is affected in particular ways by weather, geography, culture, or ethnicity.

A large segment of this setting consists of local health departments, state agencies, and voluntary health agencies such as the American Heart Association, the American Lung Association, and the American Diabetes Association. The responsibilities shouldered by a person working in population settings can be quite broad. For example, there may be a need to assess health trends and risk factors for specific groups within the population, develop appropriate interventions in order to reduce these risk factors, pursue funding to implement the interventions, and then implement them. The complexity of these activities often requires a multidisciplinary team. Consequently, health and fitness professionals may work with physicians, psychologists, epidemiologists, politicians, corporations, community planners, and a number of other professionals. Competing for such jobs often requires advanced training, such as a graduate degree in kinesiology or a master of public health degree.

Roles for Health and Fitness Professionals

The types of jobs available in a health and fitness facility depend more on the specific programs it offers than on the setting into which it is categorized. All of the settings are seeing annual increases in the number of organizations offering high-quality health and fitness programs, and many types of facilities now offer similar programs.

This section describes the types of jobs available across the different health and fitness settings. It also discusses the evolution of these jobs in response to public policy, social forces, and market demands. The most important element of this evolution may be the fact that most health and fitness positions now require some expertise in multiple areas. In other words, someone hired as a personal trainer may also be asked to lead group exercise classes or perform some duties of a health and wellness coach. In addition, as discussed in chapter 11, it is critical that you fully understand what it means to be a professional; therefore, the discussion emphasizes this common aspect of these jobs. The take-home message here is that it is in your best interests to be as knowledgeable and skilled as possible in a wide array of health and fitness occupational areas and to develop a professional mind-set in approaching these occupations.

Group Exercise Instructor

Once referred to as "aerobics instructor," the position of group exercise instructor has changed in recent decades. Group exercise instructors continue to lead aerobic exercise classes, but they now serve a broader population and provide instruction in a variety of other activities as well. For example, they teach outdoor activities; aquatic fitness; and exercise classes for specific populations, such as prenatal and postnatal women, seniors, children,

Rise of Noncommunicable Diseases

Perhaps the most alarming trend that will drive future growth in the health and fitness industry is the incidence of noncommunicable diseases (NCDs). This is a global problem. The World Health Organization (2016) estimated that NCDs kill 38 million people each year. The leading killer is cardiovascular disease, which contributes to 46 percent of these deaths; it is followed by cancer, respiratory disease, and diabetes. These four diseases account for 82 percent of all NCD deaths, and a disproportionate number of these deaths occur in low- and middle-income countries. Apart from the tragic losses on the personal and familial levels, the socioeconomic impact is staggering. For example, Brazil, Russia, India, and China together lose more than 20 million productive life-years annually to chronic disease, and that number is expected to increase by 65 percent by 2030. U.S. estimates indicate that nearly half of Americans suffer from an NCD; moreover, $277 billion is spent annually on treatment, and lost productivity costs the economy $1.1 trillion annually. Similar problems affect other countries—such as Canada, Australia, and the United Kingdom—that are characterized by an increasing older population, a disease-based medical model, and a shrinking taxation base (World Health Organization, 2011).

This problem is certainly too big to be solved by any one profession or industry. At the same time, it may well come to define the health and fitness profession because the expertise embodied in this profession is ideally situated to make a major contribution to reducing NCDs. This is so because critical contributors to NCDs include lifestyle issues such as tobacco use, physical inactivity, unhealthy diet, and the use of alcohol. By working to reduce such risk factors, well-trained health and fitness professionals can be tremendously helpful to their clients. Doing so, however, has been and will continue to be a challenge.

Although many people are interested in personal wellness, behavior change, and health and lifestyle improvement, government lifestyle indicators show that actual efforts to improve and maintain health are often inadequate. Indeed, 30 years of wellness programs and education initiatives have been unsuccessful, largely because the delivery has been based on risk management and clinical compliance rather than on evoking sustained behavior change. Thus, the missing component over the last few decades has been sustainable engagement, and solving this challenge is the key goal for any employer, funder (whether insurer or government), or educational initiative. This is one of the great challenges of our time, and you, as a kinesiology student and potential health and fitness professional, can play an enormous role in his work.

and clients who need medically supervised exercise. In the 1970s and 1980s, an aerobics studio was little more than an empty room with mirrors and a sound system. Those days are gone. Group exercise instructors are now more likely to use a variety of functional equipment, such as steps, spinning bikes, exercise bands, dumbbells, fitness and exercise balls, and barbells. In most places, this position exists as a part-time job with hourly pay.

The aging of the U.S. population has brought an increasing number of older and higher-risk clients into the group exercise setting. What was once an aerobics studio filled exclusively with young adults performing complicated dance routines is now a studio filled with a diversity of individuals. Consequently, group exercise instructors can develop expertise about an array of class "types" in addition to the stereotypical aerobics classes: cardiorespiratory classes (step, kickboxing, indoor cycling), strength development (kettlebells), mind–body classes (yoga, Pilates), and specialty classes that are designed for older adults (Silver Sneakers). Instructors should be dynamic, possess excellent leadership skills, and enjoy working with people in a group setting.

Recall from chapter 11 that an occupation constitutes a *profession* if it is marked by certain characteristics. Specifically, the work performed requires mastery of complex skills grounded in theory, the services are performed for clients or patients, the occupation holds a monopoly on the delivery of certain services, the work is guided by formal and informal ethical codes of conduct, and standards of professional practice are observed within the relevant working subculture.

The occupation of group exercise instructor possesses some of these characteristics but certainly not all. For instance, the activities performed in a group exercise class can be complex, but the instructor can also lead the class while remaining ignorant of the underlying scientific theory or knowledge. In other words, the instructor can know the *how* without knowing the *why*. As a result, many people learn how to be an instructor through on-the-job training—first participating in exercise classes and then gradually working into the role of instructor. In many health and fitness facilities, this work is viewed as an entry-level position and constitutes part-time, hourly employment. For example, the prerequisites for most certifications in group fitness instruction include being at least 18 years old, being current in CPR and AED registration, and having completed high school. Clearly, then, the expertise derived from earning a kinesiology degree is not needed, and these positions are not truly professional.

Personal Trainer

Personal trainers typically work one-on-one with clients and, because of this increased supervision, usually receive higher pay than group exercise instructors. Some personal trainers work independently, traveling to individual clients' homes, conducting training sessions outdoors, or subcontracting with a health and fitness facility or private gym. Trainers may bill for their services hourly, weekly, or monthly. Most often, however, a personal trainer is employed by a health and fitness facility but directly markets his or her own services to members. In this setup, the trainer and the facility split the fees charged to the client. On the other hand, some facilities hire personal trainers as independent contractors, thereby obligating the trainers to operate their own businesses, pay their own taxes, obtain their own health insurance, and forego paid vacation and sick days.

Although personal trainers can earn substantial income, doing so is not easy. A new trainer needs to build and retain an adequate base of clients. Consequently, as exemplified by many celebrity personal trainers, success may depend more on good marketing and sales skills than on professional expertise. Moreover, personal trainers work when their clients are available, which often means long work hours, including early mornings, late evenings, and weekends.

At the same time, working as a personal trainer provides a unique opportunity for the health and fitness professional to develop ongoing relationships with clients. This aspect of the job can be personally rewarding since a trainer can play a critical role in clients' success. For example, a trainer can assess individual clients' needs and quickly implement new strategies for behavior change. If qualified to do so, personal trainers can also work with clients on a number of health issues, such as weight management, chronic disease management, general fitness, and sport conditioning.

Current trends in this occupation include specialization in working with specific populations (e.g., older adults, children, clients with medical concerns) and working toward specific client goals (e.g., improved appearance, injury rehabilitation, participation in a 10K race or triathlon). Increasingly, personal trainers cater to these market segments by offering not just individual services but also small-group specialty fitness programs; in this way, specialty fitness programs bridge the gap between one-on-one personal training and group exercise. Such programs are offered to small groups (two to six people each) for a defined time and are usually focused on a specific goal (e.g., weight loss). These programs have grown in popularity because they allow people access to personalized training within a small group, thus making the service appealing and affordable to a wider population.

In order to succeed, personal trainers need a variety of skills. Besides having a strong knowledge base in fitness assessment and exercise prescription, they must also be excellent communicators and counselors; possess strong business skills in sales, marketing, administration, and time management; and be able to work with a diverse client population. Personal trainers are the artists of the health and fitness professions. Besides evaluating individual clients' needs and designing individualized programs, they must continually present new and exciting ways to keep clients on track. This approach helps keep clients interested and motivated to adopt and maintain new healthy behaviors as part of their lifestyle.

The qualifications needed by a personal trainer vary dramatically across different types of health and fitness facilities. In some facilities, personal trainers are not required to have a relevant college degree or relevant certifications. In fact, they may be little more than "hard bodies" or "fitness models" who have learned some skills as part of their personal exercise avocation but do not possess substantive scientific knowledge. At the other end of the spectrum, the personal trainers at some facilities are truly professionals. They are required not only to possess a college degree in their field and be certified by a nationally recognized association but also to have specialized training and many years of experience. An increasing trend in commercial health and fitness facilities is to classify personal trainers based on skill, experience, sales volume, education, and certifications held. In this case, a more advanced trainer may be titled a "master trainer." Experienced trainers are usually able to charge higher rates for their services and therefore earn higher rates of pay from their facilities.

Over the past decade, the U.S. public has become more knowledgeable about the types of credentials and certifications that are appropriate for the personal trainer position. Increasingly, simply being fit is not an acceptable qualification. Unfortunately, however, most certifications for personal trainers use the same minimal prerequisites used for a group exercise instructor. However, it is becoming more common to see personal trainers who possess a bachelor's degree in kinesiology. Regardless, in order to be marketable, personal trainers must also possess extensive experience in working with clients on a variety of health and fitness issues. They must also keep their skills current by attending conferences, workshops, and other continuing education programs.

As you can see, personal trainers carry complex responsibilities and the need to possess a robust scientific knowledge base in order to meet these responsibilities. As a result, this occupation should be viewed as a profession, as defined in chapter 11. Recognize, however, that some individuals who work as personal trainers do not have the appropriate background and training to fulfill the responsibilities described here. Therefore, when seeking employment as a personal trainer, it is to your advantage to assess the qualifications of the personal trainers working in the facility in order to help you determine whether it is the type of facility you want to be part of.

Working with a qualified personal trainer is a good way to get an individualized training program.

Health and Fitness Specialist

The health and fitness specialist position is typically a professional position for individuals who have completed a bachelor of science degree in kinesiology or exercise science as well as additional training and certifications. Job duties include conducting fitness and functional assessments and designing comprehensive individualized exercise programs that incorporate strength, flexibility, and aerobic fitness components and maximize safety and long-term results. This job provides opportunities to introduce diverse individuals to the benefits of exercise. For many sedentary adults who are joining their first health and fitness facility, the health and fitness specialist is the initial contact person who can either excite or discourage them in their endeavor to become fitter and healthier. Most reputable health and fitness facilities provide new participants with a health screening or fitness assessment. In this process, the health and fitness specialist assesses such things as blood pressure, body composition, aerobic capacity, muscular strength, muscular endurance, and flexibility. The results are used to design an individualized program that meets the client's specific health and fitness needs.

In many health and fitness facilities, these responsibilities are performed by the same employees who function as personal trainers or health and wellness coaches. Moreover, the term, *health fitness specialist* is not universally used in the health and fitness industry; common alternatives are *health fitness instructor* and *exercise physiologist*. Although the latter term is common, its use in this context is controversial and unclear because exercise physiology is one subdiscipline of kinesiology and the term *exercise physiologist* has traditionally referred to faculty with doctoral-level training who perform research in the field of exercise physiology.

Health and Wellness Coach

The position of health and wellness coach may be the most exciting growth area in the health and fitness professions. Health and wellness coaches provide guidance to diverse populations on a broad range of health and fitness topics and are in increasingly high demand in worksite wellness programs. In some facilities, this position has either replaced that of personal trainer or health and fitness specialist or been "hybridized" with it; in other facilities, it exists as a new category of employee.

The following job description has been provided by the International Consortium for Credentialing Health and Wellness Coaches.

> Health and Wellness Coaches partner with clients seeking self-directed, lasting changes, aligned with their values, which promote health and wellness and, thereby, enhance well-being. In the course of their work, health and wellness coaches display unconditional positive regard for their clients and a belief in their capacity for change, and honoring that each client is an expert on his or her life, while ensuring that all interactions are respectful and non-judgmental.

Because *health and wellness* is such a broad term, the job duties of these coaches can also be quite broad. In addition to prescribing exercise programs, they can include working with clients on behavior change, stress management, relaxation techniques, time management, smoking cessation, and weight management. Of course, health and wellness coaches must be able to conduct health screenings and fitness assessments, but these are just some of the professional tools that they use in assisting clients. They may not, however, apply all of their techniques during an initial appointment; instead, they may first assess clients' readiness for change (Prochaska & DiClemente, 1986), help them set short- and long-term goals, and then help them move at an appropriate pace toward a state of optimal health.

In other words, health and wellness coaches help facilitate attitudinal and behavioral changes in their clients in order to help them succeed. This approach differs from the approach often used by personal trainers. In many trainer–client relationships, the trainer is the expert who tells the client what to do, and the client then follows the instructions. A health and wellness coach is also an expert but tries to guide the client toward making his or her own healthy choices. In this approach, the client is ultimately in charge, and this collaborative coaching approach is far superior if the goal is to evoke behavior change.

Health and wellness coaches must possess appropriate educational credentials and certifications and participate in continuing education. They need to be organized, understand principles of behavior change, possess excellent communication skills, and be effective at marketing and promoting programs and services. They must also be able to provide counseling on a broad range of health topics and engage people.

As with the other health and fitness occupations, health and wellness coaches must choose from a host of academic and nonacademic training programs purporting to help them develop expertise. As with the other occupations, some of these programs are very good; others are not. Unlike individuals in the other occupations, health and wellness coaches have far fewer options for certification from well-regarded professional organizations with appropriate expert support. One of them, the American Council on Exercise (ACE) Health Coach certification, is accredited by the National Commission for Certifying Agencies. Another, offered by Wellcoaches, is endorsed by the American College of Sports Medicine and approved by the National Consortium for Credentialing Health and Wellness Coaches. This group has also partnered with the National Board of Medical Examiners to launch a national certification program for health and wellness coaches in 2017.

It is highly recommended that students with an interest in working in health and fitness facilities also consider pursuing a health and wellness coach certification. There is a tremendous need for health and wellness coaching skills in order to bridge the gap between the chronic health conditions (e.g., obesity, heart disease, hypertension, diabetes) that affect many clients and the lifestyle changes needed to improve these conditions. The Affordable Care Act and subsequent legislative efforts to revise health insurance nationally may promote the use of preventive services to maintain patients' health, thus helping to create a health care environment that is highly conducive to employment opportunities for health and wellness coaches. Consequently, in addition to having job options in commercial fitness facilities, health and wellness coaches may also be hired by health insurance providers or worksites in order to improve employee health and decrease medical expenditures. Kinesiology-trained graduates commonly provide similar preventative physical activity services supported by health insurance in Canada and various European countries.

Specialty Positions

Many health and fitness settings now incorporate specialty positions into their programs, including the following (see also chapter 13):

Case Study in Transitioning From School to a Work Setting

Will was hired several weeks ago as a personal trainer at a commercial gym that specializes in training athletes. He had earned his bachelor of science degree in kinesiology from a program that emphasized the importance of fitness assessments, health screening, and safe exercise techniques in working with adults who want to become fitter and healthier. During his first few weeks on the job, he noticed that many of his fellow trainers were promoting nutritional supplements to their clients. He knew from reading the trainer biographies posted on the lobby wall that, although most of the trainers in this facility had a formal degree in kinesiology, none of them had a degree in nutrition, advanced training in dietetics, or a license as a registered dietitian nutritionist (RDN).

Will was worried that his fellow trainers were not qualified to give nutritional advice to clients, much less recommend specific supplements. As a consequence, they were being negligent in providing evidence-based nutrition recommendations, thus potentially wasting clients' money and, worse, putting them at risk for adverse physical performance or health complications. Furthermore, he was concerned that without including an RDN in these dietary programs, the personal trainers and the health and fitness facility were putting themselves at risk of being sued for practicing dietetics without a license. Will was thinking about approaching his manager to discuss this concern.

Your Decision

What do you think Will should do in this situation?

Resolution

Will decided to talk to his manager and express his concerns. His manager listened very carefully. She then explained that the selling of supplements was a major contributor to the facility's profits and the personal trainers' incomes. When Will expressed concern that everyone involved could be sued, she said it was worth the risk. Will did not want to work in a facility that engaged in these unethical practices, and he began looking for a new job.

- Clinical exercise physiologist
- Physical therapist
- Athletic trainer
- Sport coach
- Registered dietitian
- Population specialist (e.g., working with older adults, children, people with chronic disease)
- Public health educator

Specialty positions provide services that go beyond those provided by other health and fitness professionals. These jobs can be exciting and challenging and allow for increased earning potential. However, many of them are licensed professions. They require graduate education and additional qualifications beyond what is provided by a bachelor's degree in kinesiology or other discipline related to health and fitness. In other words, a bachelor's degree, even with strong grades and additional experience, may serve only as a starting point in becoming prepared for one of these professions.

Here are three examples: (1) Becoming a physical therapist requires competitive admission to, and the completion of, a 3-year doctor of physical therapy degree program. (2) Within the next several years, becoming an athletic trainer will require a master's degree in athletic training, with competitive admission based on an undergraduate degree in a relevant field such as kinesiology. (3) To become a registered dietitian, one must graduate from an academic degree program that meets the requirements set forth by the Academy of Nutrition and Dietetics and also earn competitive admission to an accredited program in dietetics. The rigor of many kinesiology undergraduate programs also allows strong graduates to be competitive for admission to medical, physician assistant, dental, and other professional graduate programs.

What differences exist between the health and fitness professions and these other professions? The distinction can sometimes be difficult to discern. Consider the provision of nutrition information to clients in a health and fitness facility. In this environment, health and fitness specialists, health and wellness coaches, and personal trainers are often

The health and fitness field encompasses many types of careers.

© Human Kinetics

asked to provide dietary advice. These professionals can ethically provide basic, general nutrition information to healthy adult clients. However, because of a lack of formal training, they are not qualified to prescribe a specific type of diet or recommend special dietary supplements to anyone or, in the case of clients with specific health conditions, make dietary recommendations. In fact, doing any of these things is unethical and, in a court of law, could be considered malpractice in the form of practicing dietetics without a license.

Attaining the requisite expertise requires additional course work and internship experiences that are not normally provided with an undergraduate kinesiology degree. It also requires passing the national examination provided by the Academy of Nutrition and Dietetics. Therefore, only a registered dietitian (RD or RDN) possesses the appropriate education and qualifications to legally and ethically make such medically oriented dietary recommendations.

As a result, 46 states have enacted legislation to regulate the practice of dietetics. Unfortunately, anyone, no matter how inadequate his or her educational background or training, can legally adopt the label of "nutritionist." (A web search for the term "celebrity nutritionist" will reveal many examples.) The distinction between a dietitian and a nutritionist is difficult for clients to understand; it is also confusing for many in the health and fitness industry. For this reason, the Academy of Nutrition and Dietetics has recently begun allowing RDs the option of being credentialed as a registered dietitian nutritionist, or RDN. An RD and an RDN do not differ in their qualifications; this change was implemented entirely to facilitate clients' understanding that "all Registered Dietitians are nutritionists but not all nutritionists are Registered Dietitians" (Commission on Dietetic Registration, 2016).

Of course, clients may also ask RDs and RDNs about exercise programs. In the absence of any specialized training, these professionals should limit their advice to basic, general exercise information. However, if they did overstep the bounds of their expertise and provide information that fell within the expertise of health and fitness professionals, they would not be viewed in the United States as practicing kinesiology without a license. This is true simply because neither licensure nor registration with a national accrediting body is required to be allowed to dispense exercise-related information. In contrast, in Canada, some provinces offer licensing of kinesiology professionals that does provide some protection of the physical activity domain.

Injury rehabilitation provides another example of the boundaries between various health and fitness professions and specialty professions in a health and fitness facility. Personal trainers should obtain a full injury history for each client and consider this information when designing an exercise program. An appropriately designed exercise program can help people prevent injuries, but personal trainers should not design programs for specific injury rehabilitation. Instead, physical therapists and athletic trainers are the specialists with the appropriate knowledge and qualifications to provide injury rehabilitation services to clients.

In an ideal scenario, a personal trainer would work in partnership with a physical therapist. The personal trainer would supervise specific aspects of the program and communicate with the therapist regarding the client's progress. However, the personal trainer does not have the expertise to replace the therapist in guiding the client's rehabilitation. As with RDs and RDNs, licensure for physical therapists is required in all 50 states; for athletic trainers, it is required in 44 states.

Many specialty positions have arisen that focus on particular populations, and many of these specializations have led to additional certification options. Due to the special needs involved in providing exercise training for a specific population, these positions often require training or experience beyond what is provided as part of a typical bachelor's degree in kinesiology. Most of these certifications do not, however, require an additional academic degree. Here are a few of the many certification possibilities:

American College of Sports Medicine

- Certified Clinical Exercise Physiologist (clients with cardiovascular, pulmonary, and metabolic diseases and disorders)
- Registered Clinical Exercise Physiologist (clients in a clinical or research environment with a variety of chronic diseases or who are physically limited)
- Exercise is Medicine Credential (facilitation of collaboration between health care providers and health and fitness professionals)
- ACSM/NCHPAD Certified Inclusive Fitness Trainer (clients with physical, sensory, or cognitive disability)
- ACSM/ACS Certified Cancer Exercise Trainer (clients who have been affected by cancer)
- Physical Activity in Public Health Specialist (promotion of physical activity in public health at local, state, or national level)

Kinesiology Colleagues

Kevin Heiberger

Kevin Heiberger is an assistant football performance coach at the University of Virginia whose primary responsibilities involve teaching speed dynamics. After earning a bachelor's degree in marketing, he realized that his passion was found in strength and conditioning. He subsequently earned bachelor's and master's degrees in kinesiology while working as an unpaid strength and conditioning intern at a Division I university. He then completed two other unpaid internships before finally being hired as a performance coach at IMG Academy in Bradenton, Florida.

Photo courtesy of UVA Media Relations

Over the 4 years he spent at IMG, Heiberger developed expertise as a speed and movement specialist while also working with well-regarded professionals in the field. His expertise, along with the connections he made along the way, resulted in his now having the job he had long wanted. Although Heiberger's path to his dream job was circuitous and challenging, it was not at all atypical for those who seek a career in strength and conditioning.

American Council on Exercise

- Health Coach (health and wellness coach certification)
- Medical Exercise Specialist (clients who have common noncommunicable diseases such as cardiovascular, pulmonary, metabolic, and musculoskeletal conditions)
- Behavior Change Specialist (working with clients to evoke behavior change)
- Sports Conditioning Specialist (clients with a focus on athletic performance)
- Weight Management Specialist (clients who need long-term weight management strategies)
- Senior Fitness Specialist (clients age 55 or over)
- Youth Fitness Specialist (working with adolescents and teenagers)
- Functional Training Specialist (clients desiring to improve balance, stability, or movement efficiency)
- Mind Body (holistic approach to exercise and wellness using activities such as yoga and Pilates)

One specialty position that is increasingly popular among kinesiology students is that of strength and conditioning coach. Although some duties of this position overlap with those of a personal trainer, the focus here is on training athletes with the primary goal of improving athletic performance. Course work in most kinesiology degree programs will help prepare students for this setting, but it will not be entirely sufficient. Even for entry-level positions, job applicant needs to possess a wealth of practical experience in a strength and conditioning environment.

In most college settings, this often means working as an intern (often unpaid) in the athletic department's strength and conditioning program.

Individuals can further enhance their employability by gaining additional relevant experiences by working with athletes in other settings, participating in college-level athletics, or being mentored by well-respected professionals in the field. Appropriate certifications are also desirable. As with other health and fitness professions, many certifications are available. One well-respected entry-level certification is that of Certified Strength and Conditioning Specialist (CSCS), which is offered by the National Strength and Conditioning Association. CSCS certification requires a bachelor's degree, typically in kinesiology.

As mentioned earlier, many specialty positions require formal education beyond a bachelor's degree in kinesiology. If you are interested in pursuing a specialty position, check with your academic advisor to determine how you should proceed with your academic program.

Public Health–Oriented Careers

Historically, kinesiology has been viewed in the context of professions in physical education or athletics. This has begun to change, largely because noncommunicable diseases have become an enormous contributor to premature death and disability in the United States. Sedentary lifestyles and physical inactivity are important factors in this epidemic and, as such, constitute major public health problems. Consequently, over the last 25 years, many kinesiology programs have returned to their health promotion roots by focusing more on the public health aspect of physical activity.

On the undergraduate level, this shift may be seen in some of the courses available to you. For example, you may have the opportunity to complete courses related to principles of public health, health behavior change, and physical activity epidemiology (Ainsworth & Hooker, 2015). There may also be a degree concentration, minor, or option in the role of physical activity in public health. Besides preparing you for a career in the long term, course work can prepare you in the short term for the Physical Activity in Public Health Specialist certification offered by the American College of Sports Medicine in collaboration with the National Physical Activity Society.

Historically, students who obtained an undergraduate degree in kinesiology and wanted to continue their education pursued a graduate degree in kinesiology. Today, students with a strong interest in health promotion are increasingly choosing to pursue a master of public health (MPH) degree. This degree is offered with specialization in physical activity and public health by several kinesiology departments that are situated within a school or college of public health. More generally, more than 160 MPH programs are now accredited by the Council on Education for Public Health. These developments have resulted in opportunities for health and fitness professionals to learn how to navigate the legislative process; partner with local,

Professional Issues in Kinesiology

Evidence-Based: Access for Everyone

Most health and fitness facilities strive constantly to recruit and retain members. Even so, this process often overlooks one segment of the population—namely, individuals with disability, be it physical, intellectual, or otherwise. The U.S. Census Bureau estimates that about 19 percent of the U.S. population has some form of disability. Thus, it makes good business sense to create facilities that offer inclusive programming, equipment, and accessibility for individuals who are not ambulatory. However, a survey of community fitness facilities found that, for members who use a wheelchair, the most common challenges were access to and free movement around exercise equipment, as well as full access to restrooms (Dolbow & Figoni, 2015). Other common challenges included not having equipment adapted for wheelchair users and not having staff with adequate training. These challenges are rectifiable. If you may be interested in this area, you can gain insight into working with differentially abled individuals by accessing the resources provided by the National Center on Health, Physical Activity and Disability and by pursuing a specialty certification such as that of ACSM/NCHPAD Certified Inclusive Fitness Trainer.

Citation style: APA

Dolbow, D.R., & Figoni, S.F. (2015). Accommodation of wheelchair-reliant individuals by community fitness facilities. *Spinal Cord*, *53*, 515–519.

state, and federal agencies; design, implement, and evaluate programs; and promote physical activity as an essential public health component.

Besides pursuing an MPH, training as a public health educator may also enable you to fill this role and design programs for implementation at the community level. Job settings for public health educators include local health departments, non-profit organizations, schools, colleges, hospitals, and health insurance companies. For those wishing to pursue a position as a public health educator, the credential of Certified Health Education Specialist (CHES) is offered by the National Commission for Health Education Credentialing.

Health and Fitness Director

The health and fitness director is a key position in most health and fitness settings. In larger facilities, the health and fitness director may be part of the general management team and supervise a team of managers who handle specific responsibilities within departments. Examples of these management positions include directors for group exercise, yoga, Pilates, personal training, aquatics, athletics and recreation, and youth activities.

KEY POINT

Health and fitness directors must understand the ways in which social and cultural forces affect their industry and how to respond to these forces in order to keep their programs viable.

The job description of a health and fitness director includes hiring, training, and providing support for a diverse staff of health and fitness professionals, as well as support personnel such as custodians and equipment maintenance personnel. In addition, health and fitness directors are responsible for conducting business planning, managing budgets, planning facility renovations, selecting equipment, designing and marketing programs, and forecasting trends in health and fitness programs. The director must possess a broad, interdisciplinary education that includes business planning, budgeting, marketing, and staff management, as well as a strong foundation in kinesiology.

Health and fitness directors also need to be effective leaders. Besides managing day-to-day operations, they must be capable of articulating a vision and motivating other individuals to work toward specific goals. They are both visionaries who keep current with the field of health and fitness and mentors who support staff members in moving toward individual goals and aspirations.

Educational requirements for the health and fitness director position emphasize academic degrees more than certifications. Kinesiology is an appropriate undergraduate degree, and many facilities also require the health and fitness director to have a graduate degree in kinesiology or another discipline related to health and fitness. It may also be desirable to have a business-related degree or master's in sport management, because the director is focused primarily on the administrative and business aspects of the facility. Other requirements often include experience in a position such as health and fitness specialist or personal trainer. Through experience in such positions, directors gain a full understanding of individualized program design and strategies for enabling behavior change. In addition, these job experiences provide valuable insight into the management responsibilities of hiring, training, and supporting staff.

Certification and Continuing Education

To be hired for a position in the health and fitness professions, you will need to fulfill specific requirements in your education and training. Once you are hired, you will need to continue seeking information and developing new skills, because the health and fitness field is dynamic and new information is continually becoming available.

Corporate, commercial, community, and population-based organizations typically prefer to hire health and fitness instructors, health and fitness specialists, and personal trainers who hold at least a bachelor's degree in either kinesiology or another discipline related to health and fitness. Indeed, many organizations require such a degree. This is a good thing because club members, employees, and patients now demand high-caliber, educated professionals who possess extensive knowledge of health and fitness.

Still, for better or for worse, many people employed in the health and fitness industry do not have a kinesiology degree. Some are not qualified to work in this industry, whereas others may have extensive work experience and a great deal of continuing education and practical training. Because of the prevalence of poorly trained and self-proclaimed fitness experts, most employers now require individuals to hold a certification from a nationally recognized organization as a condition of employment.

Hundreds of certifications are available for professionals in the health and fitness industry.

Some certifications truly identify you as being knowledgeable about kinesiology and current standards of professional practice; others are little more than moneymakers for the agency that provides them. Some are highly regarded; others are not. Consequently, it can be a daunting challenge to decide which certification(s) to pursue and, due to the costs involved, an expensive mistake if handled poorly.

With this pitfall in mind, the International Health, Racquet & Sportsclub Association recommends that facilities hire personal trainers with certifications from organizations with third-party accreditation. Accreditation is the process by which a nationally recognized accrediting agency evaluates a credentialing program against standards expected in a legitimate credentialing process. Consequently, a good first step toward identifying a good potential certification is to determine whether it has been accredited by the National Commission for Certifying Agencies (NCCA). You can find out by doing a web search for the phrase "NCCA accredited certification programs." Currently, 16 organizations catering to health and fitness professionals have been accredited:

Skills and Competencies Required in the Health and Fitness Profession

Job title	Evolution	Job duties	Skills and competencies
Group exercise instructor	This position no longer exclusively involves teaching aerobic dance classes; rather, it now teaches a broad range of classes to a diverse population of individuals.	Lead a wide variety of group exercise classes for various population groups.	Musts: Certification by a nationally recognized organization Strong teaching skills Helpful: Additional certifications or training in specialized areas
Personal trainer	This position has changed from exclusively providing individualized exercise programs to providing services on a broad range of health topics to both individuals and small groups.	Provide ongoing support and guidance to a diverse population of clients on topics such as physical fitness, weight management, stress management, and sport conditioning.	Musts: Certification by a nationally accredited organization Counseling and teaching skills Business, marketing, sales, and promotion training Skills and experience with fitness testing and exercise prescription Preferred: Bachelor's or graduate degree in a field related to health and fitness Helpful: Additional certifications or training in specialized areas
Health fitness specialist	To meet the needs of clients who may have health concerns, many facilities now require their fitness professionals to have a formal degree in their field, as well as nationally recognized certification.	Design customized exercise programs to meet the needs of diverse participants. Conduct fitness assessments to inform program design and monitor progress. Provide expert guidance to help clients meet their personal exercise-related goals. Design and implement programs.	Musts: Bachelor's or graduate degree in a field related to health and fitness Certification by a nationally recognized organization Preferred: Skills and experience in fitness testing, exercise prescription, and program planning Helpful: Additional certifications or training in specialized areas

- Academy of Applied Personal Training Education
- ACTION Certification
- American College of Sports Medicine
- American Council on Exercise
- Collegiate Strength and Conditioning Coaches Association
- Cooper Institute
- International Fitness Professionals Association
- National Academy of Sports Medicine
- National Council for Certified Personal Trainers
- National Council on Strength and Fitness
- National Exercise and Sports Trainers Association
- National Exercise Trainers Association
- National Federation of Professional Trainers
- National Strength and Conditioning Association
- Pilates Method Alliance
- PTA Global, Inc.

Skills and Competencies Required in the Health and Fitness Profession *(continued)*

Job title	Evolution	Job duties	Skills and competencies
Health and wellness coach	Having evolved from the more traditional fitness instructor position, the health and wellness coach provides counseling on a broad range of health topics in addition to conducting fitness assessments and designing exercise programs.	Provide guidance to a diverse population of individuals in areas such as behavior change, stress management, smoking cessation, weight management, and exercise programming.	Musts: Bachelor's or graduate degree in a field related to health and fitness Certification by a nationally recognized organization Skills in counseling, behavior change, cultural diversity, teaching, and exercise prescription Preferred: Specialized health and wellness coach certification Marketing and promotional skills
Specialty positions	The number and types of such positions have expanded in recent years and moved from exclusively clinical settings to a variety of health and fitness settings.	Provide specialized health and fitness services to clients with special needs. Examples include physical therapist and public health educator.	Musts: Graduate degree (possibly required) Other specific experiences and skills required for each type of specialist position Preferred: Additional certifications and licensure (possibly required to practice in certain states)
Health and fitness director	This position has evolved from a fitness instructor with additional administrative responsibilities to a full-time manager who actively participates in all aspects of personnel and facility management.	Manage all aspects of a health and fitness department. Handle responsibilities including departmental leadership, staff management, program planning, and all aspects of business administration.	Musts: Bachelor's or graduate degree in a field related to health and fitness Additional skills in business administration, management, marketing, and promotion Preferred: Previous experience as a health and fitness specialist or in a similar position

If you are considering a certification that has not been accredited, the second step in deciding whether to do so is to assess the organization providing it. This step can also be challenging because certifications are offered by numerous organizations, most of whom purport to provide the "best" or "most comprehensive" or "most respected" certification available. In addition, some legitimate certifications for specialty positions or skills are highly desired but not yet accredited. Recall from the previous section of this chapter that it is recommended to pursue the Certified Strength and Conditioning Specialist certification for a career in the strength and conditioning field. This certification has been accredited; however, additional certification may also be desired from unaccredited agencies, such as the Strength and Conditioning Coach Certified (SCCC) offered by the Collegiate Strength and Conditioning Coaches association for positions as a collegiate strength and conditioning coach. This certification, while not accredited yet, might be necessary to apply for certain strength coach positions.

Consequently, the third step in identifying a potential certification is to gather expert opinions by talking with your professors and with professionals in the field. These individuals should be able to help you determine the legitimacy of a possible certification and decide whether it will enhance your marketability to potential employers. This advice is more important than evaluating certifications based on their cost. The cost varies widely and may be high because an organization is for-profit rather than because it maintains rigorous standards and provides professional support. Any organization can claim to be important and charge a lot for training, test preparation, and testing. The rigor and reputation of the certification are the important factors to consider in seeking a professional certification. Some of the most widely recognized and respected certifications come from the American College of Sports Medicine (ACSM), the American Council on Exercise (ACE), the National Academy of Sports Medicine (NASM), and the National Strength and Conditioning Association (NSCA).

As you can see, you must do your research when applying for health and fitness positions so that you will know what educational certifications and programs are credible and most desired by specific health and fitness facilities. In addition, be aware that certification organizations require you to continue your education by taking recognized continuing education units (CEUs) in order to maintain certification.

It is also professionally essential for you to stay current with new scientific material even after you have completed your undergraduate degree. Being part of a professional association can help you do so. Joining a professional association is therefore another important step in becoming a health and fitness professional. Such associations typically sponsor regional and national conferences and provide journals, educational materials, and training. Many also offer discounted membership rates for students. Important organizations in this profession are listed at the end of this chapter. To find out about membership, publications, workshops, and regional and national conferences, contact any one of them.

Trends and Opportunities in Health and Fitness

In every professional field, the leaders are those who keep up with the latest research in the field and with relevant social and cultural trends. They continually ask questions about how the world around them affects their particular organization. This chapter is designed primarily to give you an understanding of the various positions available in the health and fitness professions. Now that you have this understanding, you are encouraged to broaden your view by adding an understanding of the trends that affect these professions.

KEY POINT

Health and fitness programs increasingly address dimensions of wellness beyond physical fitness. They draw from a more holistic, multidimensional model that includes physical, intellectual, emotional, social, and spiritual well-being.

As a future health and fitness professional, you should be aware of the ever-changing context in which these professions exist. This is a dynamic time in the health and fitness field. Personal, social, cultural, and political issues exert powerful influences on the ways in which we view health and fitness and the ways in which we work in the profession. Although it is beyond the scope of this chapter to outline all of these forces, this section discusses three major issues currently influencing these professions: increasing interest in a multidimensional model of wellness, emerging demographic trends, and health care reform. The rise of noncommunicable diseases is an important fourth issue, but it has been previously described in this chapter.

Multidimensional Model of Wellness

It is a trend across all health and fitness settings that program offerings are being expanded to address additional dimensions of wellness. This expansion reflects increasing appreciation of a more holistic definition of health—namely, that being healthy involves far more than simply being physically fit. In this view, wellness is exemplified by the pursuit of comprehensive well-being, not merely the absence of disease, infirmity, or deficiencies in physical fitness. The multidimensional model of wellness addresses at least five dimensions: (1) physical—the ability to conduct daily tasks with vigor; (2) emotional—the ability to control emotions and express them appropriately and comfortably; (3) intellectual—the ability to learn, grow from experience, and use intellectual capabilities; (4) spiritual—a guiding sense of meaning or value in life that may involve belief in some unifying or universal force; and (5) social—the ability to have satisfying interpersonal relationships and interactions with others (Dintiman & Greenberg, 1986) (see figure 12.1).

In accordance with this model, worksite programs have evolved to provide services that not only help employees make changes but also support them in maintaining healthy lifestyle behaviors. Thus, besides offering employees state-of-the-art fitness facilities, many worksites now offer family activities, volunteer service opportunities, and programs that facilitate personal growth and improvement of communication skills.

As facilities expand their program offerings, they also expand their job descriptions. If you elect to focus exclusively on the physical dimension of wellness, you will continue to find many exciting job opportunities. You will find even more, however, if you embrace the challenge of becoming able to deliver services that address other dimensions of wellness.

As discussed in this chapter, many health and fitness positions have already expanded to involve a broader range of responsibilities than in the past. Thus, personal trainers and health and wellness coaches continue to design exercise programs but also help clients with stress management, personal growth, and involvement in social activities. In short, industry leaders in all settings are looking for bright people who can deliver a broad range of services and integrate concepts of exercise and fitness into a broader definition of health.

Demographics

Demographic trends are radically redefining the meaning of healthy lifestyles and healthy aging. According to the U.S. Census Bureau, the median age of the total U.S. population increased from 23 years in 1900 to 30 years in 1950, 35 years in 2000, and nearly 38 years in 2015 (U.S. Census Bureau, 2015). In the 2015 census, people under the age of 18 made up 23 percent of the total population, people aged 18 to 64 accounted for 62 percent, and people aged 65 or older made up 15 percent. Looking to the future, the distribution of ages in the

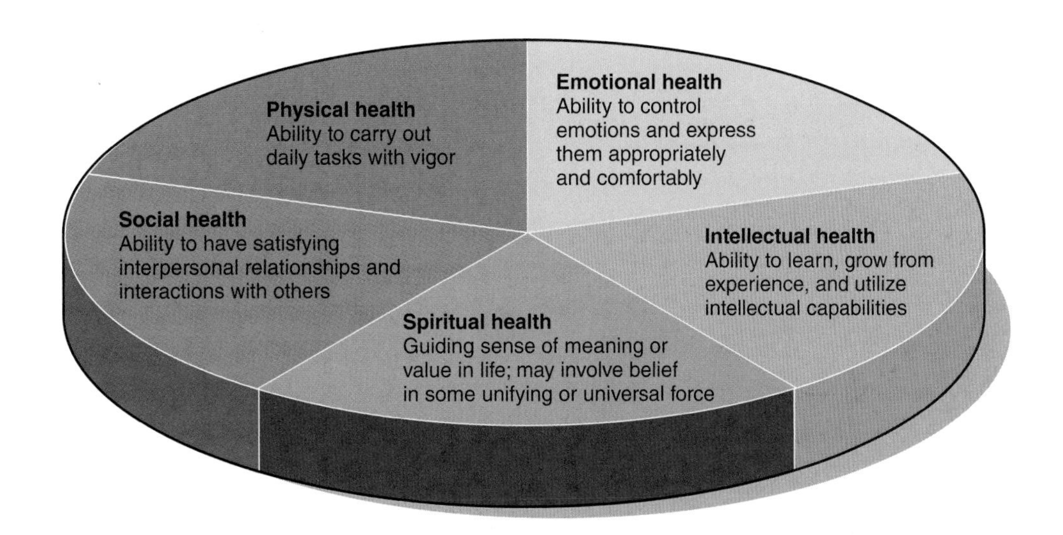

FIGURE 12.1 Dimensions of wellness.

Multidimensional Approach to Fitness and Healing

Cancer patients at the University of California, San Francisco (UCSF), have the opportunity to meet with cancer exercise trainers certified by the American College of Sports Medicine. The trainers demonstrate simple exercises that can be done at home, design personalized exercise programs, and refer patients to cancer-specific yoga and group exercise classes in the community. The service, known as the Cancer Exercise Counseling program, is provided at the UCSF Helen Diller Family Comprehensive Cancer Center and funded by grants and donations. The exercise trainers have worked with UCSF's department of physical therapy and rehabilitation services to write a manual titled "Moving Through Cancer: A Guide to Exercise for Cancer Survivors." Cancer exercise programs continue to increase in popularity as new research studies confirm the relationship between physical activity and improved quality of life for individuals diagnosed with cancer (Bulmer et al., 2012).

United States is expected to change dramatically. Over the next 45 years, which will likely span the entirety of your working career, every age category will decline *except* for people of age 65 or older (see figure 12.2). A substantial driver of this increase is the aging of the baby boom generation, which consists of the 77 million people born in the United States between 1946 and 1964. Because of its sheer number, this generation has had a huge effect on the marketplace for the past quarter century. This influence will continue for decades.

What might this "age wave" (Dychtwald, 1990) mean for the health and fitness fields? What does it mean when people say that baby boomers not only will grow older chronologically but also will "age youthfully"? Will middle age be reinvented? What will happen when future customers, clients, and patients retire at 65 years of age and still have another 25 years of living? Consider the fact that the percentage of people over age 85 is expected to increase by 30 percent from 2015 to 2030 and to increase 140 percent by 2060. These demographic shifts will surely challenge our definition of physical activity as it relates to sport, competition, adventure, movement, pleasure, and health.

The demographic patterns shaping the United States are likely to produce many job opportunities. In addition to the age-related changes, we will also be a far more diverse nation in the coming years. For instance, according to the 2010 census report, more than half of the 28-million increase in the U.S. population between 2000 and 2010 was due to an increase in the Hispanic population. However, the fastest rate of growth during that time was in the

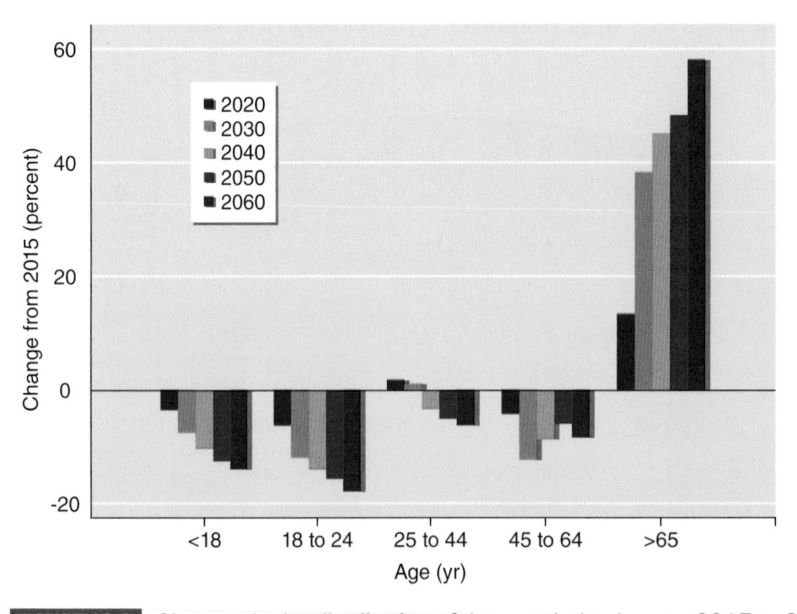

FIGURE 12.2 Changes in the distribution of the population by age, 2015 to 2060.

Data from https://www.census.gov/programs-surveys/popproj.html

Asian population, which increased by a remarkable 43 percent (Humes, Jones, & Ramirez, 2011).

The task of defining and categorizing the U.S. population into specific racial and ethnic groupings is subject to increasing methodological, empirical, and theoretical difficulty. At the same time, such trends are exerting an enormous influence on the ways in which many U.S. health and fitness businesses are developing their strategic plans for the next decade. Many settings are already taking advantage of this demographic shift in their hiring, programming, and marketing practices. As a result, new kinesiology graduates will likely be working in multigenerational and multiracial work teams as the need for new allied health professionals continues. For this reason, you are strongly encouraged to gain expertise in understanding and working with diverse populations. Doing so could entail, for example, taking academic courses related to diversity or pursuing an academic minor in gerontology.

Health Care Reform

Health care costs, which have skyrocketed over the past decade, underlie many of the decisions made by companies about their worksite health programs. As a result, these costs may affect the types of job opportunities you will have when you begin your career.

The total cost of health care in the United States is considerably higher than it is in other industrialized nations; in fact, it is unsustainable. In a publication titled *Health Care Costs: A Primer*, the Kaiser Family Foundation (2012) provided the following facts: "In 2010, the U.S. spent $2.6 trillion on health care, an average of $8,402 per person; the share of economic activity (GDP) devoted to health care has increased from 7.2 percent in 1970 to 17.9 percent in 2009 and 2010; and half of health care spending is used to treat just 5 percent of the population" (p. 1). Despite the large annual dollar amount, many people in the United States have not received the health care they need.

Clearly, health care reform has been needed, and it has needed to be sizable. Consequently, and with much fanfare and debate, then-president Barack Obama signed into law the Affordable Care Act in March 2010. The controversial law put into place comprehensive health insurance reforms, with most changes taking place by 2014. One positive consequence of the Affordable Care Act has been a decline in the number of uninsured Americans; specifically, about 16 percent of Americans were uninsured in 2011 (DeNavas-Walt, Proctor, & Smith, 2012), but this percentage had decreased to about 10 percent by 2014 (Smith & Medalia, 2015).

Even with reforms as extensive as these, perhaps the best way to control one's health care costs is to stay healthy. Of course, a good way to stay healthy is to adopt and maintain healthy lifestyle behaviors, including daily physical activity. However, both the rising costs of medical care and the manner in which health care is paid for in the United States have slowed the progress toward offering more preventive services. Health care organizations are so busy developing cost-effective ways to treat people who are sick that they seem unable to focus on developing better strategies for prevention.

Providers are often more motivated to invest money in preventive services when they are confident that they will reap the benefits in future cost savings. For example, consider a person who has a myocardial infarction (heart attack) and is treated with a percutaneous coronary intervention. This scenario is relatively common and typically results in a hospitalization cost of more than $19,000 (Afana, Brinkikji, Cloft, & Salka, 2015). If the person is employed, then a significant portion of their insurance is paid by their employer. Both the health insurance provider and the employer want to minimize the employee's health care expenses. It is becoming increasingly appreciated by most employers and insurance providers that it is cheaper to keep employees healthy than to pay for their treatment when they get sick.

Accordingly, many health plans now offer their members an extensive menu of health and fitness programs free of charge. The rationale is simple: Healthy members are less expensive than unhealthy members. In some cases, this effort pursues the goal of improving the overall health of potential subscribers in a geographic area. In other cases, it is a marketing strategy aimed at increasing the number of people who enroll in the health plan.

Today, for the first time, physical activity is increasingly seen as an integral component of the nation's health care delivery system. The American College of Sports Medicine is working with the American Medical Association to promote physical activity as another "vital sign" that physicians monitor in checkups with patients. It is possible that a perspective more focused on health promotion and wellness will dominate the health care system in the future, thus giving kinesiology graduates improved career opportunities. Perhaps the model hospital of the future will feature on-site fitness facilities and a staff of health and fitness counselors and personal trainers in addition to traditional medical personnel.

Of course, health and fitness professionals must recognize that the addition of insurance-sponsored prevention programs is not the only answer. Creative solutions require not just providing individuals with access to preventive programs and services but also facilitating their participation in these programs. Encouraging people to adhere to good diets and other healthy habits, such as adopting a physically active lifestyle, is just the beginning. The key is to develop and implement effective, outcome-driven programs.

Advice for Health and Fitness Students

Today's health and fitness professional needs to obtain a broad education across several scientific, behavioral, and humanities-based disciplines (see the job descriptions in the Skills and Competencies sidebar earlier in the chapter). To develop the necessary skills and competencies, seek a combined degree in kinesiology and health with additional courses from a wide range of areas, such as teaching; behavior change psychology; neuroscience; social and personal media; computer technology; gerontology; marketing; and the sociology of race, gender, and ethnicity. You will also benefit from course work in theoretical kinesiology, and you can round out your professional practice knowledge with other specialized courses related to health and fitness.

In addition, you need to gain practical experience in counseling clients on a variety of health and fitness topics. You should become able to relate to a diverse population of clients from different age groups and ethnic backgrounds. You can involve yourself in a health and fitness job setting through an internship, cooperative education experience, or work-study program to keep up to date on what is happening in the field.

Take special care in choosing elective courses in your degree program, and talk with your advisor for assistance in doing so. If your department does not offer a course in teaching or counseling skills, then ask what departments on campus do offer it. Alternatively, perhaps a faculty member with teaching or coaching expertise can supervise you in an independent study. You might also benefit from business and marketing classes. Marketing skills are important for health and fitness professionals in our media-saturated culture; after all, a big part of your job will be to motivate people to take part in the programs that you offer and to promote the idea of making healthy lifestyle changes. Again, if your department does not offer a specific class in health and fitness marketing, look for another course on campus that provides general marketing information that you can apply to your discipline. What will matter most when you interview for a job is that you have the knowledge and skills to do the work effectively. Do your homework early and make sure that you obtain this expertise along the way.

New graduates often show up for interviews excited and with degrees in hand but unprepared for the types of jobs available. In some cases, students interviewing for specialist positions in health and fitness report that they do not have any teaching experience because they chose an area of emphasis that deals strictly with the biophysical sphere of physical activity. Because teaching courses fall within another area of emphasis (i.e., the behavioral sphere), these students did not believe that they needed them. Learn from these mistakes—don't limit yourself to one area of emphasis at the expense of gaining skills that will help you be successful in the career you want.

Doing well academically by learning what is taught and earning good grades is important, but it is not enough. Your academic efforts will help prepare you as a professional by giving you the scientific, theoretical, and practical knowledge you need. However, when you graduate from college, you will compete for jobs with people who have similar academic credentials as well as extensive practical experience. For this reason, it is absolutely critical that you obtain practical skills as part of your education. Seek out volunteer and extracurricular activities that give you hands-on, real-world

New Technology and Innovations in Health and Fitness Products

With the advent of cloud and GPS technology, myriad products and apps have been made available to measure physiological variables, sleep, and the effects of exercise or physical activity. Consider how these products may affect the fitness and health industries now and in the future. Do the potential benefits in terms of motivation, wellness, and performance outweigh privacy issues related to storing personal information on the cloud?

experiences. If you are uncertain how to do so or where to begin, talk with your academic advisor, kinesiology faculty members, and professionals in the careers you are seeking.

In addition, read the relevant health and fitness journals and industry publications. Many of these publications are available through the associations listed at the end of this chapter. Most are also available to students at substantially reduced rates, and some are even provided free of charge to health and fitness professionals. A number of online resources are available for health and fitness professionals who wish to share knowledge and ideas. Many have a membership subscription fee but in return offer continuing education and community workshops.

Given the competitive nature of the job market, you must maximize your educational experience while in college. Work with your advisor to create a degree plan that will prepare you for the job market now and in the future. Besides taking the courses required by your major, select elective courses that prepare you for the type of job you want. Take advantage of as many internship and community service learning experiences as possible. Again, a degree in a health and fitness discipline alone will likely be insufficient to get you the job you want. In addition to having the appropriate credentials on paper, you must demonstrate that you can apply the skills you have learned.

This is a dynamic and exciting time to become a health and fitness professional. Maximizing the opportunities presented by your college course work and associated extracurricular activities will help you progress toward becoming a professional in this field. Upon graduation, if you have taken these steps and exhibit a passion for helping others become healthier, then you will be well prepared for a rewarding career.

KEY POINT

To maximize your educational experience in college, select a broad spectrum of elective courses in your degree program.

Wrap-Up

This chapter describes the health and fitness field and summarizes the types of settings and jobs that are available for health and fitness professionals. It also gives insight into current and future trends in health and fitness and describes ways in which these trends do and will affect the field. Return now to the situations described in the chapter opening. Consider how individuals in health and fitness careers can help each of the individuals described there. The world of health and fitness is ever changing, which is partly what makes this profession dynamic and exciting.

If you are considering a career in the health and fitness professions, learn as much as you can about them. It is never too early to begin to tailor your education to your personal passion. Consider industry trends as you complete this phase of what should be an ongoing educational process. This discussion about the health and fitness professions should serve only as the starting point from which you launch your process of discovery. Enjoy the journey.

MORE INFORMATION ON CAREERS IN HEALTH AND FITNESS

Organizations

American College of Sports Medicine

American Council on Exercise

American Public Health Association

IDEA Health & Fitness Association

International Health, Racquet & Sportsclub Association

Medical Fitness Association

Medical Wellness Association

National Academy of Sports Medicine

National Commission for Health Education Credentialing

National Strength and Conditioning Association
National Wellness Institute
SHAPE America (Society of Health and Physical Educators)
Society for Public Health Education
Wellness Coaches USA

Journals

ACSM's Health & Fitness Journal
American Journal of Health Behavior
American Journal of Health Education
American Journal of Health Promotion
American Journal of Public Health
Exercise and Sport Sciences Reviews
Health Education & Behavior
The Health Educator
Health Promotion Practice
International Journal of Sport Nutrition and Exercise Metabolism
Journal of Nutrition
Journal of Physical Activity and Health
Journal of Strength and Conditioning Research
Medicine & Science in Sports & Exercise
Research Quarterly for Exercise and Sport
Strength and Conditioning Journal
Translational Journal of the American College of Sports Medicine

REVIEW QUESTIONS

1. Identify program objectives and target markets for each of the following health and fitness settings: (a) worksite, (b) commercial, (c) community, (d) clinical, and (e) population.
2. Compare the sales-based and retention-based business models for commercial health and fitness facilities.
3. Describe typical job duties for the following health and fitness positions: (a) group exercise instructor, (b) fitness instructor, (c) health and fitness specialist, (d) health and wellness coach, (e) personal trainer, and (f) health and fitness director.
4. How will demographic trends in the U.S. population (e.g., aging baby boomers) affect the health and fitness professions?
5. List the types of knowledge and the skills and abilities that you need to obtain in order to be competitive in the health and fitness job market. How do certifications fit into these qualifications?
6. List past and future trends in health and fitness programs, equipment, resources, and settings.

Go online to complete the web study guide activities assigned by your instructor.

13

Careers in Therapeutic Exercise

Chad Starkey

CHAPTER OBJECTIVES

In this chapter, we will

- acquaint you with the wide range of professional opportunities in the sphere of therapeutic exercise;
- familiarize you with the purpose and types of work done by professionals in therapeutic exercise;
- inform you about the educational requirements and experiences necessary to become an active, competent professional in the area; and
- help you identify whether one of these professions fits your skills, aptitudes, and professional desires.

While going in for a lay-up, a basketball player has her legs taken out from under her. The athletic trainer arrives on the scene and finds that the player has a fractured right wrist. From here, the athletic trainer will guide the immediate care of this athlete, refer her to a physician, and develop an intervention program to safely return her to sport following initial medical treatment.

Following a stroke, a 54-year-old chief financial officer is recovering from surgery. Along his road to recovery, he will meet physical therapists, occupational therapists, and cardiovascular rehabilitation specialists. Each of these professionals will assist in restoring his function, ultimately allowing him return to his career.

Surgeons repair an infant's club foot, but the child exhibits a delay in walking upright. The physician refers the child to a physical therapist for the rehabilitation necessary to achieve normal development.

A 30-year-old man is planning a career in law enforcement. To help him meet the physical requirements—which include a timed run, a demonstration of strength, and other measures—he seeks the services of a strength and conditioning specialist.

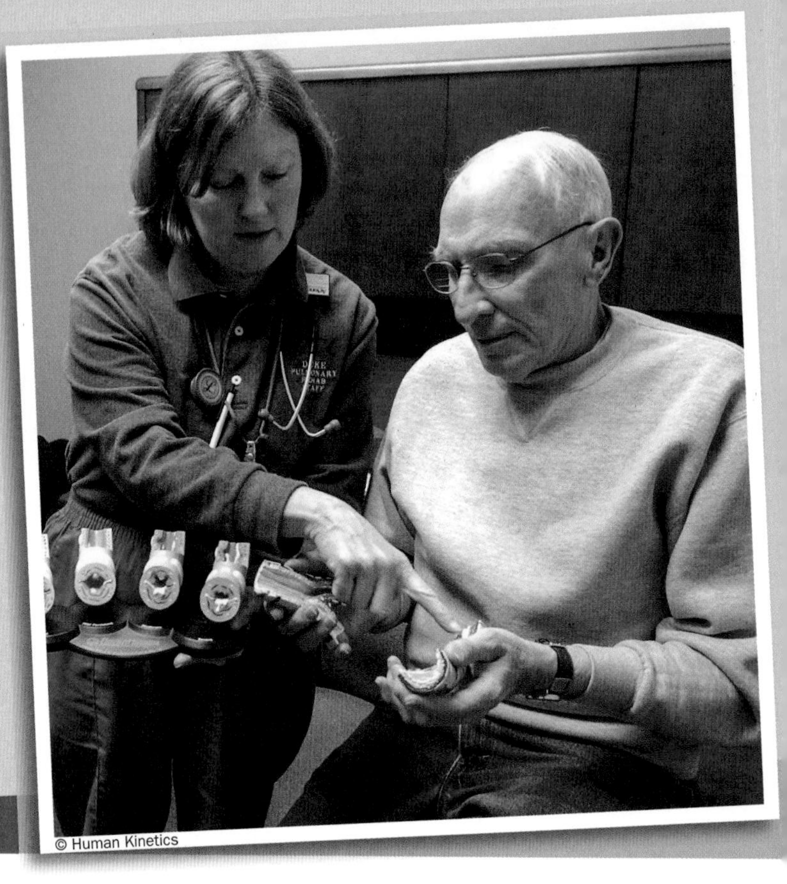

These scenarios depicts situations faced by therapeutic exercise professionals on a daily basis in work settings such as hospitals, laboratories, clinics, and sports medicine facilities. Although their patients and clients may differ, these professionals—including athletic trainers, clinical exercise physiologists, occupational therapists, physical therapists, therapeutic recreation specialists, and strength and conditioning specialists—all apply exercise and movement experiences to improve a person's physical functioning. Physical activity professionals in this sphere help people restore lost function (through rehabilitative therapeutic exercise) or acquire skills and functions considered normal or expected for advanced fitness (through habilitative therapeutic exercise).

What Is Therapeutic Exercise?

Exercise therapists design and implement movement experiences to restore or improve motor function to a level that enables people to reach personal or career goals unencumbered by physical limitations. This work requires knowledge of human anatomy, human physiology, exercise physiology, neurology, and kinesiology in structured activity programs. To develop therapeutic goals, clinicians must call on their knowledge of the effects of exercise on the muscular, nervous, skeletal, and cardiovascular systems and relate those effects to the patient's needs and expectations. Depending on the patient, the workplace, and the conditions being treated, therapeutic goals may include restoring muscular function and strength, joint range of motion, proprioception, cardiovascular and pulmonary function, or metabolic function so that the patient can participate in activities that her or she deems important.

A set of core competencies for individuals who provide patient care has been developed by the Health and Medicine Division (formerly the Institute of Medicine) of the National Academies of Sciences, Engineering, and Medicine (Greiner & Knebel, 2003). The competencies are as follows:

- **Provide patient-centered care:** As described in the next section, caregivers should respect patients' differences, their values, and their need to improve function and overall quality of life.
- **Work in interdisciplinary teams:** Patients are treated by a variety of medical and health care personnel, and this team should communicate and work in unison in order to provide continuous and reliable care.
- **Employ evidence-based practice:** Caregivers should incorporate patient care techniques that have been scientifically validated.
- **Apply quality improvement:** Caregivers should evaluate the health care delivery process to decrease errors and improve the efficiency of care.
- **Use informatics:** Caregivers should gather, process, and analyze a range of data (e.g., patient records, patient outcomes, satisfaction surveys) to support individual and operational decision-making.

You will probably be introduced to most of these core competencies during your undergraduate education. You will become more fully immersed in them when you train for and enter a therapeutic exercise profession.

Reassessing "Disability"

Historically, health care and medicine tended to focus on what patients and clients were unable to do as the result of injury, illness, or inactivity. This focus on limitations, reflected in the term *disability*, carries a stigma that emphasizes the negative aspects of the person's state of well-being. In 2002, the World Health Organization (WHO) developed a system that places the emphasis on what the patient *can* do—the **International Classification of Functioning, Disability, and Health** (ICF) (WHO, 2002). The ICF system incorporates the medical aspect of resolving pathology and the social model of reducing the condition's negative effect on the person's life (figure 13.1). This model illustrates the relationship between function and

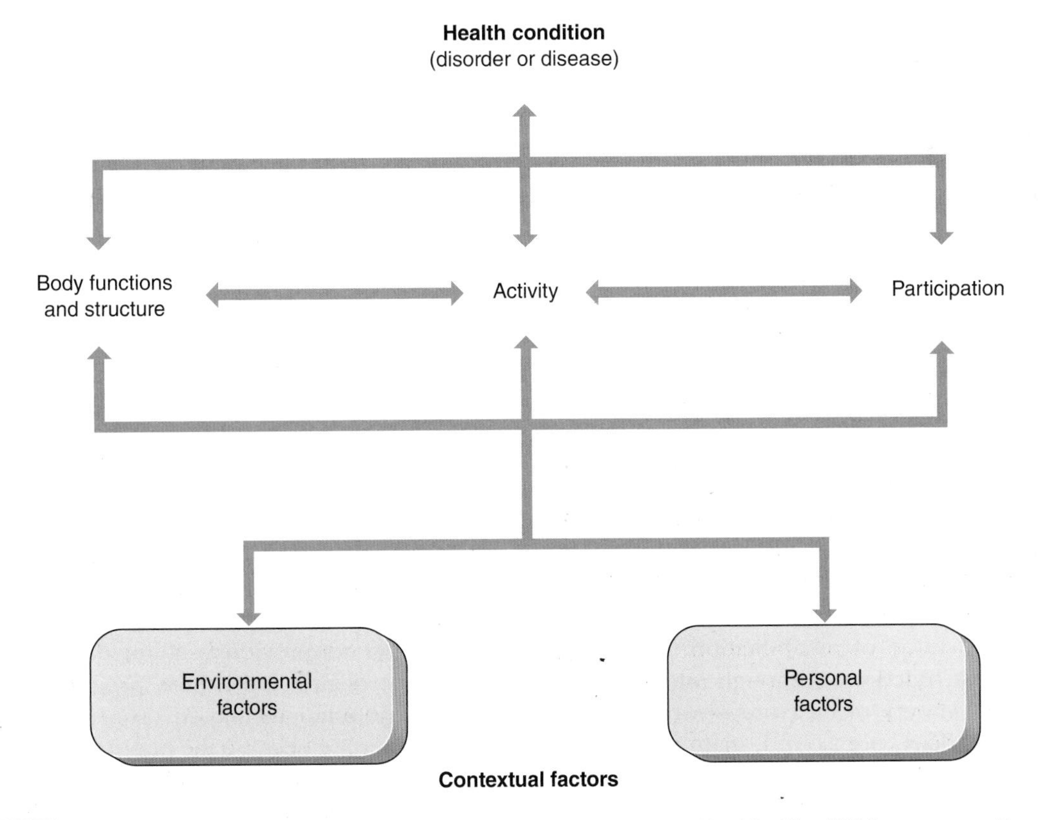

FIGURE 13.1 International Classification of Functioning, Disability, and Health. The ICF focuses on the person's level of function and identifies interventions that maximize function and lead to a more individualized course of care. Thus, activity (function) is the core of the model.

Reprinted from *Towards a common language for functioning, disability and health: The international classification of functioning, disability and health,* page 18, Copyright 2002, World Health Organization. Available: http://www.who.int/classifications/icf/training/icfbeginnersguide.pdf

disability as the result of the interactions between health conditions (injury and disease) and contextual factors. Examples of contextual factors include external environmental factors (e.g., social attitudes), physical characteristics of the places where the person lives and works, and personal factors (e.g., gender, profession, life experiences).

Therapeutic exercise specialists either help otherwise healthy people reach the level of activity they desire or need in order to function (**habilitative therapeutic exercise**) or help individuals who are ill or injured to return to their prior level of function (**rehabilitative therapeutic exercise**).

Therapeutic Exercise, Rehabilitation, and Habilitation Defined

What is therapeutic exercise? In technical terms, it is an intervention that uses systematic and scientific application of exercise and movement experiences to either *develop* or *restore* muscular strength, endurance, or flexibility; neuromuscular coordination; cardiovascular efficiency; or other health and performance factors. In practice, therapeutic exercise is programmed physical activity aimed at improving or restoring the quality of life. We draw a distinction between these two aspects of therapeutic exercise in this chapter in order to help you understand the different types of professional work associated with this area.

KEY POINT

Therapeutic exercise can be classified as either rehabilitative (restoring lost function) or habilitative (helping to acquire normal function).

Broadly speaking, **rehabilitation** describes processes and treatments (interventions) that restore skills or function that were previously acquired but have been lost because of injury, disease, or behavioral traits such as voluntary inactivity. If you have ever torn a muscle or broken a bone, then you know the value of rehabilitation in helping you regain lost functions. Although interventions can involve a variety of measures—for example, ice, heat, electricity, ultrasound, manual therapy, and psychological counseling—we focus in this chapter on interventions that rely heavily on physical activity.

Rehabilitation specialists require a thorough knowledge of the pathological aspects of injury and disease, the limitations that they impose on human performance, and the types of treatments required to meet the patient's functional needs. Because people are more than just muscles and bones, rehabilitation specialists must also take into account the psychological and social effects of the injury, the patient's personal goals and expectations, and the course of care leading to recovery. One example of rehabilitative therapeutic exercise can be found in the regimen used by an athletic trainer to restore a running back's knee so that he can return to competition.

Generally speaking, the concept of rehabilitation includes the notion of **habilitation**, or processes and treatments leading to the acquisition of skills and functions that are normal and expected for an individual based on both age and status (Olivares, Gusi, Prieto, & Hernandez-Mocholi, 2011). The standards or expectations that signal a need for habilitation may differ vastly for individuals of the same age. For example, we would expect different physical performance standards of a lawyer than we would of an athlete. A lawyer who is physically fit according to the definition presented in chapter 3 is probably not in need of habilitation; her state of fitness is desirable for good health, and she does not need special physical abilities for any rigorous occupation. In contrast, an athlete who is fit but lacks the advanced cardiovascular endurance or strength to perform the tasks needed in his sport is a candidate for habilitation involving intensive conditioning, training, and other skill-specific exercise.

In both rehabilitative and habilitative therapeutic exercise, the caregiver must consider the individual's physical status in terms of any permanent disability or impairment, such as blindness, amputation, or paralysis. For example, a paraplegic person's functional loss of use of the lower extremities would not be a cause for rehabilitation, but habilitative care would be required if the person lacks upper body strength relative to what is considered normal with a spinal cord injury. A physical therapist who is attempting to correct congenital postural problems is also practicing habilitative therapeutic exercise because it involves bringing the client to a level of functioning not previously attained.

Clearly, health and fitness programs designed for unfit populations may involve habilitative therapeutic exercise because the populations served are usually performing at levels below what is considered normal and may not have previously attained the level of health or fitness desired. Because the professional opportunities in fitness and health promotion were addressed in chapter 12, we do not further discuss them here. We do, however, consider other types of professional services that include habilitative and rehabilitative therapeutic exercise.

Each of the professions described in this chapter emphasizes the importance of preventing injury and disease. An athletic trainer, for example, helps identify musculoskeletal problems such as muscle weakness, flexibility limitations, balance disturbances, and medical conditions that could predispose an athlete to injury or even death. Similarly, a physical therapist may assess a person's posture, or evaluate a worker's technique for lifting heavy objects, in order to prevent job-related injuries. Preventive medicine is also practiced by a clinical exercise physiologist who performs electrocardiograms and exercise stress tests in order to identify people at risk of experiencing a heart attack. By developing appropriate exercise programs to remedy such problems, these specialists help reduce the likelihood of specific injuries and diseases.

KEY POINT

"An ounce of prevention is worth a pound of cure." In this spirit, both rehabilitative and habilitative therapeutic exercises focus on developing the body's systems so that injury or disease is less likely.

Although preventive therapeutic exercise programs are usually habilitative, they may sometimes be considered rehabilitative. Habilitative therapeutic exercise is preventive because bringing a person's physical state up to expected standards reduces his or her risk of injury or illness. Rehabilitative therapeutic exercise may also be preventive; specifically, by restoring a patient's physical skills and functions, rehabilitation reduces his or her risk of reinjury and of other types of injuries.

Rehabilitative Therapeutic Exercise

Physical dysfunction is characterized by inability to use one or more limbs or the torso in a manner that is meaningful and useful to the person. If you have ever broken an ankle or sprained the ligaments of your wrist, then you have experienced physical dysfunction. When an individual experiences physical dysfunction due to traumatic injury, congenital defect, or disease, a physical rehabilitation specialist can help the person regain use of the affected body part or compensate for its disability. Physical rehabilitation specialists include physical therapists, occupational therapists, and athletic trainers; they work closely with other health care professionals such as physicians, nurses, and dietitians.

KEY POINT

Restoring physical function often involves both passive intervention (e.g., heat, cold, manual therapy, electrical stimulation) and active intervention (e.g., exercises to improve strength, range of motion, and balance).

Exercise Therapy for Rehabilitation of Musculoskeletal Injuries

The muscles and the nervous system can be prevented from functioning properly by a stroke, spinal cord injury, back problem, or other type of injury, whether due to an accident in an automobile or in an industrial, sport, or home setting. Physical activity, of course, carries the risk of trauma—for example, when repetitive motions lead to neuromuscular trauma. Injury to a muscle, joint, or bone not only affects the traumatized tissues but also may lead to immobilization and decreased physical activity, which can cause more widespread problems in neuromuscular function and cardiovascular endurance. For example, when an injury inhibits a limb's functionality, the resulting disuse may affect both the limb itself and the activity of the entire body, thus perhaps degrading the efficiency of the heart and lungs. Perhaps you have experienced this effect after injuring a bone or joint; after regaining full motion you may have noticed that running up

Considering a Career in Therapeutic Exercise?

All therapeutic exercise professions discussed in this chapter must work closely with people. These professionals interact daily with a number of clients or patients and routinely discuss intervention plans with colleagues and other physical activity professionals. Do you enjoy contact with people? Do you have good communication skills? Good problem-solving skills? If you are considering a career in therapeutic exercise, what are the areas in which you might need to seek out further development?

a flight of stairs or playing basketball left you more winded than it would have before your injury.

Therefore, the goal of orthopedic rehabilitation programs is twofold: to restore symptom-free movement and to restore the function of the cardiopulmonary system. Restoring the function of a limb consists of increasing joint range of motion, increasing muscular strength and endurance, reeducating neuromuscular pathways, and building cardiovascular endurance. Clinicians involved in this type of rehabilitation use both passive and active exercise to restore limb function. The next step involves strength and muscular endurance training, which protects the limb. These exercises may be augmented by various forms of heat, cold, electrical stimulation, therapeutic ultrasound, and manual therapy. Throughout this process, the clinician attempts to maintain the patient's level of cardiovascular endurance. Most types of neuromuscular rehabilitation occur in outpatient physical therapy clinics, hospitals, and athletic training clinics.

Exercise Therapy for the Rehabilitation of Athletic Injuries At some point or another, you have either experienced or witnessed an athletic injury. Perhaps you broke a finger while participating in a sport or anxiously waited during a time-out to find out how seriously your favorite football player was injured. Almost immediately following a sport injury, the athlete's rehabilitation begins, and restoring the body part to its prior level of function is only one part of the process. Once the limb begins to regain strength, range of motion, and neuromuscular coordination, the athlete must be reintegrated into athletic activity. This sport-specific functional progression is a unique aspect of athletic rehabilitation. It requires that the traditional rehabilitation protocol be merged with the skills and tasks needed to compete in a particular sport.

Sports medicine is an aspect of therapeutic exercise that is exclusively dedicated to preventing, treating, and rehabilitating athletic injuries. It is distinguished from other aspects of therapeutic exercise in that it uses rehabilitation programs that are more aggressive than those used for the general population. The spectrum of sports medicine is broad and encompasses a range of medical and allied health professions that exceed the scope of this text. Students who are interested in sports medicine careers should explore the sections of this chapter that address athletic training and physical therapy, as well as other subdisciplines of kinesiology and medical fields that contribute to sports medicine teams.

KEY POINT

Sports medicine **is a generic term that refers to the practice of medicine, the art of rehabilitation, and the sciences that relate to preventing, treating, and rehabilitating athletic injuries (Prentice, 2015).**

No longer unique to the sports medicine setting, a new form of therapeutic exercise has emerged from this area: **prehabilitation**. Whereas rehabilitation occurs following an injury or after surgery, prehabilitation occurs following the injury but before surgery. When a physician deems that surgery is necessary for an orthopedic injury such as an ACL tear, part of the rehabilitation process must focus on restoring lost function. Prehabilitation in particular involves developing as much strength and range of motion as possible before surgery. As a result, the patient emerges from surgery having already regained some function on which to build. This approach is also used in some cases to strengthen the heart and improve the circulation of people who will undergo cardiac surgery.

Exercise Therapy for Rehabilitation of Postsurgical Trauma Although surgery is performed to restore a person's health and function, the process of surgery itself has detrimental effects on the body. The incision affects the involved muscles, nerves, and tissues, and the decreased activity associated with recovery leads to decreased muscle mass and function and decreased cardiovascular efficiency. Although this effect is most evident when surgery involves the limbs, the heart and lungs are also affected. Just as skeletal muscle wastes away when it is not used, cardiac muscle responds similarly. Therefore, in most cases, people recovering from long-term disability must participate in neuromuscular rehabilitation to restore normal cardiopulmonary function.

Exercise Therapy in Cardiopulmonary Rehabilitation

Diseases of the cardiopulmonary system include coronary artery disease, arrhythmia, hypertension, heart attack, and emphysema. Collectively, these diseases constitute the leading cause of death and long-term disability among adults in the United States. Undiagnosed heart disease is also a leading cause of death among young athletes (Maron, Haas, Ahluwalia, Murphy, & Garberich, 2016). Early identification of people at risk for these conditions, followed by intervention, reduces the likelihood

Professional Issues in Kinesiology

Overload: Selecting Progressive Training for Recovery From Injury

The overload principle in rehabilitation refers to the intensity of motion and exercise used to safely stress the body's tissues, especially the muscles. Healthy, active people require a greater amount of stress than those who are injured or who lead a sedentary life. Historically, the initial management of acute orthopedic injuries involved immobilizing the body part, a technique that caused atrophy and reduced range of motion. Recognizing that immobilization is often counterproductive, the trend now is to encourage placing gentle loads on the tissues immediately following an injury.[1] In this case, simple movement of the affected joint provides the amount of stress needed in order to produce overload. As the patient progresses through the healing phases, the amount of stress placed on the tissues is increased, thus encouraging increased strength and range of motion.

Citation style: AMA

1. Bleakley CM, Glasgow P, MacAuley DC. PRICE needs updating, should we call the POLICE? *Br J Sports Med.* 2011;1. doi:10.1136/bjsports-2011-090297

of their occurrence and increases the chance of survival following an episode.

Working closely with a physician, an exercise therapist participates in planning, implementing, and supervising physical activity programs designed to help restore individuals to normal function. Therapeutic exercise specialists must be aware of certain conditions that can make rehabilitative exercise a potentially deadly activity. Asthma, for example, can be dangerous because the lungs cannot exchange oxygen and carbon dioxide at the rate needed to support the body's metabolism during exercise. Because the risk is greatest when exercise occurs in a hot, dry environment, therapeutic exercise specialists who treat asthmatic clients prescribe regimens that the client can tolerate more easily, such as aquatic exercise routines.

KEY POINT

Because of the nature of the diseases being treated, a physician supervises most cardiopulmonary rehabilitation programs.

Cardiovascular exercises that improve the efficiency of the heart and lungs include breathing exercises, walking, running, swimming, and resistance exercises such as weight training. These exercises are prescribed at progressive levels of intensity and duration. The patient's progress is monitored by means of diagnostic tests such as stress testing, $\dot{V}O_2max$ testing, electrocardiograms, and echocardiograms, which allow caregivers to adapt the level of activity as needed. Stress tests and other diagnostic tests are usually conducted in a hospital or other medical environment. A cardiovascular exercise programs is usually implemented in a clinic, health club, or wellness center; the patient is also given an independent program prescription for home use. For a high-risk patient in a **cardiopulmonary rehabilitation** program, a licensed physician must supervise all diagnostic tests and exercise prescriptions.

Exercise Therapy for Rehabilitation of Older Populations

Humans are living longer, and the percentage of the population aged 65 or over continues to increase. As people age, the natural decrease in strength and flexibility and the increase in age-related skeletal diseases (e.g., arthritis) often hinder the quality of life. As we saw in chapter 2, the loss of ability to perform activities of daily living can exert profound physical and emotional effects on people's lives. In addition, as a person enters the later years, the body's aerobic ability decreases and the percentage of body fat increases, thus reducing one's capacity to do physical work.

Various forms of therapeutic exercise designed to increase strength, slow the loss of flexibility, and improve cardiovascular condition can delay the onset of aging-related problems and minimize their effects, thereby helping seniors regain or maintain independent, active living and improve cognitive function (Laitman & John, 2015). Programs that involve physical activity can decrease blood pressure, control cholesterol, and reduce the risks of stroke and heart attack; they may also lower the risk of certain types of cancer.

At the same time, exercise in the older population is often a double-edged sword. Although physical activity is required in order to maintain an appropriate level of fitness, the body is less capable of withstanding the forces of exercise. Therefore, clinicians working with clients who have, for example, arthritis, must remember that their joints cannot tolerate the physical stresses associated with extensive walking, jogging, or running. To account for this reality, these clinicians must modify exercise programs to minimize the amount of force exerted on the joints. In such cases, water aerobics can be used to maintain cardiovascular fitness without further injuring the joints.

Exercise Therapy for Rehabilitation of Psychological Disorders

The physical benefits of exercise should now be apparent to you. However, one aspect of physical activity is often overlooked—namely, its psychological and emotional benefits. Exercise is a wonderful way to reduce stress and unwind from the toils of life. People who engage in regular physical activity tend to sleep better and suffer fewer emotional disorders than those who do not engage in such activity. In fact, 20 to 30 minutes of aerobic exercise can produce changes that are on a par with standard forms of psychotherapy (Sims, Hill, Davidson, Gunn, & Huang, 2006). In the healthy population, aerobic exercise can prevent the onset of some types of psychological maladies; for people who have such conditions, exercise can often be considered a form of treatment.

KEY POINT

Exercise is as important for the mind as it is for the body. Improving one's physical health through exercise also improves one's mental state.

Therapeutic interventions can take place in many settings with varied populations.

Kinesiology Colleagues

Dustin Grooms

Dustin Grooms had the opportunity to serve as an intern athletic trainer for a professional football team. During this experience, he assisted with the rehabilitation of a quarterback who had sustained a torn anterior cruciate ligament (ACL). He noticed that although the intervention focused on the knee, subtle differences appeared in how the patient moved his leg. He began to wonder if injury to the ACL affects how the brain processes information to and from the knee.

While earning his doctorate in neuroscience, Grooms worked with colleagues to study how the brain processes information from the knee in people who have intact ACLs versus those who have torn their ACL. Using a special form of magnetic resonance imaging (MRI), the team identified definite changes in how the brain processes information following an ACL tear. The results of this research will improve the rehabilitation (and subsequent function) for people who suffer ACL tears by reeducating and remolding their brains to function as they did prior to injury. This strategy may also be an important step toward preventing ACL injuries.

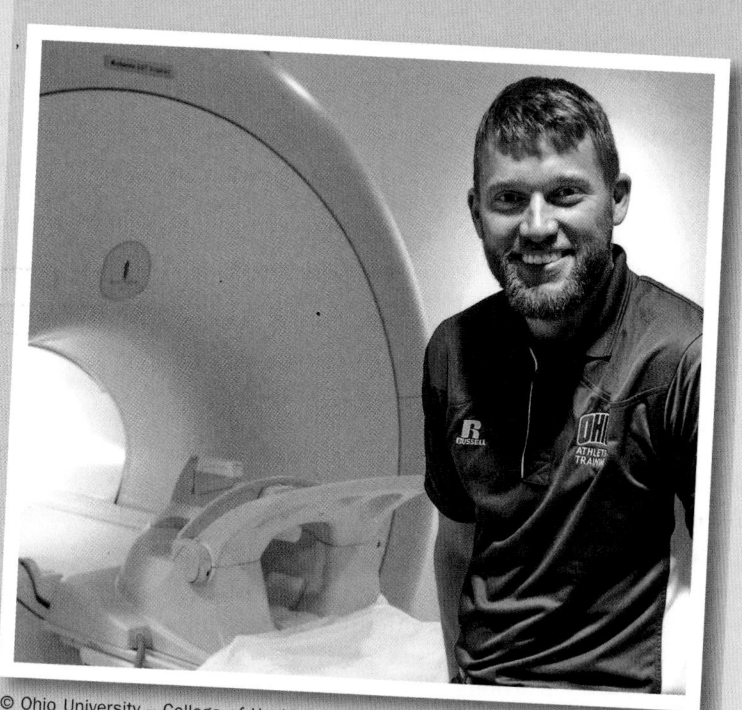

© Ohio University – College of Health Sciences and Professions/Lauren Dickey with permission

Providers of mental health care often prescribe exercise regimens as part of their patients' therapy programs. The older population referred to in the preceding section can also gain psychological benefits from exercise. Elderly people who engage in physical activity exhibit decreased signs of depression, enhanced self-image, better physical health, and improved morale, all of which lead to better quality of life (Knapen, Vancampfort, Moriën, & Marchal, 2014). Exercise is also included in many smoking and alcohol cessation programs, although scientific data have yet to show that this is effective (Ussher, Taylor, & Faulkner, 2012).

Habilitative Therapeutic Exercise

Habilitative exercises help bring people in line with established physical standards such as those presented in chapter 3. The goal of this type of therapeutic exercise is to help people reach the expected level of physical fitness for their demographic classification. Throughout this text, we have described the relationship of physical activity to health, as well as the kinesiologist's role in improving health through physical activity. Many of the health and fitness professions described in chapter 12 also meet the definition of habilitative therapeutic exercise. The following sections describe some of the areas in which therapeutic exercise is used for habilitative purposes.

Exercise Therapy for Specialized Habilitation

Although not recognized as such, specialized habilitation is probably the single largest role of habilitative therapeutic exercise. Specialized habilitation involves bringing specific groups of people in line with standards that exceed rather than merely meet those of the general population. These standards might include, for instance, running a 5-minute mile, having a body mass index within a certain range, or being able to lift a certain amount of weight. Specialized habilitative therapeutic exercise takes place in settings such as preseason sport training camps, military boot camps, and police and firefighter academies.

Exercise specialists who work in this area must fully understand the muscular and cardiovascular capabilities needed by people in these special groups. Often, the exercise specialist has a professional background in the specialty area. For example, a former baseball player with a degree in kinesiology might specialize in organizing habilitative exercise regimens for spring training camps. Such experience can help keep the exercise program in line with physical expectations for the players involved.

Exercise Therapy for Habilitation of Obese Populations

Physical fitness, exercise, and weight loss are controversial American obsessions. Many people overemphasize weight loss, and others ignore all aspects of healthy eating and exercise habits. Both approaches are hazardous and potentially deadly. Fad dieting, exercise addiction, and distorted body image have reached almost epidemic proportions among females of high school and college age. Medical conditions associated with such obsessions are increasing at alarming rates; this is especially true for anorexia and bulimia (Erskine, Whiteford, & Pike 2016). At the other end of the spectrum, the overweight, sedentary portion of the U.S. population practices lifestyles that predispose individuals to cardiovascular disease and diabetes. Meanwhile, the mass media bombard us with conflicting and confusing information about which diets do—and do not—work. These contradictions leave many Americans unclear about what is meant by healthy eating.

Therapeutic exercise that is used to control weight and body mass is considered habilitative because of its role in bringing people in line with established standards. Promoting a healthy lifestyle and decreasing the incidence of disease also requires proper nutritional and exercise counseling. Therefore, specialists such as exercise physiologists, strength and conditioning specialists, and personal trainers work in cooperation with physicians, registered dieticians, counselors, and other professionals to develop and implement programs that help people return to normal, functional, healthy lifestyles.

Exercise Therapy for Habilitation of Children With Developmental Problems

In the not-too-distant past, children born with certain physical conditions faced life with functional limitations and participation restrictions. Now, however, early identification of specific conditions can be followed by appropriate intervention to provide countless people with the opportunity to have a more active life.

The goal of this type of habilitative exercise is to help the child adapt to, or compensate for, functional anatomical and physiological deficits. In certain cases, the underlying condition may have been surgically corrected. Other cases may involve teaching the child how to use a prosthetic device or how to perform basic skills such as rolling over, walking, or eating. For example, exercise therapy may be used to strengthen the muscles in children with cerebral palsy, thereby improving their walking performance.

Working with children provides therapists with a unique set of rewards and presents them with a unique set of challenges. Issues unique to this population include communication between therapist and child, home care problems, and schooling issues. Regardless of the challenges facing the therapist, the ultimate goal of habilitative regimens for this population is to promote physical, social, and cognitive development at a rate near the norm for the child's age group.

Exercise Therapy for Habilitation for General Fitness

With all the benefits of exercise described in this text, a question logically arises: Why doesn't everyone participate in exercise? Although more people are now regularly exercising than ever before, most of the U.S. population lives in a sedentary manner. Although no single factor is fully to blame for this lack of exercise, contributors include the longtime popularity of television and the recent growth in computer use.

KEY POINT

During the industrial age, a larger proportion of the population engaged daily in strenuous physical activity. Today's information age has created a relatively sedentary population. Even casual exercise can offer a more balanced lifestyle.

Research has shown that exercise can help prevent disease and enable people to reach their maximum life expectancy. Therapeutic exercise professionals play an important role in helping the general population achieve healthy standards of physical fitness by introducing them to exercise through hospital-based fitness centers, commercial gymnasiums and fitness centers, employer-based wellness centers, and community-based nonprofit agencies.

Settings for Therapeutic Exercise

Therapeutic exercise professionals work in a variety of settings, including inpatient facilities (e.g., hospitals, residential facilities), outpatient clinics (e.g., physical therapy offices, athletic training clinics), and even clients' homes. Some are employed by other people, and some run their own private practice. Although the decision to pursue a given profession in the sphere of therapeutic exercise may dictate the setting in which you work, you may still have opportunities to work with professionals from a wide range of work settings. Most health care facilities use a team approach in which representatives from several professions collaborate to plan and deliver patient care.

Inpatient Facilities

Rehabilitation hospitals provide specialized care aimed at returning individuals to their maximum level of function. People who have a severe condition (e.g., brain or spinal cord trauma) or are recuperating from severe disease may require a long-term stay in a rehabilitation facility. **Custodial care facilities** such as nursing homes provide services to help with activities of daily living and to meet people's specific medical needs. In these two types of **inpatient facilities**, the level of a patient's function usually dictates the degree of coordinated effort that must be arranged by physicians, rehabilitation specialists, and social services personnel.

In addition to having the other skills described in this chapter, therapeutic exercise professionals who work at inpatient facilities must often possess knowledge of ambulation and patient transfer techniques, prevention of bedsores, and other needs unique to the long-term care of sedentary or immobile patients.

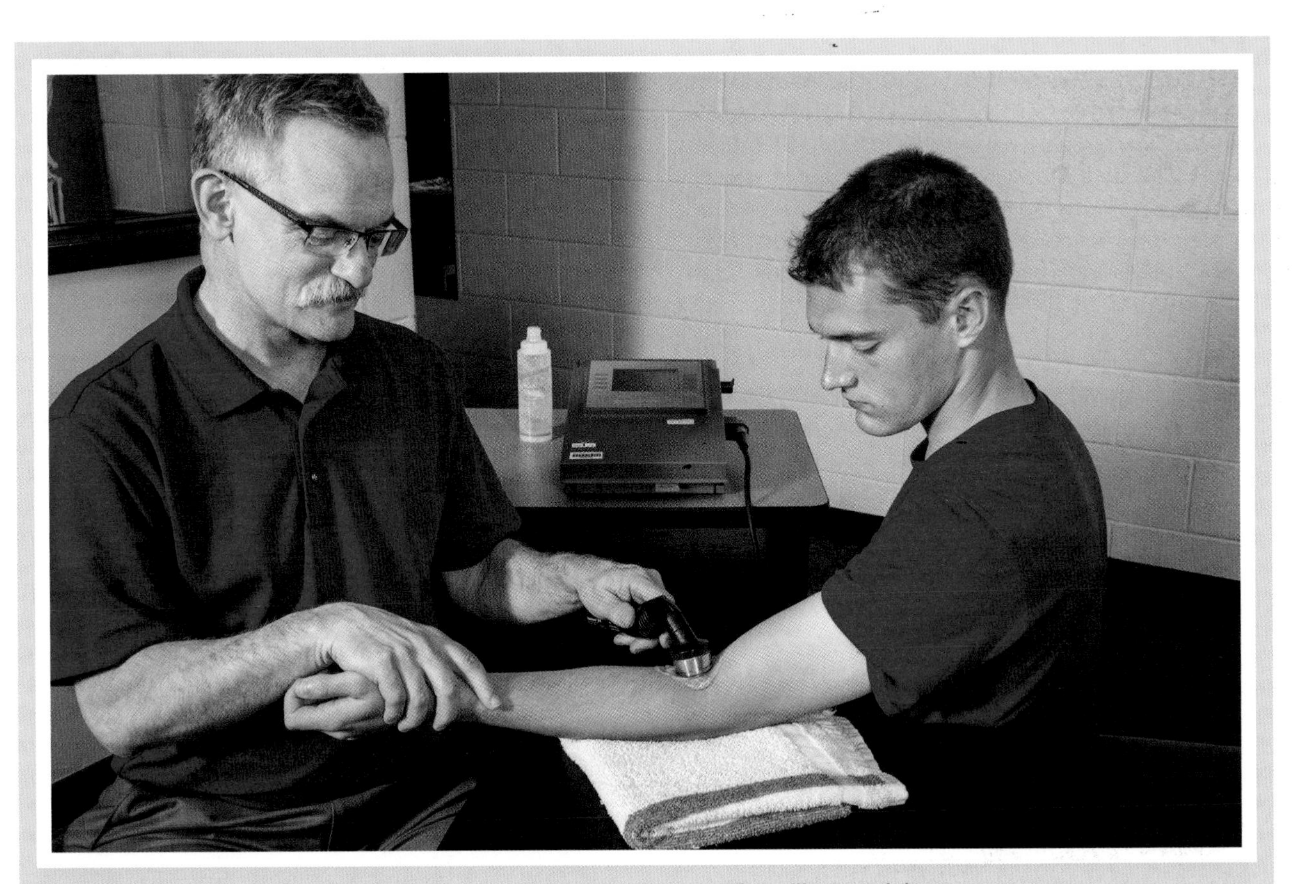

Therapeutic exercise professionals provide care for people recovering from illness or injury.

Outpatient Clinic Settings

Outpatient clinic settings are characterized by short-term patient visits (patients do not stay overnight). These types of facilities, which comprise the most diverse settings for therapeutic exercise professionals, include physical therapy clinics, sports medicine clinics, and cardiac rehabilitation facilities. Outpatient facilities may have a wide range of specialized equipment to meet the needs of their clientele—the more diverse the range of people being treated, the broader the range of equipment. Most outpatient clinics consist of examination areas, specialized treatment and rehabilitation areas, hydrotherapy pools, and open-space exercise areas. Some outpatient clinics are specialized and accept only certain patients—for example, people with cardiovascular disease, hand injuries, or spinal conditions.

People may be hindered from traveling regularly for treatment at an **outpatient facility** by social, economic, or other factors. As a result, many outpatient clinics send therapists to patients' homes to provide both rehabilitative and habilitative home care. The quality and quantity of home health care can be improved through the use of technology in the form of apps and web portals (Lee & Billings, 2016).

Sport Team Settings

Athletic training clinics constitute a specific type of outpatient facility; they are typically located in high schools, colleges, and professional team settings. The first athletic "training rooms," as it were, consisted simply of a table and a whirlpool tank located in the corner of the locker room. Today, although the size and complexity may vary, athletic training facilities at some high schools and at major colleges and universities are state-of-the-art health care clinics.

Facilities typically include treatment and rehabilitation areas, examination rooms, hydrotherapy areas, and space dedicated to injury prevention methods such as taping, wrapping, and bracing. Most professional sports medicine facilities, and many of those in colleges and universities, also have X-ray rooms and space where minor medical procedures can be performed. Many sport team settings also offer weight rooms, cardiovascular fitness centers, nutrition centers, swimming pools, and exercise science laboratories. These types of facilities expand the scope and breadth of the therapeutic exercise programs that can be offered to habilitate and rehabilitate athletes.

Private Practice

Many people are intrigued by the thought of being their own boss. **Private practice** is an entrepreneurial venture in which a professional establishes his or her own place of work. In order to launch a practice, the professional needs a source of funding (e.g., business loan, personal wealth, investors) because physical therapy and occupational therapy are capital-intensive professions that require expensive equipment and a lease on a physical facility. Other expenses that must be considered include utilities and support staff.

Most professionals do not begin their careers in their own practice; instead, they usually make the transition into private practice after working at an established clinical practice. This approach allows beginning professionals to learn about the nuances of the business world and build a patient base and a professional reputation.

Identifying Whom You Would Like to Work With

People exploring a career in therapeutic exercise often first identify the population with whom they wish to work. Take a moment to think about your favorite populations. Put a check mark by any of the following groups that seem appealing to you.

____ Infants
____ Children
____ Adults
____ Senior citizens
____ People with injuries
____ People with limited function
____ Athletes or other physically active people
____ Nonathletes
____ Healthy people
____ Recovering patients
____ People requiring long-term care
____ Other: ____________________

Roles for Therapeutic Exercise Professionals

Because kinesiology students develop a background in the anatomical and physical principles of human movement, they are well prepared to enter health care professions that use therapeutic exercise as part of habilitation or rehabilitation treatment regimens. These professions include athletic training, cardiac rehabilitation, occupational therapy, physical therapy, and strength and conditioning. This section describes these professions, the settings in which they take place, and the educational requirements and credentials needed to practice them.

The professional roles discussed here often overlap. Indeed, many people pursue multiple credentials, such as athletic training and physical therapy, or physical therapy and strength and conditioning (table 13.1). The need for multiskilled and multicredentialed individuals will increase as competition in the health care industry increases and the health care job market tightens.

Many therapeutic exercise professions, such as physical therapy and occupational therapy, require a postbaccalaureate education to enter the profession. Physical therapists in the United States, for instance, must graduate from a clinical doctorate in physical therapy (DPT) degree program. In another example, as of 2022, the professional athletic training degree will be at the master's level. In contrast, the entry-level education requirement for physical therapy assistants and occupational therapy assistants is currently an associate's degree.

An undergraduate education in kinesiology should provide you with comprehensive knowledge of human movement, the effects of exercise, and normal anatomical and physiological function. An undergraduate kinesiology degree is viewed as an ideal background for advanced studies in many graduate-level therapeutic exercise programs. You should always check the current educational requirements for any profession that you intend to pursue and use elective courses to fulfill any specific programmatic requirements.

The educational requirements of these professions vary depending on their focus, but most have the common traits of a strong science base and an active clinical education component. You will also benefit from course work in theoretical kinesiology (covered in part II of this text), and you can round out your professional knowledge by taking other

Table 13.1 Skill Matrix

	Athletic trainer	Clinical exercise physiologist	Occupational therapist	Physical therapist	Therapeutic recreation specialist	Strength and conditioning specialist
Habilitation						
Fitness assessment		X				X
Analysis of function	X	X	X	X	X	X
Neuromuscular conditioning	X	X	X	X		X
Diagnostic skill	X		X	X		
Rehabilitation						
Developing intervention programs	X	X	X	X	X	
Restoring range of motion	X		X	X		
Restoring strength	X	X	X	X		X
Improving cardiovascular function	X	X	X	X		X
Restoring activities of daily living	X	X	X	X	X	X
Restoring work or sport function	X	X	X	X	X	X

specialized courses and clinical practicums. Your background in kinesiology will provide you with a well-rounded education including many of the skills that will be expected of tomorrow's health care professionals.

The therapeutic exercise professions are characterized by differing levels of regulation and required preparation. Therefore, health care professionals must pay close attention to state licensure requirements, especially when changing location. To explore these professions in more detail, see the resources listed at the end of this chapter, which include key journals and professional organizations.

Athletic Trainer

High school, college, and professional athletes become injured at staggering rates; for example, as many as one-third of the 1 million high school football players in the United States suffer an injury each year (King et al., 2015). Combining the excitement of athletics with the demands of health care, athletic training is the only profession that falls entirely within the realm of sports medicine. The title *athletic trainer* can be misleading. These professionals are not coaches, personal trainers, or general performance-improving personnel. Rather, athletic training is a health care profession that addresses prevention, risk management, clinical diagnosis, immediate care, treatment, and rehabilitation of injuries and other conditions experienced by athletes and other physically active individuals. An athletic trainer (AT) works under the direction of a licensed medical (MD) or osteopathic (DO) physician. Besides attending to the direct health care of athletes and others engaged in physical activity, an athletic trainer coordinates referrals to appropriate medical and health care specialists.

Employment Settings

Traditionally, athletic trainers have been employed by high schools, colleges and universities, and professional sport teams. Now, however, new roles are emerging for athletic trainers in hospitals, sports medicine clinics, military clinics, industrial rehabilitation clinics, and other allied medical environments. Thus, an athletic trainer's roles and responsibilities vary depending on the work setting.

States are gradually beginning to mandate athletic training coverage of high-risk sports at the high school level. In addition, a trend toward hiring full-time athletic trainers is increasing, although many secondary schools that use the services of athletic trainers do so by contracting them through local sports medicine clinics and hospitals. Colleges and professional teams hire their own full-time athletic trainers. Athletic training is now practiced in this traditional athletic stetting and in other clinical settings.

› **Athletic setting:** Athletic trainers are uniquely positioned to implement procedures for preventing or decreasing the occurrence of sport injuries. They have the opportunity to work with the athlete before an injury, examine the injury to form a clinical diagnosis, and then work with the patient during the rehabilitation process.

› **Clinical setting:** Community recreation leagues, road races, and personal workout regimens have exposed millions of Americans to the possibility of incurring athletic injuries. Add the fact that many high school athletes do not receive the direct services of a full-time athletic trainer, and you can understand the increase in the number of sports medicine and physical therapy clinics. The emergence of athletic trainers in these settings has spurred growth in other rehabilitation settings, including corporate and industrial clinics, hospitals, and private practice. Athletic trainers' roles in these emerging settings may differ from those of athletic trainers in more traditional settings. These roles may also vary across states.

Education and Credentials

As of 2022, athletic trainers seeking national certification will be required to graduate from an entry-level master's degree program accredited by the Commission on Accreditation of Athletic Training Education (CAATE). Academic course work is supplemented by clinical education that permits students to affiliate with athletic training clinics at high schools, colleges, and professional sport teams and gives them the opportunity to learn in private clinics and hospitals. Many athletic trainers also go on to pursue postprofessional education in the field through additional graduate programs or training in additional clinical treatments.

One ideal way to meet the prerequisite course requirements for an entry-level master's degree program is to obtain an undergraduate degree in kinesiology or a related field; such a degree also gives you exposure to the required foundational knowledge.

National certification testing of athletic trainers is conducted by the Board of Certification (BOC). BOC certification is required to work as an athletic trainer at a major college or university, with a professional sport team, or with the U.S. Olympic

Athletic Training: Prerequisite Courses and Foundational Knowledge

Prerequisite Courses

- Anatomy
- Biology
- Chemistry
- Physics
- Physiology
- Psychology

Foundational Knowledge

- Biomechanics and pathomechanics
- Epidemiology
- Exercise physiology
- Human anatomy
- Pathophysiology
- Public health
- Research design
- Statistics

Committee. It is also required by 49 U.S. states as a prerequisite for practicing athletic training (Board of Certification, 2016).

In Canada, athletic trainers (referred to as *athletic therapists*) must complete an academic program accredited by the Canadian Athletic Therapists Association (CATA), as well as 1,200 hours of clinical experience under the supervision of a certified athletic therapist, or CAT(C). CATA and BOC have a reciprocity agreement for certification between the United States and Canada.

Employment Opportunities

The U.S. Bureau of Labor Statistics (BLS) projects that jobs for athletic trainers will grow at a much faster rate than for other professions through 2024 (U.S. BLS, 2016a). Part of the reason is the fact that school districts are replacing part-time athletic training positions with full-time positions and adding more athletic trainers to existing sports medicine departments in order to follow a collegiate model of athletic health care. Employment opportunities for athletic trainers may also increase at hospitals, physical therapy clinics, and sports medicine clinics, where athletic trainers work as part of a rehabilitation team. Additional job growth is predicted in military and industrial settings (where athletic trainers care for line employees) and in physician practices and hospitals (where they extend the functions of a physician). The average salary in 2014 for an athletic trainer who held a master's degree was $53,162 (National Athletic Trainers' Association, 2015).

Clinical Exercise Physiologist

Exercise supervised by a certified exercise physiologist (ACSM EP-C) is an effective treatment for people with chronic cardiovascular, pulmonary, and metabolic diseases. Certified exercise physiologists administer and interpret fitness assessments and from those data develop appropriate exercise programs. The ACSM EP-C administer cardiopulmonary exercise tests in hospital laboratories, administer fitness testing, and implement and deliver cardiovascular conditioning programs for individuals who are apparently disease free or those with medically controlled diseases such as diabetes.

The next level of exercise physiologist is the certified clinical exercise physiologist (ACSM CEP). Exercise testing such as graded exercise tests and exercise prescription and fitness counseling for those individuals with cardiovascular, respiratory, or metabolic disorders all fall in the ACSM CEP's domain.

The registered clinical exercise physiologist (ACSM RCEP) is the highest level of certification for exercise physiologists. A hospital may choose to offer a cardiac rehabilitation exercise program for people recovering from heart surgery. The registered clinical exercise physiologist and the physician, working as a team, collect the client's medical history and evaluate her cardiopulmonary function. Using these data, the two clinicians develop a structured exercise program, whereupon the client works with an exercise specialist to progress through the program. The patient's progress is evaluated at regular intervals by the program director, physician, and exercise specialist. This type of exercise physiologist is also responsible for administering the rehabilitation center and educating the cardiac rehabilitation staff and is often engaged in research.

KEY POINT

Credentials similar to the American College of Sports Medicine's ES and RCEP certifications are available to people who want to work in the fitness setting, such as health and fitness instructors.

A hospital may choose to offer a cardiac rehabilitation exercise program for people recovering from heart surgery. The Registered Clinical Exercise Physiologist and physician, working as a team, collect the client's medical history and evaluate his or her cardiopulmonary function. Using these data, the two clinicians develop a structured exercise program, whereupon the client works with an exercise specialist (ES) to progress through the program. The patient's progress is evaluated at regular intervals by the program director, physician, and exercise specialist.

Employment Settings

Clinical exercise physiologists and exercise specialists are employed in hospitals, specialty clinics, and health and fitness centers. Most ACSM RCEPs work in rehabilitation centers where patients are recovering from injury, disease, or surgery. Others work in stand-alone cardiac rehabilitation centers where specialized exercise equipment and rooms are available in an outpatient facility. Some hold exercise classes at YMCAs, schools, or other external facilities. Generally, clinical exercise physiologists work closely with nurses, nurse practitioners, physicians, and physician assistants. Exercise specialists play an important role in rehabilitation programs and may also work in prevention programs. In 2016, the average salary for an exercise physiologist was $49,135 (Salary.com, 2016).

Education and Credentials

Typical course work required of people pursuing a career in clinical exercise physiology includes anatomy, physiology, nutrition, biomechanics, exercise testing and prescription, and cardiac testing and prescription. Students should also pursue clinical rotations and fieldwork to supplement their learning experiences.

The American College of Sports Medicine offers certification for exercise physiologists, clinical exercise physiologists, and registered clinical exercise physiologists. Certification as an exercise physiologist requires a bachelor's degree in exercise science, exercise physiology, or kinesiology and must currently be adult CPR/AED certified. Clinical exercise physiologist certification requires the same education as the exercise physiologist certificate, but it also includes 400 hours (accredited program) or 500 hours (nonaccredited program) of practical (clinical) experience and certification in basic life support or CPR for the professional rescuer. To become a registered clinical exercise physiologist, a master's or doctoral degree in physiology is required along with 600 hours of clinical experience and certification in basic life support or CPR for the professional rescuer. Please refer to the ACSM's website for the current requirements.

Employment Opportunities

The U.S. Bureau of Labor Statistics does not collect specific data on exercise physiologists, but the aging of the U.S. population will continue to create demand for those involved in cardiovascular rehabilitation. The bureau (U.S. BLS, 2016b) projects a 24 percent increase in the demand for cardiac rehabilitation specialists through 2018. One's employment opportunities are increased when status as a Registered Clinical Exercise Physiologist is combined with another degree or credential—for example, in nursing.

Occupational Therapist

An occupational therapist (OT) helps people with physical, emotional, or mental disability to restore or develop as much independence as possible in daily living and work throughout their lives (the term *occupational* has roots in the word *activity*). Some of the physical care rendered by occupational therapists closely mirrors some of the rehabilitative exercises used by physical therapists. Indeed, in many rehabilitation centers, a physical therapist and an occupational therapist work together on a single patient's case.

As with many health professions, the roots of occupational therapy can be traced to postwar rehabilitation of veterans; specifically, pioneers in the profession taught craft skills to soldiers with disability (Ambrosi & Barker-Schwartz, 1995). Today, occupational therapists specialize in functional bracing and the modification of everyday items for the special needs of their patients. Functional bracing is the use of a supportive or assistive device that allows a joint to function despite anatomical or biomechanical limitations. Examples include knee braces and splints of various kinds, such as wrist and hand splints that allow people to eat with a fork. Occupational specialists also work with people to improve their concentration, motor skills, and problem-solving ability.

Occupational therapists may specialize in task-specific, work-related rehabilitation in order to help clients reacquire the motor and cognitive skills needed to return to work. This type of activity may range from teaching basic skills such as coordinated movement to specific skills such as hammering, typing, or driving a car. To assist people with disability, an occupational therapist may be called on to evaluate the layout of schools, homes, and workplaces in order to suggest methods of eliminating functional barriers. An occupational therapist often employs a Certified Occupational Therapy Assistant (COTA) as a clinical assistant in carrying out rehabilitation plans.

Employment Settings

Occupational therapists and occupational therapy assistants (OTAs) work in hospitals, rehabilitation centers, nursing homes, and orthopedic clinics and provide outpatient service in secondary schools and colleges. In 2015, an estimated 114,600 occupational therapists were in the U.S. work force (U.S. BLS, 2016a, 2016c). They instruct people with disabilities (e.g., muscular dystrophy, cerebral palsy, spinal cord injury) in the use of adaptive equipment such as wheelchairs, walkers, and aids for eating and dressing. They may also design special equipment for home or work use and cooperate with employers or supervisors to modify the work environment for a patient.

Schools often hire occupational therapists to work with a particular age group or a group with a particular disability. Occupational therapists may evaluate a child's abilities, modify equipment, and help the child participate fully in school programs and activities. Other occupational therapists work in mental health settings or drug or alcohol rehabilitation settings. These individuals work in large rehabilitation centers in cooperation with physical therapists or provide home health care services.

Education and Credentials

Occupational therapists generally attain their education at the master's degree level, and doctoral-level programs are also an option. Certification or licensure as an occupational therapy assistant requires successful completion of a 2-year program.

Both occupational therapists and occupational therapy assistants must be licensed in the states where they practice. The basic requirement for licensure is to complete an accredited program and pass the National Board for Certification in Occupational Therapy examination. Those who pass the examination earn the title of Occupational Therapist Registered (OTR) or Certified Occupational Therapist Assistant (COTA).

Professional reciprocity exists between the United States and Canada for occupational therapy. Graduates from Canadian-accredited occupational therapy programs are eligible to sit for the U.S. examination; students in the province of Quebec must also complete three Test of English as a Foreign Language (TOEFL) examinations. Graduates of accredited programs in the United States are eligible to practice in Canada.

Employment Opportunities

According to the U.S. Bureau of Labor Statistics, the employment picture for occupational therapists looks promising through 2024 due to an expected 27 percent increase in employment opportunities (U.S. BLS, 2016c). Demand for occupational therapist services will increase as the baby boom generation moves into middle age, during which the incidence of heart attack and stroke tends to increase. In addition, the steady growth in the population over age 75 will ensure that demand for therapeutic services remains high. Thus, hospitals will continue to employ a large number of occupational therapists, including in their outpatient rehabilitation clinics.

The market for school occupational therapists also looks promising as the population of schoolchildren increases and children with disability continue to receive specialized attention. The median annual income of occupational therapists in 2015 was $80,150 (U.S. BLS, 2016c).

Physical Therapist

The physical therapist (PT) is educated to provide rehabilitative care to a diverse patient population with a wide range of injuries, illnesses, and diseases. Patients include accident victims; people with low-back pain, arthritis, or head injury; and those with a congenital or acquired disease state such as cancer. A physical therapist may treat individuals ranging from infants to seniors with conditions such as joint injury, burns, cardiovascular disease, and neurological deficits. In addition, physical therapists are often in daily contact with other health professionals, such as nurses and physicians.

KEY POINT

Physical therapists administer the patient's program and perform the required functional evaluations. Physical therapy assistants function under a physical therapist's supervision and assume much of the hands-on patient care.

Therapists work approximately 40 hours per week, including some evenings and weekends. Physical therapists combine diagnostic tests with passive interventions, active interventions, and manual techniques such as joint mobilization to rehabilitate or habilitate their clients. The therapy used depends on the condition being treated. Therapy for orthopedic injuries relies heavily on resistance training and proprioception activities (i.e., activities that improve knowledge of the position, weight, and resistance of objects in relation to the body). In contrast, therapy designed to treat people with disease states and neurological conditions tends to emphasize cardiovascular aspects and neuromuscular control.

Entry-level education prepares students to enter the workforce as generalists, but once a person begins to practice professionally, the tendency is to begin to specialize in one of seven areas (see the sidebar titled Physical Therapy Specialty Certifications).

The practice of physical therapy has been influenced by changes in the health care system, such as managed care and the limitation of reimbursement. Specifically, more of the responsibility for patient treatment (i.e., the use of therapeutic modalities and therapeutic exercise) is moving to the physical therapy assistant (PTA), whereas the physical therapist performs patient evaluations and assumes a broader administrative role.

Employment Settings

The employment settings for physical therapists and physical therapy assistants are as diverse as their patient base. Because they tend to specialize in their practice, physical therapists may work in specialized private clinics or serve specialized (e.g., pediatric, neurological) roles in hospitals, although some professionals still fill the generalist role.

About two-thirds of physical therapists work in a hospital, skilled nursing facility, health and wellness center, or clinic. Physical therapists may also be in private practice or provide home care services. In addition, they may be employed by school systems, colleges, and universities to provide on-site services to students. Some physical therapists hold more than one job—one in private practice and one in a health care facility.

Education and Credentials

The wide range of potential patients and the broad scope of conditions that physical therapists treat require a vigorous and demanding academic program. The program typically includes clinical affiliations that expose the student to several work settings and patient types. Entry-level physical therapy education takes place at the master's degree level, but virtually all entry-level programs are clinical doctorates. Physical therapy programs are accredited by the Commission on Accreditation in Physical Therapy Education (CAPTE).

KEY POINT

Physical therapists are educated as generalists but tend to develop specialties while in the workforce.

The national professional organization for physical therapists is the American Physical Therapy Association (APTA). PTs are licensed to practice on a state-by-state basis, and the APTA offers voluntary certification for each specialty area. Licensed professionals may also seek advanced education in physical therapy via residency or PhD programs. Physical therapists who specialize in sports medicine have a strong orthopedic interest and may choose to pursue specialty certification in orthopedic or sport physical therapy. These individuals may

Physical Therapy Specialty Certifications

- Cardiopulmonary: treatment of patients with acute or chronic diseases of the cardiovascular or respiratory system
- Clinical electrophysiology: measurement of normal and abnormal electrical activity in the human body
- Geriatrics: treatment of conditions related to aging or other problems associated with older members of the population
- Neurology: treatment of patients with injuries or diseases of the brain and nervous system
- Orthopedics: treatment of patients with injuries or diseases of the muscles, bones, and joints
- Pediatrics: treatment of children during development from birth through adolescence
- Sport physical therapy: treatment of an athletic population, usually people who have incurred injuries as the result of competition

also have education in athletic training, strength and conditioning, or exercise physiology.

Becoming a physical therapy assistant involves completing a 2-year associate's degree accredited by CAPTE, the same organization that accredits physical therapy programs. In many states, though not all, physical therapy assistants must be licensed in order to practice. Regardless of whether they are licensed, physical therapy assistants work under the supervision of a physical therapist to carry out the prescribed protocol. Patients who are under the care of a physical therapy assistant must be reevaluated regularly by the physical therapist.

Licensure of physical therapists and physical therapy assistants is contingent on completing the appropriate program (i.e., physical therapist or physical therapy assistant) and passing an examination recognized by the state. Historically, the physical therapist's patients were required to have a referral from a physician, as well as physician oversight. Most states now permit direct physical therapy access, thus eliminating the need for a referral. However, either a referral or physician consultation is often required for insurance reimbursement.

Regulations for physical therapists educated outside of the United States vary from state to state. In general, these individuals must complete a course of study that is substantially equivalent to that of U.S. physical therapy education and must be proficient in the English language. For more information, contact the Foreign Credentialing Commission on Physical Therapy.

Employment Opportunities

The job outlook for physical therapists in the United States is encouraging. Employment is expected to grow much faster than the average for all occupations through 2020. As with occupational therapists, the demand for physical therapists is likely to increase as the age of the general population increases.

With advances in medical treatments, a higher percentage of accident victims are likely to survive, which increases the demand for rehabilitative care. In addition, employers, eager to reduce work-related injuries, are likely to employ more physical therapists to serve as worksite evaluators and as safety instructors to limit the number of injuries (U.S. BLS, 2016e). Pending federal legislation imposes limits on reimbursement to therapists and may affect job growth in the short term, but the long-term employment picture remains quite healthy.

The need for physical therapists is expected to grow by 34 percent, thus creating 71,800 new jobs, through 2024 (U.S. BLS, 2016e). In 2015, the median salary for physical therapists was $84,020 (U.S. BLS, 2016e). On average, physical therapy assistants earned $42,980. The need for physical therapist assistants is expected to increase by 40 percent through 2024 (U.S. BLS, 2016d).

Therapeutic Recreation Specialist

The physiological and psychological benefits of exercise and leisure activities are described throughout this text. Therapeutic recreation specialists, or recreation therapists, use these activities—as well as dance, drama, arts and crafts, animals, and community reintegration outings—to help people with physical, cognitive, emotional, or behavioral disabilities restore function, reduce or eliminate the effects of disability, and develop independence.

Recreation therapists work cooperatively with others on the treatment team to integrate functional activities into an overall rehabilitation program. Depending on the setting, the treatment team may consist of physical therapists, occupational therapists, speech therapists, physicians, dietitians, and mental health counselors. As part of this team, the recreation therapist focuses on leisure activities to restore function. For example, recreation therapists can use leisure activities such as fishing, painting, or singing to restore limb dominance, to decrease anxiety, to introduce positive leisure values and alternatives, or simply to promote the individual's overall sense of well-being.

In many programs, the goals are to achieve independent function and promote social integration of the patient (Wardlaw, McGuire, & Overby, 2000). For example, a recreation therapist working in an outpatient program for people with severe and persistent mental illnesses may have a strong emphasis on working toward social skill development and leisure education. In contrast, a person working with older adults who have dementia may use leisure activities to help slow the decline in cognitive function.

Those considering a career in recreation therapy have an opportunity to work with a wide variety of people in many different settings. Personal foundations for entering this career include a strong belief in the role played by recreation in creating a healthy lifestyle and an acceptance of people who have different ability levels. Recreation therapists learn diverse techniques for helping people meet their personal goals.

Employment Settings

Historically, recreation therapists have worked in inpatient and outpatient settings in psychiatry and physical rehabilitation. As with other health care professionals, recreational therapists are working increasingly with older adults across their continuum of care in facilities such as subacute care settings, adult day care centers, and assisted living residences.

Many recreation therapists work in community settings, most often in community recreation but also in programs for people with continuing disabilities, such as cerebral palsy, intellectual disability, or developmental disability. In these settings, recreation therapists often help people continue their treatment programs after discharge from an inpatient program; they also introduce people with disability to new or modified leisure options. Recreation therapists in community settings work both in segregated programs created specifically to meet the needs of people with disability and as inclusion specialists, wherein they help include people both with and without disability in typical recreation programming.

Recreation therapists are also employed in substance abuse rehabilitation centers, residential facilities for people with intellectual disability or developmental disability, camps for children or adults with disability, and public schools. They usually incorporate community settings into their patients' treatment plans. In inpatient settings, recreation therapists are often the members of the treatment team who determine readiness for discharge by observing the patient's functioning in a community setting during an outing toward the end of a hospital stay. Approximately one-third of recreational therapy jobs are based in hospitals, and another third are based in nursing care facilities; relatively few recreational therapists are self-employed (U.S. BLS, 2016f).

As with all allied health professionals, recreation therapists assess patients to determine their needs and desires, create treatment plans in conjunction with patients, and implement and evaluate the plans. These professionals must be familiar with the limitations associated with various medical conditions and accurately document medical records. They also need good interpersonal skills.

Education and Credentials

Entering the therapeutic exercise profession usually requires a bachelor's degree in therapeutic recreation or a degree in recreation with an emphasis in therapeutic recreation. A limited number of institutions offer stand-alone therapeutic recreation programs; most commonly, therapeutic recreation is offered as an emphasis area in a general recreation program. The entry level for therapeutic recreation professionals usually requires a bachelor's degree in therapeutic recreation, a term (quarter- or semester-long) internship under a certified recreation specialist, and completion of the national certification examination. A Certified Therapeutic Recreation Specialist (CTRS) is credentialed by the National Council for Therapeutic Recreation Certification (NTRC). The NTRC has also established alternative training and internship requirements so that graduates from other majors, such as kinesiology, can sit for the CTRS certification.

Employment Opportunities

The job market for recreation therapists is expected to increase by 12 percent through 2024. However, hospitals, which employ one-third of recreation therapists, are expected to reduce their staffs because of cost containment pressure, and some of these cuts will undoubtedly affect recreation therapists. Thus, the most promising opportunities for recreation therapists are likely to be found at nursing facilities, assisted living residences, and adult day care centers because of the increasing population of older adults. The median salary is $45,890 (U.S. BLS, 2016f).

Strength and Conditioning Specialists

People need proper strength and conditioning in order to achieve maximum physical performance, reduce the frequency of injury, and decrease the possibility of cardiovascular disease. Strength and conditioning specialists design weight training programs and cardiovascular conditioning programs based on the demands inherent in specific sports and on the particular needs of individual athletes. In athletic settings, such as high schools, colleges, and with professional teams, strength and conditioning specialists are often referred to as strength coaches. In these settings, strength and conditioning specialists may develop individualized programs in conjunction with athletic trainers, physical therapists, or physicians for athletes who have specific deficits or who have completed their rehabilitation programs.

Employment Settings

Strength and conditioning specialists are employed by university athletic departments, professional sport teams, health clubs, and corporate fitness centers. Universities with Division I-A athletic teams have invested heavily in strength and conditioning programs for athletes based on their recognition of the contribution of training programs to athletic success. These programs are commonly housed in facilities of more than 15,000 square feet (1,400 square meters) and may feature a staff of four or more strength and conditioning coaches, each of whom is responsible for one or more sports. Given the large number of athletes participating in a variety of sports, these facilities tend to be in use throughout the year. An increasing number of high schools are also recognizing the value of having strength and conditioning specialists on staff.

Education and Credentials

Individuals who are certified as strength and conditioning specialists often hold other professional credentials in areas such as athletic training, exercise physiology, physical therapy, and general medicine. Credentialing of strength and conditioning specialists is not associated with a degree program, but a candidate must have a bachelor's degree and current cardiopulmonary resuscitation (CPR) certification in order to take the examination, which is sponsored by the National Strength and Conditioning Association (NSCA). A degree in kinesiology is preferred. To receive certification, candidates must complete a two-part examination.

KEY POINT

A teaching degree is useful for gaining employment as a high school strength coach; college athletic departments often look for candidates with a degree in exercise physiology or a related field.

Passing both sections of the examination designates the person as a Certified Strength and Conditioning Specialist (CSCS). Kinesiology professionals who become certified as strength and conditioning specialists can compete more strongly in the job market and increase their salaries. Most NSCA members hold a degree beyond the level of a bachelor's degree.

Individuals who are interested in becoming a strength coach for a collegiate or professional team should also possess a bachelor's degree in a related field such as exercise physiology or kinesiology. Although no uniform course of study is offered at the undergraduate level, excellent preparation can be obtained by choosing an emphasis in fitness leadership with courses in performance assessment, program organization, and exercise leadership, in addition to the scientific core of courses normally offered in a kinesiology undergraduate program. Of course, it is useful to participate in an internship in a department of athletics conditioning facility and to engage in appropriate volunteer experiences. In addition, certification as a teacher enhances the possibility of being hired as a high school strength coach. One's marketability can also be improved through a graduate assistant position or internship position as a strength and conditioning coach.

Employment Opportunities

In 2015, the median pay for strength and conditioning specialists was $39,621 (Payscale, 2017), although many head strength coaches at large universities with football programs earn several hundred thousand dollars per year.

Trends and Opportunities in Therapeutic Exercise

Many people are now living longer than at any point in human history. Even so, obesity is prevalent in the United States; due largely to inactivity, it increases the risk of other conditions, primarily diabetes. Sedentary living also carries other consequences, such as cardiovascular disease. To combat the physical toll and financial costs of unhealthy lifestyles, increased emphasis is being placed on the habilitative aspects of therapeutic exercise.

Thus, in the years to come, all of the professions described in this chapter will place increased emphasis on identifying physical barriers to activity (e.g., issues of strength, range of motion, and cardiovascular limitation) and biomechanical factors

Are You Clinically or Technically Oriented?

Do you prefer to devise solutions on your own, or would you rather follow a structured sequence of steps? Clinicians devise and develop therapeutic exercise plans and are therefore called on to solve problems and make decisions. Technicians, in contrast, are experts at performing specific sets of skills. Technical professions include physical therapy assistants and occupational therapy assistants. Clinical professions include athletic training, physical therapy, occupational therapy, and higher levels of cardiac rehabilitation.

that may increase the risk of injury. In addition, the aging of the population will require professionals to be educated regarding the unique needs of older individuals. Already, courses in geriatric health care are being added to undergraduate and postbaccalaureate curriculums, and this topic is often emphasized in postprofessional continuing education courses as well.

Perhaps the biggest trend in future health care, however, will be increased emphasis on remote home health care. This Internet-based approach allows clinicians to prescribe, monitor, and evaluate therapeutic exercise programs by means of video and voice connections, web-based forms, and electronic monitoring devices. Perhaps the most intriguing aspect of this approach lies in its use of video gaming systems, such as Xbox, to entertain patients and clients during exercise sessions while also providing valuable medical feedback.

Advice for Therapeutic Exercise Students

By now, you understand that those who work in the sphere of therapeutic exercise must possess keen knowledge of human anatomy, human physiology, and kinesiology and of the effects of exercise on both healthy and unhealthy individuals. These areas, however, represent only a partial list of the skills and knowledge required to practice in this sphere.

Exercise is therapeutic only when it is performed in a safe and appropriate manner. The work of developing and delivering appropriate therapeutic exercise routines requires the ability to solve problems, collect information regarding the patient's or client's condition, identify the therapeutic goals, and determine the appropriate course of action. To solve problems, therapeutic exercise professionals must be able to access, manage, and interpret various forms of information and apply them to the current case. This task often requires the use of computers and an understanding of how to access appropriate web sites and perform effective literature searches.

Health care professionals must also be able to communicate effectively across demographic, sociocultural, and professional boundaries. Certain settings or roles require additional areas of knowledge. For instance, professionals who work in fee-for-service environments need to become familiar with record keeping and insurance billing systems. A business background would be useful for practitioners in private practice, and a person in any type of leadership position would benefit from course work in management and administration.

Therapeutic exercise professionals must adhere to basic ethical and professional principles that are defined by state laws, professional membership organizations, and certification and licensure agencies. Although the particulars vary from profession to profession and from state to state, some themes are common across the professions. For one thing, practitioners must adhere to the profession's scope of practice—the legal parameters that define the profession. Another tenet holds that the services provided must be in the patient's or client's best interest, including use of the best scientific evidence in the patient's care. Practitioners must also maintain the confidentiality of medical records. Furthermore, when services are billed to a third party, such as an insurance company, the service descriptions must accurately represent the care provided. Although your instructors and supervisors will identify these issues for you, it is ultimately your responsibility to seek out relevant information and ensure that you remain in compliance. Failure to do so could result in criminal liability.

How does one choose an area in which to work? Although reading can provide you with an overview of these professions, the only way to discover what a profession is really like is to see professionals at work. Many professionals allow students to shadow them throughout the course of a day. You may be able to make such a connection with the help of a professor or other acquaintance. You may also have an opportunity to do an internship or apprenticeship or find a part-time or summer job in one of these areas.

Occupational Skills Analysis

If you are interested in entering a therapeutic exercise profession, you will need to work toward a degree in kinesiology and become certified in a particular area. What other skills do you need to learn? Do you have good problem-solving and communication skills? Will you enjoy the business side of these professions—billing, record keeping, and so on? In what ways can you test the waters of these professions in order to see whether they are a good fit for you?

Case Study in Ethics

Suppose that you are an athletic trainer or physical therapist for Costello College. A family member refers a friend to you for examination and care. It seems that the friend, Pat, sprained an ankle while playing in a pickup basketball game. Immediately following the injury, Pat went to the emergency department for care. Several hours later, the physician ruled that there was no fracture. The laws in your state that define the scope of practice for an athletic trainer allow you to treat "anyone who is competing in, or preparing for, competition in organized athletics." The physical therapy laws in your state allow treatment only by physician referral.* You explain to Pat that you cannot legally provide care. Pat says that he understands the law but views this as helping out a family friend.

Your Decision

Although friendship is important, the laws governing both physical therapy and athletic training preclude you from treating your family friend. The law specifies that the athlete must be injured during organized athletic participation in order for an athletic trainer to render services. It also states that a physician must specifically prescribe that a patient be treated by a physical therapist. Furthermore, since you would be providing care to Pat at Costello College, that would place the college at legal risk and potentially jeopardize your employment. Would you provide care to Pat?

Resolution

Your team physician at Costello College has a local private practice. To keep your care legal, you help Pat set up an appointment with your team physician. She can then refer Pat to you for legal treatment, thus keeping everything within the scope of the state laws regulating the practices of athletic trainers and physical therapists.

*These laws are rare but do exist.

Wrap-Up

Professions arise to fulfill a societal void. The therapeutic exercise professions fill a void by either helping people obtain their desired level of physical fitness (habilitation) or helping injured individuals regain lost function (rehabilitation). Part of the attractiveness of this sphere of professional practice derives from the wide range of the population served. Highly skilled athletes, newborns, and the geriatric population—all may rely on therapeutic exercise specialists.

Diversity also characterizes the settings in which therapeutic exercise professionals work. From the structured environments of hospitals and laboratories to the hectic atmosphere of athletic training rooms, these work settings can accommodate a wide range of interests.

This chapter covers some of the more prominent therapeutic exercise professions, such as athletic training, cardiac rehabilitation, occupational therapy, and physical therapy. If you are interested in any of these professions, we encourage you to seek more information from your professors and explore this area in greater depth.

MORE INFORMATION ON CAREERS IN THERAPEUTIC EXERCISE

Organizations

American Association of Cardiovascular and Pulmonary Rehabilitation

American College of Sports Medicine

American Kinesiotherapy Association

American Occupational Therapy Association

American Physical Therapy Association

American Therapeutic Recreation Association
Canadian Association of Occupational Therapists
Canadian Athletic Therapists Association
National Athletic Trainers' Association
National Strength and Conditioning Association

Journals

American Journal of Occupational Therapy
Annual in Therapeutic Recreation
Athletic Training and Sports Health Care
Canadian Journal of Occupational Therapy
Clinical Kinesiology
International Journal of Athletic Therapy and Training
Journal of Athletic Training
Journal of Cardiopulmonary Rehabilitation and Prevention
Journal of Clinical Exercise Physiology
Journal of Orthopaedic & Sports Physical Therapy
Journal of Sport Rehabilitation
Journal of Strength & Conditioning Research
Physical Therapy
Physical Therapy in Sport
Sports Health
Sports Medicine
Strength and Conditioning Journal
Therapeutic Recreation Journal

REVIEW QUESTIONS

1. Describe how therapeutic exercise promotes healthy lifestyles.
2. Discuss the similarities and differences between habilitation and rehabilitation.
3. Identify the common overlaps among the therapeutic exercise professions described in this chapter.
4. What skills or attributes are unique to the therapeutic professions presented in this chapter compared to other kinesiology professions?
5. To make themselves more marketable, many people obtain multiple credentials or specializations. Describe some possibilities for dual credentials and cross-training based on the descriptions of professions given in this chapter. What benefits are provided by dual credentials and cross-training? What professions *not* described in this chapter would also lend themselves to dual credentials?

Go online to complete the web study guide activities assigned by your instructor.

Careers in Teaching Physical Education

Kim C. Graber and Thomas J. Templin

Special appreciation is extended to Darla Castelli, Ben Kern, and Sara Russell for their contributions to this chapter.

CHAPTER OBJECTIVES

In this chapter, we will

- describe what a physical education teacher does,
- provide insights into exciting research,
- explain how to remain current in the field,
- describe settings in which physical educators teach,
- describe highly effective teachers, and
- provide information about career options.

As Ms. Schmidt sat among dozens of other teachers at Peter Robert High School during orientation day, she could hardly believe that she had landed her dream job as a physical education teacher and volleyball coach in the same school at which she had completed her student teaching. She had also completed a practicum experience at the school and had been so impressed by the quality of facilities and the teaching that she had requested the school for her student teaching experience.

During student teaching, Ms. Schmidt had spent many hours planning high-quality lessons that engaged students in high amounts of physical activity. She had managed her classes well, and the students had responded eagerly to her enthusiasm. Her professionalism and commitment to effective teaching had caught the attention of both the department head and the principal, and she had been invited to apply for a permanent position after another teacher retired. Now, sitting among her colleagues, she was proud of herself for landing this job.

Ms. Schmidt had entered college intending to become a physical therapist. She had felt that it would be great to help others recover from injury and reestablish a physically active lifestyle. Then, during an introductory kinesiology class, she had worked with a professor in the physical education teacher education program who talked about research as it related to effective instruction and the importance of engaging children in physical education as one mechanism to offset childhood obesity. She had also learned about quality physical education programs and the need for teachers who were committed to teaching skills and engaging students in high levels of physical activity. In addition, she had read research about equality in teaching and how appropriate practice can help all students, even those who are low skilled, acquire motor competence.

After working in the research labs of both an exercise physiologist and a teacher educator, Ms. Schmidt had begun to realize that physical education was the career best suited to her interests and skills. Now, sitting next to her new colleagues, some of whom possessed many years of experience and others who had none, she was nervous yet excited about the possibilities her new career would offer.

© Human Kinetics

As a participant in a physical education class, you may have been concerned about improving your performance and fitness levels, playing well during competition, or simply enjoying the experience. You may have felt proud of your ability, hard work, and dedication when you turned in an especially successful performance. The truth is that if your performance improved, and if you enjoyed the experience, then your teacher was largely responsible.

Effective teachers understand their subject and know how to convey it appropriately. For example, expert teachers in river rafting know which techniques should be used to propel a raft when the river is smooth as opposed to when it is at its most dangerous. They understand when to expect a change in the conditions of the river and how to maneuver safely and effectively from one type of water to another. Good teachers also know which techniques are most appropriate for novices and which are best for experts. They convey instructions in a manner that is understandable and provide the best practice opportunities to help learners develop new skills. In addition, good teachers integrate technology into lessons when appropriate; they also know how to diagnose movement technique and provide meaningful feedback to help students improve.

Successful teaching requires a background in and knowledge of the curriculum and pedagogy (teaching) of physical education. Without a background in pedagogy, physical activity instructors tend to "wing it," an approach that can sometimes work out in the short run but often leads to wasted time and effort—and stress for the teacher. In the opening scenario, Ms. Schmidt was hired because she demonstrated effective teaching behaviors during her student teaching. More specifically, she exhibited knowledge of curriculum development, understood effective pedagogy, and put these qualifications to good use with her students.

Before you read further, consider these questions:

- If you become a physical education teacher, how will you get students to become interested in the subject matter of physical education?
- How will you convey your instructions to students?
- Should physical education class emphasize skill development or physical activity?
- How will you structure class to ensure that all students can be successful?
- How will you motivate students to engage in physical activity outside of the school setting and to make nutritious food selections?

A person with a background in pedagogy knows the answers to these questions. A knowledgeable teacher knows that the best way to get and keep students' attention is to establish rules and procedures for good behavior in the gymnasium and on the playing field. An effective teacher with a background in pedagogy knows that practicing a skill correctly and in gamelike situations is more important than listening to a 20-minute lecture on basketball techniques.

An effective teacher also understands the critical importance of physical activity to children's health. The teacher knows that 17 percent of youth in the United States are considered obese (Ogden et al., 2015) and that many do not receive adequate physical education or engage in adequate physical activity (Society of Health and Physical Educators [SHAPE America] and American Heart Association [AHA], 2016). The teacher also believes that these negative trends can be reversed in part through effective physical education.

Pedagogy is the art, science, and profession of teaching. Effective pedagogy focuses on teaching behaviors and producing learning in students. The distinction between teaching and producing learning boils down to focusing on the student instead of on the teacher. Rather than approach each class with the question, "What am I going to teach today?," the most effective teachers instead ask themselves, "What are my students going to learn today?" This focus on learning can improve the interaction between the learner and the teacher and lead to better education for students. This transition of focus from teaching to learning is one of the key discoveries of the early research in pedagogy and education. Ideally, then, individuals who pursue a career in teaching are concerned with producing learning in students.

Whether you want to become a physical education teacher, a physical activity director, or a high school coach, you need to learn the skills associated with effective pedagogy. Prospective teachers typically gain pedagogical knowledge through an accredited teacher education program at a college or university. In some cases, they acquire knowledge by working with a mentor, such as a cooperating teacher in a public school; by attending a teachers' conference; or by reading a professional journal.

What Is the Teaching of Physical Education?

This chapter introduces you to the teaching of physical education, sometimes referred to as **sport pedagogy**. As you read this chapter, think of your educational experiences thus far and ask yourself whether you received effective instruction—and why or why not. Consider how you might use some of the research findings discussed in this chapter, particularly if these findings contradict your

How Would You Rate Your Experiences in Physical Education Class?

Reflect on your own experience by answering the following questions:

- Who stands out in your memory as the best physical education teacher you ever had?
- What specific qualities made this teacher effective?
- Who stands out in your memory as the worst physical education teacher you ever had?
- What weaknesses did this teacher possess, or what qualities made him or her ineffective?
- What do you think are the most important traits for a teacher to possess? Why?

experiences as a student. Because most of the research on pedagogy focuses on teaching in public schools, much of this chapter addresses public school physical education. However, we also illustrate the application of pedagogy to other types of physical activity instruction—for example, instruction delivered by community recreation directors, coaches, and exercise instructors for older adults.

Experts in physical education teaching have many career options. Most elect to obtain certification and become a teacher or coach in a K-12 school. Many choose to work at the elementary level because they like interacting with young children and appreciate the enjoyment that youngsters experience while engaged in physical activity. Others decide to work in a middle school or high school because they appreciate that age group or wish to coach. As a means of supplementing their salary, some teachers also work in the role of referee, swim instructor, or recreation leader.

Some people who enter a department of kinesiology in order to become a physical activity instructor do not intend to teach in a public school. Instead, they decide that they are better suited to teaching in a private school or community recreation center. Some elect to become a specialist, such as a wilderness expert or golf pro. Others choose to work at a college or university as an instructor or professor. Most, however, elect to become an instructor of physical activity because they are interested in helping people improve their health and wellness levels, engaging people in activities that they can continue throughout their life span, and working with children in schools.

Influence of Experience on Your Beliefs

Research has long demonstrated that undergraduate students who are interested in teaching often believe that they already know all there is to know about teaching and have little more to learn (Lanier & Little, 1986). In a kinesiology major, this belief may relate to experiences as a student in K-12 physical education, as an athlete on a team, or as a participant in a park district program. Consider that you have spent approximately 13,000 hours as a student (Lortie, 1975) and many other hours in other learning environments. Through these experiences, you have developed many strong beliefs about teaching.

KEY POINT

All instructors are influenced by their experiences in K-12 schooling and other educational settings. The best instructors, however, ground their programs and teaching methods in evidence-based pedagogical knowledge acquired during their professional preparation.

Despite the logic of emulating someone whose qualities you admire, relying only on experience has some drawbacks. Because teachers influence different students in different ways, you cannot assume that everyone had the same experiences with a particular instructor. Nor can you assume that a teacher is effective simply because everyone liked him or her. For example, if you were a starter on your high school's lacrosse team, your physical education teacher may have allowed you to skip class on game days, asked you to demonstrate lacrosse skills to classmates, and given you more opportunities than others to play lacrosse during class.

Such personal experiences might lead you to believe that your teacher's pedagogy was sound. You might believe that athletes *should* be released from class before a game, that better players *should* demonstrate their skills for the benefit of others, and that higher-skilled students *should* receive more opportunities for activity than do lower-skilled students. Despite your personal experience, however, these beliefs may be mistaken, because they are based on unsound, unjust, or unproven teaching principles.

Imagine how the scenario might have been different. If your teacher wanted all students to have a positive response to the sport of lacrosse, he or she might have spent many hours planning to ensure that everyone had an opportunity to improve their skills and experience success. In addition, your teacher wouldn't want you to miss class on game days, because physical education is another opportunity to improve your skill level. This teacher's practices would have been grounded in effective pedagogy.

Although it is human nature to accept and imitate the behavior of people you respect—and to reject the behavior of those you do not—personal experience alone is insufficient for becoming an effective teacher. A better approach for learning how to teach is to analyze the available research about teaching and learning and use this analysis as the basis for your behavior and values. To some degree, you will always refer to personal experience, but it should never be your only criterion.

Certification and Continuing Education

The majority of students admitted to a kinesiology or physical education certification program intend to teach in a K-12 school. Others may wish to become a recreation specialist, fitness trainer, coach, or outdoor education expert. Regardless of their career path, students must meet many program requirements before gaining formal admission to a teacher education program and before achieving certification. These requirements typically involve the following:

- **Grades or grade point average:** In most teacher education programs, students must maintain a certain minimum grade point average. In fact, most programs require a designated minimum grade point average just for admission into the program. In addition, students must often maintain a grade of B or higher in their professional education courses in order to continue in the program and graduate with an emphasis in teacher education.
- **Praxis I:** Before admission into certain courses (e.g., methods courses) or a particular phase in a program, students are required to complete preprofessional skills testing (PPST or Praxis I). This testing entails a series of tests that assess basic proficiency in reading, writing, and math. Students must achieve minimum scores in each area in order to progress in the program.

To be an effective teacher, you must rely on more than your own personal experiences.

- **Praxis II:** Before admission into certain courses or student teaching, students must pass Praxis II. This exam evaluates specialty content knowledge and pedagogical knowledge (methods and curricular knowledge). It is administered by the Educational Testing Service.

Although certification requirements vary from state to state and country to country, they usually involve a series of exams. In addition, the program in which you are enrolled is responsible for ensuring that you have developed adequate knowledge and certain competencies. The curriculum in such programs is periodically evaluated by state officials. If a teacher education program fails to meet established state standards, it is placed on probation and may lose its ability to provide teacher certification.

After a student meets all program and state testing requirements, he or she may apply for a teaching certification or license. In many cases, reciprocity between states enables a graduate from a program in one state to teach in one or more other states throughout the country. Some states require teachers to complete continuing education courses periodically in order to maintain a teaching license. This requirement helps ensure that teachers remain current and knowledgeable about the most effective teaching practices. For example, teachers may be required to take 12 credits of continuing education courses every 4 years in order to remain certified. They can usually meet this requirement by taking courses at a college or university, enrolling in a master's degree program, or attending approved professional conferences; however, requirements can vary considerably from state to state.

With regard to coaching, some schools mandate that coaches be certified teachers; others require only an undergraduate degree in any subject matter, and the remainder expect coaches to have specialized training or an endorsement from a group such as Human Kinetics Coach Education. Similarly, fitness professionals are often expected to become certified as a personal trainer. Some students enter colleges and universities for their training, whereas others obtain certification through online courses.

KEY POINT

Teaching and learning journals provide knowledge about advances in the field, innovative teaching strategies, and research results. To remain current, teachers should subscribe to and read as many journals as time permits.

In the United States, the easiest way to remain current and informed is to join the Society of Health and Physical Educators (SHAPE America). The group's website allows you to access information about research, conferences, grants, professional standards, and effective teaching practices. In exchange for your annual dues, you will receive a monthly subscription to the *Journal of Physical Education, Recreation and Dance* and will be able to attend the national convention at a reduced rate. You will also be able to log on to a members-only portion of the website, where you can access the most recent and important updates in the field. The best professionals who are the most committed to effective teaching know that membership in this organization is essential for remaining current in the field and contributing to its development in the United States.

In Canada, you can remain informed by joining Physical and Health Education Canada (PHE Canada). This association, represented in all provinces, is committed to advancing the field of physical education. It offers the *Physical and Health Education Journal*, which contains much useful information for teachers. PHE Canada members and SHAPE America members often attend the same national conferences. For more information, log on to the PHE Canada website. Similar national professional organizations for physical educators are found in most countries.

Considering the cost of your undergraduate education, the annual fee for joining these associations, especially at student member rates, is a cost-effective means of remaining current. Indeed, becoming a member is a professional responsibility that informed professionals take seriously. Other professional associations and journals are listed at the end of this chapter.

Research on Effective Teaching

We can learn much about what teachers do by examining the wealth of research that has been conducted in the area. Research on teaching pertains to the teaching and learning process that occurs in educational settings such as public schools or private schools. Researchers in a typical study might try to determine whether students learn better through one method of instruction or through another. These researchers might look at how many times an instructor provides feedback to individual students during class and whether that feedback is specific or general. Or, researchers might investigate whether students who receive more practice opportunities during class learn more than those who receive fewer opportunities. Many of the findings about effective teaching are transferable from one setting (e.g., school) to another setting (e.g., recreation center).

Research on teaching is concerned with the scientific study of the processes of teaching and learning and is deeply rooted in what happens in schools. Over the past two decades, scholars have undertaken many types of investigations and examined many aspects of teaching. For instance, researchers have studied the characteristics of effective teachers and their classrooms. Although most research examining the effectiveness of physical activity instruction has been conducted in the realm of physical education, many of the findings are easily transferable to other educational settings. Therefore, it can inform other types of physical activity instructors.

KEY POINT

Teaching experience alone does not guarantee expertise.

Appropriate Practice Experiences for Students

Your early experiences in physical education, recreation, or camp settings may have been similar to those of many of your age peers. If so, your teachers and counselors probably had you play dodgeball, kickball, "steal the bacon," and other such games. As you progressed into the middle and high school grades, you may have received a few days of instruction before playing softball, basketball, football, volleyball, soccer, or the like. If you are now interested in pursuing a career in the physical activity field, you probably had above-average skills in these games and sports; for instance, you may have enjoyed eliminating other students in dodgeball or scoring the winning goal in soccer. Such memories probably remain satisfying for you and likely influenced your interest in a physical education career.

Unfortunately, it is unlikely that all of your classmates shared your positive experiences of physical activity. For students who were less skilled—the ones that you quickly eliminated from games such as dodgeball—physical activity time was at best supervised recess, and at worst an opportunity

Expertise

David Berliner (1988) posited that teachers progress through a series of stages as they gain expertise as educators. Moreover, some teachers progress through these stages with little difficulty, whereas others may never progress to the expert stage. Although experience can enhance teacher effectiveness, it doesn't guarantee that a teacher will acquire expertise (Dodds, 1994; Graber, 2001). By analogy, Michael Phelps, Simone Biles, Aaron Rodgers, and Serena Williams have high levels of performance expertise in their respective sports of swimming, gymnastics, American football, and tennis. Other athletes, however, may practice hard for many years yet never achieve the same level of expertise.

Physical education teachers who possess a high degree of skill in one or more sports or activities may have acquired content area expertise. That alone, however, does not guarantee teaching expertise. In addition to possessing content expertise, they also need to be able to convey content to their students through appropriate pedagogy before they can be considered to have acquired teaching expertise.

Unfortunately, some teachers become bored or burned out with teaching and lose interest in learning about ways in which they can improve their teaching or influence student leaning. These teachers no longer learn from experience; nor are they interested in reading professional journals, taking graduate classes, or attending professional conferences. In contrast, expert teachers—and all of those who continue to learn about teaching by reading and by attending professional conferences—are less likely to experience burnout.

To learn more about teaching expertise, we suggest that you explore the following literature:

- Manross, D., & Templeton, C.L. (1997). Expertise in teaching physical education. *Journal of Physical Education, Recreation & Dance,* 68(3), 29–35.
- O'Sullivan, M., & Doutis, P. (1994). Research on expertise: Guideposts for expertise and teacher education in physical education. *Quest, 46,* 176–185.

for embarrassment. For most of those students, no instruction was available; thus, they had little chance of improving their skills. Think about it: Did you acquire your athletic expertise by participating in physical education class? Or did you acquire it through free play or through extracurricular activities organized by coaches (or parents) who were determined to help you improve your playing ability?

Indeed, it is sometimes difficult to defend physical education classes and unstructured recreation programs as opportunities for participants to improve their skill levels. For example, if a student spends 1 or 2 days practicing a volleyball bump pass before being expected to perform the pass correctly during a game, how much chance does the student have to succeed? Consider that serious volleyball players spend hundreds of hours developing their skills. For students to succeed at any skill, they must be exposed to *appropriate practice*. Thus, when designing activities, instructors need to consider whether drills simulate game situations. For example, the most appropriate method of practicing a volleyball set is to practice setting the ball to a target area from a position in which the volleyball is bump-passed to the setter—not tossed. This drill closely resembles a game situation.

Active Learning Time

Do you recall the many times as a student when you had to wait in line for a turn? If you're like many students, you sometimes had to wait through half the class for an opportunity, say, to bounce once or twice on a trampoline. Or maybe you played volleyball with 11 others on your side of the net, waiting minutes at a time for a chance to touch the ball. If this sounds familiar, you're not alone in your experience. For various reasons, thousands of students each year receive inadequate opportunities to practice and learn new skills.

Research conducted in the classroom (Brophy & Good, 1986) has clearly linked student achievement to time on task. Also called **engaged time**, time on task consists of the time that students spend *doing* physical activity or sport. Unfortunately, engaged time has been low in many physical activity settings. In fact, pedagogy research has reported that students in traditional physical education classes spent only about 30 percent of class time engaged in physical activity (Anderson & Barrette, 1978; Metzler, 1979, 1989). What are they doing instead? Historically, students have spent 50 percent to 70 percent of their class time waiting, transitioning, and being managed by the teacher. Fortunately, today's teachers are being trained to engage students in activity for at least 50 percent of the time, which is what experts recommend (Siedentop & Tannehill, 2000). Increased emphasis is also being placed on ensuring that students are *appropriately* engaged—that is, performing correctly with frequent success. Such time is often called **academic learning time** or **functional learning time**.

For some students, physical education class is the only opportunity throughout the entire school day to engage in physical activity. Therefore, teachers must keep students active (and successful). This need is given additional urgency by the obesity epidemic and the high rate of physical *inactivity* that characterizes many people throughout the world. Thus, physical education teachers must be committed to keeping students moving in appropriate learning activities for the majority of the class period.

Effective Class Management and Discipline

How do you handle a student who refuses to take a time-out? How do you deal with a student who ridicules or bullies other students? Such questions about class management and discipline tend to top the list of common concerns among new teachers. *Class management* involves organizing students in such a way that learning is most likely to occur, whereas *discipline* involves establishing rules, enforcing them, and rewarding exceptional behavior. Regardless of their experience level, teachers can never prepare for all of the situations they will encounter. Moreover, physical education teachers have a particular need to be experts in class management and discipline because their classes are much larger than those of most classroom-based teachers.

Are You Suited for a Career in Teaching Physical Education?

Do you enjoy helping others learn? Do you like working with children? How do you feel about long but fulfilling work days? Would you like to work in an environment where you can share your enjoyment of physical activity with others? If you responded positively to these questions, then you may be well suited for a career as a physical education instructor or physical activity instructor.

KEY POINT

When teaching rules and routines, instructors can best help students by setting high expectations, being firm but warm, developing clear rules, and describing how rules will be enforced.

The best thing instructors can do in this regard is to take proactive measures that tend to minimize potential difficulties. Graham (2001) has suggested ways to create a positive learning environment. In general, he suggests using the first few classes of a new academic year or unit to establish and implement rules and procedures for good behavior. For example, in a physical education class, students need to learn the new teacher's signals, such as start and stop signals and other signs to look or listen for during class. These signals and signs tell students to behave in a certain way or let them know that a particular behavior choice is good or poor. If students learn such lessons early, then the entire school year will run more smoothly for the class.

In a study of seven effective elementary specialists, instructors were observed teaching students the stopping and starting signals for class by using the signals a total of 346 times during the first few days of instruction (Fink & Siedentop, 1989). After the first few classes, most students were behaving appropriately, and teachers promptly reprimanded those who did not follow the signals. By implementing this routine during the first few days of every unit, these teachers quickly constructed an environment in which learning could take place.

Students also learn how to behave in class through clearly stated and consistently enforced expectations. One instructor characteristic that promotes appropriate student behavior is called **with-it-ness**. Teachers demonstrate their with-it-ness by knowing what's happening in the learning environment and displaying their awareness through oral or other communication with students. By holding students accountable for their actions, these teachers reduce the likelihood that students will try to manipulate the learning environment.

Assessment

Effective teachers understand the importance of assessing student achievement. They use formative assessment to determine whether students are learning the content of a unit as it progresses; then, at the end of the unit, they use summative assessment to determine whether students have acquired the relevant skills and knowledge. Without formal assessment, it is impossible to determine whether children have learned. Unfortunately, some teachers do not assess students to determine whether they have learned as a result of class. Instead, these teachers base students' grades primarily on dressing out, attendance, and good behavior in class. Although you might believe that it isn't fair to grade students in physical education because some are naturally less skillful than others, do you believe that students in math or science should not be graded because the curriculum is less intuitive for some than for others?

Many states have established learning standards that specify what students should know and be able to do as a result of physical education. Many of these standards are based on SHAPE America National Standards and Grade-Level Outcomes (Society of Health and Physical Educators, 2014). As schools are increasingly being held accountable for standards-based student learning, a greater number of teachers are beginning to use formal assessments in physical education. To aid in this process, SHAPE America has published a series of assessments, called PE Metrics, that match the national standards and are designed to inform teachers, students, parents, and others about whether students are meeting the national standards for their age level (Society of Health and Physical Educators, 2015a, 2015b). These assessments have undergone vigorous testing to ensure their validity and reliability. Fitnessgram is another measurement tool that is widely used by teachers to accurately assess students for health-related fitness (Cooper Institute, 2016).

Student assessment is linked to teacher evaluation. Presently, about two-thirds of U.S. states require that teachers be evaluated on two or more measures, such as student learning metrics and classroom observation. The practice of evaluating physical education teachers on the basis of student skill development, fitness development, or personal-social development is grounded in teacher accountability. Various parties—including legislators, parents, and school administrators—want to make sure that teachers are performing at a high standard and that children are learning in physical education classes. Therefore, future teachers in physical education teacher education (PETE) programs should understand that once they become certified, they will be evaluated annually and must possess the tools to assess student learning in order to demonstrate their own effectiveness.

Case Study in Assessment

After being offered several teaching positions, Sarah Smith accepted her first job as a physical education teacher at Chavez Middle School. She did so because Mr. Carson, the principal, believed strongly in the importance of physical education.

Both before and during her college years, Ms. Smith had noticed that many physical education teachers did not care about assessment despite the existence of new state-mandated teacher evaluation systems that assessed both teacher performance and student learning outcomes. In contrast, since learning about PE Metrics and Fitnessgram in her PETE program, she viewed it as extremely important to assess her students.

Ms. Smith decided to grade her students on their effort, skill improvement, attitude, knowledge, and fitness; in addition, each area would be weighted according to its importance. The effort portion of a student's grade would include dressing appropriately for activity and participating during class. Skills would be tested during each unit and graded only on improvement; more specifically, students would take a skill pretest before each unit and a posttest at the end of the unit based on PE Metrics. Student attitude and knowledge development would be assessed by means of cognitive exams. Finally, Ms. Smith decided to use Fitnessgram to assess her students' fitness levels. Although this assessment would not affect their grades, it would help her determine how they were progressing in various dimensions of health-related fitness.

When Ms. Smith met her colleague, Mr. Jones, a veteran of 20 years at Chavez, she explained her approach to assessment. She suggested that her model might help Mr. Jones since he was going to be assessed within the new teacher evaluation system. After a few minutes, however, she could tell that the interaction wasn't going well. Before long, Mr. Jones' face turned bright red, and he barked, "No way are you going to do that here! If the kids show up and dress, they get an A. That's the way it's going to be. I don't care about the new evaluation system, and our principal won't worry about it either." Ms. Smith was shocked and slowly retreated as she said, "Let's see what Mr. Carson has to say about this."

Your Decision

Do you think Ms. Smith should have made such a fuss over the assessment process she had chosen for her class? Based on your experience in high school physical education classes, how likely is it that other veteran physical educators might react similarly to Mr. Jones? Do you think Ms. Smith should have threatened to go to the principal? How would you have handled the situation?

Resolution

Ms. Smith, Mr. Jones, and Mr. Carson met to discuss the assessment challenge. Even though Mr. Carson had been a colleague of Mr. Jones for more than 10 years, he was emphatic in his support for Ms. Smith. Moreover, he told Mr. Jones that perhaps it was time for the "old dog to learn some new tricks" from this beginning teacher. He went on to explain that because of the new state standards related to teacher evaluation, it was time for all of the teaching staff at Chavez to implement state-of-the-art techniques, including student assessment. With some hesitation, Mr. Jones consented, and Ms. Smith felt relieved, albeit certainly cautious about future interactions with her new colleague.

KEY POINT

Learning how to implement new procedures, such as the state-mandated teacher evaluation system, is difficult for some teachers because they already have many responsibilities and are often more comfortable with procedures that are familiar.

Feedback

In addition to providing clear instructions, teachers must also provide clear, specific, and immediate feedback (Rink, 2006). Unfortunately, physical education teachers sometimes provide too many instructions, whereupon students either get bored and stop listening or try to listen but fail to remember everything the teacher says. In contrast, during practice, teachers seldom provide students with an adequate amount of feedback that is correct, prompt, and specific.

Research and our own experiences indicate that teachers make the following mistakes when providing feedback.

- Feedback is often incorrect. For example, a teacher may not see the fundamental error committed by a student and may focus instead on a secondary feature of performance.
- Teachers sometimes focus on an aspect of performance that does not require feedback while neglecting an area that does require feedback. For example, Stroot and Oslin (1993) found that teachers neglected to provide feedback on a component of the overhead throw because they were unable to diagnose deficits; instead, they provided feedback about aspects that students had already mastered.
- Some teachers do not time their feedback so that the learner receives prompt help when practice trials are defective in form or when they produce unsuccessful results. Practicing incorrect performance for an extended period produces predictably negative results!
- Teachers provide less feedback during game play. However, feedback provided during game play can be valuable; it gives all students information about ways to improve.

Alternative curriculums offer exciting possibilities to students who may not be interested in traditional activities.

KEY POINT

Teachers can make instruction more likely to be effective by providing appropriate learning activities; maximal active learning time; and feedback that is correct, prompt, and specific.

Alternative Curriculums

You and most of your classmates were likely exposed to a traditional curriculum in which instructors designed activities with little student input and used a teacher-dominated approach. Although a traditional approach can be effective if used by a creative teacher, alternative curriculums offer strong promise and can be implemented in many different environments.

Physical education today is much more exciting than ever before. Although the obesity epidemic is a serious health concern, it has at least motivated teachers to improve their physical education programs. You may have heard the term *new physical education*. This movement is made up of teachers who incorporate novel and exciting learning ideas into their classes—and especially those who emphasize fitness and nontraditional activities such as cycling, rock climbing, and in-line skating. In fact, these teachers are sometimes profiled in the national media as having changed the face of physical education.

Physical education teachers can structure their curriculum in many ways. Here are different forms of curriculum that are used in K-12 schools:

- The *elective* curriculum allows learners to choose one activity from a wide selection of options and encourages them to develop skills that will transfer into a lifetime interest in that activity. This approach is offered by many large-school programs. Some high school programs take an approach similar to that of a fitness club by allowing students to select classes that match their fitness and developmental needs.

› The *fitness curriculum* emphasizes cardiorespiratory efficiency, muscular strength and endurance, flexibility, and body composition, particularly at the secondary level. Students participate in health-related fitness activities such as running, swimming, circuit training, cycling, in-line skating, and aerobics (Corbin, Lindsey, Welk, & Corbin, 2001). This type of curriculum is becoming increasingly popular because it is perceived as a way to address such health concerns as the obesity epidemic. If implemented appropriately, a fitness curriculum gives students a chance to engage in enjoyable moderate to vigorous physical activity. Physical educators must not forget, however, that it is equally important to teach skills to students so that they can engage successfully in multiple activities throughout their lives.

› The *sport education* model appears to hold promise particularly as a means of improving secondary physical education (Siedentop, Hastie, & van der Mars, 2011). In this approach, teachers treat students as athletes. Students are formed into teams and not only learn how to play but also assume some responsibility for roles as managers, coaches, trainers, officials, statisticians, and tournament administrators.

› The *wilderness and adventure education* curriculum introduces learners to such activities as canoeing, backpacking, camping, white-water rafting, skiing, first aid, ropes courses, climbing walls, and new and cooperative games. Evidence suggests that wilderness and adventure education can be highly appealing to students (Dyson, 1995).

Professional Issues in Kinesiology

Equality in U.S. Public Education

Schools and classrooms are diverse settings. Regardless of gender, race, age, social class, sexual orientation, ability, or any other demographic characteristic, each student has the right to an education free of bias and discrimination.

In 1972, Title IX was legislated by the United State Congress. It mandates that girls and women have the same educational opportunities as boys and men. Specifically, it states in part, "No person in the United States shall, on the basis of sex, be excluded from participation in, be denied the benefits of, or be subjected to discrimination under any educational program or activity receiving Federal financial assistance…."

Despite measures such as Title IX, some students continue to be disadvantaged for reasons beyond their control. In physical education settings, common forms of discrimination include bias against students who possess less ability or are considered obese. In elimination games, for example, students with the lowest levels of skill are usually the first eliminated, although they are the ones most in need of practice. Issues of equity and inclusiveness have been addressed by considerable research in physical education, and we encourage you to read the following landmark publications:

Citation style: APA

Davis, K. (2003). Teaching for gender equity in physical education: A review of the literature. *Women in Sport and Physical Activity Journal, 12*(2), 55–82.

Griffin, P.S. (1984). Girls' participation patterns in a middle school team sports unit. *Journal of Teaching in Physical Education, 4,* 30–38.

Griffin, P.S. (1985). Teachers' perceptions of and responses to sex equity problems in a middle school physical education program. *Research Quarterly for Exercise and Sport, 56*, 103–110.

Staurowsky, E., Hogshed-Makar, N., Kane, M., Wughalter, E., Yiamouyiannis, A., & Lerner, P. (2007). Gender equity in physical education and athletics. In S. Klein (Ed.), *Handbook for achieving sex equity through education.* (pp. 381–410). Routledge: New York.

- The *social development model* is concerned with teaching students self-control and responsibility and draws heavily on the pioneering efforts of Hellison (2011). This model is particularly effective in environments in which students have not had opportunities to develop personal responsibility and positive social skills.
- The *teaching games for understanding* model has received increasing attention from both teachers and researchers. Whereas the traditional way of teaching games seeks first to ensure that students have acquired the motor skills necessary for game play, this model emphasizes teaching tactical awareness, game strategies, and game appreciation while placing less emphasis on initial motor skill development (Griffin, Mitchell, & Oslin, 1997).

KEY POINT

Effective teachers seek to implement curricular models that are interesting to students and produce the greatest opportunity for student learning.

Role Conflict, Burnout, and Resilience

Teaching is a very demanding job, and the rigors of the teaching role must be managed on a daily basis in combination with other roles (e.g., coach, student club sponsor, school committee member, cafeteria monitor, bus supervisor). As a result, your ability to fulfill multiple roles while teaching physical education may determine your ability to perform the teaching role competently. Moreover, beyond the roles that are specific to school, a teacher may also hold a second job, shoulder family responsibilities, or volunteer with a civic organization. As you can see, then, although the teaching profession is sometimes viewed as less than demanding, teachers actually work long hours; they also sometimes face additional stress due to the pressure to excel. The exact nature of these demands depends on where one teaches and on the various parties with whom one interacts, such as students, parents, administrators, and other teachers.

As discussed later in this chapter, some school settings are much more desirable than others. Teachers are subject to varying degrees of stress and burnout, and their ability to be resilient in the face of school demands remains important throughout their teaching career (Richards, Levesque-Bristol, Templin, & Graber, 2016; Richards, Templin, Levesque-Bristol, & Blankenship, 2014). Burnout negatively affects performance and can be manifested in emotional exhaustion, reduced personal accomplishment, and a sense of depersonalization, wherein the teacher feels disconnected from colleagues (Maslach & Jackson, 1986).

In contrast, resilient teachers are better able to resist burnout and persist through stressful events while continuing to deliver effective instruction (Gloria, Faulk, & Steinhardt, 2013). In fact, teachers with high levels of resilience are able to cope even in contexts that are rife with stressful working conditions. These individuals are more likely to enjoy their jobs and remain in the teaching profession for their entire working life.

KEY POINT

Both teaching and coaching are satisfying career choices. Individuals who choose to engage in both simultaneously must be careful to fulfill the obligations of each role.

Another role that many teachers of physical education assume is that of coach for one or more sport teams at the school. Many physical education teachers competed on interscholastic or intercollegiate teams themselves before becoming a teacher, and some elect to teach specifically because of a strong desire to coach. Researchers have studied the dual roles of teaching and coaching over the last decade, and they have found that for some teacher-coaches, a conflict arises in trying to perform both roles (Konukman et al., 2010). Specifically, the research has focused on how one can be an effective teacher of physical education while also striving for success as a coach.

Role conflict occurs when a person takes on two or more roles that are difficult, if not impossible, to perform simultaneously. If we examine the many characteristics linked to teaching and coaching, we can see both compatibility and incompatibility between the two roles. If you recall your own observations of former teachers and coaches and reflect on each role, you may see how successfully performing them both can be difficult for some people. As a result, many individuals withdraw into one role that is preferred over the other; this phenomenon is referred to as *role withdrawal* or *retreatism*. When this happens, one role suffers, and the clients (in this case, either students or student-athletes) likely suffer the consequences of a less than satisfactory experience.

You will occupy many roles in your life, both personal and professional. As you prepare for them and take them on, think about your ability to perform them concurrently and about the implications for your own success and that of others. If, as a teacher-coach, you put as much time into planning good lessons for your teaching as you do in planning good practice sessions for your athletes, then you will have a good chance of succeeding in both roles.

What Instructors Need to Know About Effective Pedagogy

Although research has the potential to inform teachers about how to teach most effectively, most teachers do not subscribe to research journals. Instead, they often derive the knowledge that informs their teaching practice through their experiences as students in the public schools and in other physical activity settings and through certification programs, informal discussions with other instructors, and in-service workshops and conferences.

KEY POINT

Although professional practice knowledge is a powerful source of information, it cannot substitute for thoughtful consideration of the available research literature on effective teaching.

In addition to what they learn from experiences outside of instruction time, teachers also acquire much of their pedagogical expertise in the form of professional practice knowledge. Over hours, weeks, and years of instructional experience, teachers learn (often the hard way!) which techniques work well with students and which do not. Instructors and coaches often refer to their professional practice knowledge when explaining their instructional decisions to others.

Nevertheless, as valuable as personal experience can be, it is a serious mistake for teachers or coaches to look to personal experience as their only guideline for how to teach or coach. Sufficient scholarly research has now been done that any teacher, no matter how experienced, can learn from research. This is particularly important because pedagogical research has less bias and more potential for effectiveness than does personal experience, which is of course limited to the students and situations that a given teacher has encountered.

Research sometimes seems inaccessible to instructors because it can be difficult to understand or does not apply directly to their situations. Even so, scholarly study has influenced the ways in which instructors teach physical education. For example, instructors now have a vocabulary for discussing issues such as academic learning time, management time, off-task student behaviors, and quick lesson transitions. Thus, teachers are gradually becoming aware of the research on effective teaching and making attempts to incorporate the results into their own practice.

Physical Activity and Academic Performance

As a prospective physical activity expert, you will be asked to expound on the benefits of your work with participants. The evidence-based answer, according to a worldwide consensus of scholars (Bangsoo et al., 2016), is that students and clients

Summary of Key Evidence-Based Pedagogical Principles

- Begin to develop expertise by acquiring teaching experience and new knowledge.
- Provide appropriate practice.
- Provide a large amount of academic learning time.
- Always be concerned about class management and discipline.
- Assess student learning.
- Provide clear, specific feedback.
- Develop knowledge about alternative curricular models.
- Ensure an equitable learning environment that addresses the individual needs of all learners.
- Be mindful of the teacher–coach role conflict.

likely benefit in 21 major ways related to fitness and health, cognitive function, and psychological and social well-being. The research on the cognitive benefits is particularly important for our focus here. Although the primary goals of physical activity leaders are to meet objectives related to skill development, fitness, and activity, the movement experiences they facilitate also enhance cognitive health, which in children is measured as academic achievement.

The relationships between physical activity, fitness, cognition, and academic achievement are outlined in a systematic review sponsored by the American College of Sports Medicine (Donnelly et al., 2016). Donnelly and colleagues defined academic *achievement* as the degree to which educational goals were achieved by an individual, which could be expressed in terms of a standardized test score, grades, or school attendance. A relationship also appears to exist between engagement in physical activity and academic *performance*, which can be defined as having the following three components (Rasberry et al., 2011):

1. *Cognitive skills and attitudes:* basic cognitive abilities such as executive function, memory, attention, information processing, motivation, and self-concept
2. *Academic behaviors:* on-task behavior, organization, planning, attendance, emotional control
3. *Academic achievement:* test scores, other formal assessment results, grade point average

Although improving test scores should not be the primary goal of physical education class, participation in physical activity is important for daily growth and development and for maintenance of cognitive functioning. Several comprehensive reviews of the literature on this topic have found illustrative positive associations between physical activity and academic performance (e.g., Castelli et al., 2014; Howie & Pate, 2012):

- Ongoing or chronic physical activity is mostly beneficial to academic performance.
- There is no adverse effect on academic performance of spending additional time in physical education rather than more time in other subjects. Thus, children can be offered opportunities to be active across the school day without sacrificing academic performance.
- Being physically fit and having a lower body mass index (BMI) appear to be positively associated with academic performance.
- Extending physical education is associated positively with higher math, reading, and writing test scores.
- Although the research is inconsistent, some studies reveal a positive relationship between participation in interscholastic sport and academic performance.
- Variables such as gender, socioeconomic status (SES), and age all reveal mixed (positive or negative) association between physical activity and academic performance. For example, a more positive association is typically revealed for children with higher SES than for those with lower SES.
- Opportunities for physical activity before, during, and after the school day should be encouraged by principals and teachers in all subjects and should become a part of ongoing school policy.

This area of research is still evolving. Although there are many positive trends related to the relationships between physical activity, physical education, cognitive development, and academic performance, you should read the research literature to remain informed about future research in this area. Demonstrating that physical education can enhance cognitive performance helps teachers advocate knowledgeably for their programs when communicating with people who are skeptical or who believe that academic performance is affected negatively by time spent in physical education.

Settings for Teaching Physical Education

In an ideal world, we all would have been exposed to excellent physical education settings. These settings typically have good facilities and outstanding teachers. They are a joy to experience, and they contribute to participants' quality of life. In these settings, learning is fun, and participants come away with an expanded set of physical skills. Teachers who work in this type of setting are highly motivated and thrive in a workplace that promotes their effectiveness. In short, these settings are characterized by the factors that promote high-quality work life: respect from students, teacher participation in decision making, collegial support and stimulation, a sense of efficacy among teachers, resources that

The quality of the teaching environment affects teachers' enjoyment of their jobs.

enable effective performance, and a common vision throughout the school (Stroot, 1996).

The quality of the instructional environment in a school often determines the success and satisfaction of both teachers and students. Therefore, it is critical for a teacher of physical education to understand the influence of the workplace and the cultural context in which he or she is employed (Flory & McCaughtry, 2011). Like other workplace settings, schools contain a host of contextual factors that one must understand and cope with in order to be successful. These factors—social, psychological, political, economic, and other organizational factors—may emerge in the individual work setting or class. For example, it is critical for a teacher to have the moral and financial support of colleagues and administrators.

In addition, what happens in the workplace may be affected by larger influences from outside the workplace—for instance, issues related to family, population demographics, civil rights, health care, drugs and violence, educational reform and policy, and advances in information technology (Tyson, 1996). The ways in which instructors cope with these contextual factors inevitably influence their success as instructors and, ultimately, the success of learners.

KEY POINT

Many research findings are interesting to all instructors of physical activity, regardless of the setting in which they work. Some findings are positive, whereas others offer unflattering portraits of the school environment.

Life and Work in the Physical Education Setting

Teacher effectiveness and satisfaction are often linked to the characteristics of the work setting. Thus, it is important to understand the nature of a specific workplace. One important early research study (Locke 1975) yielded the following results:

1. The work life of some physical education teachers is characterized by diverse activity and isolation from other adults.
2. Teacher effectiveness is influenced by differences between students.
3. The curriculum of physical education differs distinctly from those of other subjects in terms of space, activities, and relationships.

Researchers have illustrated both the positive and the negative aspects of school physical education—in short, what life in the gym is like for teachers and students. Research has shown that although elementary physical education is in a relatively healthy state, secondary physical education suffers from a variety of difficult social and political problems (Graber, Locke, Lambdin, & Solmon, 2008; McKenzie & Lounsbery, 2009).

Some scholars have also vividly depicted the malaise that existed in U.S. society relative to the role of secondary physical education in the mid-1980s. Cases of struggle appeared to be more commonplace than profiles of excellence. The various obstacles to excellence at that time were summarized as follows by Griffin (1986):

- Lack of teacher or program evaluation
- Lack of formal incentives or rewards for good teaching

- Lack of professional support for teacher and program development activities
- Inadequate facilities, equipment, and scheduling
- Failure to include teachers in decision making
- Higher value placed on compliance and smooth operations than on teaching competence
- Acceptance of mediocrity
- Isolation

In discussing these obstacles, Griffin stated, "In most cases, the schools did not intentionally limit what could be accomplished. . . . Instead, they practiced benign neglect. The tendency of everyone to ignore physical education presented what seems to have been the most formidable obstacle to excellence" (p. 58). In essence, physical education as a subject area at the secondary level seems to be perceived as superfluous to student learning; accordingly, the physical education teacher holds an occupational status that has been described as marginal.

KEY POINT

Research on school physical education has revealed both positive and negative dimensions of the school environment.

Influences on life and work in the physical education setting include political, organizational, personal, and social factors. For example, a school's economic conditions often influence what facilities and equipment are available to teachers and students. As a result, some schools are able to offer a wide variety of curricular opportunities for students, whereas other schools are more limited. Figure 14.1 illustrates the factors most likely to influence workplace conditions.

In the school context, teachers' success and satisfaction are linked to many variables. Class size, for example, has the potential to influence what material a teacher is able to cover, the amount of feedback that he or she can provide to individual students, and the degree to which students have sufficient opportunities to practice. In addition, school policy may dictate how often students receive instruction or how they will be graded. Thus, the setting in which a teacher works can affect his or her career satisfaction and ability to teach effectively.

Expanding Physical Education

There is reason for optimism about creating a positive work and learning environment for teachers and students in physical education. This hope stems in part from legislative acts such as Public Law 108-265, which requires every school in the United States that participates in the National School Lunch Act to implement a school-wide plan addressing physical activity and nutrition. In fact, physical education teachers have never been better positioned to help lead school initiatives to improve students' health, wellness, and fitness (Graber & Woods, 2007; Woods & Graber, 2007). Though formerly marginalized, physical education teachers are now encouraged to assume roles such as school physical activity director (Castelli & Beighle, 2007).

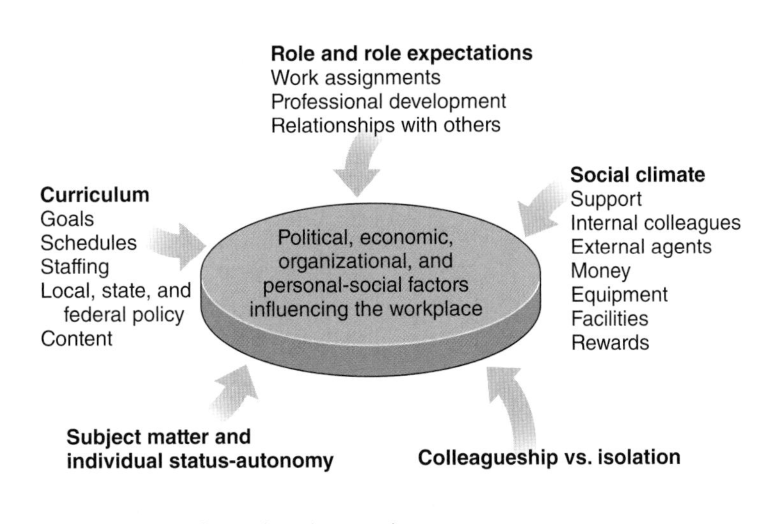

FIGURE 14.1 Workplace conditions influencing the teacher.

One of the most promising new models is the comprehensive school physical activity program (CSPAP), which has the potential to change how physical education is valued in school communities. The model uses a multidimensional approach to engage students in physical activity by encouraging them to participate in physical education, in before- and after-school programs, and in other physical activity programs offered at school. It partners physical education teachers with other members of the school community (e.g., classroom teachers, principals, local professionals) and encourages parental and community engagement in order to foster opportunities for students to participate in at least 60 minutes of moderate to vigorous physical activity during the school day (Society of Health and Physical Educators, n.d.).

The CSPAP model is occasionally criticized because some perceive that its strong focus on physical activity comes at the expense of motor skill development during physical education. However, many scholars believe that a strong CSPAP program can achieve both goals. That is, students can still acquire important motor skills while they are engaged in moderate to vigorous physical activity for the majority of the physical education period.

Roles for Physical Education Professionals

As physical educators become increasingly active in addressing public health problems such as the obesity epidemic—and as greater emphasis is placed on physical activity across the curriculum—it is conceivable that physical education teachers will become less marginalized. In fact, they may come to be perceived as school leaders in promoting physical activity and health through high-quality school programs (Graber & Woods, 2007; Woods & Graber, 2007). Of course, program improvement is not easy, and it must be led by dedicated individuals who possess the ability to negotiate a variety of obstacles and employ effective strategies for youth physical activity interventions (Lau, Wandersman, & Pate, 2016). Physical education teachers who succeed with CSPAP are likely to do the following:

- Use innovative instructional strategies.
- Develop novel curriculums.
- Integrate physical education with other subject matters.
- Develop unique ways to promote learning progressions.
- Use exemplary classroom management strategies.
- Become involved in professional development activities.
- Model athletic skill and fitness.
- Implement after-school programs for students and adults.
- Partner with others, both inside and outside of the school community, to foster additional physical activity opportunities for children throughout the school day.

In addition, successful teachers are likely to have support in the following forms:

- High-profile public relations programs
- Supportive colleagues and administrators
- Adequate funding
- Administrators who promote equitable learning settings

These teachers demonstrate a sincere interest in and enthusiasm for teaching, a genuine concern for students, and a continued desire to grow and develop as teachers. The settings in which they work are not unlike other physical activity settings (e.g., fitness clubs, private golf and tennis clubs, recreation centers) in which a high priority is placed on customer service and on the welfare of participants.

Typical Teaching Responsibilities

Before becoming a certified teacher, most people spend many hours in schools as part of practicum and student teaching experiences, during which teachers in training observe and assist experienced teachers. Although no experience can completely prepare new teachers for all that they will encounter once they enter the instructional setting as certified teachers, practicum and student teaching experiences provide teachers in training with good exposure to students that will ease the transition into the teaching role.

Teachers can expect to meet with students for about 180 days per school year. Although most schools operate on a 9-month calendar, some school districts convene throughout the year. In year-round schooling, students attend school for the same number of days as those who attend for 9 months, but their schedule includes breaks that occur frequently throughout the year.

Teachers at the elementary level generally teach eight to ten classes per day in 30- to 45-minute periods. These teachers meet with students from one to five times per week and sometimes split their time between two schools. Teachers at the secondary level teach fewer classes (five to seven) per day but in longer (45- to 60-minute) periods. These teachers usually meet with students three to five times per week at the same school. Whereas physical education teachers at the elementary level may be the only physical activity specialists at their schools, those in secondary settings often have several colleagues in physical education.

U.S. recommendations indicate that elementary school students should receive at least 150 minutes of weekly physical education and secondary students at least 225 minutes of physical education per week for the entire school year (SHAPE America & AHA, 2016). In most states, however, students receive far less school-based physical education instruction than is called for in these recommendations. For example, the percentage of students who receive physical education in the United States declined from 42 percent to 28 percent between 1991 and 2003. Fortunately, however, because of increased concern about children's health, 44 states have introduced legislative acts since 2005 to improve or increase the physical education (Cawley, Meyerhoefer, & Newhouse, 2006).

With regard to content, teachers at the elementary level tend to teach short units of instruction that emphasize basic skill development and introduce children to concepts such as speed, force, and direction. Teachers at the middle school level tend to teach longer units that introduce students to a variety of sports, games, and fitness and exploratory activities. Those at the high school level typically teach the longest units as a means of helping students acquire adequate skill to participate proficiently in a few activities throughout the life span. Teachers at all levels may also be required to teach classes specifically designed for children with special needs or include them in their regular classes. As a result, most teacher education programs offer courses that provide future teachers with knowledge about such needs, as well as appropriate ways of structuring the curriculum to serve them.

Although teacher salaries are usually modest, the benefits are generally good, and the personal rewards of teaching can be great. Witnessing the joy of a child who accomplishes a new skill or seeing a previously inactive student become an active participant—such moments offer satisfaction that is difficult to find elsewhere.

Physical education teachers help students learn movement skills, sport skills, and lifespan physical activities.

Rewards for Outstanding Teaching

Teachers who are hardworking and competent—particularly those who remain current in the field—may be recognized for their contributions to the profession at the local, state, or national level. For example, they may receive recognition for being among the best in their profession by displaying special characteristics linked to excellent teaching, outstanding curriculums, and the promotion of student learning. You may have wondered who these people are. Where do they teach? How do they teach? What led to their development as great teachers? We have all known great teachers in our own lives. Who are the outstanding teachers who made a difference in your life?

For every subject area, teachers across the United States receive recognition on an annual basis for excellence in teaching. These teachers are selected through various award programs. For example, one program has been offered since 1952 by the Council of Chief State School Officers (CCSSO). According to its nomination process, candidates for this award should

Kinesiology Colleagues

Photo courtesy of William G. Russell

Sara Russell

Each year, SHAPE America honors physical education and health teachers across the nation for exemplary performance. In 2016, the award for High School Teacher of the Year went to Sara Russell of Ravensdale, Washington. A graduate of Central Washington University and a teacher for 12 years, she loves being around students, wants to give back to her community and profession, and promotes well-being in others.

Russell's main goal for students is to gain the knowledge and skills they need in order to live an active and healthy lifestyle. She believes that students should leave class knowing not just what they did but also why they did it and what they learned. To that end, she created a year-long, standards-based class called Foundations that is required for first-year students. The class focuses on health and fitness knowledge, and the activity portion introduces the physical activity electives from which students can select for future classes.

Russell believes that physical education is one of the most important subjects in a student's education. She recognizes the many challenges in teaching and appreciates the many rewards. For instance, she enjoys seeing a student smile when expressing interest or experiencing success in class. Her suggestion to individuals who plan to teach physical education is to be a committed professional and lifelong learner by viewing the best in the field as models.

- inspire students of all backgrounds and abilities to learn;
- be respected and admired by students, parents, and colleagues;
- play an active, useful role in the school and in the community; and
- be poised, be articulate, and possess the energy to withstand a taxing schedule.

In another example related more closely to our discipline, SHAPE America conducts a teacher of the year (TOY) program with sponsorship from Sportime. The first TOY award in this program went to Regina McGill of Bettendorf, Iowa. The selection process begins at the state level, and state winners are then considered for district honors. From there, a special panel of judges considers district winners for the national awards, which are now presented at three levels: elementary, middle, and high school. TOY nominees on all three levels are expected to meet the following criteria:

- Conduct a quality physical education program that reflects appropriate standards of practice.
- Use various teaching methods and plans to provide innovative learning experiences that meet the needs of all students.
- Serve as a positive role model for personal health and fitness, enjoyment of activity, fair play, and sensitivity to the needs of students.
- Engage in professional development.
- Provide service to the profession through leadership, presentations, or advocacy.

In a study of four teachers who received prestigious teaching awards from the state of Georgia, DeMarco (1999) discovered that these teachers maximized student engagement in motor tasks and managed the physical education class with great effectiveness (i.e., students were not disruptive). DeMarco found that all four teachers held a strong belief in the value of their work and in their ability to do that work. They also loved students and loved teaching. Furthermore, they were greatly admired by students, fellow teachers, administrators, and parents. Finally, all four teachers were active in professional organizations, including local teaching organizations and the state and national levels of SHAPE America.

Most students who major in physical education decide to teach in the public schools, but various other options are also available (Graber, Templin, Haag, & O'Connor, 2011). Most teachers, particularly at the high school level, also choose to coach. Some individuals pursue graduate education and obtain a master's degree, which enables one to be an instructor at a college or university; others obtain a doctorate degree so that they can teach and conduct research in a higher education setting. Some elect to work in a fitness facility or community recreation center. Others may simply decide to volunteer their time as a youth sport coach.

Teaching Physical Education in a School

Most certified teachers elect to work in a public or private school at the elementary, middle, or high school level. Their college course work will have taught them both about the *process* of teaching (i.e., instructional strategies, teaching styles) and about the *product* (i.e., curriculum). In most states, prospective teachers must pass one or more state certification exams designed to ensure that they possess basic skills in subjects such as reading and math, have acquired knowledge about the subject matter they plan to teach (in this case, physical education), and are competent to teach

SHAPE America National Standards for Initial Physical Education Teacher Education

Standard 1: Content and Foundational Knowledge

Physical education candidates demonstrate an understanding of common and specialized content, and scientific and theoretic foundations for the delivery of an effective preK-12 physical education program.

Standard 2: Skillfulness and Health-Related Fitness

Physical education candidates are physically literate individuals who can demonstrate skillful performance in physical education content areas and health-enhancing levels of fitness.

Standard 3: Planning and Implementation

Physical education candidates apply content and foundational knowledge to plan and implement developmentally appropriate learning experiences aligned with local, state, and/or SHAPE America National Standards and Grade-Level Outcomes for K-12 Physical Education through the effective use of resources, accommodations and/or modifications, technology, and metacognitive strategies to address the diverse needs of all students.

Standard 4: Instructional Delivery and Management

Physical education candidates engage students in meaningful learning experiences through effective use of pedagogical skills. They use communication, feedback, and instructional and managerial skills to enhance student learning.

Standard 5: Assessment of Student Learning

Physical education candidates select and implement appropriate assessments to monitor students' progress and guide decision making related to instruction and learning.

Standard 6: Professional Responsibility

Physical education candidates demonstrate behaviors essential to becoming effective professionals. They exhibit professional ethics and culturally competent practices; seek opportunities for continued professional development; and demonstrate knowledge of promotion/advocacy strategies for physical education and expanded physical activity opportunities that support the development of physically literate individuals.

effectively. In order to help university professors design an undergraduate curriculum that adequately prepares future teachers, SHAPE America (2017) has developed a set of national standards that most U.S. teacher education programs strive to achieve (see the sidebar listing the SHAPE America national standards). These standards relate to the knowledge and skills that you will be expected to apply as a teacher of physical education.

Teaching in Higher Education Settings

Some individuals choose to teach at a college or university. If you obtain a master's degree, you are qualified to work as a college-level instructor, in which position you might be responsible for teaching skills courses (e.g., basketball, aquatics, racket sports) or supervising student teachers. Although some people elect to remain employed as an instructor throughout their career, others view this option as too limited because, without a doctorate, they cannot teach or advise graduate students and may be unable serve on important committees.

By obtaining a doctoral degree, which is the precursor to the professorial ranks, people gain access to additional career responsibilities that are often considered desirable. Most professors of pedagogy have also worked at one point or another as a teacher in a public school. Many elect to work at the college or university level because they want to improve teacher education. An increasing number of professors also want to engage in research and mentor graduate students. They design studies intended to help us better understand the work of teaching and how to promote instruction that leads to meaningful student learning. Professors also present papers at professional conferences; participate on national committees; collaborate with teachers in the public schools; remain informed about state certification requirements; and serve on departmental and other committees.

KEY POINT

If you decide to become a professor of physical education pedagogy, you should acquire public school teaching experience. This background gives you greater credibility in the eyes of both your students and public school teachers.

Regardless of whether you elect to pursue graduate study in physical education teacher education, you should obtain public school teaching experience. Would an instructor or professor in a university teacher education program have greater credibility in your eyes if he or she had several years of experience as a teacher? In most cases, your answer will be yes. Public school experience can also open the door to other intriguing career possibilities such as recreation instructor, whitewater rafting guide, or other areas involving physical activity. In many cases, it is a requirement if you wish to be hired at the college or university level as a teacher educator.

Adapted Physical Education

Adapted physical education consists of a program of skill instruction and exercise that has been modified so that all students can participate. In the United States, federal Public Laws 94-142, 101-476, and 105-17 stipulate that students with disability will be provided with a program of physical education that includes motor skills; fundamental skills such as running, catching, and throwing; and skills in games, sports, and aquatics. These laws cover a huge range of disabilities, including Down syndrome, hearing disorders, spina bifida, muscular dystrophy, and a host of others. If a child has unusual restrictions, the instruction must be provided through an individualized program developed by the child's teacher, parents, and other interested parties. Where the disability permits, children must be mainstreamed into the regular classroom or gymnasium as a way of meeting Public Law 94-142, which means that physical education teachers must be prepared to teach a curriculum to children with a diverse range of abilities. In larger school districts, students with similar disabilities may be grouped for physical education instruction.

One cannot presume to teach students who have serious disabilities without understanding something about disability and how it can affect a child's performance. To meet this need, some colleges and universities offer certification or endorsement in physical education special education; these programs are often modeled on the Adapted Physical Education National Standards (APENS) and the Certified Adapted Physical Educators (CAPE) national certification. Your state probably requires at least one course in adapted physical education as part of your teacher certification program; you might also consider specializing in this exciting area.

Beyond the public schools, specialized schools usually hire a staff of adapted physical education teachers to work with their chosen populations, such as children with intellectual disability, a severe orthopedic disorder, or limited or no capacity for

Kinesiology Colleagues

College of Applied Health Sciences, University of Illinois at Urbana-Champaign

Ben Kern

Ben Kern initially decided to become a physical education teacher because it matched his interests in exercise science, teaching, and coaching. After teaching for 7 years in a high-poverty, rural public high school in southern Colorado, he realized that the work of physical educators involves much more than just exercise and sport; it is also a critical part of developing young peoples' knowledge, skills, and attitudes for living an active and healthy life. Kern also served as director of strength and conditioning and as an adjunct instructor at Adams State University in Alamosa, Colorado, before being offered a unique opportunity—to help design and implement a program using evidence-based physical education to support children's health in high-poverty areas. This work was part of a research initiative funded by the University of Colorado Denver, and his role was to support physical education teachers as an instructional coach.

The success of the program convinced Kern that he could help others achieve a healthier life through research, teaching, and service in kinesiology. Thus, he completed his PhD at the University of Illinois at Urbana-Champaign and is now a faculty member at the University of Louisiana at Lafayette. There, he continues to conduct research and teach others about ways to improve quality of life through physical education and school-based physical activity.

seeing or hearing. If working with children and adults who have been challenged by disease or injury is a career possibility that excites you, give some thought to obtaining a master's degree in adapted physical education; this background will prepare you for the challenges of such a career.

Coaching

Many people who enter teaching do so because they desire a career in coaching. That is, they are less interested in teaching and more interested in coaching. Although a career in teaching physical education offers entry into coaching, a teacher-coach can become less committed to teaching than to coaching (see the earlier discussion of teacher–coach role conflict). In such cases, students who are in the classes of these teacher-coaches often suffer.

Of course, many teacher-coaches do an outstanding job of balancing the workload of teaching and coaching. These individuals believe that physical education is critical because it offers an opportunity for students to acquire skills regardless of ability. They carefully plan lessons that promote skill development in all students. These teacher-coaches also believe that athletics are important because they offer students an opportunity for advanced competition. Therefore, they carefully manage their time to enable success in both roles.

Coaching is not an easy job, and coaches may confront situations that are challenging and unexpected. Those who elect to coach spend many hours at their schools each day. They often arrive earlier than other teachers do in order to plan classes and practices. Although some coaches receive an extra release period during the day for planning purposes, many spend their evenings and weekends planning practices and scheduling competitive events. Immediately before or after school, they spend several hours practicing with students, engaging them in drills and fitness activities that will enable them to compete successfully against other students. Some coaches handle only one sport during the year, whereas others elect to coach year-round and even during the summer months. Some become so engaged in the coaching role that they offer summer clinics or camps for students.

Although most physical education teachers elect to coach in the public schools, some prefer

Case Study in Adapted Instruction

Town Middle School had a great adapted physical education program that helped students with disability stay active. The teacher responsible for the wonderful program was Ms. Kelly, and students with disability were uniformly placed in her adapted physical education class. In contrast, students with disability were never placed in the other physical education class, which was taught by Coach Stevens.

Then, one day when Coach Stevens was taking attendance, he called the name "Joe Henninger," only to have no one respond. He assumed that the student was absent that day, until a few other students explained to him that Joe was hearing impaired. At first, Coach Stevens thought it was a student prank, but he soon discovered that the students were telling the truth. He then watched as a student walked up to him while speaking and signing at the same time. With a worried look on his face, Coach Stevens thanked Joe for introducing himself.

During the class activity, which was knockout basketball, Joe played the game well and even won a few times. Coach Stevens was surprised that Joe was athletic and exhibited great skills. However, these factors did not change Coach Stevens' mind about asking the principal, Mr. Hicks, to place Joe in Ms. Kelly's care in the adapted physical education class. After a meeting between Mr. Hicks, Coach Stevens, and Ms. Kelly, Mr. Hicks agreed to move Joe back into the adapted class. Joe was extremely disappointed and appealed to his parents to change the situation. He felt that his rights were being violated.

Your Decision

Do you think Coach Stevens' reaction to Joe was fair? Based on your experience with students with disability, how likely is it that other experienced teachers would respond in a similar way? How might Ms. Kelly (or you) advocate for Joe to remain in Coach Stevens' class? What role does federal law have in the local school district's decision about Joe's placement?

Resolution

In a meeting involving Joe's parents, their attorney, Mr. Hicks, Coach Stevens, Ms. Kelly, and the school district's attorney, the parties concluded that, under federal Public Law 94-142, Joe was entitled to be placed in Coach Stevens' class and that Coach Stevens should make every effort to accommodate Joe in promoting a meaningful learning environment.

coaching at a college or university. Most begin their careers as graduate teaching assistants enrolled in master's degree programs. As they gain experience and acquire name recognition, a few of these people obtain an assistant or head coaching position in higher education. At the most competitive level, Division I, these individuals only coach, and their career advancement is contingent on winning. As a result, they face enormous pressure to recruit and produce winning teams. For more information on coaching at a competitive level, see chapter 15.

Trends and Opportunities in Teaching Physical Education

If you intend to become a teacher of physical education, your prospects for finding employment are high. In many states, schools have a desperate need for physical education teachers. In Illinois, for example, students at all grade levels are required to take physical education daily; as a result, severe teacher shortages exist in many areas of the state. In some states, the need for teachers is so severe that undergraduate students can apply for a loan that will be forgiven if they agree to teach in that state for a certain length of time after they become certified. Rapidly growing states and counties also have a tremendous need for good physical education teachers. If you are willing to move to a part of the country that needs physical education teachers, you are likely to find employment quickly.

You can improve your employability by ensuring that you acquire as many skills as possible while enrolled in a teacher education program as an undergraduate. You may know that physical education teachers who are also good coaches are always in demand, but you may not know that physical

Case Study in Coaching

Before the season started, rumors started filling the hallways of Eden High School. The students were talking about "hazing." Overhearing them, Bonnie, a physical education teacher at the school, briefly wondered whether the buzz was about a recent incident at a nearby school, which had been in the news, or was just gossip. Her attention quickly moved to other subjects, however, because she had never seen a problem with hazing at Eden High. She definitely didn't think it was a problem in her classes or with her girls' basketball team.

Soon, however, Bonnie learned that the problem was, in fact, happening at Eden. To her dismay, she learned that hazing was an annual ritual of the boys' basketball team—and one that was known by the team's head coach. In the ritual, the team's seniors made the first-year members (some of whom were in her physical education class) sneak out of their homes at midnight and meet in a secluded area behind the local mall in subfreezing weather. Forced to strip down to only their underwear, these boys were subjected to an hour of strenuous physical exercises. Afterward, the first-years were sworn to secrecy about the "private meeting" and told to go home. This time, however, one of them had "blown the whistle" after becoming very ill and having to go to the emergency room.

Your Decision

What is the responsibility of the head coach of the boys' team in this situation? What is the role of the school principal and athletic director? Should rituals be accepted in interscholastic sports? If so, what types of rituals? What do you think should happen to the seniors at Eden? To the head coach?

Resolution

Separate meetings were called with the seniors and the first-years, as well as their parents. The meeting participants included school administrators, the athletic director, the boys' basketball coach, and representatives of the school faculty. The situation was presented by the school principal, testimony was obtained from all parties, and the school hazing policy was reviewed. Subsequently, the evidence was evaluated by a school panel established to rule on the case. The panel ruled that the seniors should be suspended from school for a week and dismissed from the team. The coach was permanently dismissed from the coaching role and placed on probation for the remainder of the year as a teacher.

education teachers often find employment because they also possess the following qualifications:

- Certification as an athletic trainer
- Certification to teach health education
- Ability to speak a second language fluently
- Endorsement to teach in additional high-demand subject areas
- Certification to teach driver's education

Schools that do not currently offer physical education are likely to add it as a required curricular area in the future. This likelihood relates to the fact that the obesity epidemic has created greater awareness of the need for students to engage in physical activity on a daily basis. For some students, school is the only place where they have an opportunity to be active. As more schools add physical education or increase the time devoted to the subject, the need for teachers will increase. Thus, there has never been a better time to enter the profession.

For individuals who wish to teach or coach outside of the K-12 setting, the education they receive while enrolled in a teacher certification program will place them in good standing for whatever teaching-related position they may eventually pursue. For example, have you ever observed community soccer coaches who allow students to stand in long lines waiting for a turn while one student at a time kicks the ball? This problem would not occur in a program run by someone who was educated in a teacher certification program, regardless of whether that person ever taught in a K-12 school. Those who have been trained in pedagogy understand that students cannot learn without practice opportunities and that standing in a long line does not facilitate student learning.

Selecting a Career

List the advantages and disadvantages of each of the following career options.

- Teaching physical education in a public school
- Teaching in a higher education setting
- Teaching adapted physical education
- Coaching in a public school
- Coaching at the college level (e.g., Division I school)

Advice for Physical Education Students

When effective teachers are asked to describe the point at which they really learned how to teach, their response is that they have never stopped learning to teach. That is, they view learning to teach as a lifelong process that continues well after one becomes certified and leaves the influence of a teacher education program. In short, learning to teach is an ongoing process. Therefore, we encourage you to engage in professional discourse with colleagues, attend professional conferences, and become familiar with knowledge generated by research investigations. Just as you expect your physician to remain current in the medical field, parents and students will expect you to remain current in yours.

Wrap-Up

This chapter introduces you to the knowledge base for the pedagogy of physical education. Exciting developments have occurred over the past 2 decades, and new knowledge will continue to emerge as researchers ask more sophisticated questions and develop new methods for collecting data. The information presented here is but a small sample of the types of knowledge currently available to those who work in pedagogy.

We hope that you will aspire to be like one of the teachers of the year—someone who is committed to student learning, engages in effective teaching practices, works toward developing subject-matter expertise, and remains professionally involved and current in the field. We also hope that you will receive the same satisfaction in your career that has been experienced by the thousands of teachers who enter the workplace every day in the belief that they can make a difference in the lives of children.

If you are considering embarking on a teaching career, you will need to know much more about pedagogy than we are able to discuss here. You will find greater success if you consistently and objectively reevaluate your current beliefs in light of new, research-based knowledge about effective teaching. As a member of the next generation of teachers, you can make a difference.

MORE INFORMATION ON CAREERS IN PHYSICAL EDUCATION

Organizations

American Educational Research Association

Association Internationale des Ecoles Supérieures d'Education Physique

Federation Internationale D'education Physique

National Association for Kinesiology and Physical Education in Higher Education

Physical and Health Education Canada

Society of Health and Physical Educators

Research on Learning & Instruction in Physical Education

Special Interest Group of the American Educational Research Association

Journals

Action in Teacher Education
American Educational Research Journal
Educational Technology
Elementary School Journal
High School Journal
International Journal of Physical Education
Journal of Educational Research
Journal of Physical Education, Recreation and Dance
Journal of Research and Development in Education
Journal of Teacher Education
Journal of Teaching in Physical Education
Kinesiology Review
Medicine & Science in Sports & Exercise
Physical Educator
Quest
Research Quarterly for Exercise and Sport
Sport, Education and Society
Strategies: A Journal for Physical and Sport Educators
Teaching and Teacher Education

REVIEW QUESTIONS

1. How do your experiences as a student of physical activity connect to the examples of effective and ineffective teaching discussed in this chapter?
2. How can you develop expertise in one or more of the content areas of physical education?
3. How might you convince others that physical education class does not detract from student learning in other academic subject matters? What would you say?
4. Describe three ways in which a teacher you have observed could increase the amount of learning time available to students.
5. To reduce management time, what routines are most important for students to learn? Which routines would you emphasize? Describe how you would implement those routines.
6. Which curricular framework would you choose to implement as an instructor? Why?
7. How can you avoid the pitfalls of teacher–coach role conflict?
8. Describe your high school physical education program and how you would work to improve the quality of that program.

Go online to complete the web study guide activities assigned by your instructor.

15

Careers in Coaching and Sport Instruction

Joseph A. Luxbacher and Duane V. Knudson

The authors acknowledge the contributions of Shirl J. Hoffman to this chapter.

CHAPTER OBJECTIVES

In this chapter, we will

- acquaint you with the wide range of professional opportunities available in the sphere of coaching and sport instruction;
- familiarize you with the work of, and qualifications for, coaching and professional sport instruction;
- inform you about the educational requirements and life experiences necessary to become a qualified and successful coach or sport instructor; and
- help you determine whether one of these professions matches your skills, aptitudes, and professional desires.

Rob, a former Division I soccer player who earned a graduate degree in kinesiology, coached soccer at a nationally ranked university. His team had suffered through a couple of down years, and Rob knew that retaining his job could depend on qualifying for the NCAA tournament. He was counting on the team to make a run through the conference tournament and earn an automatic bid to the NCAA tournament. Rob's hopes were based in part on the talents of his star player, Javier, who led the conference in goals scored, assists, and total points. The team won its first-round playoff match 2-1, with Javier scoring both goals, then won its semifinal match to advance to the conference championship. Javier also scored a goal in the semifinal but had to leave the game shortly after halftime when he clashed heads with an opponent while trying to head the ball.

During a light training session on the following day, Javier complained of a headache, upset stomach, and dizziness. Rob realized that these symptoms could be indications of a concussion. As a coach, he had to make a decision—inform the training staff of Javier's symptoms, which would probably keep him out of the championship game, or keep the information to himself. The next day, at the pregame walk-through, Rob asked Javier how he felt. Javier said that he still had a headache and had not slept much the night before, but he attributed that to nervousness about the game. He insisted that he was fine—just sore—and wanted to play.

Rob weighed everything in his mind and came to a conclusion. He informed the training staff of Javier's symptoms, and Javier was pulled from the final. The team went on to lose a tightly contested game in overtime and did not receive an NCAA bid. Rob was disappointed with the outcome but content with his decision. Not long after, he was informed by the athletic director that his contract would not be renewed and that he should begin searching for a new job.

Coaching can be a satisfying and enjoyable profession; it can also pose philosophical challenges and dilemmas. Coaches at the highest levels of competition are expected to compete for championships, but the pursuit of that goal is often affected by factors and circumstances that lie beyond the coach's control. By its nature, coaching also brings into close contact individuals with different personalities who may have different expectations of the sport experience, whose short- and long-term goals may conflict, and who may perceive the world of athletics in different ways.

The previous chapter focused on careers in teaching physical education. Although teaching is a major professional track in kinesiology, some students set their sights on a career as a coach or as a sport instructor in a nonschool setting. Of course, physical education teachers in schools often take on coaching positions as additional duties, but there are also increasing opportunities to coach in nonschool settings. These opportunities will only continue to grow as more swimming, tennis, and golf facilities open; as sport camps and instructional sport academies continue to expand; and as public interest in community nonprofit sport initiatives increases. These careers serve as the focus of this chapter.

All of the professions in this sphere involve helping participants develop either modest or high levels of motor skill and sport tactics. The participants served by coaches and instructors vary tremendously, from young children to university students and athletes to professional sport figures. There is also great variation in the settings in which these individuals work—for example, public and alternative schools, community agencies and organizations, colleges and universities, sport clubs and academies, community agencies, and professional organizations that are operated as businesses. Although institutions and organizations employ most coaches and instructors, many are freelancers; that is, they are self-employed instructors (e.g., in gymnastics, golf, baseball, tennis, swimming) who operate their own small companies.

What Are Coaching and Sport Instruction?

Coaching and sport instruction have a long and storied tradition, extending back as far as the ancient Greeks and Romans. We know, for example, that coaches were hired by early Greek municipalities, and supervised by public officials called *gymnasiarchs,* for the job of teaching young people how to compete in various forms of athletics. Although held in high esteem, these coaches could be demanding taskmasters. One notable Greek vase depicts a trainer (or coach) flogging two *pankratiasts* (athletes who performed a vicious form of wrestling popular in the early Olympic Games) for gouging each other in the face while wrestling.

Now, many centuries later, coaches continue to be popular and well regarded, and many continue to have reputations as demanding taskmasters (although none flog their athletes!). Some ply their trade in the anonymous quarters of community gymnasiums and school athletic fields; others work in the glaring spotlight of the collegiate and professional ranks. Some coaches earn very little; others earn millions. Regardless of fame or fortune, coaching and sport instruction can be a fascinating and rewarding career. This chapter helps you determine whether it might be one for you to consider.

Coaching and sport instruction are similar in many ways. Both are performed by professionals who have mastered knowledge about physical activities, as well as the skills and techniques required to transmit that knowledge to other people. Both also require expertise in designing practice experiences that stimulate learning and conditioning experiences to enhance performance.

But coaching and instruction also differ significantly; in particular, they tend to teach people at different skill levels. An instructor's efforts are usually, though not always, directed toward novices or those lacking a high level of proficiency—for example, young children in a community youth sport development program, middle-aged people in a beginner swimming class at a community pool, and people of all ages who want to improve their swing by consulting a teaching professional at the local driving range.

In contrast, coaches at the high school, college, and professional levels typically direct their planning and efforts toward an already skilled population. Teams at the high school level are composed primarily of students with above-average physical proficiency for their age and gender. College teams tend to be populated by an even more select group culled from the best of the high school teams, and, of course, professional and elite (i.e., international-level) athletes comprise an even more highly skilled group. As the level of competition increases, the pool of eligible performers gets smaller, and the pressure to win typically grows.

Even youth sport programs, whether inadvertently or by design, tend to weed out those who are less gifted in their age groups, leaving a population of relatively skilled youngsters as compared with others of similar age. Sad to say, this phenomenon appears to have become the rule rather than the exception. As a result, kids who are physically less mature are typically eliminated early in the selection process and therefore do not have an equal opportunity to continue with the sport.

Despite these differences, teaching and coaching share many functions. Both involve instructing and interacting with people on a personal basis; in fact, the dictionary definition of *coaching* is close to that of *teaching*. For example, both a gymnastics coach and a gymnastics teacher try to impart knowledge, help gymnasts develop skill at certain routines, instill in them a love of the activity, and help them improve their performances. Improvement comes both through practice and conditioning and through explanations, instructions, and feedback (verbal as well as visual, in the form of video replays and demonstrations).

So, how do we determine what distinguishes instruction or teaching from coaching? One good way to address the question is to distinguish between the acts of teaching and coaching and the professions of teaching and coaching. The acts of teaching and coaching may both be designed as attempts to alter the thinking, feelings, or behavior of a particular clientele by systematically exposing

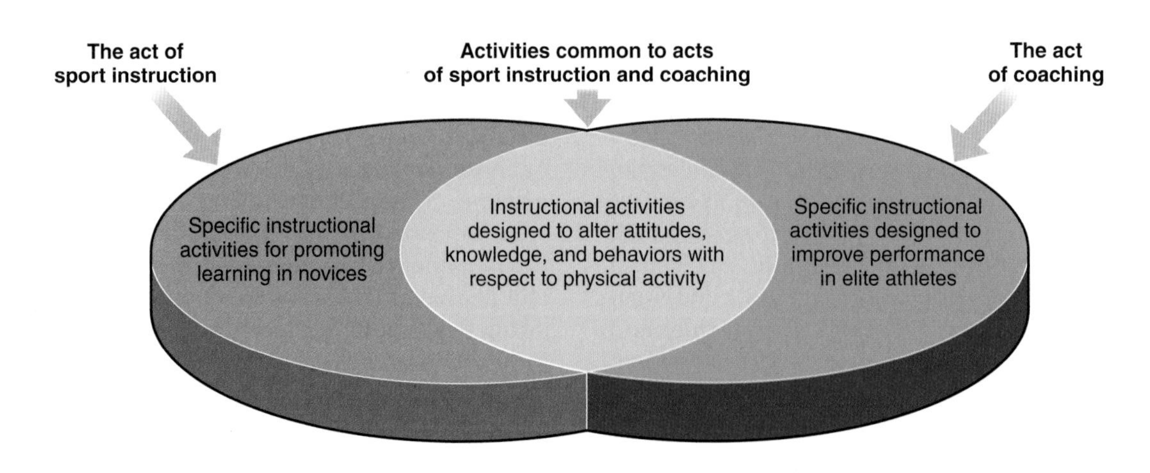

FIGURE 15.1 The acts of sport coaching and of sport instruction are characterized by considerable overlap.

them to physical activity experiences, along with appropriate verbal and visual experiences, to bring about predetermined outcomes. Figure 15.1 shows the overlap in the acts of instruction and coaching, as well as their uniqueness; the areas that do not overlap tend to be those directly related to the clients served.

As mentioned earlier, instruction focuses primarily on novices, whereas coaching focuses on more advanced performers. As a consequence, instruction typically focuses on skill development, whereas coaching encompasses skill *refinement* coupled with learning about tactics (strategy). Coaching is usually directed toward select or advanced populations who, at least to some extent, have already acquired the skills, knowledge, and attitudes essential for performance. Thus, the overriding emphasis for coaches, particularly at higher levels of competition, must be placed on individual and team performance, because success is typically measured by wins and losses. As a result, coaches, especially those in the college and professional ranks, tend to direct much of their effort toward improving skills already learned, rather than teaching new or basic skills, and then meshing that improved technical ability with sport-specific tactics and a high level of muscular fitness, aerobic fitness, or both.

Overall, coaches may spend a disproportionate amount of time motivating and conditioning athletes and refining and retaining acquired skills. Sport instructors, on the other hand, are likely to focus their efforts on helping their clients acquire new skills and learn how to apply them in real-life settings.

But making too fine a distinction between the acts of teaching and coaching is probably not useful because the roles associated with these positions can switch back and forth. Indeed, it is safe to say that most coaches consider themselves to be, first and foremost, teachers; conversely, sport instructors, particularly at the youth level, commonly envision themselves as coaches. Youth sport coaches, for example, spend much of their time teaching children how to perform basic skills; their emphasis is not, and should not be, on who wins the games but rather on maximizing each player's level of expertise. However, coaching at the high school varsity or college level usually involves more of the unique aspects of coaching identified in figure 15.1.

Sometimes, professional sport instructors serve in a role that we might define as coach-instructor—for example, when they supply instruction on a one-to-one basis to professional golf or tennis players. In fact, some of the more successful golfers regularly visit such instructors (also called *teaching professionals* or "swing doctors") to correct flaws that have caused their performance to deteriorate. Many professional tennis players include staff in their entourage who perform the same function. Regardless of whether one refers to these individuals as teaching professionals, golf or tennis instructors, or coaches, they play fundamentally different roles—and work in quite different settings—than do coaches of scholastic athletics or college-level athletes or those who supply sport instruction to large groups in municipal recreation settings.

Despite this considerable overlap between the *acts* of coaching and instruction, the same is not true for the *professions* of coaching and sport instruction. By *profession*, we refer to the entire range of duties carried out by a coach or instructor beyond the direct acts of coaching and instruction per se. Instructors and coaches must carry out different professional

Case Study in Determining Practice Schedule

Mohammed was the newly appointed coach of the high school track team. He had plenty of experience as a runner, having competed at the international level, but very little experience in performing or coaching field events. He was concerned about his lack of knowledge of the weight events, particularly the shot put. He had read books about the skill but wanted to take a scientific approach that incorporated all that he had learned in kinesiology classes. He reflected on what he had learned in motor learning and performance, biomechanics, and exercise physiology to come up with a basic practice plan.

Bringing his analytical ability to bear, Mohammed recognized that the shot put is a closed skill (see chapter 3) that requires improvement in both physical performance capacity (strength, mainly) and skill development (specifically, a coordinated series of perfectly executed movements). He reviewed his notes from his classes and believed that he now had a good handle on the mechanics of the event and understood the best way to develop athletes' strength. The question of how to arrange practices, however, remained unanswered.

Your Decision

How would you design practices for this event? Would you vary the practice conditions by changing the distance and direction of the athletes' throws? Would you have the athletes perform the movements without the shot during early trials? Would you have them practice the skill repetitively—say, at least 25 successive trials in each practice session?

Resolution

Given that the shot put is a closed skill, Mohammed decided to have his athletes focus on developing a stereotyped, effective movement pattern rather than varying the practice conditions from trial to trial. He realized that expert performance of this skill ultimately depends on executing a power-producing series of movements and that the weight of the ball might cause his athletes to develop mechanically incorrect technique in early trials. Therefore, he decided to move slowly by first having athletes work on technique and then introducing the ball. Finally, he recognized that "massed practice trials" would lead quickly to fatigue and faulty performance patterns, so he adopted a "distributed" schedule in which his athletes performed only 15 trials per practice. Moreover, the throws were interspersed with walking to retrieve the shot and with rehearsing the movements without the ball.

responsibilities in different occupational subcultures. Figure 15.2 shows the differences and similarities in the instruction and coaching professions. As you can see, the similarities here are considerably fewer than the similarities between the acts of instruction and coaching shown in figure 15.1. There are two major differences:

1. Instructors tend to spend more of their time involved in on-task duties—that is, in disseminating knowledge and molding students' behavior. Coaches spend less of their time on this work and more on the off-task duties of recruiting, scouting, reviewing video, planning strategy, addressing compliance (eligibility) issues, scheduling, budgeting, and fundraising.
2. The nature of the off-task duties also differs between the two jobs. Whereas instructors spend time maintaining records, repairing and maintaining equipment, advertising classes, and (if teaching in an institution) attending to institutional demands, coaches are more likely to become absorbed in the off-task duties described in figure 15.2.

KEY POINT

The *acts* of sport coaching and sport instruction are more similar than distinct; the *professions* of sport coaching and sport instruction are more distinct than similar.

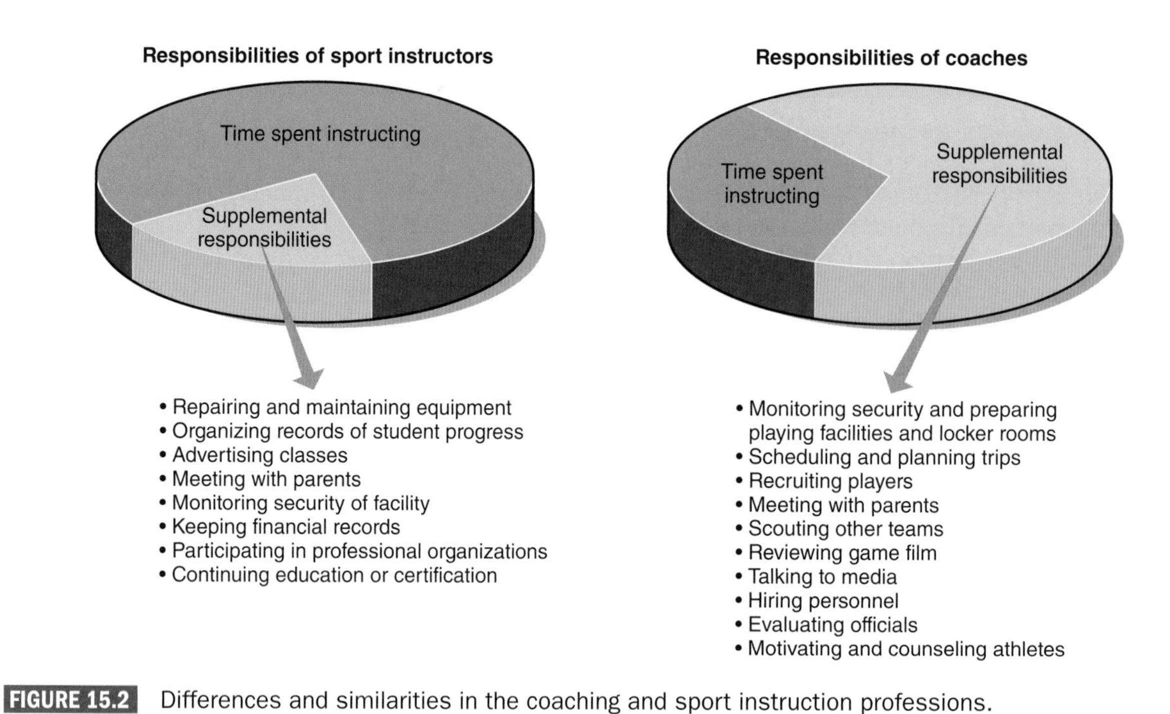

FIGURE 15.2 Differences and similarities in the coaching and sport instruction professions.

Based on the study of G.M. DeMarco 1999, "Physical education teachers of the year: Who they are, what they think, say, and do," *Teaching Elementary Physical Education* 10(2): 11-13.

Clearly, then, in the most fundamental analysis, the *act* of instruction is essentially the same as the act of coaching. To some extent, all teachers are coaches, and all coaches are teachers. However, when we consider the *professional* responsibilities of the two roles, we see that the jobs are really quite different. Even when a person plays both roles in the same job—for example, when a school or college physical education teacher also coaches—the demands of the two roles are distinct. Therefore, although we recognize similarities in many of the professional responsibilities of coaches and teachers, we also need to keep in mind that each operates in a distinctly different occupational subculture and requires a unique set of professional knowledge and skills. In addition, the ultimate goals to be achieved are not always identical.

Settings for Coaching and Sport Instruction

A work setting usually defines the job expectations and duties of those who work there. Coaches and sport instructors work in many different settings, and their duties and responsibilities vary accordingly. Most of the settings fall into one of the following three categories: community based, institutional, and commercial.

Community Settings

Community organizations offer a variety of settings for coaching and sport instruction. Because most of the organizations offering sport programs in these settings are nonprofit, many of the coaching positions are filled by volunteers, although some youth and adult sport leagues hire part-time coaches. Parks and recreation departments, for example, usually sponsor athletic leagues for both youth and adults. Similar opportunities are provided by other organizations, such as the YMCA and YWCA, the Young Men's Hebrew Association (YMHA) and Young Women's Hebrew Association (YWHA), the American Youth Soccer Organization (AYSO), the U.S. Volleyball Association, and the U.S. Fencing Association.

Community sites for coaching and sport instruction may also be overseen by municipal recreation authorities. Although recreation has developed into a specialized profession in its own right, the career aspirations of recreation graduates and kinesiology graduates sometimes overlap. If a career in municipal recreation interests you, explore the possibility of obtaining an undergraduate minor in recreation or taking some courses in recreation leadership and management.

Community organizations are civic minded; indeed, their first goal is to improve the social, phys-

Profile of Boys & Girls Clubs of America

- Overview: safe and fun places for youth during out-of-school hours; diverse programming for participants
- Scope: national (4,200 locations in 50 states)
- Locations: urban, suburban, rural
- Setting: community-based organization, public schools, recreation centers
  - 1,600 school-based clubs
  - 490 affiliated youth centers on U.S. military installations worldwide
  - 970 clubs in rural areas
  - 300 clubs in public housing
  - 170 clubs on Native lands (largest youth development provider for Native youth)
- Participants: kindergarten through high school students
- Funding: public grants, corporations, foundations, investment income

ical, and moral development of the local community. Think back to the discussion, in chapter 11, of the two value systems that influence professional conduct. Coaches and instructors working in community programs are usually guided by a social trustee, civic professionalism orientation. For funding, community organizations rely on municipal funding, user fees, dues, private and public contributions from agencies such as the United Way, and business or corporate sponsorships of athletic teams. Some organizations are sponsored directly by tax money; most are funded through donations and corporate sponsors. The Boys & Girls Clubs of America, for example, serve 4.5 million boys and girls at 4,200 locations in 50 states with a trained staff of 49,000—and they do so through funding (approximately $230 million) provided by major corporations. Another prominent organization is the Joy of Sports Foundation in Washington, DC, which is one of the largest community programs to serve at-risk young people through sport; it uses both paid and volunteer staff.

Instructors in community settings may work in pool areas teaching swimming classes, in gymnasiums or weight rooms facilitating adult group exercise programs, in testing rooms conducting stress tests and fitness evaluations, or in conference rooms engaging in health and nutrition consulting. Typical worksites for coaches of community-based sport teams include gymnasiums; skating rinks; soccer, football, softball, and baseball fields; and tracks. Generally, games are played on the organization's property or on property owned by municipalities or schools.

Institutional Settings

Most of the sport instruction in institutional settings takes place in physical education class (see chapter 14). As an arm of education, physical education is part of the formal curriculum in public and private schools, colleges, and universities. However, sport also is an important part of the extracurricular program

Settings for Coaching

Community settings	Institutional settings	Commercial settings
YMCA, YWCA	Middle schools	National and Olympic facilities
Boys & Girls Clubs of America	High schools	Private clubs, camps, and sport academies
Municipal recreation facilities	Private secondary schools	Professional sport organizations
National youth sport programs	Community colleges	Freelance and entrepreneurial settings
Various nonprofit organizations	Colleges and universities	

in educational institutions. As a high school student, you probably gained an appreciation of the importance that people in your school and community placed on interscholastic sports. As a college student, you are undoubtedly aware of how important sport programs have become for institutional recognition and recruitment. Many universities are identified more by the success of their football and basketball programs than by their academic excellence.

High School and Middle School Settings

In public and private schools, coaches usually serve on the teaching faculty and perhaps in an administrative capacity as well; alternatively, some districts hire part-time coaches if fewer full-time teachers opt to take on these additional responsibilities. These part-time coaches may have expertise in their chosen sport but sometimes have little training in kinesiology or education. Fortunately, school boards and superintendents are increasingly recognizing the need to hire only fully qualified coaches to work with student athletic teams. Still, in most cases, coaching is an add-on to the individual's primary responsibility as a teacher of physical education or another subject.

KEY POINT

For middle and high school coaches, directing the team is merely one of many responsibilities they have at the school. For most, the primary responsibility is teaching classes.

Scholastic coaches are often in the public eye, particularly when they coach football or basketball. Because athletic teams are viewed as representatives of the school to the broader community, coaches must pay careful attention to the conduct of their athletes and, of course, their own conduct. Practices and games typically take place on school property; of course, travel to other schools is necessary for away games. Coaches usually work during after-school hours, and in some sports the season can last more than 5 months. In many private schools, every student is required to be a member of at least one interscholastic team, and coaches must accept all willing and committed students as part of a given team, regardless of their athletic talent. Consequently, the range of talent among team members in private schools can vary greatly. In these situations, the coach must be able to deal effectively with both highly skilled and less skilled players.

College and University Settings

Work settings at the college level vary according to the type of institution and the level of athletic competition. Coaches at community colleges usually coach one sport and teach kinesiology classes, although the setup can vary. Teaching and coaching take place in the same facility, which generally consists of an office complex, gymnasiums, weight rooms, swimming pools, and an athletic training room. Because the programs at many community colleges and technical colleges cater to part-time adult students, some schools do not place strong emphasis on athletic programs. At these schools, athletic programs and facilities may be modest, although that is not always the case. As you might expect, larger schools that emphasize sport tend to have larger, better-equipped facilities. Athletic programs at larger community colleges usually participate as members of the National Junior College Athletic Association (NJCAA), which sponsors national championship competitions each year in a variety of sports. The NJCAA has more than 500 member institutions and offers sports in Divisions I through III to more than 45,000 athletes.

The worksites at some small 4-year colleges are similar to those of community colleges. Coaches at small colleges are likely to hold faculty posts in kinesiology in addition to their coaching responsibilities. College coaching linked to a faculty positon almost always requires a master's degree; coaches in some small institutions gain tenure as faculty members. Coaches who serve as instructors in the physical activity academic program must juggle the responsibilities assigned by the head of the kinesiology department with those assigned by the athletic director. In addition, coaches at smaller colleges often become very involved in the social and academic life of the school. Usually, the outside obligations and pressure to win are not as burdensome at small (Division III) colleges as they are for coaches at the highest level (Division I), although self-imposed pressure to win is a potential cause of stress at all levels of competition.

Among larger U.S. colleges and universities, coaches are commonly employed full-time to coach a single sport. Moreover, at universities whose teams rank among the top in the country, a head coach often operates more as a CEO than as an instructor, overseeing a team of assistant coaches and a vast operation involving high-pressure recruiting and intense publicity. In fact, this is the rule rather than the exception in Division I football and in Division I basketball (men's and women's). Coaches at large public universities are often the highest-paid public employee in their state, earning several times more than the university president or the governor. The large salary comes with heightened expectations—if head coaches are not highly successful, they are quite likely to be replaced.

For example, the Football Bowl Subdivision (FBS) includes 128 member teams, and after the 2015 season ended there were 28 head coach openings; all were filled by January 2016 (Barnett, 2015). Thus, about 22 percent of FBS head coaches lost their jobs! To compound the situation, when a head coach is fired, most of his or her assistants and support staff are also fired when the new head coach arrives.

Coaches at 4-year colleges and universities work in athletic facilities consisting of numerous offices, gymnasiums and practice centers, weight rooms and conditioning rooms, swimming pools, sport science centers, dietetic and dining centers, and athletic training rooms. Schools that offer football commonly have a separate stadium and office complex for the football staff. These athletic facilities typically house not only coaches and athletic administrators (e.g., athletic directors and their assistants) but also compliance officers, academic tutors and counselors, sport marketers and promotion personnel, media relations officers, facility directors, and event directors. Although the structure of university athletic departments varies, all athletic faculty report to an athletic director, who in turn reports directly to a vice president or to the university president.

Other Institutional Settings

Opportunities for coaching are also available in many other institutional settings, including residential and nonresidential schools for children with disability (e.g., schools for people with deafness or blindness), residential juvenile institutions for youth who are troubled or disadvantaged, detention facilities, and military settings. In addition, opportunities are now available for coaches, instructors, and administrators in some large churches that have excellent facilities for sport and exercise programs. In all of these settings, the institution's philosophy and mission guide coaches and coaching practices. Generally, sport in such settings is intended to accomplish an educational or recreational mission rather than provide entertainment for spectators. If you are considering a coaching job in such a setting, make sure that you feel comfortable working in the environment of the specialized institution and serving its mission.

Surging Salaries for Division I Football and Basketball Coaches

Success (i.e., winning) translates into big bucks in Division I athletics. If you want to work at a major university and make a tremendous amount of money, then plan to become a big-time basketball or football coach. According to a *USA Today* report (USA Today, 2017) on the 2016 salaries of NCAA Division I basketball coaches,

- Rick Pitino (Louisville) was the highest-paid coach at more than $7 million,
- 8 NCAA men's basketball coaches earned more than $3 million—more than the average for chancellors at the same universities, and
- 41 men's NCAA basketball coaches made more than $1 million each.

On the women's side, the salaries are not as extreme, but coaches can still rank among the highest-salaried individuals at an institution. Coincidentally, the highest-paid women's coach is a man, Geno Auriemma (Connecticut), who was paid just over $2 million in 2015 and is scheduled to receive $2.4 million for the 2017–2018 season—not including income from his camps and endorsements (USA Today, 2017). Auriemma is arguably the most successful active coach in either men's or women's NCAA basketball; therefore, it is evident that the top women's coaches, though paid well, still do not earn on the same level as the men's coaches. Five women's coaches were paid more than $1 million dollars per year in 2016.

According to the same report, 36 NCAA Division I football coaches were paid more than $3 million in 2016. The list was topped by Jim Harbaugh (Michigan), whose total compensation was more than $9 million. In most instances, the salaries paid to Division I football and basketball coaches are substantially higher not only than those of the presidents at their respective universities but also more than that of the president of the United States. Do you think the salaries of top coaches and professional athletes are fair in terms of their value to an institution or to society as a whole? Why, or why not?

Professional Issues in Kinesiology

Expertise: Predicting Future Talent

Coaches of advanced and elite athletes must often try to identify athletes with potential. It is an extremely challenging task to assess expertise and identify potential for outstanding performance. Expertise in a sport involves a complex combination of sport-specific, fitness-related, and psychological factors. This complexity is heightened by the uncertainties inherent in developing such abilities, as well as genetics, injuries, training issues, and motivational factors.

Thus, you can see why predicting outstanding future expertise poses a difficult problem. In fact, measuring and understanding the variables involved in expertise and talent development constitute specialized subdisciplines in kinesiology (see the section later in this chapter on the growth of sport statistics and performance analytics), and coaches must integrate them in order to succeed at identifying talent. One caveat: A given sport metric that best describes elite performance may not be most important factor at lower levels of competition or in athlete development.

Commercial Sport Settings

Commercial sport settings are those in which sport instruction or coaching is offered as a product or service in a business venture, usually by self-employed professionals. These settings range from the huge arenas, stadiums, and practice facilities of professional teams to tennis, swimming, or squash facilities at swanky country clubs and health spas. An increasing number of human performance clubs are focusing attention on development of sport skills and improved athletic performance, in addition to exercise. For-profit sport team clubs are fairly new developments in the United States but are quite common in Europe and Australia, where there are fewer school-based sport teams.

Coaches and teaching professionals who work with elite athletes often have the most diverse worksites of all of their colleagues. Some coaches work at a fixed facility, whereas others teach and coach at a variety of locations. Some itinerant golf or tennis teachers work out of their home or car trunk. Other sport instructors own and operate a martial arts studio, gymnastics school, or swimming club. Of course, coaches of professional sport teams in football, baseball, soccer, hockey, and basketball work in fixed settings as employees of sport organizations. Professional coaches work at permanent facilities similar to those in colleges and universities; such facilities contain office space, practice and game areas, weight rooms, and athletic training rooms. In this type of organization, coaches usually report to a general manager and team owner.

Elite athletes hire specialized coach-instructors who typically have their own experience as athletes and coaches in the sport. These professionals tend to freelance and are often on the road, traveling to their clients' hometowns. For example, figure skating coaches work with high-caliber competitive figure skaters at the skaters' home rinks. On the other hand, professional coaches who are employed by sport academies have the luxury of remaining at a fixed location. In the United States, these coaches may not hold a degree in kinesiology; in most other countries, elite coaches do hold bachelor's and master's degrees in kinesiology or sport science.

KEY POINT

In advanced sport settings, some coaches have a stable worksite, whereas others coach at a variety of facilities; still others are self-employed and may work out of their home.

A few sport instructors who work with elite athletes have the luxury of a stable worksite—usually a school or academy with residences on or near the campus where budding teenage Olympic or professional athletes can stay for several months of the year. For example, affluent parents who believe that their children have a promising career in basketball, baseball, ice hockey, golf, soccer, or tennis might consider enrolling their children in the IMG Academy or another such organization. Aspiring young golfers might also enroll full-time at the International Junior Golf Academy in Hilton Head, South Carolina, where they spend much of the day practicing and learning golf from highly skilled instructors and attending campus-based preparatory schools.

When we think of coaching and sport instruction settings, we usually think of the popular sports that we see every day on television. However, coaches

Sport Instruction on a Large Scale: IMG Academy

IMG Academy is the world's largest, most successful institution dedicated to multisport training and education. Two of IMG's founders were kinesiology professors who transitioned into commercial sport. Average annual enrollment consists of more than 12,000 junior, collegiate, adult, and professional athletes from more than 80 countries. Although IMG may be best known for its famous alumni—including Andre Agassi (tennis), Vince Carter (basketball), and Adam Dunn (baseball)—you don't have to be a serious competitor to take advantage of the programs. IMG Academy is situated on a gated, 400-acre campus in Bradenton, Florida, that consists of golf driving ranges and practice facilities, tennis courts, baseball fields, turf training fields, indoor basketball facilities, sport performance facilities, gyms, clubhouses, and dormitories. It also has two cafeterias (one for adults and professionals and one for students) and operates an on-campus restaurant for students in the evening. It offers youth sport camps in the following areas:

- Tennis
- Golf
- Soccer
- Baseball
- Basketball
- Lacrosse
- Football
- Peak performance training

Peak performance training encompasses a holistic approach to improving performance that includes physical, nutritional, mental, and leadership training. Young athletes stay in campus dormitories and have access to a highly competent teaching and coaching faculty. For all of this, their parents pay more than $30,000 per year. Tuition for the on-campus boarding school and fees for private lessons and other amenities can cost a grand total of $70,000 per year or even more. IMG also offers peak performance training for college and professional athletes and boasts a number of former students who competed in the 2016 Rio Olympics.

work in a variety of settings. For instance, instructors in rowing, kayaking, and canoeing work on rivers. With competitive rock climbing becoming more popular, we find instructors at indoor climbing walls, in the great outdoors scaling mountains, and on stone or brick buildings in towns and cities. Similarly, as interest continues to grow in ecotourism and outdoor adventure experiences, the need for instructors and coaches in outdoor recreation pursuits is expected to increase accordingly. In addition, each summer, camp directors, some with large budgets and impressive facilities and equipment, hire expert sport instructors to oversee instructional operations; sport instructor positions are also available at large resorts that sponsor extensive sport programs.

Roles for Coaching and Sport Instruction Professionals

Professional sport instructors usually work with relatively small numbers of highly motivated learners, athletes, or clients under optimal conditions in which attention can be focused exclusively on one skill or activity. These peak performance experts typically work with individual or small groups of elite athletes but may also work with teams in concert with the team coach or manager to maximize both individual and team performance. Let's examine more closely the duties and responsibilities of sport instructors and coaches.

Community Physical Activity Program Instructors

Physical activity classes in community-based programs are offered by municipalities through departments of recreation and other nonprofit agencies. For the most part, nonprofit organizations work on the front lines in an effort to strengthen communities by providing needed services to individuals and families. In meeting this challenge, an important role can be played by physical activity program leaders. These leaders also teach older people in physical activity programs offered by community centers, recreation facilities, retirement centers, and other nonprofit agencies. Adult programs may include soccer, softball, tennis, swimming, golf, adult exercise, yoga, dance, and martial arts. Working in nonprofit agencies such as these can be rewarding for professionals who are interested in serving an

Coaches who work with underserved children not only help them develop athletic skill, but they also teach them to be personally and socially responsible young people.

older clientele. As with employment in all nonprofit settings, financial compensation tends to be less than that of commercial or other institutional settings, but knowing that your services are valued can be a form of compensation in its own right.

KEY POINT

The majority of physical activity programs in community settings are offered by nonprofit organizations and parks and recreation departments.

Community Youth Sport Development Leaders

One particularly exciting area of community work focuses on underserved youth. Many communities have experienced damaging social, economic, and political influences, and these pressures have directly affected the ways in which human services are made available today.

Throughout the United States, communities are looking for solutions to the crises facing America's youth. The typical approach has been to focus on what is wrong with young people and then develop strategies and programs to solve the problem. This traditional approach is well intended but focuses on kids' deficiencies and problems rather than on their capacities, skills, and assets. A newer approach assumes that young people have knowledge and positive qualities that can be enhanced, and this asset-based model is increasingly being used by community organizations in their physical activity settings for youth.

Physical activity is a particularly effective way to focus on kids' capacities, skills, wellness, and assets. The unique energy, human interactions, and creativity that are harnessed by physical activity allow community-based leaders of physical activity programs to see kids' potentials and personal qualities that might otherwise go unnoticed. Furthermore, youth development philosophy holds that young people who are having difficulty in school or at home, or even those involved with the juvenile justice system, can, when given the proper opportunities, beat the odds and become contributing members of society.

Because of the various needs of community-based organizations, physical activity professionals at this level perform a range of tasks that go beyond instruction. Typically, they direct physical activity programs, which includes marketing and budgeting the program. They may also perform a variety of managerial duties, as well as tutoring, mentoring, and overseeing literacy programs and a host of other community services. If you are a kinesiology major who wants to work with youth but not in a public school, then you may be interested in becoming a community physical activity instructor. Salaries for some of these community youth development positions tend to be lower than those of public school teachers. However, they offer the reward of making a difference in kids' lives and knowing that you are contributing to your community.

Professional Sport Instructors

Teaching professionals and professional sport instructors offer services (private or group lessons) in a specialized physical activity in a professional setting. Swim, golf, and tennis professionals who work either as freelance entrepreneurs or at clubs may also hold positions as members of the physical activity instruction faculty at colleges and universities. Sometimes they teach beginners, but they may also coach highly skilled individuals who are competing in leagues and tournaments. Outside the university setting, teaching professionals can be found at private, semiprivate, and public teaching facilities, where they teach beginning, intermediate, advanced, or elite players. Golf professionals, for example, may work at a municipal golf course, giving lessons to the public and conducting golf camps or managing the pro shop or the clubhouse. Teaching professionals have often attained national or international sport-specific certifications, some of which require continuing education and offer opportunities for specialization at advanced levels.

KEY POINT

Professional sport instructors, who offer services (private or group lessons) solely in an area of expertise, are more like coaches than are any other teachers, yet they are free from the administrative duties of coaching.

Professional sport instructors typically are self-employed and are hired as independent contractors. Therefore, their employment security depends on the satisfaction of their clients and of the owners or managers of the instructional facility by whom they are contracted. Like professionals in other fields, certified professional sport instructors have a high degree of control over their careers. Their continued employment depends simply on their ability to provide services that their clients find valuable and are willing to pay for. Unlike coaches at institutions, however, professional instructors are usually not covered by group insurance policies or retirement plans.

More employment opportunities are available at the beginner or intermediate levels than at the elite or professional levels. Although many people want to learn to swim, golf, rock-climb, and skydive, only a select few (in relative terms) want to work hard and pay an instructor to become proficient enough to perform at an elite level. Incomes for private instructors can range from only a few thousand dollars per year (e.g., for a golf instructor whose office is the trunk of her car) to more than $100,000 per year for an instructor-entrepreneur (e.g., owner, operator, and instructor of a successful golf instruction school).

Certification and Education of Instructors

Do you recall the discussion in chapter 11 about professionalism and the importance of certification procedures to ensure that only the most highly qualified people can serve in a professional capacity? In keeping with this approach, various types of credentials are required of the instructors who lead community physical activity and sport programs.

Certification for Community Physical Activity Instructors

No certification programs exist specifically for youth sport development work focused on underserved children; however, a bachelor's degree is required for most positions, preferably from a program that offers a minor specialization in youth sport development. In this vein, an increasing number of U.S. colleges and universities are beginning to offer bachelor's and master's degrees in kinesiology with a concentration in what is termed either *community youth sport development* or *community leadership* (Hellison et al., 2000). These programs teach students how to develop, implement, and evaluate community-based physical activity programs for young people. Such academic programs also include courses on youth agency administration, principles of community organization, grant writing and fundraising, social work, and community development. Elective courses may also be helpful in areas such as social work, counseling, and management. Another important part of these programs consists of internships that involve working with children on a daily basis.

Certification for Professional Sport Instructors

Although a college degree is usually sufficient for coaches of the most popular sports (e.g., basketball, football, baseball) in the United States, certification can be important when the coach or instructor works in a noninstitutional setting, such as a sport club. Some sort of instructor or coaching certification is offered in most organized sports. For example, most coaches of age-group teams in large soccer clubs are certified (licensed) by the United States Soccer Federation (commonly referred to as U.S. Soccer), the National Soccer Coaches Association of America (NSCAA), or both. Likewise, in order to teach swimming or work as a lifeguard, you will most likely be required to secure American Red Cross certification as a Water Safety Instructor. This certification enables you to teach swimming lessons and coach swim or synchronized swim teams at a private club, YMCA, or community center. If you are interested in teaching scuba diving, you will need to acquire certification from one or more of several organizations; for example, the Professional Association of Diving Instructors offers certification from the assistant instructor level to the open-water scuba instructor level.

In another example, the United States Fencing Coaches Association (USFCA) offers a rigorous certification program through its Coaches College Academy, in which would-be instructors move through five levels. At levels I and II, they master skills in using three weapons (épée, saber, foil); at levels III through V, they specialize in one of the weapons. To pass the first part of the certification program, they must write a paper, propose a 1-year training cycle, and serve as an intern at Coaches College Academy. At the highest level, they must develop a fencer under their supervision. Certification is granted by the USFA only when the fencer places in the United States Fencing Association ranking list or among the top 12 NCAA fencers.

Instructor certification programs are also offered by the men's Professional Golf Association (PGA) and the Ladies Professional Golf Association (LPGA). One becomes a golf professional under either PGA or LPGA auspices only by passing an arduous education and testing program that can take up to 8 years to complete. The PGA certification program requires study, training, experience, and testing in every aspect of golf—not only playing aptitude and teaching technique but also club repair, analysis and correction of swing flaws, tournament supervision, and the business side of golf (including golf shop operations).

Certification programs also exist for many other sports (see the section titled Certification and Continuing Education of Coaches later in this chapter). If you plan a career as an instructor or coach in one of these sports, you should begin working toward certification in that sport as soon as possible. Finally, kinesiology professionals sometimes find a career in the national office of a certification or sport professional organization.

Coaching

We have all known people who touched our lives as no one else could. Often, our memories of these people is associated with emotions—respect, awe, deference, and admiration—that seem as vibrant as when we first experienced them. For many current and former athletes, a coach is in the forefront of their mind and heart.

KEY POINT

As with being a teacher, being a coach involves building relationships that can change the life of another person for the better.

Many opportunities are available to those who want to coach. Coaches work in many different settings with a wide variety of people, ranging from young children to adults and from recreational athletes to elite performers. Some college and professional coaches are paid hundreds of thousands or even millions of dollars, whereas others coach simply for the joy it brings. Coaches can be found in sport programs in communities, secondary schools, colleges and universities, and professional sports. These different programs carry different coaching requirements, philosophies, and expectations, and the coaches in these programs often differ in educational background and coaching experience.

Youth Sport Coaches

Municipal recreation departments usually offer a wide range of competitive youth and adult sports, which are the responsibility of the director or supervisor of recreation. The range of activities offered depends on the ingenuity of the director and the available budget and facilities. Activities may include age-group swimming programs conducted in public swimming pools, as well as highly organized competitive programs in football, baseball, softball, soccer, basketball, tennis, and other popular sports. Generally, but not always, these programs tend to deemphasize winning while emphasizing inclusiveness and fair play. They are usually staffed by volunteers who are recruited and trained by the program director. The director is also responsible for hiring and training officials to oversee the contests. Games usually take place during evening hours, on weekends, or during the summer.

Organizations that rely on volunteer coaches often face a difficult task in finding qualified, certified coaches. Because many states do not have specific standards or certification requirements for volunteer coaches, anyone interested in coaching is often deemed qualified. Furthermore, community organizations are often short of volunteers, so anyone who wants to coach at this level can usually find a coaching position. Often, a parent of one of the participants assumes the role of coach. Coaches who have had formal education in kinesiology may serve as administrators of large community-based youth sport leagues or as instructors for coaching clinics offered to laypersons who coach in a youth sport league.

If you are interested in pursuing a career in coaching, you can obtain excellent ground-floor preparation by volunteering to coach a local youth sport team during your undergraduate years. The experience will also help you determine whether coaching is the right profession for you. Consider, however, that the essentials of coaching at the youth level differ from those required at the high school and college levels.

One relatively recent development in youth sport involves the rise of private sport programs in which, for a fee, children join "travel teams" or "club teams." As the first name implies, these teams travel around a county or state to experience competitive interleague play. Proceeds from team fees are used to pay the coach. Some civic government programs offer competition on a regional or national scale. For example, age-group soccer and swimming programs compete at local, regional, and national levels. In theory, the more gifted competitors are selected for the traveling teams, which offer the players opportunities to compete with others of the same ability. In practice, however, many of these programs are marked by inherent problems, such as player selection and the added competitive pressures imposed on young athletes that is not always in their best long-term development. One of the most prominent examples of such a program is the annual Little League baseball tournament, which culminates in the nationally televised Little League World Series in Williamsport, Pennsylvania.

Some of these programs hire professional coaches. Two examples in the United States are (Pop Warner) football and the Amateur Athletic Union (basketball), both of which are private national organizations that are more competitive than community-based programs. Coaches at this national competitive level usually spend more time performing the nonteaching duties associated with coaching, such as scouting and recruiting new talent, viewing game films, and organizing team travel.

Large soccer clubs field teams in several age groups (U-10 through U-18 for boys and for girls). Even at young ages, these teams face intense pressure to win. Winning clubs attract more players, which in turn generates more money for those clubs. Coaches are usually experienced players themselves or are licensed by U.S. Soccer or NSCAA. The coaching staff is typically hired and coordinated by the club's director of coaching, who oversees the entire program. As a consequence, coaching directors of large soccer clubs often earn a salary commensurate with that of most college soccer coaches and are held to the same high expectations. If their teams are successful (i.e., if they win), then these coaches are likely to keep their positions; if not, they may experience the same fate as college and professional coaches who do not win.

KEY POINT

Community coaches volunteer in community-based sport programs that range from educational, inclusive, and relatively noncompetitive programs to those that emphasize competition, performance, and winning.

Coaching in Institutional Settings

Institutional coaches work in educational institutions such as secondary schools (i.e., middle, junior high, and high schools), colleges, and universities. As mentioned earlier, they may also serve in specialized residential institutions for young people who have disability or are socially disruptive; some coaching opportunities are also available in the armed services. In each case, the coach strives to evoke each athlete's potential and guide him or her in overcoming personal challenges and achieving personal success within his or her limitations. Coaching duties include teaching physical skills; keeping individual and team statistics; scheduling practices, games, and tournaments; and managing and maintaining equipment.

Working at an institution requires physical activity professionals to relinquish some autonomy in carrying out their duties. Generally, coaches working in an institution do not enjoy the freedom of freelancers; on the other hand, they do reap the benefit of a steady and reliable income. Coaches who work in an institution must comply with institutional expectations and conduct themselves in accordance with established codes of conduct. Ultimately, these coaches serve at the behest of the institution's chief administrator or board of directors.

Coaching in Middle and Secondary Schools In 2016, the number of participants in high school sport in the United States reached an all-time high of nearly 8 million, which marked an increase of nearly 62,000 from the previous year (National Federation of State High School Associations, 2016). The task of coaching these athletes requires an enormous number of people. For those who love sport and love being around young people and serving as a positive role model, coaching can be an exciting and rewarding career. At the same time, you should factor a certain dose of realism into your decision about whether to pursue coaching as a career. The

When Commercialism and Professionalism Threaten Youth Sport

People who are committed to the educational benefits of youth sport programs face a growing problem in the tendency for organizations, media, and corporations to exploit them for commercial gain. For instance, some people question whether the Little League World Series, televised around the world, is appropriate for young children. Should preteens be placed on a world stage where they are critiqued, ridiculed, and, in the opinion of some, exploited for financial gain? The final game in the 2016 edition of the event was decided on a passed ball as New York defeated South Korea 2-1. Imagine the overwhelming pressure experienced by the young player whose error ultimately cost the team a world title in front of an international audience.

Others question the increasing trend for professional sport organizations to sign athletically promising young children to professional contracts and commercial endorsements. Sponsors and professional clubs have signed young children to lucrative multiyear contracts in the hope that they will mature into professional standouts. For example, in 2003, Nike signed Freddy Adu, an exceptionally skilled American soccer player who was just 13 years old, to a $1 million endorsement contract. In 2004, Adu became the youngest U.S. athlete to sign a major league contract in more than 115 years when he joined D.C. United as the number one overall pick in the Major League Soccer SuperDraft. Thus, at an early age, Adu was thrown into the competitive spotlight, where he experienced the pressures and expectations that come with fame. To date, Adu has yet to achieve the level of prominence in the soccer world that was expected.

That cautionary tale has not stopped other top professional teams from copying Nike's move. Bayern Munich, Germany's richest and historically most successful professional soccer team, signed youth prodigy Mehmet Ekici when he was 8 years old, then sold his rights to Werner Bremen for €5 million. In 2011, Real Madrid, a top Spanish club, signed young prodigy Leonel Angel Coira to its youth academy after seeing him in tryouts. He was 7 years old! Lionel has not disappointed. The talented Argentine youngster was incorporated into Madrid's famous youth academy and since then has continued to attract interest after scoring more than 300 goals for Los Merengues. Even so, he recently signed with Spanish League rival Valencia because he was not enjoying his time in Madrid. In any event, Leonel's signing at such an early age reflects the emerging philosophy of top teams that try to invest in fresh talent as early as possible to avoid paying huge transfer fees when their talent blossoms. For these young athletes, then, soccer is no longer merely a "youth sport," and the pressure imposed on them is intense.

most important point may be the fact that coaches work many hours for relatively low compensation. In high school, even the coaches of major sports such as football often earn only a small stipend on top of their regular teaching salary. The committed coaches who love what they do will tell you that the money is merely icing on the cake; the real compensation for them consists of the rewards of getting close to the kids on their teams, playing a substantial role in their development, and earning the respect of the community at large. Weigh these considerations before you make your decision.

KEY POINT

Secondary-level school coaches typically spend their days teaching and receive an additional stipend for coaching.

Middle, junior high, and high school coaches are usually certified public school teachers who teach a full schedule and receive a stipend for taking on the extracurricular activity of coaching. Certified, part-time coaches who aren't members of the school faculty may be hired to coach specific teams. Although the qualifications for coaches in the public schools vary by state, they typically require a teaching or coaching certification and a background check. Coaches are among their school's most visible representatives. Unlike teachers, who work within the space of the classroom—where their performance is assessed only by students, parents, and the principal—coaches perform in front of all of the preceding people as well as the community at large. Their successes and failures may be covered in the daily newspaper. Their performance is often the center of conversation at local diners and coffee shops. Everyone is allowed to hold an opinion (and to voice it!) about the effectiveness of a coach.

In public secondary school athletics, the number of contact hours between coaches and athletes varies depending on the governing body in question. Typically, however, secondary school coaches spend 8 to 10 hours per week in after-school practices. When competitive play begins, the number of practice hours may decrease to 8, whereas game hours may total 4 to 6 hours. Some coaches have early-morning practice schedules; others have late-afternoon or early-evening schedules.

In addition to spending time at practices and games, coaches shoulder many responsibilities behind the scenes (see figure 15.2). To varying degrees, all coaches are involved in organizing and scheduling team meetings and practices, ordering equipment, maintaining and overseeing inventory, checking athletes' eligibility, arranging transportation to and from events, talking with members of the media, preparing for home events, fundraising, and organizing end-of-season banquets. The extent of coach involvement in these activities depends in part on the size of the coaching staff and the number of support personnel. For example, some schools have a full-time athletic director (AD), who is responsible for organizing home contests. Other schools may have a part-time AD, who must delegate that responsibility to a coach. Likewise, some sports (especially football) require many assistant coaches, to whom head coaches can delegate specific duties.

The differences in coaching at the various educational levels hinge on the athletes' developmental differences. Some middle school athletes are "diamonds in the rough" who have the potential to be excellent athletes but are physically undeveloped (i.e., prepubescent and small) as compared with others on the team. Some may have just gone through a growth spurt and need to relearn a given skill in their resized body.

The Parent Problem

One growing concern among sport scientists, program administrators, and parents stems from the fact that an overwhelming amount of parental involvement in youth sport is detrimental to the development and experiences of the young athletes. Parents who desperately want their children to succeed in athletic competition can make serious problems at the site of an event. Recent events have included a father being sent to jail for beating and berating a coach because the coach took his 11-year-old son out of a game; a father (a dentist) who sharpened the nose guard of his son's football helmet so that the son could slash opponents; and a police officer who gave a Little League pitcher $2 to hit a 10-year-old player with a fastball during a game. In Massachusetts, a hockey player's father who was supervising a practice session was beaten to death by another father because of a disagreement about the rough play allowed during practice.

Such incidents have prompted youth sport leaders to promulgate codes for sporting behavior among parents. In some states, parents must sign the code before their children are permitted to play. Efforts to educate parents and coaches about this issue have been aided by organizations such as the Institute for the Study of Youth Sports (ISYS) at Michigan State University and the National Alliance for Youth Sports (NAYS). Two codes that have been widely adopted by youth sport programs across the United States are the Sport Parent Code of Conduct, provided by NAYS, and the Bill of Rights for Young Athletes, written by Dr. Vern Seefeldt and Dr. Rainer Martens and distributed by ISYS.

The increasing reports of parents engaging in physically violent or verbally abusive behavior toward young athletes, coaches, officials, and opposing spectators have led many organizations to clearly define the role of parents in youth sport. For example, the American Youth Soccer Organization requires parents of players below the age of 8 to attend classes addressing sporting behavior. AYSO and similar youth sport organizations emphasize that parents must remain mindful of why kids participate in sport—in most cases, to have fun, according to a study of more than 25,000 children from across the United States (Ewing & Seefeldt, 1989). The next most popular reasons were to learn new skills, be with friends, and experience the thrill of competition. Children did list winning as a reason for playing sport, but it was not one of the most popular reasons; yet many parents and youth sport coaches erroneously believe that winning is the number one reason that children want to play. Parents who become preoccupied with winning and losing place an unreasonable amount of pressure on their child and increase the probability that he or she will drop out of sport and physical activity altogether.

Middle and junior high school coaches have the unique responsibility of introducing young athletes to the world of interscholastic sport, which requires athletes to balance the attention they give to their academics with their focus on sport. The primary tasks for middle school coaches are

- to help students interested in athletics develop a positive view of themselves as athletes and
- to teach them important, complex physical skills and the basic strategies needed to become competitive student-athletes.

These coaches also must introduce their athletes to a concept of competition in which winning takes on greater importance than it often does at the community level.

High school coaches are primarily involved with students who are interested in athletics and want to invest the time needed to become good

Long Shot: Probability of Competing in Athletics Beyond High School

A recent NCAA survey (NCAA, 2016) provides some sobering numbers for individuals who envision playing at the college or professional level. Although many high school players dream of playing at the next level, these data show the low probability of fulfilling such dreams.

Student-athletes	Men's basketball	Women's basketball	Football	Baseball	Men's ice hockey	Men's soccer
High school student-athletes	541,054	433,344	1,093,234	482,629	35,393	417,419
High school senior student-athletes	154,587	123,813	312,353	137,894	10,112	119,263
NCAA student-athletes	18,320	16,319	71,291	33,431	3,976	23,602
NCAA first-year roster positions	5,234	4,663	20,369	9,552	1,136	6,743
NCAA senior student-athletes	4,071	3,626	15,842	7,429	884	5,245
NCAA student-athletes drafted	47	32	255	638	60	72
% high school to NCAA	3.4	3.8	6.5	6.9	11.2	5.7
% NCAA to profes-sional	1.2	0.9	1.6	8.6	6.8	1.4
% high school to professional	0.03	0.03	0.08	0.53	0.04	0.06

Percentages based on estimated data.

©2016 National Collegiate Athletic Association. *2016-2017 guide for the college-bound student-athlete.* NCAA Eligibility Center. Available: http://www.ncaapublications.com/productdownloads/CBSA16.pdf

In a nutshell, the NCAA estimates that an athlete's chance of competing in his or her chosen sport at the college level is very low (NCAA, 2016). For example, only 3.4 percent of high school senior men basketball players play NCAA-sponsored basketball, and only 3.8 percent of senior high school women go on to play NCAA basketball. These figures do not take into account players competing in the National Association of Intercollegiate Athletes (NAIA) or the National Junior College Athletic Association (NJCAA).

athletes. They also have the opportunity to build on the foundation laid by middle and junior high school coaches as they teach more complex tasks and game strategies. This level is also marked by increased focus on competition and winning, which requires coaches to choose their teams wisely. They also spend more time reviewing game film, developing strategies specific to a given opponent, and scouting and recruiting potential team members.

Coaching in Colleges and Universities Clearly, winning matters a great deal to all intercollegiate coaches; the pressure to win at this level, especially at large universities, is much greater than that at the high school level. In fact, the enormous financial stakes of big-time college athletics can reduce it from an enjoyable, educational experience for everyone to a brutal exercise in survival. As anyone who reads the sport pages knows, the coach who doesn't win his or her share of games is soon fired.

KEY POINT

College coaches aim to maximize the athletic potential of already-skilled athletes but also have many other responsibilities, including team and facility management, budgeting, monitoring of academic progress, compliance issues, and recruiting.

The quest of college coaches is to maximize the athletic potential of already-skilled athletes by building on the foundation established in high school and fine-tuning each athlete's ability in order to move him or her toward becoming a member of the elite. Most college coaches have the advantage of working with a relatively small number of highly skilled athletes in well-equipped facilities. Given the popularity of basketball and football in U.S. society, these two sports tend to receive most of the resources in an athletic department. Coaches of less prominent sports—such as cross country, volleyball, softball, soccer, tennis, and golf—usually have to make do with modest budgets and resources.

Like secondary-level coaches, college coaches carry many responsibilities behind the scenes. To varying degrees, all coaches are involved in team, facility, and equipment management and in budgeting. At the college or university level, other responsibilities also take on greater significance, such as talent scouting, recruiting, fundraising, public relations, and athletic eligibility. The degree of significance often depends on the type of college in which the coach works.

Employment opportunities at the college or university level are not as plentiful as they are at the secondary level. Because there are no official national or state qualifications or requirements for coaching at the college or university level, most colleges and universities look for coaching applicants who have an established track record in the relevant sport. Typically, establishing a name requires college playing experience, coaching experience, a history of successful seasons, and being known by many other coaches and administrators. Many Division I coaches began their careers at the secondary level and moved through the lower divisions of college athletics. Others managed to secure an internship during their undergraduate years or sign on as team manager and, with a great deal of effort, work their way up to an assistant coaching position.

Smaller colleges, most notably Division II and Division III institutions, offer many opportunities for coaches to get close to their players and to meet and interact with them in many settings. Although the pay is generally not high, the rich quality of life and reduced pressure to win offer compensation of a different sort. Coaching in larger institutions where coaches' jobs, and their assistants' jobs, depend on the team's win–loss record usually requires an enormous commitment in time and energy, sacrifice of one's family and social life, and the fortitude to withstand the anxiety in years when the team fails to win many games. If you happen to be one of the few perennially successful coaches, however, coaching can be a lucrative and exciting career.

Coaching in Professional Sport

Who comes to mind when you think of professional coaches? The question may evoke images of Joe Maddon and Buck Showalter (baseball), Greg Popovich and Billy Donovan (basketball), and Bill Belichick and Mike Tomlin (football). Although the number of professional sport coaches is small when compared with the number of coaches at secondary schools and colleges, professional sport coaches typically have national prominence.

KEY POINT

A professional sport coach usually has many nonteaching and noncoaching duties, including administration, recruitment, and media appearances. Continued employment depends heavily on producing a winning team.

As with other professional coaching positions, facilitating athlete performance is central to the job of the professional sport coach. The position also includes elements of administration, athlete recruitment, and public relations events and media appearances. Generally, the professional sport coach has more nonteaching duties than does the teaching professional discussed earlier. For example, some sports, such as football and hockey, require coaches to spend considerable time participating in coaching meetings and reviewing game films. In addition, the win–loss record in such sports is more important than it is in sports such as golf and tennis; in fact, the coach's continued employment depends heavily on team performance. Finally, the professional sport coach deals solely with elite athletes, whereas the teaching pro often interacts with beginning and intermediate athletes.

Employment opportunities for the professional sport coach are few and far between. The number of professional sport teams is limited, but many individuals are qualified to coach at the professional level. As in the college and university setting, no particular national or state qualifications are needed for coaching in the professional ranks. Prospective coaches do need an established name in the chosen sport, and most acquire it by coaching successfully at the college or university level.

Job security for the professional sport coach is tenuous at best. When you coach a team sport at the professional level, you have entered an industry where success relies in large part on public support. In professional sport, however, public support waxes and wanes in direct relation to a team's win–loss record, as well as the whims of the media and the mass of uninformed spectators on social media. Just as the organization's success depends on producing a winning team, so does your coaching job. Many professional sport coaches thrive on this aspect of the job. They bask in the challenge of creating or maintaining a winning tradition, and accomplishing that goal brings them great rewards and personal satisfaction.

Certification and Continuing Education of Coaches

Coaching has been described as both an art and a science. Whatever the balance between the two may be, coaching is a highly complex process of applying knowledge about human performance, human development, motor learning, and exercise. It also involves high-order cognitive skills to integrate sport strategy and diagnose sport skills. All of these skills are applied by using sport psychology skills that enable the coach to communicate and motivate athletes. Thus, all successful coaches have excellent interpersonal skills and the ability to teach the techniques, tactics, and strategies of a given sport based on motor learning principles.

Coaches must also possess various other kinds of knowledge and skill. For instance, they must understand the fundamentals of exercise as it relates to training and conditioning. In addition, managing any sport enterprise entails planning, organizing, staffing, leading others and monitoring progress toward team goals, and understanding the legal liability and risk factors associated with making decisions. Furthermore, in the last 2 decades, coaches have taken on concerns related to first aid and drug education. Moreover, as sports medicine and athletic training have become more established as fields, the coach's role has expanded to include injury prevention and facilitation of

Life as a Professional Sport Coach: Few Jobs, Less Security

If your ultimate goal is to land a head coaching position at the professional level, be prepared for the most difficult job search imaginable. Consider the following: The five major professional sports in the United States have a total of only 142 head coaching positions (National Football League, 32; Major League Baseball, 30; National Hockey League, 30; National Basketball Association, 30; Major League Soccer, 20). Even if you are fortunate enough to land a pro head coaching job, your probability of keeping it for any extended time is questionable. For example, from 2012 through 2015, 22 NFL teams replaced their head coach, and 11 did it twice!

On the other hand, although the tenure of a professional sport coach may be brief, openings do occur on a regular basis. It is also true that coaches who are already in the professional coaching community are often first in line to fill the most recent vacancy. Indeed, a sort of informal fraternity appears to exist among professional coaches, which is evidenced by the fact that several coaches in each of the major leagues have coached two or three teams in their league. Thus, they get fired or leave one job only to fill a similar position with another club.

emergency care. When injuries occur, the coach should be knowledgeable enough to help trained medical personnel.

Certification of Youth Sport Coaches

Certification is becoming increasingly important as the problems afflicting youth sport become more serious. Never before has there been such an obvious need for highly qualified coaches who operate on the basis of a sound coaching philosophy. The coaches for most youth sport teams are parents and others who volunteer to serve. The kinesiology graduate is more likely to be not a coach but the program organizer and administrator. As a result, you should recognize the value of certification for your coaches. A number of certification programs are available for youth sport coaches. Excellent educational programs for coaches are offered by the National Alliance for Youth Sports (NAYS), formerly known as the National Youth Sports Coaches Association (NYSCA), and the Human Kinetics Coach Education Center. The United States Sports Academy also offers coaching courses and coaching certificates that are available online, on-site, or through correspondence. Most of these programs require a modest tuition fee although a few of the introductory courses are free. Various sport-specific certification agencies (e.g., U.S. Soccer) also offer options for equipping youth sport coaches with the skills and knowledge required for successful coaching.

Education and Certification of Middle and High School Coaches

Your department may offer a curriculum—usually as a minor rather than a major—to prepare you for coaching. However, the United States has yet to establish consistent research, journals, or curriculums on coaching in the way that some countries have done in Europe and Asia. In the United States, most regional school districts and local coaching societies sponsor continuing education programs and annual clinics for coaches at the middle school, high school, and college levels. In some states, these clinics attract thousands of coaches and feature presentations by some of the most popular college and professional coaches. Some coaches, eager to learn more about the scientific side of coaching, enroll in master's degree programs that offer a concentration in coaching or a related kinesiology subdiscipline. For instance, coaches in a technique-centric sport such as gymnastics might benefit from an advanced degree in biomechanics, whereas those in a training-centric sport such as running might benefit from a master's degree in exercise physiology.

KEY POINT

The Human Kinetics Coach Education Center and the Coaching Association of Canada offer certification programs for community and scholastic coaches covering sport psychology, sport physiology, coaching philosophy, sport pedagogy, sport management, sport first aid, and drug education.

Coaching education experiences are available through various national coaching associations. One good example is the certification program provided by the National Soccer Coaches Association of America (NSCAA), which sponsors a coaching

U.S. Soccer National Coaching Certification Program

Millions of boys and girls under the age of 19 play organized soccer in the United States, which creates a huge demand for coaches at all levels. In an effort to educate coaches and set a standard for coaching young soccer players, U.S. Soccer has instituted a national coaching certification program that offers both state and national coaching licenses for various levels of expertise and experience. Certification begins at the state level, which focuses on elementary principles of coaching and prepares coaches for the 36-hour State D License course. National A, B, and C courses consist of 7 days of instruction and 2 days of extensive oral, written, and practical examinations.

U.S. Soccer also offers the National Youth License. This coaching license provides youth coaches, club coaching directors, physical education teachers, and soccer administrators with the knowledge to successfully structure soccer environments for children aged 4 to 12. The National Youth License course explores the role of the coach as a facilitator; the physical, mental, and emotional needs and capabilities of 4- to 12-year-old players; lessons from developmental psychology; and the art of teaching. To facilitate analysis of their performance, candidates are video-recorded during live training sessions.

NSCAA Nonresidential Coaching Programs

The National Soccer Coaches Association of America (NSCAA) offers a variety of coaching courses designed to accommodate all levels of coaching experience, ranging from youth coaches through professional team managers. The introductory Coaching Development Courses are nonresidential and can be offered at local soccer clubs throughout the country. These courses are designed primarily for youth soccer coaches and focus on player development through small-sided games. Three courses are offered based on the age and ability of players being coached.

1. **Coaching 4v4 Diploma:** This 3-hour, in-person course is designed for coaches working with players aged 8 or under. The curriculum is packed with age-appropriate training activities based on a 4v4 game structure. The time commitment for this diploma is only 3 hours, and no testing is required.
2. **Coaching 7v7/9v9 Diploma:** This 9-hour, in-person curriculum focuses on coaching players aged 9 to 12. Course content addresses player development, principles of play, technique, and tactics with small-sided games. No testing is required.
3. **Coaching 11v11 Diploma:** This curriculum is designed for intermediate coaches who work with players aged 13 and older. Course topics include the modern numbering system of players, formations and systems of play, player competencies, and defensive and offensive tactics. This diploma requires a 12-hour commitment but no testing.

academy that offers residential and nonresidential courses. The residential academy offers course work toward National, Advanced, and Premier Diplomas.

The primary objective of the National Diploma is to educate the coach about the needs of soccer players and provide an organized methodology of coaching those needs. The central theme of the course is basic technical and tactical knowledge in a 7v7 model. The course includes sessions on the teaching process, psychological skills training, and "laws of the game." The National Diploma consists of 45 hours of instruction and 6 hours of testing and evaluation. The Advanced Diploma is a 45-hour course for coaches with at least 10 years of experience. The Premier Diploma is a 45-hour course that examines different systems of play, nutrition, sporting behavior, ethics, and personal and professional development. To enroll for the Premier Diploma, a candidate must have passed the Advanced Diploma with a grade of "distinguished pass." These courses are very informative; they can also be quite costly.

Education and Certification of College and Professional Coaches

Although coaching at the high school level may require preparation in kinesiology, coaching positions at the highest levels (college and university or professional teams) may require very little in the way of formal academic preparation. Duke basketball coach Mike Krzyzewski, for example, holds a B.A. from the U.S. Military Academy; former University of Maryland basketball coach Gary Williams earned a B.S. in business; and New England Patriots coach Bill Belichick holds a bachelor's degree in economics. Thus, none of these coaches availed themselves of a formal education in kinesiology. (In a side note, the academic credentials of some of our most illustrious college coaches remain buried in university personnel files, unavailable for public inspection—a bit odd for institutions that supposedly take academic qualifications very seriously.)

Thus, if your goal is to coach football or basketball at a very high level in the United States, then you are likely to benefit less from obtaining highly polished academic credentials than from getting your foot in the door to the coaching establishment at a fairly young age and showing that you can produce winning teams. Experience as a head coach in a major college sport can also set the stage for a career in athletic administration once the fire to continue coaching has been extinguished. In fact, a number of prominent football and basketball coaches have moved to administration later in their careers. The experiences gained while serving, in effect, as the CEO of a major college sport can be beneficial in the administrative role as well.

Ethics and Coaching

In the recent past, a prominent coach at a major university was fired for allegedly knowing that play-

Kinesiology Colleagues

Courtesy of Chris Gilman

Chris Gilman

Have you ever had the desire to work with high-level coaches and elite athletes at a major Division I university? Chris Gilman hadn't planned on such a career, but a series of opportunities and decisions provided a path for him. Initially, he had a keen interest in kinesiology and intended to follow in his father's footsteps by becoming a health and physical educator. He graduated from Edinboro University in Pennsylvania with a B.S. in health and physical education, then taught and coached in the public school system while competing nationally in the sport of Olympic weightlifting. During that time, he began to realize the depth of his interest in strength training and coaching.

After much thought, Gilman decided to follow his passions and accepted a graduate assistant football position at Clarion University, where he coordinated the football strength and conditioning program. To further his knowledge and experience, he spent a summer as an intern coach at the University of North Texas, whereupon he decided that coaching collegiate strength and conditioning was the path that he wanted to follow. Two internships and one part-time position later, he landed a full-time position at the University of Pittsburgh.

ers had violated NCAA rules but not reporting the violations to his superiors. In another celebrated case, a college basketball coach was terminated for allegedly lying to NCAA investigators during a probe of recruiting violations. The issues at Penn State University with former football assistant coach Jerry Sandusky are well documented. Such ethical issues are not limited to the highest levels of amateur sport. Every year, a number of middle school and high school coaches are also fired for inappropriate conduct ranging from unethical relationships with students to psychological or physical abuses of the athletes for whom they are responsible.

Ethical issues related to coaching conduct now form an important consideration at all levels of coaching. Whether it involves encouraging players to cheat, pressuring other faculty members to give passing grades to a star player, or performing antics during a game that incite players and spectators against officials, unethical behavior should have no part in the professional's life. To ensure that athletics are integrated with the total educational program and that coaches act with their athletes' welfare in mind at all times, the National Youth Sports Coaches Association (NYSCA) has promulgated a code of ethics for coaches (see sidebar). If you plan to coach in a middle school or high school, memorize this code.

In the final analysis, ethical conduct flows from a sound personal philosophy. As addressed in chapter 4, it is vital for professionals to have a solid base of values on which to make ethical decisions that affect the lives of those for whom they work. Without developing such a base and applying it to coaching, you stand a good chance of making decisions by the seat of your pants rather than from the larger perspective of the values that you hold sacred. Have you developed a personal philosophy that can guide you in making decisions as a coach?

Trends and Opportunities in Coaching and Sport Instruction

The outlook for securing a coaching job is somewhat promising. According to the U.S. Bureau of Labor Statistics, (2015), employment of coaches is projected to grow 6 percent from 2014 to 2024, about as fast as the average for all occupations. Increasing participation in high school and college sports will boost demand for coaches and scouts. In addition, coaching was listed seventh in the 2016 ranking of best education jobs by *U.S. News & World*

Report ("Sports Coach," 2016), although the salaries for many positions remain fairly low. Specifically, the median annual salary for sport coaches, taking into account all levels, was $30,640 in 2014, according to the U.S. Bureau of Labor Statistics (U.S. Bureau of Labor Statistics, 2015). Elementary and secondary schools are the largest employers of sport coaches. Major spectator sports (college and professional football and basketball) provide the best compensation; they are also by far the most difficult jobs to get and retain due to the intense competition for positions.

The job outlook is less clear for sport instructors and coaches at various commercial sport academies. One might logically expect, however, that as long as public interest in sport remains high, job opportunities in these areas will remain available.

One emerging job market in U.S. sport is that of statistical analyst. According to the U.S. Bureau of Labor and Statistics (2015), career opportunities for stats analysts in competitive sport will increase steadily over the next decade. Sport stats analysts collect data and then create mathematical models to evaluate players and teams in order to project performance. They use statistical models to analyze the data and simulate how various changes in the game might affect chances of success. For example, professional baseball teams commonly use infield shifts with certain hitters based on the frequency with which the batter hits the ball to specific areas of the field. The same approach can be applied to pitch selection for a certain hitter depending on statistical analysis of which pitches the player hits best or worst. This approach was popularized by Michael Lewis' book *Moneyball*, which was turned into a 2011 movie by the same name.

Thus, a kinesiology major with a keen interest in sport and statistics may find a niche in the arena of highly competitive sport. Kinesiology has always had research on sport stats, or analytics, but historically this work has been most appreciated and most used in the United Kingdom and Australia. In contrast, it is a recent development in the United States. Therefore, kinesiology graduates who are interested in this career need to leverage the available research from throughout the world and develop their mathematical skills in order to compete with graduates who hold any of various degrees in business, computing, or math.

Advice for Coaching and Sport Instruction Students

The first piece of advice that any coach or sport instructor will give to students who want to coach

NYSCA Code of Ethics

I hereby pledge to live up to my certification as a NYSCA Coach by following the NYSCA Coaches' Code of Ethics:

- I will place the emotional and physical well-being of my players ahead of a personal desire to win.
- I will treat each player as an individual, remembering the large range of emotional and physical development for the same age group.
- I will do my best to provide a safe playing situation for my players.
- I promise to review and practice basic first aid principles needed to treat injuries of my players.
- I will do my best to organize practices that are fun and challenging for all my players.
- I will lead by example in demonstrating fair play and sportsmanship to all my players.
- I will not cheat or engage in any form of unethical behavior that violates league rules.
- I will provide a sports environment for my team that is free of drugs, tobacco, and alcohol, and I will refrain from their use at all youth sports events.
- I will be knowledgeable in the rules of each sport that I coach, and I will teach these rules to my players.
- I will use those coaching techniques appropriate for all of the skills that I teach.
- I will remember that I am a youth sports coach, and that the game is for children and not adults.

Are You Suited for a Career in Sport Instruction or Coaching?

If you are considering a career in coaching or sport instruction, ask yourself which setting appeals to you most. Will you enjoy working in a community, commercial, or institutional setting? Are you attracted to working with underserved youth in community youth development programs or with skilled or budding athletes in specialized sport programs? Do you enjoy being "on the job" on weekends, when most games are played and most recruiting takes place? Do you have an entrepreneurial spirit that will support a venture as an independent sport instructor or teaching professional? Are you comfortable marketing yourself and your expertise?

Consider also whether you like dealing with people as much as you like participating in your favorite sport. Are you an effective communicator? Do you have what it takes to be a leader? Do you like to plan and organize activities? Do you like to engage in strategy? Are you organized? How committed are you to the sport you want to coach and to the young people who play it? Do you have good judgment? Will you be happy experiencing both your successes and your failures in the public eye? Will it bother you to have others evaluate your performance, second-guess you, and criticize you? Do you have what it takes to look beyond the win–loss record and focus on the growth and development experienced by the athletes who play for you? Are you willing to experience the emotional ups and downs that typically characterize a career in coaching?

Answer these questions truthfully and you will get a better idea of whether you are emotionally and intellectually suited for a career in sport instruction, coaching, or both.

is to make sure that they are suited for a career in coaching. Take a moment to review the questions listed in the sidebar titled Are You Suited for a Career in Sport Instruction or Coaching? If, after going through this exercise, you believe that you are suited for a career in this area, then consider this second piece of advice: Carefully lay out a plan for developing a career in coaching. This process entails the following four steps: doing appropriate academic work, gaining coaching experience, staying current in the field, and building a coaching network.

The first step is to take coaching courses in your department or at neighboring colleges and universities. If your primary program of studies will accommodate a coaching minor, arrange to do it.

The second step is to gain coaching experience, which is a very important element in the overall process of preparing for a career in coaching. On-the-job learning, particularly at the collegiate and professional levels, is often viewed as more important than the completion of coaching courses or certification licenses. Opportunities for coaching internships abound in most communities. Explore opportunities to become a volunteer coach in a local youth sport organization; sport associations and recreation departments are always looking for volunteer coaches. You probably will start out as an assistant coach, so be patient—other opportunities will develop. If you have a special interest in swimming, tennis, or martial arts, inquire about volunteer opportunities at local clubs or schools. If golf is your passion, make inquiries at local country clubs or golf courses. Opportunities to gain experience may also be available in athletic programs in local school districts. Many coaches welcome offers of help from coaches-in-training. Also check with private schools in your area; these schools often fill coaching positions with college students who are preparing for coaching careers.

KEY POINT

Successful coaching requires skills that go beyond scientific knowledge. Effective teachers and coaches also develop communication, leadership, and organization skills that enable them to develop rapport with their participants and foster a sense of community.

The third step toward developing a career in coaching is to stay up-to-date on the latest trends and tactics in your chosen sport. As a coach, you are regarded as an expert in your sport. To live up to that standard, attend clinics, watch webinars, and become a voracious reader of materials related to the sport you wish to coach. Read books, articles, and newsletters on coaching; many of these materials are available from national or local coaching associations for your sport of interest. Attend conferences and other events for coaching education in your sport. Earn coaching licenses and certifications.

Selected teaching and coaching resources are listed at the end of this chapter, along with organizations and associations that can help you develop as a coach. With some diligent website navigation (beginning with the sport that interests you), you will find many more organizations.

The fourth step is to begin, while you are still an undergraduate student, building a network of active coaches. You will find no quicker way of joining the fraternity of the coaching ranks than cultivating contacts who can serve as mentors. Do not underestimate the importance of being connected in the coaching community; networking is essential.

As you progress through the steps to becoming a career coach, it is important to periodically take stock of your personal abilities and develop those areas in which you detect weaknesses. For example, coaching success depends a great deal on communication. A course in communication skills can help, but it cannot take the place of experience. Teachers and coaches can be highly knowledgeable about the principles of communication but not be effective communicators. Similarly, they may know the principles of leadership but not be effective leaders. Being a successful teacher or coach requires that you possess not only knowledge but also the skill to share that knowledge in such a way that people listen and act accordingly.

This view is reflected in the fact that the greatest athletes on the field do not necessarily make the greatest coaches. Although it is difficult to pinpoint why this is true, part of the reason may lie in the fact that sport came easily to them and they never developed the communication skills needed to explain the reasons for their success to less talented individuals. Conversely, athletes endowed with less natural ability may have had to study the sport in greater detail and depth in order to achieve competitive levels of performance. They have experienced adversity in sport, can understand the challenges confronting the average athlete, and can therefore relate and communicate more effectively with the group as a whole.

Successful coaching also puts a premium on organizational skills because key components of the job include detailed planning of in-season practice sessions, off-season training and conditioning programs, and game preparation. Remember, too, that for most middle school and high school coaches, and for some college coaches, coaching constitutes only one part of an individual's overall job responsibilities. Fitting it all together requires a person to be organized and possess sound time management skills.

Do you possess good leadership skills? Coaching effectiveness also hinges largely on your ability to inspire athletes to higher levels of performance. This undertaking requires leadership. Hundreds of books on leadership are on the market, including many written by business executives and successful coaches. Begin a course of reading that will help you develop effective leadership skills.

Examining your own style of communication, organization, and leadership will enable you to identify the weak spots and develop a plan for remediation. Simply put, the more you do, the more you learn. Take advantage of these opportunities to enhance your knowledge and skills, thus making yourself a more competent and effective teacher or coach when you enter the profession.

Wrap-Up

This chapter presents basic information about the coaching and sport instruction professions. It also provides an overview of possible careers in teaching and coaching, as well as the work settings in which these careers take place. Both professions involve instruction and offer opportunities to serve a wide range of people across many ages and ability levels. In fact, few professions offer such a wide array of populations from which to choose. A great range and number of opportunities are available to those people who decide to enter a teaching or coaching career—opportunities not only to capitalize on personal interest but also to promote physical activity programs and serve society. Ask yourself the following question: Do you want to make a positive difference in someone's life? If the answer is yes, then a career in sport instruction and coaching may be right for you.

MORE INFORMATION ON CAREERS IN COACHING AND SPORT INSTRUCTION

Organizations

Amateur Athletic Union
American Baseball Coaches Association
American Swimming Coaches Association
Black Coaches & Administrators
Boys & Girls Clubs of America
Coaching Association of Canada
Canadian Swimming Coaches and Teachers Association
Institute for the Study of Youth Sports
Joy of Sports Foundation
National Association for the Education of Young Children
National Association of Intercollegiate Athletics
National Collegiate Athletic Association
National Federation of State High School Associations
National Junior College Athletic Association
National Soccer Coaches Association of America
SHAPE America
United States Olympic Committee
Women's Sports Foundation
YMCA of the USA
YWCA of the USA

Journals

Applied Research in Coaching and Athletics Annual
Coaching: An International Journal of Theory, Research, and Practice
International Journal of Sport Sciences
Journal for the Study of Sports and Athletes in Education
Journal of Intercollegiate Sport
Journal of Science and Medicine in Sport
Journal of Sports Sciences
International Journal of Performance Analysis in Sport
International Sport Coaching Journal
Research Journal of Sport Sciences
Soccer Journal (NSCAA)
The Sport Journal

REVIEW QUESTIONS

1. What distinguishes sport instruction from sport coaching?
2. Describe the differences and similarities between the act of instruction and the act of coaching.
3. Contrast the professions and subcultures of coaching and those of sport instruction.
4. Outline the variety of work settings for coaches and sport instructors.
5. List the three professions in coaching and sport instruction that are most appealing to you. Examine and discuss the educational requirements and qualifications for these careers.
6. Consider two professions from this chapter—one in sport instruction and one in coaching. Describe the duties and responsibilities for each, including their primary purposes.

Go online to complete the web study guide activities assigned by your instructor.

10

Careers in Sport Management

G. Clayton Stoldt and Mark C. Vermillion

CHAPTER OBJECTIVES

In this chapter, we will

› acquaint you with the wide range of professional opportunities in sport management;

› familiarize you with the purposes and types of work done by professionals in sport management;

› inform you about the educational requirements and experiences necessary to become an active, competent professional in the field;

› apprise you of current trends and associated opportunities in the field; and

› help you identify whether one of these professions fits your skills, aptitudes, and professional desires.

ider the following scenarios representing challenges that sport managers routinely
t.

- n account executive for a professional sport franchise keeps busy during the off-season by elling season tickets, group ticket packages, and promotional packages to area businesses and individuals. Revenue generated by these sales serves as the lifeblood of the franchise, ensuring a successful upcoming season regardless of team performance.
- Staff members and interns in a college athletic department's communication office activate numerous social media channels in support of a student-athlete's candidacy for a prestigious national award. By motivating the department's fan base and alumni to vote in an online poll, they give the student-athlete a better chance of being selected.
- An events coordinator for a sport-specific international governing body plans and oversees a range of tournaments around the world each year. Key tasks include coordinating the efforts of the event staff at each tournament, dealing with site and venue issues, and ensuring that event staff and volunteers are adequately trained.
- The sport director for a community recreation commission develops a schedule for the upcoming boys and girls soccer season. With 600 children registered, 250 games to be played, and six fields available, scheduling poses a significant challenge.

In each of these scenarios, sport managers and employees of sport organizations handle challenges and make decisions that serve the consumer while enabling the sport organization to meet its financial goals.

© Human Kinetics

Students with an interest in sport and physical activity may find a career in sport management appealing. Whether the job involves the use of marketing, program management, financial, or other administrative skills, managerial positions in sport usually prove to be both challenging and rewarding. Although competition for jobs is often fierce, numerous career options are available, and the field continues to grow.

Plunkett Research (n.d.) estimated the size of the sport industry in 2015 to be $472 billion in the United States and $1.5 trillion worldwide. Their analysis included factors such as ticket sales, licensed products, sport video games, sporting goods, and facilities income. Using statistics from the U.S. Bureau of Labor statistics, Plunkett also estimated that 1.2 million people in the United States work in some facet of the amusement, sport, and recreation industry. Because of the many job opportunities available in sport management, it makes up one of the spheres of professional practice centered in physical activity.

KEY POINT

The thriving sport management industry offers numerous viable options to students looking for a career in sport.

The career opportunities in sport management are many, varied, and expanding. Job opportunities exist in professional sport, amateur sport, sport participation, sporting goods, and support services. Therefore, graduates with a well-rounded degree can choose from a variety of roles, ranging from event management to human resource management to marketing management. We hope this chapter helps you decide whether a career in sport management is right for you.

What Is Sport Management?

The **sport industry** is composed of several distinct sectors, but, overall, it is clearly big business. Sport management scholars have defined the term *sport industry* in a number of ways (Li, Hofacre, & Mahony, 2001; Meek, 1997; Milano & Chelladurai, 2011; Pitts, Fielding, & Miller, 1994; Plunkett Research, n.d.), which accounts for the variation in estimates of its size. The analysis conducted by Plunkett Research included recreation, fitness, and sport in educational settings, along with sport entertainment and other sectors of the industry.

More specifically, **sport management** is "the study and practice of all people, activities, businesses, . . . [and] organizations involved in producing, facilitating, promoting, or organizing any sport-related business or product" (Pitts & Stotlar, 2007, p. 4). The sport management sphere of professional practice centered on physical activity can intersect with career opportunities in three of the other spheres: health and fitness, coaching and sport education, and therapeutic exercise.

Physical activity professionals are expert manipulators of physical activity experiences. In other words, their expertise lies in manipulating physical activity experiences in order to improve performance and health. They may serve as personal trainers, teachers, athletic trainers, or coaches. Many sport management professions relate indirectly to the manipulation of physical activity itself (e.g., fitness center administration), and others focus on manipulating elements that support spectatorship (e.g., promotion, marketing, public relations) in order to maximize customer satisfaction. Generally, these professionals provide support services, facilities, and other amenities to make physical activity and spectatorship possible. For example, a sport manager may be employed as one of the following:

- Marketing coordinator for a speedway (see the section titled Sport Entertainment)
- Program coordinator for a YMCA (see Sport Participation)
- General manager for a golf course (see Sport Participation)
- Athletic director for a high school (see Sport Entertainment)
- Development officer for a college or university athletic department (see Sport Entertainment)
- Event manager for a triathlon (see Sport Services)
- Manager of a golf equipment retail store (see Sporting Goods)

Given the overlap between sport management and other spheres of professional practice, students interested in building a career as a sport manager sometimes find themselves competing for jobs with students who have studied in related areas, such as recreation, leisure, sport tourism, and business. Although similarities exist between these areas, so do distinctions.

Recreation or leisure may involve physical activity that results in personal enjoyment, excitement, and fulfillment. One form of recreation is, of course, sport (Hurd & Anderson, 2011). Students in recreation programs often prepare for careers in corporate fitness, nonprofit recreation, military recreation, parks and recreation, campus intramurals, and other forms of leisure services. Sport management students may have similar career interests, as is detailed in the later section of this chapter titled Sport Participation. Graduates of sport management programs do sometimes land jobs with parks and recreation departments, nonprofit fitness providers, and campus recreation programs.

Sport tourism involves people whose primary motivation in traveling outside of their communities is to participate in physical activity, watch sport events, or visit sport attractions—for example, a Hall of Fame (Gibson & Fairley, 2014). Students in sport tourism programs often prepare for careers as special event managers, travel coordinators, and sport commission administrators. Here again, sport management students may pursue similar career paths and land jobs in sport tourism.

What distinctions exist between these areas of study? It is difficult to define fixed differences,

Sport Management Versus Recreation and Leisure Studies

The following comparison of course descriptions illustrate distinctions between recreation, sport tourism, and sport management. The first set of courses is part of an undergraduate curriculum in the recreation, park, and leisure studies program at the University of Minnesota. The second set is drawn from the Tourism and Hospitality Management specialization within the Tourism, Recreation, and Sport Management department at the University of Florida. The third set comes from the sport management program at Wichita State University.

Selected Courses in Recreation, Park, and Leisure Studies (University of Minnesota)

- Orientation to Leisure and Recreation: opportunities to explore the field of recreation and the role it plays in society and human development; visits to recreation facilities representing public, quasi-public, and for-profit agencies; overview of the recreation field and foundation for continuing on to more advanced recreation courses.
- Recreation Programming: methods, skills, and materials needed for planning, developing, implementing, and evaluating professional recreation programs for diverse populations in various settings.
- Leisure and Human Development: exploration of issues associated with roles of leisure throughout the life span; principles and procedures for designing programs, services, and facilities relative to individual values, attitudes, identity, culture, age, and gender.

Selected Courses in Tourism and Hospitality Management in the Department of Tourism, Recreation, and Sport Management (University of Florida)

- Principles of Travel and Tourism: overview of the travel and tourism industry; coverage of historical, behavioral, societal, and business aspects of travel and tourism.
- Fundamentals of Tourism Planning: examination of the planning of tourism services and facilities, including the identification, planning, and use of the physical, social, and economic resources necessary to develop and support tourism.
- Tourism and Hospitality Marketing: exploration of a wide range of subjects relevant to tourism and hospitality marketing from both macro (global) and micro (organizational) perspectives.

Selected Courses in Sport Management (Wichita State University)

- Legal Aspects of Sport (I): knowledge, understanding, and application of how legal issues influence the sport industry; specific coverage of the legal system, statutory law, risk management, tort law (negligence and intentional torts), contracts, and employment-related issues in the sport industry; enhancement of decision-making and problem-solving ability regarding legal issues in sport and physical activity.
- Human Resource Management in Sport: introduction to the administration of sport in public schools, institutions of higher education, community recreation, and commercial and professional sport organizations; coverage of various components of sport administration and how to apply managerial decision-making and leadership theories in a complex and diverse environment.
- Technology in Sport Management: applications of current technology in sport management, including fundamentals of computers and their use, application of commercial software to sport management, and ethical issues faced by sport managers in using computers to conduct research and work with various social media platforms in sport settings.

Data from University of Minnesota 2016; University of Florida 2016; Wichita State University 2016.

given the diverse ways in which sport management, recreation and leisure programs, and sport tourism programs are structured. Still, we can say that recreation programs commonly offer greater depth in terms of specific forms of recreational programming and related considerations. Sport tourism programs address how to develop and manage sport-related travel opportunities and determine the effects of such initiatives. Sport management programs, in contrast, tend to focus on the sport-specific application of business principles such as management and marketing in varied for-profit and nonprofit settings.

Although the academic study of sport management is relatively new, management of the sport industry has a long history. In the early 1800s, people planned how to promote boxing and billiards, and by the mid-1800s, those responsible for the promotion of baseball games scrutinized the placement of baseball fields outside of urban areas. The 1880s and subsequent years brought about distribution issues that required the input of sport managers. During the 1890s, sporting goods manufacturers commonly hired professional advertising agencies (Fielding & Miller, 1996). By 1920, the structure of organizations in the sport industry was well established: management, marketing, legal issues and regulations, financial structure, and distribution.

KEY POINT

The sport industry has grown considerably since the 1800s. This blossoming has been caused in part by technological developments, increased discretionary money and time, evolving demographics, consumer needs, and other societal influences.

Settings for Sport Management

The sport industry can be divided into three segments: sport entertainment, sport participation, and support services. All three segments share functional roles within the industry (see figure 16.1). For instance, management and marketing are required by both professional sport franchises and nonprofit sport organizations. Likewise, financial and legal services are required by both nonprofit organizations (e.g., YMCA) and for-profit organizations (e.g., ice-skating arena). In fact, as illustrated later in this chapter, a great deal of interdependence exists between the various sport industry segments. Furthermore, the task listings for the various functional areas detailed in figure 16.1 (e.g., event management, financial management) are not exhaustive. Human resource management, for example, could also include motivating, compensating, retaining, and evaluating.

KEY POINT

Sport managers are more involved in the activities and job responsibilities surrounding an event than they are in the sport or activity itself.

The number of tasks handled in-house or outsourced varies both across the industry and within individual sport organizations. Similarly, industry segments differ in how each function is allocated among employees. Larger sport organizations often assign specialized responsibilities to individuals based on areas of expertise. In contrast, people in smaller organizations are likely to be responsible for performing a variety of job functions.

Examining Your Interactions With Sport

To consider the prevalence and popularity of sport, you need only examine your own interactions with sport and leisure during a typical day. Sport is an integral part of many peoples' lives, and that is good news for sport managers because consumer demand for sport and leisure activities results in job opportunities for sport management students.

Think about your average week. How much time do you spend watching sport programming on television? How much time do you spend reading about sports online, discussing it on social media platforms, or participating in a fantasy sport? How much time do you spend engaging in sport activities, whether it be golfing, cycling, or even e-sport? How much money do you spend on apparel featuring team logos or brand marks of sporting goods companies? How much time do you spend in sport-related conversation?

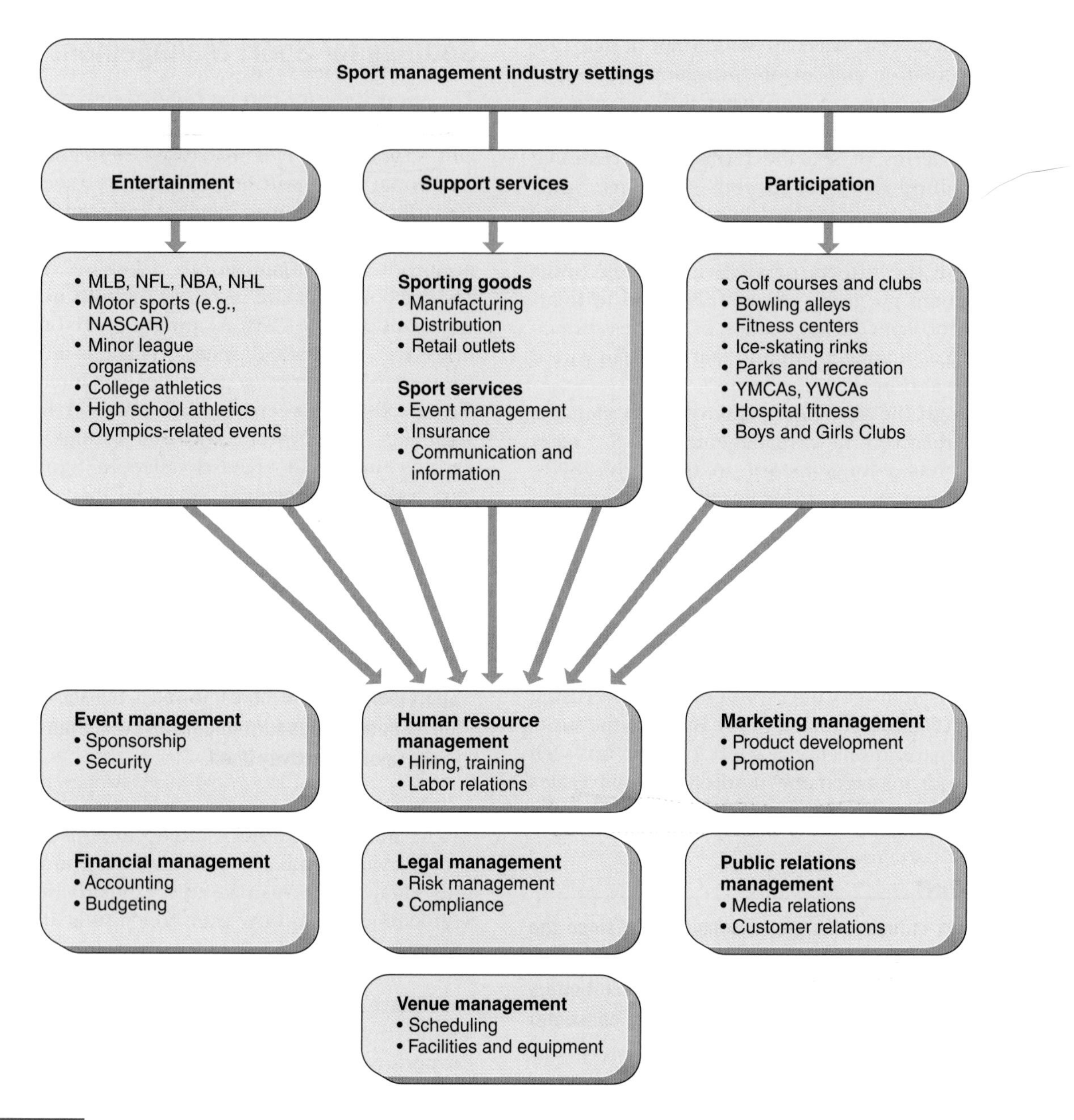

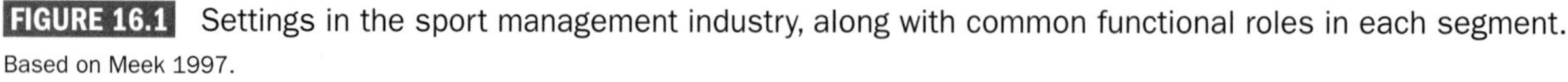

FIGURE 16.1 Settings in the sport management industry, along with common functional roles in each segment.
Based on Meek 1997.

Sport Entertainment

The proliferation of **sport entertainment** opportunities has created an abundance of jobs. At the professional level, fans are entertained by a multitude of major and minor league teams, as well as motorsports, golf, tennis, and mixed martial arts events. The options are increased further by expansion teams, new professional leagues, and the professionalization of alternative sports (e.g., action sports). Each entertainment production (i.e., game or event) represents a tremendous undertaking, combining the skills, competencies, and knowledge of many people working in separate functional areas. As indicated in figure 16.1, each professional sport entertainment organization provides career opportunities in the following areas:

- Event management
- Financial management
- Human resource management
- Legal management
- Marketing management

- Public relations management
- Venue management

These characteristics of professional sport are mirrored in many ways by the amateur sport entertainment sector. As with professional sport, the entertainment alternatives are vast and the career opportunities abundant in settings such as Olympics-related events and college athletics. Career opportunities in this industry segment often include the following:

- Development
- Public relations
- Marketing
- Operations
- Ticket sales
- Compliance
- Student services
- Facilities

Sport Participation

Whereas managers in the entertainment sector enable consumers to enjoy the sport performances of others, those in the participation sector provide opportunities for customers to engage in sport activities themselves. The sport participation sector includes both for-profit and nonprofit organizations.

For-Profit Participation Segment

For-profit sport participation organizations include health and fitness clubs, bowling alleys, roller-skating rinks, miniature golf courses, golf courses, ice-skating rinks, sport parks, and more. The increasing presence of such organizations is exemplified by the growth in the number of health and fitness clubs over the last decade, which includes the economic recession of 2008 to 2011. Specifically, according to the International Health, Racquet and Sportsclub Association (IHRSA, 2016), the number has grown during this decade by nearly 10,000 to a total of more than 36,000 clubs operating in the United States at the start of 2016. As indicated in earlier chapters, the health and fitness industry has many facets. An individual pursuing a career as a personal trainer, for example, needs to prepare for different professional responsibilities and therefore requires different educational training than does a person aiming to work in a sport management capacity.

Of particular interest to potential sport managers is the breadth of interest in sport participation. The sport industry is continuing to diversify. On one hand, according to the National Sporting Goods Association, participants continue to embrace traditional activities such as swimming, tennis, and soccer (Sports Business Research Network, 2016). At the same time, sport participation is ever evolving and includes newer activities such as fat-tire biking—a variation of mountain biking that holds particular appeal in the winter (Olmsted, 2016). Career opportunities are also available in adventure activities and events such as mud races, color races, fitness competitions, and tournaments.

The for-profit sport participation segment has also realized growth in existing product mixes by supplementing core products such as resistance training with newer offerings such as Zumba. In addition, more and more businesses cater to specific target markets, such as unfit women and time-pressed professionals. Employment opportunities vary according to the size of the organization but do exist in a variety of functional areas:

- Management
- Sales
- Marketing
- Program planning
- Human resources
- Risk management

Thus, for-profit sport organizations hold an enviable position as their target markets and product lines expand.

Considering Sport Entertainment Management

Think about the last time you attended a professional or amateur sport event or watched one on television. Identify as many as you can of the peripheral activities or extended elements (e.g., concessions) associated with the main event (i.e., the game). With this context in mind, are you interested in managing events or venues? If so, begin developing your skills in organization, time management, leadership, and communication—all of which are necessary for a successful career in sport entertainment management.

Kinesiology Colleagues

Photo courtesy of Kathleen Avitt

Kathleen Avitt

Kathleen Avitt is director of programs for the Derby Recreation Commission near Wichita, Kansas, where she oversees more than 1,200 recreation programs in areas including art, sport, special events, special populations, aquatics, fitness, and health and wellness. Prior to her arrival in Kansas, Avitt worked for the National Alliance for Youth Sports in West Palm Beach, Florida, where she served as national program director of the Parents Association for Youth Sports. There, she created one of the first youth-sport parent-education programs, which drew 2,500 parents and received national news coverage for its first event.

Nonprofit Participation Segment

As discussed in chapter 15, the nonprofit sport participation segment is replete with jobs for qualified sport management professionals. Examples include YMCAs, YWCAs, Boys & Girls Clubs, and hospital-affiliated fitness centers. As in other industry segments, individual jobs in these organizations may be specialized in one of the functional areas identified in figure 16.1 or may be more general. For example, a YMCA located in a major metropolitan area may offer job opportunities in specialized functional areas, whereas a Boys & Girls Club may employ one person to be responsible for all recreation-related areas.

Like many other areas in sport, the nonprofit sport participation sector offers considerable opportunities. Nearly 20 percent of all U.S. health club consumers are members of a nonprofit, such as a YMCA or YWCA (Knowledge@Wharton, 2014). In 2016, the YMCA had 2,700 branches with roughly 19,000 full-time employees in about 10,000 communities in the United States. The Y also has a global presence; indeed, some 45 million people in 119 countries are affected by programs, services, or facilities run by the Y (The Y, 2016).

Support Services

The support services segment includes an even wider array of sport organizations than do the previous two segments. It includes both the sporting goods industry and a variety of service providers; these vast categories are explored in the following sections.

Sporting Goods

The sporting goods industry represents a significant portion of the overall sport industry. In 2015, U.S. consumers purchased almost $65 billion of sporting goods (Statistica, 2016), and the industry employs roughly 350,000 people at either the wholesale level or the retail level (Plunkett Research, n.d.). In this context, sport managers play a vital role in ascertaining what sport products people need and want. Employment opportunities in the sporting goods industry include the following:

- Research analyst
- Retail store manager
- Manufacturing sales representative
- Accounting manager
- Licensing administrator

Sport Services

As a result of the sport industry's enormous growth, sport-related services have become more specialized. As a result, sport managers can now specialize in sport insurance, sport event management, or sport marketing; this segment of the industry is

Considering Career Options in the Sporting Goods Industry

Do you have a special interest in sports equipment or apparel? Perhaps you like to explore the newest merchandise on the market or imagine designing your own sport clothing. Or maybe you like to mine data and research trends in order to predict the newest fads in athletics. Or perhaps you're great at getting people excited about new products or services. All of these skills and interests could contribute to a career in sporting goods.

known as **sport services**. Here are a few examples of organizations that offer sport services:

- Sports & Fitness Insurance Corporation: provides insurance for health clubs, fitness centers, martial arts studios, personal trainers, and dance schools.
- Team Marketing Report: provides information services regarding sport consumers, ways to increase ticket and sponsorship revenues, individual sponsorship decision makers, and more.
- Collegiate Licensing Company: manages licensing programs for universities and colleges.
- Exclusive Sports Marketing: promotes and manages recreational sporting events such as triathlons.
- Anthony Travel: specializes in working with university athletic departments, professional sport organizations, individual athletes, and fans.

The sport services segment is likely to realize continued growth as organizations outsource job functions in order to reduce personnel costs while at the same time receiving specialized services, such as insurance, travel, marketing, or licensing services unique to sport and physical activity organizations, for seasonal sport product offerings.

Roles for Sport Management Professionals

Sport management practitioners perform many similar professional roles regardless of the sport industry segment in which they are employed. Therefore, rather than list separate professions in the sphere of sport management, as in previous chapters, we describe the sport industry in terms of functional roles and related tasks that may be assigned to one or more employees in various sport organizations.

Event Management

In all physical activity professions, the staging of an event is a tremendous undertaking. As a result, many organizations hire people with specific expertise in **venue** and **event management**. These employees are responsible for the many tasks associated with hosting an event, including risk management, security, venue setup, concessions, scheduling, game presentation, customer service, and many more. Although a single person may cover all of these responsibilities for a small-scale event, large staffs are commonly employed for a major event. For instance, Super Bowl 50 had a host committee advisory group of almost 50 people, and thousands of volunteers worked in a variety of roles to meet the minimum requirement of three shifts during Super Bowl week ("Host Committee Advisory Group," n.d.). Ultimately, the overall event attracted more than a million people to various Super Bowl-related events in San Francisco ("Super Bowl 50," 2016).

KEY POINT

Sport event management is a complex function because it requires skills specific both to securing, organizing, and executing an event and to creating partnerships with other organizations that may have a stake in the event.

Clearly, venue and event management requires sport managers to consider a multitude of factors—for example, traffic control, crowd control, security, medical care, resource availability, and management and coordination of both paid personnel and volunteers. Therefore, opportunities to work in this area are widespread and vary from the Super Bowl to an evening of Little League baseball at a local park.

Sport commissions—local entities that work to attract sporting events and franchises to their cities in order to boost the local economy—have seen an enormous increase in the United States in

Crowd Control and Security

On April 15, 2013, two terrorists planted and exploded multiple bombs at the Boston Marathon, killing three people and injuring almost 300 others. This high-profile incident highlighted the need for security and the difficult nature of implementing crowd control and security strategies at sporting events. Specifically, the unique nature of a marathon—the course is over 26 miles (42 kilometers) long—illustrated some of the different challenges in securing sporting events. The following year, security procedures for the marathon included greater police presence, more than 100 surveillance cameras, bag searches, and 400 heavily armed military police personnel (Quinn, 2014). In addition to these logistical changes, cultural changes were also implemented. For example, organizers discontinued the tradition of letting unauthorized participants run parts of the course with the official participants. Although this is a particularly large and dramatic example, adequate crowd control and security strategies should be identified prior to any sporting event, regardless of size or anticipated attendance.

Who is responsible for crowd control and security at a sporting event? What factors are involved in safely controlling a crowd?

recent years. In fact, the National Association of Sports Commissions (NASC) has grown from 15 members in 1992 to almost 750 members (NASC, n.d.). Similarly, the Canadian Sport Tourism Alliance (CSTA) now boasts more than 400 members who want to cultivate travel to their towns and cities for sporting events (CSTA, n.d.). Clearly, employment opportunities in this sector of the field are abundant and expanding globally.

Once an event is scheduled, sport commission personnel collaborate with others (e.g., team owners, employees, volunteers) in the sport industry to ensure that the event is a success. For example, a sport commission that secures a bid to bring a bowling championship to its city would work closely with practitioners in the local bowling community to ensure that adequate facilities are available, customer service is exemplary, scheduling is organized, and volunteers are gathered and committed.

Financial Management

Financial management is critical to the operation of most sport-related organizations. Financial managers may administer budgets, oversee income allocation, pursue development opportunities, and handle investments. Managing financial situations for sport organizations requires knowledge and understanding of the necessary financial tools (e.g., financial statements, budgeting procedures) and of the sport product. Without understanding the sport product, a person is not equipped to capitalize on potential revenue opportunities.

KEY POINT

High demand for sport-related products creates a positive environment for the financing of a variety of sport facilities, including health clubs, swimming pools, arenas, and stadiums.

Two areas of sport financial management that provide a considerable number of opportunities are accounting and facility finance. All sport organizations, and almost all employees in the sport industry, are involved in some manner of accounting and budgeting. Sport marketers, for example, are responsible for promotions, advertising, and sponsorships and must meet revenue goals. Similarly, sport event managers must balance the expenses associated with staging an event (e.g., facility rental, staffing, concessions purchases) with anticipated or expected revenues (e.g., ticket sales, sponsorships).

Financing facilities has become an important issue in the sport industry. More than $9 billion was spent in the United States on facility construction in 2005 (Milano & Chelladurai, 2011), and recent years have seen a large increase in the number of new facilities constructed for professional franchises and college sport programs (Muret, 2014). Sport organizations and their public constituents have used a variety of sources to fund their projects, including private-sector money (e.g., corporate sponsorships), public support (e.g., subsidies, tax abatements), and organizational revenue (e.g., personal seat licenses, luxury seating) (Howard & Crompton, 2014). The growth in facility construction has provided interested individuals with a number of finance-related jobs in the sport industry.

Kinesiology Colleagues

University of Colorado Boulder

Ted Ledbetter

Some people aspire to a successful career in college athletics. Ted Ledbetter has experienced two. As a student-athlete, he enjoyed an enormously successful baseball career at Oklahoma City University (OCU). Now he works as a sport administrator, serving as the assistant athletic director for development at the University of Colorado Boulder (CU Boulder). A two-time NAIA All-American and national player of the year while at OCU, Ledbetter pursued a range of extracurricular interests while earning his bachelor's degree in psychology. He was then drafted and spent a short stint in pro baseball.

To transition into his next chapter of life, Ledbetter began working on a master's degree in sport administration at Wichita State University. That led to an internship and then a job in athletic development at the University of Michigan, where he played a key role in securing funds for the $226 million Michigan Stadium renovation project. In 2012, Ledbetter shifted to a new job in athletic development at the University of Colorado. His responsibilities now include oversight of CU's Buff Club for annual giving, premium seating at CU's football and basketball facilities, and major gift cultivation throughout the region.

Human Resource Management

Human resource management, also known as personnel management, may be one of the most important job functions in the sport industry (Chelladurai, 2006; Miller, 1997; Weerakoon, 2016). Employees of sport organizations are arguably their organization's greatest asset. Happy and content employees are more loyal to the organization, which results in a more experienced staff and a reduction in the costs associated with employee turnover. Employees are responsible for the entire sport-product delivery process, including customer service, operations, and other crucial functions. A gap in any functional area dilutes the quality of the sport product. In turn, low-quality products eventually fall below the financial break-even point, thus causing sport organizations to go out of business.

KEY POINT

Employees arguably represent the greatest asset of any sport organization; therefore, the human resource function is critical because it selects, trains, and retains high-caliber people.

Human resource managers work to combine employees' talents and desires with the needs of the sport organization in order to create a pleasant and legally defensible work environment. These managers have traditionally dealt with the hiring, training, maintenance, and dismissal of employees. Recent trends, however, reveal a far broader scope of work associated with human resource management. For example, these managers may also organize and manage such benefits as day care for employees' children, employee stock ownership plans, professional development opportunities, fitness facilities, and counseling services—all of which are designed to attract and keep high-quality employees.

Human resource managers carry added responsibilities in sport organizations where unions are present. Unions negotiate agreements with management regarding compensation, working conditions, and other issues. As evident from previous work stoppages in professional sport, union–management relations are critical to a sport organization's profitability. Sport managers who are employed outside of professional sport may also encounter union issues when dealing with concessions and merchandising vendors, security personnel, and officials and umpires.

Considering Your Experiences as an Employee

Think of an organization where you have worked. Did that organization create a friendly work environment conducive to productivity and efficiency? What did the organization do well? What could it have done to improve the work environment? Does tackling such issues interest you? Can you see yourself being an instrument of change in an organization, recruiting others, establishing workplace policies, or ensuring company compliance with employment guidelines? If so, human resource management might be the profession for you.

Risk Management

As U.S. society has grown increasingly litigious, sport administrators have become more concerned about **risk management**. Sport managers today face many complex issues that warrant advice from attorneys and other legal experts, and all sport managers should possess the knowledge necessary to minimize risks for their organizations.

KEY POINT

Risk management involves taking steps to reduce losses of all kinds. It requires the support of all employees, from top management down to the front line.

Risk management evolved from a predominant concern about lawsuits to a more encompassing concern that includes risks of all kinds (Sharp, Moorman, & Claussen, 2010). Today, the sport industry views risk management as including, among other things, losses resulting from inadequate or improper security, food service distribution, supervision, instruction, and facility and equipment design. To protect against losses of all kinds, many sport organizations have designated risk management personnel. Risk management groups help sport entities with identifying potential hazards, suggest and implement ways to manage potential losses, evaluate existing risk management plans, and continually improve existing risk management efforts. A successful sport manager can apply risk management principles to a variety of sport-related environments, ranging from traditional sporting facilities to such unique environments as adventure sport coaching (Collins & Collins, 2013).

Risk management is especially important in the amateur and professional segments of the sport entertainment industry. Besides attending to the concerns already listed, the entertainment industry segment must also comply with complex rules and regulations established by governing organizations. Fearing losses to their athletic organizations and programs, some colleges have responded by using outside legal assistance in their compliance efforts. Many others have designated particular administrators as compliance officers.

Compliance-related failures are unfortunately common, and their consequences have been well documented by the mass media. The multitude of compliance-related failures in NCAA athletics can be illustrated by a survey of popular media investigations. For example, Ganim (2015) reviewed investigative reports from the *New York Times*, the *Ann Arbor News*, and the *Chronicle of Higher Education* and found coverage of a number of cases involving various forms of academic fraud and extra or impermissible benefits provided to student-athletes. According to a search on the NCAA's LSDBi (Legislative Services Database), 115 NCAA institutions reported separate major infraction cases across several sport programs between the fall of 2010 and the summer of 2016 (NCAA, 2016). In this context, a university compliance officer strives to keep the athletic department in compliance with the rules and regulations detailed in the lengthy NCAA compliance manual; in university policies and procedures; and in local, state, and federal laws and ordinances.

Marketing Management

Marketing management is a burgeoning area of opportunity for qualified students. Some even suggest that effective marketing is the key to success for any sport product, whether it is a new running shoe, a sport clinic, or a sporting event. Even a great sport product, offered at an affordable price and available in a favorable location, generates little revenue unless people are aware of it. Sport marketers may spend time working in a variety of marketing management areas:

- Research and development can range from sophisticated analyses of consumer opinions and economic impact studies to focus groups and brainstorming sessions on how to package and present products more effectively.

› Sport promotion involves taking action to heighten interest in sport products and increase consumption (e.g., sales, viewership). **Promotions** are tactics used by marketers to communicate with and attract buyers. They come in a wide range of forms, including product sampling opportunities, premium giveaways, price reductions, player appearances, camps, and clinics (Irwin, Sutton, & McCarthy, 2008; Mullin, Hardy, & Sutton, 2014).

› Sport sponsorship involves an exchange between a company and a sport organization. For example, a sponsoring company may provide either cash or in-kind resources (e.g., products, services) in exchange for sponsorship of an event. For example, 15 national sponsors were associated with the 2016 College Football Playoff, including Allstate, Capital One, Dr. Pepper, and Taco Bell ("National Sponsors," n.d.). Sponsorships represent powerful promotional platforms that build brand awareness for sponsoring companies. They provide an important revenue stream for sport organizations, but sport managers must be mindful of which brands make for the best and most appropriate partners for their sport organizations. This issue is illustrated in the accompanying case study.

› Advertising consists of "any paid, nonpersonal (not directed to individuals), clearly sponsored message conveyed through the media" (Mullin et al., 2014, p. 266). It is one of the most important and visible tools available to people who work in sport (Shank & Lyberger, 2015). Many sport-related organizations purchase **advertising** to promote their events or sell their services or products.

› Merchandising efforts extend far beyond apparel. Customers who visit sport specialty stores may choose from an array of licensed products, including toys, automobile accessories, and even pasta in the shape of team mascots. Sport trading cards continue to be big business, and rare or high-demand cards can command thousands of dollars. In addition, the market for e-sport continues to grow and expand globally, as is evidenced by robust financial projections nearing almost half a billion dollars ("Global Esports Market Report," 2016). Some sport organizations conduct their own

Case Study in Assessment

Randy believes adamantly in the power of sport to exert a positive effect on people. As a result, he was very excited about taking a position as director of marketing and sponsorships for the athletic department at a public (state-supported) Division II university in the U.S. Midwest. Randy saw his organization as playing an important role not only in entertaining the community but also in educating and informing citizens of worthwhile causes in the region—all while developing student-athletes into productive members of society.

Decision

Soon after starting his new job, Randy was presented with a sponsorship opportunity from a well-known beer distributor and brewery. The sponsorship would have brought much-needed revenue to an athletic department struggling with decreased donor contributions brought about by an economic recession. However, Randy was aware that alcohol abuse is prevalent on many campuses, and he wondered what kind of message is sent when a college athletic department promotes beer brands. He had many factors to weigh before making a decision regarding the sponsorship opportunity. For instance, he needed to check university policy and current sponsorship trends in college athletics while weighing the various pros and cons.

Resolution

With the athletic department's financial well-being in question, the beer distributor's proposal must be considered carefully. In accordance with the athletic department's vision (and Randy's), any signs posted in athletic facilities could be used as an awareness opportunity. For example, the posted signs, although associated with the beer distributor's name and logo, could be designed to educate community members about the dangers of excessive alcohol consumption. This resolution allows the distributor to be associated with the athletic department; brings the department a much-needed financial boost; and educates both college students and community members about the importance of responsible drinking behavior.

merchandising programs, whereas others sell licenses to other businesses to place their names and logos on various products.

› Distribution connects a sport product or service with the ultimate consumer. For sporting goods manufacturers, **distribution** includes building relationships with distributors and retailers, securing optimal shelf space, and maintaining low-cost product transportation systems. For sport managers in the entertainment or participation sector, distribution concerns may include facility design, accessibility, flexibility, and attractiveness because the sport product (e.g., game) is produced and consumed in the same location (e.g., stadium). Broadcasting arrangements may also be considered distribution-related issues because they focus on getting the sport product to consumers (Mullin et al., 2014). The NFL provides a good example of the potential magnitude of such agreements in the sport entertainment industry. The league reaps nearly $5 billion in television rights fees each year (Deitsch, 2011), and the current broadcast rights agreement that extends through the 2022 season is worth almost $30 billion (Badenhausen, 2011).

KEY POINT

Sport management students who are educated in sport marketing often hold great appeal for employers in all sport industry segments. A background in sport marketing provides both lateral and vertical career opportunities.

Public Relations Management

Public relations management in sport seeks to achieve positive relationships between the sport organization and its most important constituents (Stoldt, Dittmore, & Branvold, 2012; Stoldt, Dittmore, & Pedersen, 2014). This work involves diverse activities ranging from media relations management to community relations to investor relations.

A number of people in sport organizations work in media relations, and almost all professional and college sport organizations assign at least one person to media relations. These employees produce organizational media, manage organizational websites and social media platforms, track statistics, arrange interviews, supply story ideas, and generally service media requests. Thus, these individuals provide an integral link between the organization and its stakeholders.

Although the media may be more integrally involved in sport than in any other industry in the United States, **public relations management** involves much more than just media relations. For example, sport entertainment organizations often employ people specifically to work in fan relations, and sporting goods businesses generally have customer relations employees who provide a variety of services ranging from handling complaints to giving special instructions about product assembly.

Some sport managers also work in community relations. These public relations professionals may seek opportunities for organizational representatives to make public appearances or speeches. They may also coordinate donations to good causes,

Corporate Social Responsibility in Sport

According to Bradish and Cronin (2009), corporate social responsibility (CSR) “can be broadly understood as the responsibility of organizations to be ethical and accountable to the needs of their society as well as to their stakeholders” (p. 692). To learn about how CSR mechanisms and processes evolve over time, Heinze, Soderstrom, and Zdroik (2014) studied the CSR efforts made by the Detroit Lions. They noted that the organization’s change in approach centered on three key results and strategies.

First, the Lions decided on a strategic approach that was more direct and specific than previous broad strategies of raising money and distributing that money to a variety of charitable organizations or causes. Second, the team selected high-impact areas and used programming that focused on “sustainable community health, wellness, and development” (p. 677). Third, the organization involved the local community in a bottom-up approach in which specific communities communicated their needs to the team. Heinze et al.’s (2014) research details not only how the Lions organization affected its community but also some key processes in professional sport organizations’ continued efforts to “give back.”

promote the organization's environmental initiatives, and execute other tasks pertaining to social responsibility.

Program Management

Program management is common in all sport industry segments. Sport programs are delivered, for example, by health clubs, parks and recreation departments, athletic departments, and sport services. A parks and recreation department, for instance, may offer a youth sport program, an aquatics program, and a fitness program. In turn, each program is staffed with its own employees who deliver desired activities (e.g., lessons, clinics, tournaments).

Program management activities vary across functional areas and sport industry segments. Two of the more common task areas are scheduling and maintenance of facilities and equipment.

Scheduling

Scheduling is one of the primary concerns for program managers. For instance, many a Little League coordinator has dealt with scheduling complications stemming from too many teams with access to too few facilities. Sport entertainment administrators must also address scheduling and related considerations. Scheduling games, events, and even practices can be complicated when other events in the community affect public demand or when one facility must accommodate multiple demands. Scheduling travel can also be a complex undertaking, and some athletic programs employ administrators whose primary function is to coordinate travel.

Facility and Equipment Maintenance

Maintenance of facilities and equipment is a crucial task for program managers. Those who work in sport participation organizations must maintain and

Professional Issues in Kinesiology

Equality: Diversity and Equality in Kinesiology

For some time now, the field of sport management has emphasized the importance of understanding diversity and other issues related to equality. More than a decade ago, for instance, the NCAA noted the importance of providing diversity training and diverse opportunities for student-athletes and sport organizational personnel (NCAA, 2005). More recent examples include understanding the barriers faced by ethnic-minority students in accessing campus recreation facilities, events, and programs (Hoang, Cardinal, & Newhart, 2016); how to develop culturally sensitive sport programming in Canadian Red Cross swimming programs (Rich & Giles, 2015); and how individuals in intercollegiate athletics can improve organizational actions and policies associated with diversity (Bopp, Goldsmith, & Walker, 2014).

The commitment to diversity and equality extends to the sport management classroom, as well. Efforts are ongoing not only to develop culturally sensitive classrooms but also to educate students using descriptive and effective theories, strategies, and tools (Bruening et al., 2014). The case for valuing diversity and equality in all spheres of kinesiology has been made by Brooks, Harrison, Norris, and Norwood (2013).

Citation style: APA

Bopp, T., Goldsmith, A., & Walker, M. (2014). Commitment to diversity. *Journal of Applied Sport Management, 6*, 1–27.

Brooks, D., Harrison, L., Norris, M., and Norwood, D. (2013). Why should we care about diversity in kinesiology? *Kinesiology Review, 2*, 145–156.

Bruening, J. Fuller, R.D., Catrufo, R.J., Madsen, R.M., Evanovich, J., and Wilson-Hill, D.E. (2014). Applying intergroup contact theory to the sport management classroom. *Sport Management Education Journal, 8*, 35–46.

Hoang, T.V., Cardinal, B.J., & Newhart, D.W. (2016). An exploratory study of ethnic minority students' constraints to and facilitators of engaging in campus recreation. *Recreational Sports Journal, 40*, 69–82.

National Collegiate Athletic Association. (2005, February 25). Commitment to diversity programming. *NCAA News*, 42(5), A2.

Rich, K. & Giles, A. (2015). Managing diversity to provide culturally safe sport programming: A case study of the Canadian Red Cross's swim program. *Journal of Sport Management, 29*, 305–318.

Considering Career Options in Sport Program Management

Think of a sport organization with which you are familiar. What programs does the organization offer? Do you have the skills and aptitudes necessary to run such programs? If you are interested in program management, what skills would you need to develop? What can you do to start developing these skills? What opportunities are available to you now? What opportunities might require some research and planning to get involved in? Make a plan for your skill development.

update facilities to meet the needs and demands of consumers; this work includes keeping facilities and equipment clean, safe, and operative. In addition, sport managers must monitor the external environment (e.g., competitor actions, changes in consumer demographics and psychographics) for information that can be used to enhance an organization's consumer offerings and customer satisfaction.

Career Prospects and Opportunities in Sport Management

The sport management industry offers a plethora of job opportunities in a variety of industry segments and functional areas. In fact, it can be difficult to determine objectively which segment is the most popular, because they all appear to be hiring qualified students of sport management. Still, a review of various websites and trade publications provides insight into the available job opportunities. For example, TeamWork Online advertises administrative jobs available in most of the major professional sport leagues and in large sport entertainment companies.

KEY POINT

Job opportunities abound for qualified individuals in sport management, thanks in part to the proliferation of professional teams, societal interest in health and fitness, technological developments, and the presence of eclectic consumers with varying needs.

The vast array of job opportunities available in sport management is also illustrated by data from the U.S. Department of Labor. More specifically, table 16.1 provides information about the number of jobs and mean salaries for selected positions and sectors of the industry.

Education and Qualifications

The North American Society for Sport Management (NASSM) website lists hundreds of colleges and universities that offer sport management programs at the bachelor's, master's, and doctoral levels (NASSM, n.d.). More than 200 of those programs are located in the United States; others are found in Canada, Europe, Africa, India, Australia, and New Zealand. If you review the program information

Table 16.1 Mean Salaries for Selected Jobs Associated With Sport

Industry sector	Job type	Annual mean salary
Spectator sport	Chief executive	$217,280
	Financial manager	$137,030
	Sales manager	$113,710
	Agent or business manager for artists, performers, and athletes	$103,370
	Market research and marketing specialist	$58,640
Fitness and recreational sport centers	Chief executive	$181,170
	Sales manager	$94,798
	Recreation and fitness worker	$40,360
	Supervisor, personal services	$39,440

Data from U.S. Department of Labor, Bureau of Labor Statistics 2015.

for even a handful of the institutions listed on the NASSM website, you will likely find considerable variation among the programs in terms of focus, breadth, depth, housing (e.g., business school, colleges of education), and experiential learning requirements. High-quality programs can be structured in a variety of ways, and no single program offers the best fit for all prospective students.

The Commission on Sport Management Accreditation (COSMA) is an accrediting body whose purpose is "to recognize excellence in sport management education" (COSMA, 2016). When COSMA reviews a program, it considers, among many other factors, how well the curriculum is grounded in the program's mission, goals, and objectives, which clearly demonstrate what students are learning (COSMA, 2016). Many universities offer sport management programs, but only about two dozen provide COSMA-accredited programs. Programs seeking accreditation at the undergraduate level are expected to feature a curriculum based on standardized principles covering five broad areas (COSMA, 2016, p 12):

1. Foundations of sport: historical, sociological, and psychological
2. Foundations of sport management
   - Management concepts
   - Governance and policy
   - International sport
3. Functions of sport management
   - Sport operations
   - Sport marketing
   - Sport communication
   - Sport finance and economics
4. Sport management environment
   - Legal aspects of sport management
   - Ethical aspects of sport management
   - Diversity issues in sport management
   - Technological advances in sport management
5. Integrative experiences and career planning
   - Internship or other practical or experiential learning
   - Capstone experience

Reprinted, by permission, from Commission on Sport Management Accreditation (COSMA). 2016, *Accreditation principles manual & guidelines for self-study preparation.* Available: http://www.cosmaweb.org/accreditation-manuals.html.

Although all of the content areas listed are critical to attaining a high-quality sport management education, the internship and other integrative experiences (e.g., practicum) deserve special emphasis. Sport organizations look for people with practical, real-world sport experience. Because the sport management industry is extremely competitive, the references, knowledge, and work experience that you gain from an internship site supervisor is invaluable. Consequently, an internship or other form of integrative experience is essential to your ability not only to secure a job but also to learn the best practices in the industry.

KEY POINT

Sport management students should capitalize on available internship opportunities with credible sport management organizations.

Students can obtain internships in any of the sport industry segments discussed earlier. Responsibilities associated with internships tend to be situation specific. For example, an internship at a Division I institution might be highly specialized. In other words, the intern might work specifically in one area, such as compliance, marketing, facilities, operations, or tickets. In contrast, an internship at a Division III institution might combine roles. In this situation, the intern might work in a variety of areas, including marketing, facilities, operations, and tickets.

All of the competency areas recognized by COSMA are identified, either directly or indirectly, in figure 16.1. As you can see, each functional area is somewhat dependent on, and linked to, the others. Moreover, many benefits can be gained from an understanding of each area. For example, if you are working in a youth sport program, you will be better able to provide successful program offerings if you understand budgetary limitations, managerial issues, consumer behaviors, risk management principles, and societal implications. Similarly, if you work with a professional hockey franchise, you will gain better understanding of the intricacies involved and the communication required of individuals working in various functional areas of a large event. You will learn, for example, that executing promotions requires close cooperation among the people working in the area of facilities. Sound effects (e.g., music, clapping hands), lighting, fixtures, and so on must be coordinated so that fans are entertained, the sponsor's promotion is flawlessly executed, and promotion participants are not injured.

KEY POINT

Multifaceted individuals can make decisions that serve the best interests of the sport organization as a whole rather than benefiting only particular business functions within the organization.

When you understand how various issues and functional areas influence your product—be it a sport event, a clinic, or a line of apparel—you are better able to communicate with colleagues, employees, and customers. Your conceptual understanding makes you an invaluable asset to your sport organization. In addition, by being knowledgeable in a variety of competency areas, you enhance your marketability in the profession.

Students preparing for sport management careers follow different curricula than do students studying for such careers as teaching, coaching, personal training, and athletic training. As you have seen, sport management students are often not required to take courses in motor learning, exercise physiology, or anatomy and physiology. Course work in theoretical kinesiology, however, can significantly enhance a person's marketability in the sport participation sector. For example, students who work in the health club industry can improve their marketability by combining a sport management degree with course work in appropriate areas of theoretical kinesiology. This well-rounded education enables a sport manager to make better hiring decisions and provide better supervision.

Trends and Opportunities in Sport Management

In recent years, *Street & Smith's SportsBusiness Journal*, a weekly professional publication, has featured a series of interviews on a wide range of topics with executives in (or in partnership with) the sport industry. One of the questions posed to the executives asks what they like about the future of sport business. Here are some of their responses ("Andrew Walker," 2011; "Charlie Besser," 2011; "Conrad Smith," 2011; "Dave Butler," 2011; "Mark King," 2011; "Sarah Hirshland," 2010; "Tony Schiller," 2011).

- Sarah Hirshland, then senior vice president, Wasserman Media Group: "Consumers continue to prove how important sports are in their lives. They want more, they want it live, and they want to share it with their growing base of friends."
- Tony Schiller, executive vice president and partner, Paragon Marketing Group: "Real partnerships between ownership and players. Dollars are becoming more scarce, so more creativity in game presentation and revenue distribution are increasingly essential."
- Conrad Smith, COO, Minnesota Lynx: "Combining the in-person experience with the comforts of watching a game on television at home."
- David Butler, then CEO, Paciolan: "Fans continue to want more access, more information, and a more personal involvement, and that passion is a great sign for our industry."
- Andrew Walker, then chief marketing officer, Women's Tennis Association: "Convergence with entertainment."
- Mark King, then CEO and president, TaylorMade-Adidas Golf: "Incredible growth fueled by new media."
- Charlie Besser, president & CEO, Intersport: "The globalization of sports. . . . The world is shrinking fast!"

These comments carry an array of implications. First, as the lines blur between sport and entertainment, and as new technologies enable higher levels of direct engagement, the future of the sport business only gets brighter. For example, in the growing e-sport sector of the industry, sport management professionals who deliver compelling experiences to both spectators and participants can expect to enjoy successful careers.

The Growing E-Sport Phenomenon

When ESPN launched a website dedicated to e-sport, it was a clear indication of the media powerhouse's bullishness on this growing sector of the industry (Peckham, 2016). The e-sport market approached $500 million in 2016 and is projected to double to more than a billion dollars by 2019 ("Global Esports Market Report," 2016). The sector was fueled in 2016 by an estimated audience of 131 million avid fans and 125 million occasional fans. Little wonder that ESPN, as well as other organizations, view e-sport as a key growth area.

Engagement is no longer limited to the sport facility itself or to the delivery of sport programming via mass media. New and social media now make it possible for consumers to interact at any time and in any place. As a result of ongoing technological change, sport organizations need a strategic approach for examining technology's effect on a variety of sport relationships. For example, the NBA hired a former telecommunications expert to lead its mobile and wireless initiatives (Fisher, 2015), and the National Basketball Players Association launched an app to help players communicate more effectively with the union (Mullen, 2015).

The expert comments about the future of sport also carry a second implication—that as sport management becomes more competitive and complex, sport managers who can forge successful partnerships will be in high demand. The nature of those partnerships will vary from setting to setting. Whether the task is to negotiate mutually acceptable agreements between player personnel and management or to secure corporate support for a community running event, managers with strong collaborative skills and creative minds will enjoy advantages in the competitive marketplace.

The third implication is that the sport industry is growing more global by the day. Global sport properties, such as the World Cup, now command extravagant rights fees from media companies wanting to broadcast the matches on U.S. television. NBA stars now enjoy celebrity status in countries as far away as China. Diverse populations are now recognized for their achievement in global events such as the Paralympics. Sport managers who embrace the opportunities and challenges associated with globalization are well positioned for career growth.

Advice for Sport Management Students

As you know from reading about other physical activity professions in previous chapters, simply meeting educational requirements does not ensure your success as a sport management professional. The field of sport management is hypercompetitive, and you can enhance your career prospects by gaining as much practical experience as possible, seeking a mentor, and building a professional network.

Volunteering is a great way to build your resume and gain practical work experience.

Besides completing the required courses and obtaining your degree, you also need practical experience in your area of interest. Such experience is invaluable in helping you improve your communication skills and enhance your problem-solving skills. Understand the value of a mentor—someone in the field who can give you practical insights into your new career. In addition, keep up with the profession by noticing and being aware of sport management organizations' administrative activities and reading professional journals; resources to get you started are listed at the end of this chapter. The following sections offer details on strategies that you can pursue to prepare for a career in sport management.

Leadership Development

In a study examining the career expectations of sport management students and practitioners, Mathner and Martin (2012) found that both groups ranked leadership skills as the most critical competency. Leadership skills can be developed by just about everyone. Leadership courses, and in some cases minors, are available at many colleges and universities, and a wide range of leadership development programs and resources (e.g., books, speeches) are also available beyond college campuses. The topic of leadership may also be something to explore with a trusted mentor.

Practical Experience

You should always be at work building your resume because practical experience is a necessity in the competitive job market of the sport industry. Almost all programs have internship requirements, and we have already described the benefits of internships. The best internships, however, often go to students with the most job experience. Try to find part-time work in the sport industry and volunteer for special events (e.g., fun runs, triathlons, Special Olympics). Such experiences allows you to apply what you learned in the classroom to the job setting while simultaneously gaining valuable resume material.

Mentors

Mentors are people who work in a student's chosen profession and provide students with guidance, learning opportunities, and contacts in the field. Students who have a mentor gain insights that they cannot obtain in the classroom; they also have an advocate when they seek future jobs. Some mentor relationships occur spontaneously when someone with education and experience takes an interest in a student. You can also arrange a mentorship directly by asking someone to be your mentor. A variety of people can serve as mentors, including job supervisors, university professors, and family friends. You can even have more than one mentor, thus increasing your exposure to knowledge and opportunities.

Communication Skills

Strong communication skills are needed by sport managers in virtually every functional area discussed in this chapter. Accordingly, communication is the subject of at least one class in most undergraduate curriculums, and we encourage sport management students to take more than the minimum of communication-related courses because most jobs in the sport industry are service oriented. Therefore, it is essential that you be able to communicate effectively in interpersonal situations, write well, and know how to use new communication technologies.

Problem-Solving Skills

Sport managers who make hiring decisions want qualified employees with the educational and employment background to succeed on the job. One characteristic that often distinguishes successful practitioners is their ability to analyze a problem effectively and generate creative solutions. We encourage you to enroll in courses that help you develop such skills and to seek work experiences that require such skills. For example, by volunteering to work in a fitness center's membership campaign, you will gain experience that is more valuable than what you would gain by simply checking in members at the front desk.

Assessing Your Sport Management Preparation

Do you have the skills and aptitudes needed to pursue a career in sport management? Do you communicate clearly and confidently with people? Are you adept at problem solving? What types of experiences would be useful to pursue during your undergraduate years? Would you like to have a mentor? Have you considered seeking volunteer opportunities or part-time work in sport organizations to gain practical experience?

Technological Skills

In the aforementioned study of career expectations among sport management students and practitioners (Mathner & Martin, 2012), technological skills constituted one area where the students differed from the practitioners in their perceptions of critical competencies for career success. Specifically, practitioners indicated that technological skills were more important than students did; therefore, it behooves students to adjust their perceptions to align with those of people already working in the field. Fortunately, a wide variety of resources are available to help you address this need, such as online tutorials and college and university classes. The scope of needed skills ranges from basic ability to use computers and common office software (e.g., Microsoft Excel) to more advanced capabilities, such as using social media, programming, web design, statistics, and software related to professional specializations (e.g., PAC ticketing).

Generalization

Sport-related jobs are plentiful but in high demand. As mentioned earlier, if you obtain work experience in a variety of functional areas, you gain two advantages. First, by experiencing different facets of the field, you position yourself well to identify which areas interest you most. Second, you may expand the number of jobs for which you are qualified. Once you have established a career path or advanced to graduate school, you should work to identify a specialization.

Wrap-Up

The sport management professions offer a broad range of exciting career opportunities for qualified individuals. In areas ranging from sporting goods retail to commercial bowling, sport managers in managerial and administrative positions look to hire individuals with a sport management education, work experience, professionalism, volunteerism, networking, leadership, a good work ethic, and a positive attitude. Individuals in sport management careers often hold jobs that are interesting because of the variety they offer, exciting because of the activities they involve, and challenging because of the ever-changing nature of the field. As a result, many sport management professionals report high levels of job satisfaction. If you are looking for a fulfilling job and have a keen interest in sport, the sport management professions may be for you.

MORE INFORMATION ON CAREERS IN SPORT MANAGEMENT

Organizations

North American Society for Sport Management (NASSM)

Sport Marketing Association

Journals

Athletic Business

Athletic Management

Case Studies in Sport Management

International Journal of Sport Communication

International Journal of Sport Management

Journal of Applied Sport Management

Journal of Intercollegiate Sport

Journal of Legal Aspects of Sport

Journal of Sport Management

Recreational Sports Journal

Sport Marketing Quarterly

Street & Smith's SportsBusiness Journal

REVIEW QUESTIONS

1. Discuss the breadth of the sport industry and the related career opportunities with specific reference to the industry segments discussed in this chapter.
2. Explain how each of the following would help an individual succeed in the sport industry: (a) communication skills, (b) leadership skills, (c) practical experience, (d) a mentor.
3. What type of course work and related academic content are provided by a good sport management program?
4. Identify three functional areas in the sport industry and elaborate on the types of jobs that might be specific to those professional roles.
5. Identify two existing sport organizations for each segment of the sport industry.
6. Elaborate on the type of sport management job you would be most attracted to and why.

Go online to complete the web study guide activities assigned by your instructor.

GLOSSARY

ability—Genetically endowed perceptual, cognitive, motor, metabolic, or personality trait that is susceptible to little or no modification by practice or training.

academic learning time—Time in which students are actively participating and learning.

activities of daily living (ADLs)—Physical activities in the sphere of self-sufficiency that involve personal grooming, dressing, eating, walking, and using the toilet.

advertising—Information placed in the media by an identified sponsor that pays for the time or space.

aerobic—Involving metabolism in the presence of oxygen.

aesthetic experience—Subjective experience in which sensations appeal to a person's sense of beauty, grace, or artistic appreciation.

anaerobic—Involving metabolism in the absence of oxygen.

anemia—Condition marked by hemoglobin concentration below 12 grams per deciliter in women and below 13 grams per deciliter in men.

anxiety—Intense, unpleasant feeling that typically results from a demand or threat.

arousal—State of bodily energy or physical and mental readiness.

ascetic experience—Physical activity experience that involves either discomfort, pain, or suffering.

athletic training clinic—Health care facility that specializes in serving athletes and is usually associated with a high school, collegiate, or professional team.

attitude—Relatively stable mind-set regarding, for our purposes, physical activity.

autotelic attitude—Attitude characteristic of experiences whose value lies in the doing (i.e., in which the activity is an end in itself); typically held by individuals who are intrinsically motivated.

axiology—Branch of philosophy that examines values, the purpose of human existence, and the nature of the good life.

biochemistry—Chemistry of living organisms.

biomechanics—Subdiscipline of kinesiology that applies the principles of mechanics to document and explain the motions of living things.

burnout—State of extreme mental, emotional, and physical exhaustion that results from chronic stress.

cardiac output—Amount of blood pumped out of one side of the heart per minute.

cardiopulmonary rehabilitation—Exercise program designed to improve the function of the heart, lungs, and vascular system.

carpal tunnel syndrome—Cumulative trauma disorder that affects the hand and wrist and is usually suffered by carpenters, typists, packers, assembly line workers, and others who repeat the same movements for several hours each day.

choking—Sudden or progressive deterioration of performance below the typical and expected level of expertise for a person performing under pressure, wherein the person is seemingly incapable of regaining control over his or her performance.

clinical setting—Health or fitness facility that operates within a hospital or medical facility.

closed skill—Motor skill in which the performer must coordinate his or her movements with a predictable, usually stationary environment.

code of ethical principles and standards—Set of guidelines typically published and disseminated by a professional association to its members to ensure that the welfare of clients always receives top priority from practitioners.

cognitive skills—Human capacities that require complex modes of thought, including rational analysis and problem solving, in order to achieve predetermined goals; essential for most professionals.

cohesion—Tendency for groups to stick together and remain united in pursuing goals.

community setting—Health or fitness facility that operates within a community center, church, or other nonprofit organization (e.g., YMCA, Jewish Community Center).

competition—Principle or framework for organizing physical activity in which participants compare their performances either with each other's or with a standard for the purpose of increasing enjoyment.

concentric—A muscle action where activated muscle(s) create a torque greater than the resistance torque resulting in whole muscle shortening.

conditioning—Temporary end state of training that is reflected in the performer's possession of an adequate level of strength, endurance, and flexibility to carry out desired tasks.

cosmetic exercise—Physical training to change body shape for primarily aesthetic reasons.

cumulative trauma disorder—Injury in a muscle, tendon, nerve, or ligament brought about by repetitive motion of a body part.

custodial care facility—Long-term care facility that caters to the medical, rehabilitative, and specialized needs of patients, including assistance with activities of daily living.

dance—Involves body movements, often timed to music, that express messages, emotions, or artistic values.

deductive reasoning—Method of reflection that starts with one or more broad premises and moves toward specific conclusions.

descriptive reasoning—Method of reflection that involves looking at one example of an event and describing its essential qualities.

discipline—Organized body of knowledge considered worthy of study and usually pursued in a college or university curriculum.

disinterested spectating—Form of watching sport contests in which the observer is nonpartisan in regard to the outcome.

disposition—Short-term, highly variable psychological state that may affect one's enjoyment of physical activity.

distribution—Process of connecting a sport product or service with the ultimate consumer.

dualism—When applied to the nature of persons, a doctrine that emphasizes the radical distinctiveness and independence of mind and body.

dynamometer—A machine used to measure muscular force or torque.

eccentric—A muscle action where activated muscle(s) create a torque less than the resistance torque resulting in whole muscle lengthening.

electromyography (EMG)—System for amplifying and recording electrical activity in muscles.

emblem—Body movement (e.g., hand signal) that can be translated easily into explicit messages.

emotion—Subjective feelings, such as anger, fear, joy, and sorrow that are related to personal experiences.

employee assistance program—Program supported by business or industry to help employees with personal problems that may affect their work performance.

engaged time—Time that students spend engaged in activity.

epistemology—Branch of philosophy concerned with how we know things and the foundations on which such knowledge rests.

ergonomist—Engineer who seeks to improve the safety and efficiency of work by analyzing workers' movements and workplace conditions.

ethics—Branch of philosophy that analyzes right behavior and individuals' responsibilities toward others and themselves.

event management—Planning, organizing, execution, and evaluation of events (e.g., Super Bowl, Little League baseball tournament).

exercise—Physical activity intended to improve one's physical performance in an activity, to improve one's health, or to regain performance that has been reduced as a result of injury or disease.

exercise physiology—Subdiscipline of kinesiology that focuses on the physiology of physical activity.

experience of physical activity—Training in, observation of, practice of, or participation in physical activity.

expressive movement—Movement used in a physical activity as a way of expressing one's emotion or personality; differentiated from instrumental movement.

extrinsic motivation—Motivation derived from pursuing and obtaining rewards (e.g., money, status) outside of an activity itself.

face-to-face contact activity—Contests (e.g., football, wrestling, basketball, soccer) in which individuals interact with opponents' attempts to achieve a goal by physically manipulating their opponents' movements.

face-to-face noncontact activity—Contests (e.g., volleyball, tennis, baseball) in which individuals interact with opponents to maximize their own chances of winning but do not physically manipulate their opponents.

feedback (intrinsic and extrinsic)—Information about movement that is provided to a learner, either during or after a movement, which may derive from either an external source (e.g., instructor, video footage) or an internal source (e.g., muscles, joints, nervous system).

financial management—Financial activities such as prudent investment and use of assets, monetary development opportunities, and short-term and long-term budgeting.

flow—An optimal mental state characterized by total absorption in a task.

focus group—Group interview conducted to learn about people's shared understandings through organized discussion with selected individuals about their views and experiences; used for both sociological research and marketing purposes.

free time—Personal time that has not been encumbered with obligations; also referred to as *discretionary time*.

functional learning time—See *academic learning time*.

fundamental movement pattern—Broad category of human movement (e.g., running, throwing, kicking, jumping) used for a general purpose.

game spectator knowledge—Knowledge of elements of the game one is watching, including players, strategies, and competitive tactics.

gender—Social position based on a set of norms or expectations about how we should behave that are linked to societal understandings of sexuality and procreation.

gesture—Movement that is used to communicate one's intentions and may be classified as an illustrator, emblem, or regulator.

habilitation—Processes and treatments leading to acquisition of skills and functions that are considered normal and expected for an individual of a particular age and status.

habilitative therapeutic exercise—Processes and treatments leading to acquisition of skills and functions that are considered normal and expected for an individual of a certain age, status, and occupation.

health-related exercise—Exercise undertaken to develop or maintain a sound working body.

health-related fitness—Fitness developed through physical activity experience and characterized by capacities and traits associated with low risk of hypokinetic disease.

helping profession—Profession primarily committed to providing services.

hemoglobin—Iron-containing protein found in red blood cells that carries most of the oxygen in the blood.

holism—When applied to the nature of persons, a position that underscores the interdependence and interrelatedness of thought and physicality.

home maintenance activities—Activities in the sphere of self-sufficiency intended to improve or repair living conditions in one's home.

human agency—Concept suggesting that people are actively involved in developing or constructing their own sports.

human factors engineer—Engineer who seeks to improve the safety and efficiency of work by analyzing workers' movements and workplace conditions.

human resource management—An important area of sport management that involves recruiting, orienting, retaining, and evaluating employees; planning benefit programs; and designing and implementing internal grievance procedures.

hyperplasia—Increase in muscle mass due to splitting of muscle fibers.

hypertrophy—Increase in muscle mass due to enlargement of muscle fibers.

hypokinetic disease—Disease (e.g., obesity, high blood pressure, heart disease) associated directly with low levels of daily physical activity.

hypothermia—Condition in which body temperature falls below 95 degrees Fahrenheit (35 degrees Celsius), which causes heart rate and metabolism to slow and can be life threatening.

illustrator—Gesture used to demonstrate or complement what is said.

imagery—Invocation of the senses to create or re-create an experience in the mind.

impersonal competition—Attempt by an individual to better an established record in a physical activity (e.g., mountain climbing, long-distance swimming) that does not involve an opponent.

inductive reasoning—Method of reflection that starts with specific cases or examples and moves toward broad, general conclusions.

inpatient facility—Rehabilitation hospital, nursing home, or other institution where patients spend extended periods of time in order to receive medical or other health-related treatment.

instrumental activities of daily living (IADLs)—Activities in the sphere of self-sufficiency (e.g., shopping, telephoning, cooking, doing laundry) that are less personal than activities of daily living.

instrumental movement—Movement used in a physical activity to accomplish the goal of the action; differentiated from expressive movement.

International Classification of Functioning, Disability, and Health—Diagnostic approach developed by the World Health Organization that focuses on patient needs and considers environmental factors in order to maximize function.

internship—Culminating educational experience, undertaken by a kinesiology major who plans on a professional career, that involves extended work at one or two professional sites where the student works under supervision provided by a veteran professional.

interpersonal skill—Skill that enables one to listen to and communicate with clients, students, or patients.

intrinsic motivation—Motivation derived from an activity's inherent rewards, such as enjoyment or feelings of accomplishment.

intuitive knowledge—Knowledge that is gained through physical activity and does not depend on rational or conscious processes.

isokinetic—Muscle action involving quasi-constant joint angular velocity.

isometric—Muscle action involving tension produced by muscle groups that equals opposing resistance and does not result in any change in joint angle.

isotonic—Involving muscle action to move a constant external (usually gravitational) load.

kinematics—Branch of mechanics that describes and measures movement in terms of variables such as velocity and acceleration.

kinesiology—Discipline or body of knowledge that studies human physical activity.

kinesiology theory—Theoretical knowledge about physical activity as embodied in the subdisciplines of kinesiology.

learning—Permanent alteration in nervous system functioning that enables performers to consistently achieve a predetermined goal.

leisure—State of being in which humans find deep satisfaction and contentment, often accompanied by feelings of wonder, celebration, excitement, and creativity.

leisure activity—Physical activity that nourishes or maintains the disposition of leisure.

leisure studies—Area of study, or department in a college or university, that focuses on preparing individuals for careers in the leisure industry.

liberal studies—Area of study regarded as forming the core of higher education and inducing students to love the truth and develop independence in thought and action; also known as the *liberal arts and sciences*.

marketing management—Production, pricing, promotion, and distribution of a sport-related product or service in a way that inspires transactions between the sport organization and its consumers.

materialism—Position on human nature that views the person as being made of only one thing: atoms.

maximal oxygen uptake or $\dot{V}O_2max$—Highest rate of oxygen uptake during heavy dynamic exercise.

mechanical, market-driven professionalism—Mode of practice that prioritizes technique, methodology, profit, and prestige over clients' wants and needs.

memory drum theory—Theory developed by Franklin Henry (1960) proposing that rapid and well-learned movements are not consciously controlled but are run off automatically (as an older computer uses a memory drum to store and retrieve data).

mentor—Professional who provides less experienced professionals or preprofessionals with learning opportunities, responsibilities, knowledge, and contacts who can help them attain a successful career or continue to improve in an existing career.

merchandising—Selling of goods and services related to an organization (e.g., team apparel, coffee mugs, notepads), which gives sport managers opportunities to generate revenue and enhance brand image, brand loyalty, and customer satisfaction.

metabolic rate—Rate at which the body uses energy.

metaphysics—Branch of philosophy concerned with the nature of reality (i.e., what exists).

mindfulness—Maintaining focus on the present in an open, nonjudgmental way.

modernization theory—Theory positing that the

rise of modern sport occurred during the Industrial Revolution as U.S. society shifted away from an agricultural and local orientation and developed city-based industries rooted in science and technology.

motivation—Complex set of internal and external forces that influence individuals to behave in certain ways.

motor performance fitness—Capacity developed through physical activity experience that enables people to perform daily activities with vigor and often incorporates an element of skill.

motor skill taxonomy—Classification system that categorizes motor skills according to their common critical components.

motor skill—Human act that requires efficient, coordinated movement to achieve a predetermined goal; relied on, along with cognitive skills, by some professionals (e.g., dentists, surgeons).

movement—Any change in the position of body parts relative to each other.

muscular endurance—Ability of a muscle to repeatedly exert force over a prolonged period.

myofibril—The part of a muscle fiber that contains contractile elements.

mystical knowledge—Knowledge about another dimension of reality apprehended, for our purposes, through participation in sport and exercise.

novel learning task—Movement task with which the subject does not have prior experience and which usually involves simple movement (e.g., linear positioning or tracking).

open skill—Motor skill in which the performer must coordinate movements with an unpredictable (usually moving) environment.

outpatient facility—Short-term care facility where patients do not stay overnight.

peak experience—Type of mystical experience sometimes experienced by runners and others engaged in strenuous sport or exercise.

pedagogy—Study of teaching and learning.

perception—Meaningful construct or message based on interpretation of sensations from subjective experience.

perceptual skill—Skill that enables one to identify and recognize problems in professional settings.

personality—Unique blend of psychological characteristics and behavioral tendencies that make an individual both different from and similar to others.

phenomenology—Method of reflection that examines the content of consciousness and gives credence to differences encountered in normal subjective life.

physical activity—Movement that is voluntary, intentional, and directed toward an identifiable goal.

physical activity experience—Source of kinesiology knowledge involving direct personal experience of watching or performing physical activity.

physical fitness—Capacity developed through exercise that enables one to perform essential activities of daily living, engage in an active leisure lifestyle, and have sufficient energy remaining to meet the demands of unexpected events.

physical performance capacity—Qualities of physical capacity such as flexibility, muscular endurance, cardiovascular endurance, and strength that are developed not through learning but through training.

plasma volume—Volume of extracellular fluid in the blood.

practice—Type of physical activity experience that involves cognitive processing and leads to improvement in skill (learning); repetition of a task, often with an instructor's guidance or feedback, to promote learning.

practice theory—Knowledge about clients, methods, and outcomes that guides practitioners in performing their duties as professionals.

practitioners—Those who use knowledge to bring about predetermined objectives (e.g., professionals).

prehabilitation—Therapy that occurs following an injury but before surgery in order to develop as much strength and range of motion as possible.

private practice—Entrepreneurial venture in which a professional establishes his or her own workplace with its own client pool.

professional experience in physical activity—Source of kinesiology knowledge involving design and implementation of physical activity programs in kinesiology-related professions.

program management—Activities (e.g., scheduling, equipment maintenance) that sport organizations must perform in order to ensure operational effectiveness.

promotion—Activity (e.g., personal selling, sales promotion, advertising, publicity) designed to attract consumers to products or services.

psychoanalytic self-knowledge—Knowledge of

one's deep-seated desires, motivations, and behaviors that is gained, for our purposes, through participation in sport and exercise.

psychological inventory—Standardized or objective measure of a specific sample of behavior, typically in the form of a questionnaire.

public relations management—Tasks in areas such as media relations and community relations that are designed to achieve positive relationships between a sport organization and its most important constituents.

rational knowledge—Knowledge of facts, concepts, and theories gained through reason, logic, and analysis.

reaction time—Speed of response to a light or sound, as when pressing a single button after seeing a signal (simple) or choosing one of multiple buttons to press depending on which signal one sees (choice).

receptor—Specialized nerve ending that is found at the end of a sensory neuron and detects changes in the environment.

recreation—See *leisure studies.*

regulator—Hand or body movement used to guide the flow of conversation, as in greeting someone or parting company.

rehabilitation—Physical treatment, exercise, and educational or counseling sessions that lead a person to regain function and a personally acceptable level of independence.

rehabilitative therapeutic exercise—Processes and treatments designed to restore skills or functions that were previously acquired but have been lost due to disease, injury, or behavioral traits.

retention-based facility—Commercial (for-profit) health or fitness facility that makes a large percentage of its profit from ongoing monthly membership dues and engages in activities directed at helping members meet their goals and have a positive experience in the facility so that they will continue to pay for membership on a monthly basis.

risk management—Prevention of loss associated with inadequacy in areas such as financial planning, employee management, facility and equipment maintenance, customer service, and short- and long-term planning.

sales-based facility—Commercial (for-profit) health or fitness facility that makes a large percentage of its profit from membership initiation fees and prepaid dues and therefore lacks significant incentive to satisfy current members and, in turn, may be marked by overcrowding and lower-quality programs and services.

sarcopenia—Loss of skeletal muscle.

scholarship of physical activity—Source of kinesiology knowledge derived from systematic research and study of the theoretical and practical aspects of physical activity.

self-reflection—Process of considering the subjective experience of an activity performed in the past.

sensation—Raw, uninterpreted information collected through sensory organs.

sensation-seeking activity—Physical activity involving high speed, danger, or disorientation of the body in space.

service—Human acts intended to improve quality of life for others.

side-by-side competitive activity—Contest (e.g., golf, swim racing) in which individuals do not interact directly in striving to accomplish the goal.

skill—Quality of physical activity ability that underlies the performance of a movement.

social loafing—Decrease in individual performance within a group.

social trustee, civic professional—Professional who adheres to the creed "Healthy people and a good society first. Myself and my profession second in service of this greater good" (Lawson, 1998a).

Socratic self-knowledge—Knowledge of capacities and limitations that enables one to perform physical activity safely within the range of one's abilities.

spheres of professional practice—Categories of physical activity professions that are similar with respect to general objectives, methods, educational requirements, working environments, and other factors.

sport—Physical activity in which movement is performed in order to achieve a specific goal in a manner specified by established rules and in competitive contexts.

sport entertainment—Sport industry sector consisting of organizations that seek to attract consumers to watch athletes compete (e.g., NASCAR, NFL, NCAA institutions).

sport industry—"Market in which the products offered to its buyers are fitness, sport, recreation, and leisure related . . . [and] include goods (e.g., baseball bats), services (e.g., sport marketing, health clubs), people (e.g., professional players), places (e.g., golf courses), and ideas" (Pitts, Fielding, & Miller, 1994, p. 18).

sport management—Subdiscipline of kinesiology focused on the "study and practice of all people, activities, businesses, or organizations involved in producing, facilitating, promoting, or organizing any sport-related business or product" (Pitts & Stotlar, 2007, p. 4).

sport pedagogy—Subdiscipline of kinesiology that studies the teaching and learning of physical activity, primarily in physical education.

sport services—Sport industry sector consisting of businesses that provide needed services to other sport organizations (e.g., event management, representation services).

sports medicine—Field of medicine and therapeutic exercise that specializes in treating, preventing, and rehabilitating athletes (and others who engage in sport or strenuous exercise) and investigates training methods and practices.

sport spectacle—Staged competition designed and promoted for an audience and intended to evoke a range of human emotions by virtue of its grandeur, scale, and drama.

strength—Maximal moment of force exerted by a muscle group, typically measured in isokinetic conditions.

stress—Process in which an individual perceives an imbalance between situational demands and his or her response capabilities.

stroke volume—Amount of blood pumped out of one side of the heart per beat.

subdiscipline—Division that represents an extension of an established discipline such as psychology, physiology, or history; way of dividing the scholarly study of physical activity in order to facilitate teaching and research.

subjective experience—Individual reactions, feelings, and thoughts about events.

subjective experience of physical activity—Individual reactions, feelings, and thoughts that accompany either performance or observation of physical activity.

technical definition—Specialized meaning of a term (e.g., *physical activity*) that is used to convey knowledge to others within a scientific or technical field and differs from how people define and use the term in everyday language.

therapeutic exercise—Exercise and movement applied in a systematic and scientific manner in order to restore or develop health-related movement performance factors such as muscular strength, endurance, and flexibility; neuromuscular coordination; and cardiovascular efficiency.

tidal volume—Amount of air inhaled or exhaled per breath.

torque—Common English term for "moment of force" that is the rotation effect of a force about an axis of rotation. Joint torques measured by isokinetic dynamometers are particularly good measures of muscular strength.

training—Physical activity carried out for the express purpose of improving athletic, military, work-related, or recreation-related performance.

underwater weighing—Procedure in which a person's body weight is measured while he or she is completely submerged in water in order to determine body volume.

ventilation—Process in which gases are exchanged between the atmosphere and the alveoli of the lungs.

ventilatory threshold—Point during a graded exercise test at which ventilation begins increasing at a faster rate than $\dot{V}O_2max$ does.

venue—Facility (e.g., arena, stadium, track) in which sport competitions or activities occur.

vertigo—Sensation resulting from disorientation of the body in space, experienced often in conjunction with dangerous activities and differing from the medical conditions that result in dizziness.

wellness—An active, self-directed process of making choices to achieve a multidimensionally healthy and fulfilling life.

with-it-ness—Teacher awareness of all events transpiring in a learning environment.

workplace knowledge—Practical, mundane knowledge that is not grounded in theory and that people use to perform everyday tasks in the workplace (e.g., knowing where items are stored or how to clean, repair, or calibrate equipment).

REFERENCES

Chapter 1

American Kinesiology Association. (2017). About AKA. www.americankinesiology.org/about-us/about-us/about-aka.

Centers for Disease Control and Prevention. (2015, June 10). Glossary of terms. www.cdc.gov/physicalactivity/basics/glossary/index.htm.

Elliott, B.C., & Khangure, M. (2002). Disk degeneration and fast bowling in cricket: An intervention study. *Medicine and Science in Sports and Exercise, 34*, 1714–1718.

Herman, D.C., Onate, J.A., Weinhold, P.S., Guskiewicz, K.M., Garrett, W.E., Yu, B., & Padua, D.A. (2009). The effects of feedback with and without strength training on lower extremity biomechanics. *American Journal of Sports Medicine, 37*, 1301–1308.

Hewett, T.E., Lindenfield, T.N., Riccobene, J.V., & Noyes, F.R. (1999). The effect of neuromuscular training on the incidence of knee injury in female athletes. A prospective study. *American Journal of Sports Medicine, 27*, 699–706.

Knudson, D.V. (2013). *Qualitative diagnosis of human movement* (3rd ed.). Champaign, IL: Human Kinetics.

Mahar, M., & Crenshaw, J.T. (2015). *AKA Salary Survey 2015*. Champaign, IL: American Kinesiology Association.

Newell, K.M. (1990a). Kinesiology: The label for the study of physical activity in higher education. *Quest, 42*, 269–278.

Newell, K.M. (1990b). Physical activity, knowledge types, and degree programs. *Quest, 42*, 243–268.

Noehren, B., Scholz, J., & Davis, I. (2011). The effect of real-time gait retraining on hip kinematics, pain, and function in subjects with patellofemoral pain syndrome. *British Journal of Sports Medicine, 45*, 691–606.

Sherrington, C.S. (1940). *Man on his nature*. Cambridge, UK: Cambridge University Press.

Wojciechowska, I. (2010, August). A quickly growing major. *Inside Higher Education*, http://www.insidehighered.com/news/2010/08/11/kinesiology.

Part I

Epstein, J. (1976, July). Obsessed with sport. *Harper's Magazine*, pp. 253–255.

Sheehan, G.A. (1978). *Running and being: The total experience*. New York: Warner Books.

Chapter 2

Ainsworth, B.E., Haskell, W.L., Whitt, M.C., Irwin, M.L., Swartz, A.M., Strath, S.J., . . . Leon, A.S. (2000). Compendium of physical activities: An update of activity codes and MET intensities. *Medicine and Science in Sports and Exercise, 32*, S498–S516.

Alzheimer's Association. (2016). Alzheimer's disease facts and figures. http://www.alz.org/facts/.

Ambady, N., & Rosenthal, R. (1993). Half a minute—Predicting teacher evaluations from thin slices of nonverbal behavior and physical attractiveness. *Journal of Personality and Social Psychology, 64*(3), 431–441. http://dx.doi.org/10.1037/0022-3514.64.3.431.

Bick, A., Brüggemann, B., and Fuchs-Schündeln, N. (2016, August). Hours worked in Europe and the US: New data, new answers. IZA DP No. 10179, Institute for the Study of Labor (Forschungsinstitut zur Zukunft der Arbeit). http:/ftp.iza.org/dp10179.pdf.

Braithwaite, I., Stewart, A.W., Hancox, R.J., Beasley, R., Murphy, R., Mitchell, E.A., & ISACC Phase Three Study Group. (2013). The worldwide association between television viewing and obesity in children and adolescents: Cross sectional study. *PLoS One, 8*(9), e74263. (PMID:24086327)

California School Boards Association. (2009). California Project Lean. www.CaliforniaProjectLEAN.org.

Carr, J.H., & Shepherd, R.B. (1987). *A motor relearning programme for stroke*. Rockville, MD: Aspen.

Carroll, J. (2007). Workers' average commute round trip is 46 minutes in a typical day. *Gallup New Service*. www.gallup.com/poll/28504/Workers-Average-Commute-RoundTrip-Minutes-Typical-Day.aspx5?g_source=mn1-services.

Centers for Disease Control and Prevention. (2010a). Summary health statistics for the US population: National health interview study, 2010. www.cdc.gov/nchs/data/series/sr_10/sr10_251.pdf.

Centers for Disease Control and Prevention. (2010b). Prevalence of overweight, obesity and extreme obesity among adults: United States, trends 1960–1962 through 2007–2008. National Center for Health Statistics. www.cdc.gov/NCHS/data/hestat/obesity_adult_07_08/obesity_adult_07_08.pdf.

Centers for Disease Control and Prevention. (2014). Physical education profiles, 2012: Physical education and physical activity practices and policies among secondary schools at select us sites. Atlanta, GA: Centers for Disease Control and Prevention, US Department of Health and Human Services. www.cdc.gov/healthyyouth/physicalactivity/pdf/PE_Profile_Book_2014.pdf.

Centers for Disease Control and Prevention. (2015a). Chartbook data tables 6 [data table for figure 6]. www.cdc.gov/nchs/data/hus/2015/fig06.pdf.

Centers for Disease Control and Prevention. (2015b). Childhood obesity facts. www.cdc.gov/healthyschools/obesity/facts.htm.

Centers for Disease Control and Prevention. (2016). High blood pressure fact sheet. www.cdc.gov/dhdsp/data_statistics/fact_sheets/fs_bloodpressure.htm.

Centers for Medicare and Medicaid Services. (2014). National health expenditures 2015 highlights. www.cms.gov/research-statistics-data-and-systems/statistics-trends-and-reports/nationalhealthexpenddata/downloads/highlights.pdf.

Common Sense Census. (2015). Media use by tweens and teens. www.commonsensemedia.org/research/the-common-sense-census-media-use-by-tweens-and-teens.

Csikszentmihalyi, M. (1990a). *Flow: The psychology of optimal experience.* New York: Harper & Row.

Csikszentmihalyi, M. (1990b, January). *What good are sports?* Paper presented at the Commonwealth and International Conference of Physical Education, Sport, Health, Dance, Recreation, and Leisure, Auckland, New Zealand.

Czaja, S.J. (1997). Using technologies to aid the performance of home tasks. In A.D. Fisk & W.A. Rogers (Eds.), *Human factors and the older adult* (pp. 311–334). New York: Academic Press.

De Grazia, S. (1962). *Of time, work, and leisure.* New York: Doubleday.

Dong, N.L., Block, G., & Mandel, S. (2004, February 12). Activities contributing to total energy expenditure in the United States: Results from the NHAPS study. *International Journal of Behavioral Nutrition and Physical Activity, 1*(4). www.ijbnpa.org/content/1/1/4.

Health and Safety Ontario. (2011). Physical demands analysis [Form]. www.wsps.ca/WSPS/media/Site/Resources/Downloads/PDA_Form_FillableEx_Final.pd

Increasing employee participation in corporate wellness programs. (2013, September 1). *Occupational Health and Safety.* https://ohsonline.com/Articles/2013/09/01/Increasing-Employee-Participation-in-Corporate-Wellness-Programs.aspx?Page=2.

Johnson, L.D., Delva, J., & O'Malley, P.M. (2007, October). Sports participation and physical education in American secondary schools: Current levels and racial/ethnic and socioeconomic disparities. *American Journal of Preventive Medicine, 33*(4, Suppl. 1), S195–S208.

Katz, S., Ford, A.B., Moskowitz, R.W., Jackson, B.A., & Jaffe, M.W. (1963). Studies of illness in the aged: The index of ADL: A standardized measure of biological and psychological function. *Journal of the American Medical Association, 185,* 914–919.

Kohn, A. (1992). *No contest: The case against competition.* New York: Houghton Mifflin.

Kranz, L. & Lee, T (2015). *Jobs rated almanac: The best jobs and how to get them.* Createspace Independent Publishing Platform.

Kroemer, K., Kroemer, H., & Kroemer-Elbert, K. (1994). *Ergonomics: How to design for ease and efficiency.* Englewood Cliffs, NJ: Prentice Hall.

Kucera, K.L., Klossner, D., Colgate, B., & Cantu, R.C. (2016). Annual survey of football injury research: 1931-2015. National Center for Catastrophic Sport Injury Research. nccsir.unc.edu/reports.

Kulmala, J., Solomon, A., Kåreholt, I., Ngandu, T., Rantanen, T., Laatikainen, T., . . . Kivipelto, M. (2014). Association between mid- to late life physical fitness and dementia: Evidence from the CAIDE study. *Journal of Internal Medicine, 276,* 296–307. doi:10.1111/joim.12202.

Locke, L.F. (1996). Dr. Lewin's little liver patties: A parable about encouraging healthy lifestyles. *Quest, 48*(3), 422–431.

Martin, L. (2013). The role of women in athletic training: a review of the literature. *Skyline, 1*(1), 13. Retrieved from http://skyline.bigskyconf.com/cgi/viewcontent.cgi?article=1012&context=journal.

May, C. (2012, October). What does the way you walk say about you? *Scientific American.* www.scientificamerican.com/article/what-does-the-way-you-walk-say-about-you/.

Michigan Medicine. (2010, August). Your Child Development and Behavior Resources. www.med.umich.edu/yourchild/topics/tv.htm.

Morris, D. (1994). *Bodytalk: The meaning of gestures.* New York: Crown.

Morris, J.N., Heady, J.A., Raffle, P.A.B., Roberts, C.G., & Parks, J.W. (1953). Coronary heart disease and physical activity of work. *Lancet, 2,* 1111–1120.

National Collegiate Athletic Association. (2010). Football injuries: Data from the 2004/05–2008–/09 seasons. www.ncaa.org/sites/default/files/NCAA_Football_Injury_WEB.pdf.

National Institute for Occupational Safety and Health. (2000, June 1). Working women face high risk from work stress, musculoskeletal injuries, other disorders, NIOSH finds. www.cdc.gov/niosh/updates/womrisk.html.

Nyberg, J., Aberg, M.A., Schiöler, L., Nilsson, M., Wallin, A., Torén, K., & Kuhn, H.G. (2014). Cardiovascular and cognitive fitness at age 18 and risk of early-onset dementia. *Brain: A Journal of Neurology, 137*(5), 1514–1523.

Phenix, P.H. (1964). *Realms of meaning*. New York: McGraw-Hill.

Physical Activity Council. (2016). 2016 Participation report: Annual study tracking sports, fitness, and recreation participation in the US. www.physicalactivitycouncil.com/pdfs/current.pdf.

Pieper, J. (1952). *Leisure: The basis of culture*. New York: Pantheon Books.

Safe Kids Worldwide. (2015). Sport and Recreation Safety Fact Sheet 2015. www.safekids.org/sites/default/files/documents/skw_sports_fact_sheet_feb_2015.pdf.

Sanchez-Vaznaugh, E.V., O'Sullivan, M., & Egerter, S. (2013, December). Policy brief: When school districts fail to comply with state physical education laws, the fitness of California's children lags. San Francisco, CA. http://activelivingresearch.org/when-school-districts-fail-comply-state-physical-education-laws-fitness-californias-children-lags.

Schmidt, M.S. (2015, April 6). Battling crime and calories at F.B.I. (Fit Bureau of Investigation). *New York Times,* p. A1.

Simon, R.L. (2004). *Fair play: The ethics of sport* (2nd ed.). Boulder, CO: Westview Press.

Smith, A.D., Crippa, A., Woodcock, J., & Brage, S. (2016). Physical activity and incident type 2 diabetes mellitus: A systematic review and dose-response meta-analysis of prospective cohort studies. *Diabetologia, 59*(12), 2527–2545.

Smith, J.C., Nielson, K.A., Woodard, J.L., Seidenberg, M., Durgerian, S., Hazlett, K.E., . . . Rao S.M. (2014, April 23). Physical activity reduces hippocampal atrophy in elders at genetic risk for Alzheimer's disease. *Frontiers of Aging Neuroscience*. http://journal.frontiersin.org/Journal/10.3389/fnagi.2014.00061/full.

Stanne, M.B., Johnson, D.W., & Johnson, R.T. (2009). Does competition enhance or inhibit motor performance? A meta-analysis. *Psychological Bulletin, 125*(1), 133–154.

Strieber, A. (2010). Jobs rated 2010: A ranking of 200 jobs best to worst. CareerCast. www.careercast.com/jobs-rated/jobs-rated-2010-ranking-200-jobs-best-worst.

Thoresen, J.C., Vuong, Q.C., & Atkinson, A.P. (2012). First impressions: Gait cues drive reliable trait judgements. *Cognition, 124*(3), 261–271.

Tudor-Locke, C., Craig, C., Cameron, C., & Griffiths, J. (2011). Canadian children's and youth's predetermined steps/day, parent-reported TV watching time, and overweight/obesity: The Canplay Surveillance Study. *International Journal of Behavior Nutrition and Physical Activity, 8*(1), 66–75.

U.S. Bureau of Labor Statistics. (2011). Nonfatal occupational injuries and illnesses requiring days away from work, 2010. www.bls.gov/news.release/archives/osh2_11092011.pdf.

U.S. Bureau of Labor Statistics. (2012). American time use survey summary. www.bls.gov/news.release/atus.nr0.htm.

U.S. Department of Health and Human Services. (1990). *Healthy people 2000*. DHHS Publication No. (PHS) 91-50213. Hyattsville, MD: Public Health Service.

U.S. Department of Health and Human Services. (1996). *Physical activity and health: A report of the surgeon general*. Atlanta, GA: U.S. Department of Health and Human Services, Centers for Disease Control and Prevention, and National Center for Chronic Disease Prevention and Health Promotion.

U.S. Department of Health and Human Services. (1997, July). *Musculoskeletal disorders and workplace factors—A critical review of epidemiologic evidence for work-related musculoskeletal disorders of the neck, upper extremity, and low back*. B.P. Bernard (Ed.). Cincinnati: National Institute for Occupational Safety and Health. www.cdc.gov/niosh/docs/97-141/.

U.S. Department of Health and Human Services. (2000, November). *Healthy people 2010: Understanding and improving health* (2nd ed.). Washington, DC: U.S. Government Printing Office.

U.S. Department of Health and Human Services (Office of Disease Prevention and Health Promotion). (2012). Healthy people 2020. www.healthypeople.gov/2020/default.aspx.

United States Report Card on Physical Activity for Children and Youth. (2016). National Physical Activity Plan. http://physicalactivityplan.org/reportcard/2016FINAL_USReportCard.pdf.

Wakefield, J. (2016, May 25). Foxconn replaces "60 thousand factory workers with robots." BBC News. www.bbc.com/news/technology-36376966.

World Health Organization. (2013, July 4–5). Vienna declaration on nutrition and noncommunicable diseases in the context of Health 2020 (p. 3). www.euro.who.int/__data/assets/pdf_file/0003/234381/Vienna-Declaration-on-Nutrition-and-Noncommunicable-Diseases-in-the-Context-of-Health-2020-Eng.pdf?ua=1.

World Health Organization. (2016). Physical activity strategy for the WHO European Region 2016–2025. www.euro.who.int/en/publications/abstracts/physical-activity-strategy-for-the-who-european-region-20162025.

Chapter 3

Alberton, C.L., Pinto, C.C., Gorski, T., Antunes, A.H., Finatto, P., Cadore, E.L. et al. (2016). Rating of perceived exertion in maximal incremental tests during head-out water-based aerobic exercises. *Journal of Sports Sciences, 34,* 1691–1698.

Atkinson, G. (2004). Effects of music on work-rate distribution during a cycling time trial. *International Journal of Sports Medicine, 25*(8), 611–615.

Berkowitz, L. (1969). *Roots of aggression: A reexamination of the frustration-aggression hypothesis*. New York: Atherton Press.

Bernstein, N. (1967). *The coordination and regulation of movements*. London: Pergamon.

Black, S.J., & Weiss, M.R. (1992). The relationship among perceived coaching behaviors, perceptions of ability, and motivation in competitive age-group swimmers. *Journal of Sport and Exercise Psychology, 14*(3), 309–325.

Bouchard, C., Malina, R.M., & Perusse, L. (1997). *Genetics of fitness and physical performance.* Champaign, IL: Human Kinetics.

Brill, P.A., Burkhaulter, H.E., Kohl, H.W., Blair, S.N., & Goodyear, N.N. (1989). The impact of previous athleticism on exercise habits, physical fitness, and coronary heart disease risk factors in middle-aged men. *Research Quarterly for Exercise and Sport, 60*, 209–215.

Cauley, J.A., Donfield, S.M., LaPorte, R.E., & Warhaftig, N.E. (1991). Physical activity by socioeconomic status in two population based cohorts. *Medicine and Science in Sports and Exercise, 23*, 343–352.

Centers for Disease Control and Prevention. (2011). Glossary of terms. www.cdc.gov/nccdphp/dnpa/physical/everyone/glossary/index.htm.

Centers for Disease Control and Prevention. (2012). Physical activity facts. www.cdc.gov/healthyyouth/physicalactivity/facts.htm.

Csikszentmihalyi, M. (1990a). *Flow: The psychology of optimal experience.* New York: Harper & Row.

Csikszentmihalyi, M. (1990b, January). *What good are sports?* Paper presented at the Commonwealth and International Conference of Physical Education, Sport, Health, Dance, Recreation, and Leisure, Auckland, New Zealand.

Dishman, R.K., & Sallis, J.F. (1994). Determinants and interventions for physical activity and exercise. In C. Bouchard, R.J. Shephard, & T. Stephens (Eds.), *Physical activity, fitness, and health: International proceedings and consensus statement* (pp. 214–238). Champaign, IL: Human Kinetics.

Donnelly, P. (1977). Vertigo in America: A social comment. *Quest, 27*, 106–113.

Ebihara, O., Ideda, M., & Myiashita, M. (1983). Birth order and children's socialization into sport. *International Review of Sport Sociology, 18*, 69–89.

Fleishman, E.A., & Hempel, W.E. (1955). The relationship between abilities and improvement with practice in a visual reaction time discrimination task. *Journal of Experimental Psychology, 49*, 301–312.

Fletcher, I.M., & Hartwell, M. (2004). Effect of an 8-week combined weight and plyometrics training program on golf drive performance. *Journal of Strength and Conditioning Research, 18*, 59–62.

Fox, L.D., Rejeski, W.J., & Gauvin, L. (2000). Effects of leadership style and group dynamics on enjoyment of physical activity. *American Journal of Health Promotion, 14*, 277–283.

Garcia, A.W., Broda, M.A.N., Frenn, M., Coviak, C., Pender, N.J., & Ronis, D.L. (1995). Gender and developmental differences in exercise beliefs among youth and prediction of their exercise behavior. *Journal of School Health, 65*, 213–219.

Gentile, A.M. (1972). A working model of skill acquisition with application to teaching. *Quest*, Monograph XVII, 2–23.

Giamatti, A.B. (1989). *Take time for paradise: Americans and their games.* New York: Summit Books.

Gill, F.B. (1989). *Ornithology.* New York: Freeman.

Goldstein, J.H., & Arms, R. (1971, March). Effects of observing athletic contests on hostility. *Sociometry, 34*, 83–90.

Heltne, P.G. (1989). Epilogue: Understanding chimpanzees and bonobos, understanding ourselves. In P. Heltne & L. Marquardt (Eds.), *Understanding chimpanzees* (pp. 380–384). Cambridge, MA: Harvard University Press.

Herring, M., Jacob, M.L., Suveg, C., Rishman, R.K., and O'Connor, P.J. (2011). Feasibility of exercise training for short-term generalized anxiety disorder: A randomized, controlled trial. *Psychotherapy and Psychosomatics, 81*(1), 21–28.

Heyman, S. (1994). The hero archetype and high-risk sports participants. In M. Stein & J. Hollwitz (Eds.), *Psyche and sports* (pp. 188–201). Wilmette, IL: Chiron.

Hyland, D.A. (1990). *Philosophy of sport.* New York: Paragon House.

Jackson, S.A., & Csikszentmihalyi, M. (1999). *Flow in sports.* Champaign, IL: Human Kinetics.

Jago, R., Fox, K.R., Page, A.S., Brockman, R., & Thompson, J.L. (2010). Parent and child physical activity and sedentary time: Do active parents foster active children? *BMC Public Health, 10*(1), 194–202.

Kalakanis, L., Goldfield, G.S., Paluch, R.A., & Epstein, L.H. (2001). Parental activity as a determinant of activity level and patterns of activity in obese children. *Research Quarterly for Exercise and Sport, 72*(3), 202–209.

Kenyon, G.S. (1968). A conceptual model for characterizing physical activity. *Research Quarterly, 39*, 96–104.

King, A.C., Blair, S.N., Bild, D., Dishman, R.K., Dubbert, P.M., Marcus, B.H., . . . Yeager, K.Y. (1992). Determinants of physical activity and interventions in adults. *Medicine and Science in Sports and Exercise, 24*(6), S221–S236.

Kirby, J., Levin, K.A. & Inchley, J. (2011). Parental and peer influences on physical activity among Scottish adolescents: A longitudinal study. *Journal of Physical Activity and Health, 8*, 785–793.

Kleinman, S. (1968). Toward a non-theory of sport. *Quest, 10*, 29–34.

Kouros, Y. (1990, March). A war is going on between my body and my mind. *Ultrarunning*, 19.

Kretchmar, R.S. (1985). "Distancing:" An essay on abstract thinking in sport performances. In D.L. Vanderweken & S.K. Wertz (Eds.), *Sport inside out: Readings in literature and philosophy* (pp. 87–103). Fort Worth, TX: TCU Press.

Kretchmar, R.S. (2005). *Practical philosophy of sport and physical activity* (2nd ed.). Champaign, IL: Human Kinetics.

LaBarre, W. (1963). *The human animal*. Chicago: University of Chicago Press.

Lorenz, K. (1966). *On aggression*. New York: Harcourt, Brace, & World.

Lortie, G., Simoneau, J.A., Hamel, P., Boulan, M.R., Landry, F., & Bouchard, C. (1984). Responses of maximal aerobic power and capacity to aerobic training. *International Journal of Sports Medicine, 5*, 232–236.

Loy, J.W., McPherson, B.D., & Kenyon, G. (1978). *Sport and social systems*. Reading, MA: Addison-Wesley.

MacArthur, D.G., & North, K.I. (2005). Genes and human elite athletic performance. *Human Genetics, 116*(5), 331–339.

Madrigal, R. (1995). Cognitive and affective determinants of fan satisfaction with sporting event attendance. *Journal of Leisure Research, 27*(3), 205–227.

Martin, T.W., & Berry, K.J. (1974). Competitive sport in post-industrial society: The case of the motocross racer. *Journal of Popular Culture, 8*, 107–120.

McAuley, E., & Jacobson, L.B. (1991). Self-efficacy and exercise participation in sedentary adult females. *American Journal of Health Promotion, 5*, 185–191.

McAuley, E., Wraith, S., & Duncan, T.E. (1991). Self-efficacy perceptions of success and intrinsic motivation for exercise. *Journal of Applied Social Psychology, 16*, 139–155.

McIntyre, N. (1992). Involvement in risk recreation: A comparison of objective measures of engagement. *Journal of Leisure Research, 24*, 64–71.

McNally, M.P., Yontz, N., & Chaudhari, A.M. (2014). Lower-extremity work is associated with club head velocity during the golf swing of experienced golfers. *International Journal of Sports Medicine, 35*, 785–788.

Meehan, S.K., Bubblepreet, R., Wessel, B., & Boyd, L.A. (2011). Implicit sequence-specific motor learning after subcortical stroke is associated with increased prefrontal brain activations: An fMRI study. *Human Brain Mapping, 32*(2), 290–303.

Neulinger, J., & Raps, C. (1972). Leisure attitude of an intellectual elite. *Journal of Leisure Research, 4*, 196–207.

Ogilvie, B. (1973, November). The stimulus addicts. *Physician and Sportsmedicine, 1*(4), 61–65.

Ostrander, E.A., Huson, H.J., & Ostrander, G.K. (2009). Genetics of athletic performance. *Annual Review of Genomics and Human Genetics, 10*, 407–429.

Pate, R.R. (1988). The evolving definition of fitness. *Quest, 40*, 174–179.

Petrie, A. (1967). Individuality in pain and suffering. Chicago: University of Chicago Press.

Poulton, E.C. (1957). On prediction in skilled movements. *Psychological Bulletin, 54*, 467–479.

Public Health Agency of Canada. (2012). Healthy living. www.phac-aspc.gc.ca/pau-uap/fitness/definitions.html.

Ravizza, K. (1984). Qualities of the peak experience in sport. In J.M. Silva & R.S. Weinberg (Eds.), *Psychological foundations of sport* (pp. 452–462). Champaign, IL: Human Kinetics.

Reid, H.L. (2002). *The philosophical athlete*. Durham, NC: Carolina Academic Press.

Sallis, J.F., Haskell, W.L., Fortnam, S.P., Vranizan, M.S., Taylor, C.B., & Solomon, D.S. (1986). Predictors of adoption and maintenance of physical activity in a community sample. *Preventive Medicine, 15*, 331–341.

Sallis, J.F., & Hovell, M.G. (1990). Determinants of exercise behavior. *Exercise and Sport Sciences Reviews, 18*, 307–330.

Scanlan, T.K., Stein, G.L., & Ravizza, K. (1988). An in-depth study of former elite figure skaters: II. Sources of enjoyment. *Journal of Sport Psychology, 11*(1), 65–83.

Sheehan, G. (1978). *Running and being: The total experience*. New York: Warner Books.

Shi, J., & Ewing, M. (1993). Definition of fun for youth soccer players. *Journal of Sport and Exercise Psychology* (NASPSPA abstracts), *15*, S74.

Smoll, F.L., & Schutz, R.W. (1980). Children's attitude toward physical activity: A longitudinal analysis. *Journal of Sport Psychology, 2*, 137–147.

Storm, C. (2005). *Who's got game? State and county leaders in per capita production of professional major league athletes*. Paper presented at the Annual Meeting of the Association of American Geographers. www.sportsgeography.com/projects.htm.

Thomas, C.E. (1983). *Sport in a philosophic context*. Philadelphia: Lea & Febiger.

Torpy, J.M., Lynm, C., & Glass, R.M. (2005). JAMA patient page: fitness. *Journal of the American Medical Association, 294*, 3048. http://jama.ama-assn.org/cgi/content/full/294/23/3048#JPG1221F1#JPG1221F1.

U.S. Department of Commerce (U.S. Census Bureau). (2012). Participation in selected sports activities: 2009. Statistical Abstract of the United States, 2012. www.census.gov/library/publications/2011/compendia/statab/131ed.html.

U.S. Department of Health and Human Services. (1996). *Physical activity and health: A report of the surgeon general*. Atlanta: U.S. Department of Health and Human Services, Centers for Disease Control and Prevention, and National Center for Chronic Disease Prevention and Health Promotion.

Wankel, L.M. (1985). Personal and situational factors affecting exercise involvement: The importance of enjoyment. *Research Quarterly, 56*(3), 275–282.

Wankel, L.M., & Krissel, P.S.J. (1985). Methodological considerations in youth sport motivation research: A comparison of open-ended and paired comparison approaches. *Journal of Sport Psychology, 7*, 65–74.

Warren, T.Y., Barry, V., Hooker, S.P., Xuemei, S., Church, T.S., & Blair, S.N. (2010). Sedentary behaviors increase risk of cardiovascular disease mortality in men. *Medicine and Science in Sports and Exercise, 42*(5), 879–885.

Williams, J.F. (1964). *The principles of physical education* (8th ed.). Philadelphia: Lea & Febiger. (Original work published 1927.)

Willis, J.D., & Campbell, L.F. (1992). *Exercise psychology*. Champaign, IL: Human Kinetics.

Zaichowsky, L.B. (1975). Attitudinal differences in two types of physical education programs. *Research Quarterly, 46*, 364–370.

Zakarian, J.M., Hovell, M.F., Hofstettere, C.R., Sallis, J.F., & Keating, K.J. (1994). Correlates of vigorous exercise in a predominately low SES and minority high school population. *Preventive Medicine, 23*, 314–321.

Zillman, D., Bryant, J., & Sapolsky, B.S. (1979). The enjoyment of watching sport contests. In J.H. Goldstein (Ed.), *Sports, games, and play: Social and psychological viewpoints* (pp. 297–336). Hillsdale, NJ: Erlbaum.

Zillman, D., & Cantor, J.R. (1976). A disposition theory of humor and mirth. In T. Chapman & H. Foot (Eds.), *Humor and laughter: Theory, research, and applications*. London: Wiley.

Chapter 4

Boeck, S., & Staimer, M. (1996, December 6). NFL drug suspensions. *USA Today*, p. 1C.

Brown, W.M. (1980). Ethics, drugs, and sport. *Journal of the Philosophy of Sport, 7*, 15–23.

Csikszentmihalyi, M. (1990). *Flow: The psychology of optimal experience*. New York: Harper & Row.

Gleaves, J., & Lehrbach, T. (2016). Beyond fairness: The ethics of inclusion for transgender and intersex athletes. *Journal of the Philosophy of Sport, 43*(2), 311–326.

Husserl, E. (1900/1970). *Logical investigations, Vol. I*. Translated by J.N. Findlay. New York: Humanities Press.

Karkazis, K., Jordan-Young, R., Davis, G., & Camporesi, S. (2012). Out of bounds? A critique of the new policies on hyperandrogenism in elite female athletes. *The American Journal of Bioethics, 12*(7), 3–16.

Kretchmar, R.S. (1975). From test to contest: An analysis of two kinds of counterpoint in sport. *Journal of the Philosophy of Sport, 2*, 23–30.

Kretchmar, R.S. (1994). *Practical philosophy of sport*. Champaign, IL: Human Kinetics.

Kretchmar, R.S. (2011). Philosophic research in physical activity. In J.R. Thomas & J.K. Nelson (Eds.), *Research methods in physical activity* (6th ed.). (pp. 235–251). Champaign, IL: Human Kinetics.

Kretchmar, R.S. (2004). Walking Barry Bonds: The ethics of the intentional walk. In E. Bronson (Ed.), *Baseball and philosophy: Thinking outside the batter's box* (pp. 261–272). Chicago: Open Court.

Kretchmar, R.S. (2005). *Practical philosophy of sport and physical activity* (2nd ed.). Champaign, IL: Human Kinetics.

Lasch, C. (1979). *The culture of narcissism: American life in an age of diminishing expectations*. New York: Warner Books.

Loland, S. (2002). *Fair play: A moral norm system*. London: Routledge.

Meier, K. (1980). An affair of flutes: An appreciation of play. *Journal of the Philosophy of Sport, 7*, 24–45.

Meier, K. (1985). Restless sport. *Journal of the Philosophy of Sport, 12*, 64–77.

Metheny, E. (1965). *Connotations of movement in sport and dance*. Dubuque, IA: Brown.

Metheny, E. (1968). *Movement and meaning*. New York: McGraw-Hill.

Morgan, W. (1982). Play, utopia, and dystopia: Prologue to a ludic theory of the state. *Journal of the Philosophy of Sport, 9*, 30–42.

Morgan, W.J. (1994). *Leftist theories of sport: A critique and reconstruction*. Urbana, IL: University of Illinois Press.

Mosher, C., & Atkins, S. (2007). *Drugs and drug policy: The control of consciousness alteration*. Thousand Oaks, CA: Sage.

Nagel, T. (1987). *What does it all mean? A very short introduction to philosophy*. Oxford, UK: Oxford University Press.

Osterhoudt, R.G. (1991). *The philosophy of sport: An overview*. Champaign, IL: Stipes.

Perry, C. (1983). Blood doping and athletic competition. *International Journal of Applied Philosophy, 1*(3), 39–45.

Pieper, J. (1952). *Leisure: The basis of culture*. New York: Pantheon Books.

Simon, R., Torres, C.R., & Hager, P. (2015). *Fair play: The Ethics of Sport* (4th ed.). Boulder, CO: Westview Press.

Slusher, H.S. (1967). *Man, sport, and existence: A critical analysis*. Philadelphia: Lea & Febiger.

Suits, B. (1978). *The grasshopper: Games, life, and utopia*. Toronto: University of Toronto Press.

Teetzel, S. (2006). On transgendered athletes, fairness, and doping: An international challenge. *Sport in Society: Cultures, Commerce, Media, Politics, 9*(2), 227–251.

Thompson, P.B. (1982). Privacy and the urinalysis testing of athletes. *Journal of the Philosophy of Sport, 9*, 60–65.

Todd, T. (1987). Anabolic steroids: The gremlins of sport. *Journal of Sport History, 14*, 87–107.

Torres, C. (2000). What counts as part of the game? A look at skills. *Journal of the Philosophy of Sport, 27*, 81–92.

Torres, C. (2002). *Play as expression: An analysis based on the philosophy of Maurice Merleau-Ponty* (Unpublished doctoral dissertation). Pennsylvania State University, University Park.

Torres, C., & McLaughlin, D. (2003). Indigestion? An apology for ties. *Journal of the Philosophy of Sport, 30*, 144–158.

Wahlert, L., & Fiester, A. (2012). Gender transports: Privileging the "natural" in gender testing debates for intersex and transgender athletes. *The American Journal of Bioethics, 12*(7), 19–21.

Weiss, P. (1969). *Sport: A philosophic inquiry*. Carbondale, IL: Southern Illinois University Press.

Williams, J.F. (1930). Education through the physical. *Journal of Higher Education, 1*(5), 279–282.

Chapter 5

Adelman, M.L. (1986). *A sporting time: New York City and the rise of modern athletics, 1820–70*. Urbana: University of Illinois Press.

Adventure Cycling Association. (2016). Adventure cycling route network. www.adventurecycling.org/routes-and-maps/adventure-cycling-route-network.

American Physical Therapy Association. (2016). Physical therapist (PT) education overview. www.apta.org/PTEducation/Overview/.

Baker, W.J. (1988). *Sports in the Western world*. Urbana: University of Illinois Press.

Berryman, J.W. (1973). Sport history as social history? *Quest, 20*, 65–73.

Berryman, J.W. (1989). The tradition of the "six things non-natural": Exercise and medicine from Hippocrates through ante-bellum America. *Exercise and Sport Sciences Reviews, 17*, 515–559.

Betts, J.R. (1952). Organized sport in industrial America. *Dissertation Abstracts International, 12*(1), 41.

Blair, S.N., Mulder, R.T., & Kohl, H.W. (1987). Reaction to "Secular trends in adult physical activity: Exercise boom or bust?" *Research Quarterly for Exercise and Sport, 58*(2), 106–110.

Boyle, R.H. (1963). *Sport—Mirror of American life*. Boston: Little, Brown.

Carey, A.R., & Mullins, M.E. (1997, June 16). Toning up. *USA Today*, p. 1C.

Commission on Accreditation of Athletic Training Education. (2014-2015). www.caate.net/becoming-an-athletic-trainer/.

Dulles, F.R. (1940). *America learns to play: A History of popular recreation*. New York: Appleton-Century.

Eitzen, D.S., & Sage, G.H. (1997). *Sociology of North American sport* (6th ed.). Madison, WI: Brown & Benchmark.

Fisher, M. (2015, April 5). Baseball is struggling to hook kids - and risks losing fans to other sports. *Washington Post*. www.washingtonpost.com/sports/nationals/baseballs-trouble-with-the-youth-curve--and-what-that-means-for-the-game/2015/04/05/2da36dca.

Futterman, M. (2011, March 31). Has baseball's moment passed? *The Wall Street Journal*, p. D11.

Gerber, E.W. (1971). *Innovators and institutions in physical education*. Philadelphia: Lea & Febiger.

Gorn, E., & Goldstein, W. (1993). *A brief history of American sports*. New York: Hill & Wang.

Gorn, E.J. (1986). *The manly art: Bare-knuckle prize fighting in America*. Ithaca, NY: Cornell University Press.

Guttmann, A. (1978). *From ritual to record: The nature of modern sports*. New York: Columbia University Press.

Guttmann, A. (1984). *The games must go on: Avery Brundage and the Olympic movement*. New York: Columbia University Press.

Guttmann, A. (1991). *Women's sports: A history*. New York: Columbia University Press.

Hardy, S. (1982). *How Boston played: Sport, recreation, and community 1865–1915*. Boston: Northeastern University Press.

Hartwell, E.M. (1887). On the physiology of exercise. *Boston Medical and Surgical Journal, 16*(13), 297–302; *16*(14), 321–324.

Hartwell, E.M. (1899). *On physical training. Report of the commissioner of education for 1897–1898* (Vol. 1). Washington, DC: U.S. Government Printing Office.

Henry, F.M. (1964). Physical education: An academic discipline. *Journal of Health, Physical Education, and Recreation, 35*(7), 32–33, 69.

International Health, Racquet, and Sportsclub Association (IHRSA). (2016). Health club industry overview. www.ihrsa.org/about-the-industry/.

Johnson, W.J. (1960). *Science and Medicine of Exercise and Sports*. New York: Harper.

Karpovich, P.V., Morehouse, L.E., Scott, M.G., & Weiss, R.A. (Eds.). (1960). The contributions of physical activity to human well-being. *Research Quarterly, 31*(2), part II [Special issue].

King, B. (2015, August 10). Are the kids alright? *Street and Smith's Sports Business Journal*. p. 1.

Lapchick, R. (2007). The 2006 racial and gender report card: Major League Baseball. Institute for Diversity and Ethics in Sport, University of Central Florida.

Lapchick, R. (2016). The 2016 racial and gender report: Major League Baseball. Institute for Diversity and Ethics in Sport, University of Central Florida.

Leonard, W.M. (1998). *A sociological perspective of sport*. Needham Heights, MA: Allyn & Bacon.

Levine, P. (1992). *Ellis Island to Ebbets Field: Sport and the American Jewish experience.* New York: Oxford University Press.

Lippi, G., Banfi, G., Favaloro, E., Rittweger, J., & Maffulli, N. (2008). Updates on improvement of human athletic performance: Focus on world records in athletics. *British Medical Journal, 87,* 7–15.

Lucas, J.A., & Smith, R.A. (1978). *Saga of American sport.* Philadelphia: Lea & Febiger.

Massengale, J.D., & Swanson, R.A. (Eds.). (1997). *The history of exercise and sport science.* Champaign, IL: Human Kinetics.

Murphy, W. (1995). *Healing the generations: A history of physical therapy and the American Physical Therapy Association.* Lyme, CT: Greenwich.

National Center for Health Statistics. (2015). Leisure time physical activity. www.cdc.gov/nchs/data/nhis/earlyrelease201506-07/.

National Collegiate Athletic Association. (2015). NCAA sports sponsorship and participation report, 1981-82-2014-15.

National Federation of State High School Associations. (2016). 2014-15 high school athletics participation survey.

National Park Service. (1979). Visitor use statistics. Fiscal year visitation report - Fiscal year 10/1/1978 to 9/30/1979. https://irma.nps.gov/stats/reports/national/.

National Park Service. (2016). Visitor use statistics. Fiscal year visitation report - Fiscal year 10/12015 to 9/30/2016. https://irma.nps.gov/stats/reports/national.

North American Society for Sport Management. (2016). Sport management programs. www.nassm.com/.

Outdoor Foundation. (2016). 2016 outdoor recreation participation topline report. Outdoor Industry Association. www.outdoorfoundation.org/research.participation.2016.topline.html\.

Park, R.J. (1980). The *Research Quarterly* and its antecedents. *Research Quarterly for Exercise and Sport, 51*(1), 1–22.

Park, R.J. (1981). The emergence of the academic discipline of physical education in the United States. In G.A. Brooks (Ed.), *Perspectives on the academic discipline of physical education* (pp. 20–45). Champaign, IL: Human Kinetics.

Park, R.J. (1987a). Physiologists, physicians, and physical educators: Nineteenth century biology and exercise, *hygienic* and *educative*. *Journal of Sport History, 14*(1), 28–60.

Park, R.J. (1987b). Sport, gender, and society in a transatlantic Victorian perspective. In J.A. Mangan & R.J. Park (Eds.), *From "fair sex" to feminism: Sport and the socialization of women in the industrial and post-industrial eras* (pp. 58–93). London: Cass.

Park, R.J. (1989). The second 100 years: Or, can physical education become the renaissance field of the 21st century? *Quest, 41*(1), 2–27.

Paxson, F.L. (1917). The rise of sport. *Mississippi Valley Historical Review, 4,* 143–168.

Peiss, K. (1986). *Cheap amusements: Working women and leisure in turn-of-the-century New York.* Philadelphia: Temple University Press.

Pope, S.W. (1997). Introduction: American sport history—Toward a new paradigm. In S.W. Pope (Ed.), *The new American sport history: Recent approaches and perspectives* (pp. 1–30). Urbana: University of Illinois Press.

Rader, B.G. (1990). *American sports: From the age of folk games to the age of televised sports.* Englewood Cliffs, NJ: Prentice Hall.

Ramlow, J., Kriska, A., & LaPorte, R. (1987). Physical activity in the population: The epidemiologic spectrum. *Research Quarterly for Exercise and Sport, 58*(2), 111–113.

Regalado, S.O. (1992). Sport and community in California's Japanese American "Yamato Colony," 1930–1945. *Journal of Sport History, 19*(2), 130–143.

Rice, E.A., Hutchinson, J.L., & Lee, M. (1969). *A brief history of physical education.* New York: Ronald.

Rudolph, F. (1962). *The American college and university: A history.* New York: Vintage.

Running USA. (2015). State of the sport - Annual marathon report. www.runningusa.org/annualreports/.

Running USA. (2016). State of the sport - U.S. Road race trends. www.runningusa.org/state-of-sport-us-trends-2015/.

Shafer, R.J. (Ed.). (1980). *A guide to historical method.* Homewood, IL: Dorsey.

Smith, J.K. (1979). *Athletic training: A developing profession* (Unpublished master's thesis). Brigham Young University, Provo, Utah.

Smith, R.A. (1988). *Sports and freedom: The rise of big-time college athletics.* New York: Oxford University Press.

Staley, S.C. (1937). The history of sport: A new course in the professional training curriculum. *Journal of Health, Physical Education and Recreation, 8,* 522–525, 570–572.

Stephens, T. (1987). Secular trends in adult physical activity: Exercise boom or bust? *Research Quarterly for Exercise and Sport, 58*(2), 94–105.

Struna, N.L. (1996a). Historical research in physical activity. In J.R. Thomas & J.K. Nelson (Eds.), *Research methods in physical activity* (pp. 251–275). Champaign, IL: Human Kinetics.

Struna, N.L. (1996b). *People of prowess: Sport, leisure, and labor in early Anglo-America.* Urbana: University of Illinois Press.

Struna, N.L. (1997). Sport history. In J.D. Massengale & R.A. Swanson (Eds.), *The history of exercise and sport science* (pp. 143–179). Champaign, IL: Human Kinetics.

Swanson, R.A., & Spears, B. (1995). *History of sport and physical education in the United States.* Madison, WI: Brown & Benchmark.

Tygiel, J. (1983). *Baseball's great experiment: Jackie Robinson and his legacy.* New York: Oxford University Press.

U.S. Bureau of Labor Statistics. (1961). *Occupational outlook handbook* (Bulletin No. 1300). Washington, DC: U.S. Government Printing Office.

U.S. Bureau of Labor Statistics. (1982). *Occupational outlook handbook* (Bulletin No. 2200). Washington, DC: U.S. Government Printing Office.

U.S. Bureau of Labor Statistics. (2016). Physical therapists. In *Occupational outlook handbook* (2015, 16 ed.). United States Department of Labor. www.bls.gov/ooh/.

U.S. Department of Health and Human Services. (1996). *Physical activity and health: A report of the surgeon general.* Atlanta: U.S. Department of Health and Human Services, Centers for Disease Control and Prevention, National Center for Chronic Disease Prevention and Health Promotion.

US Lacrosse. (2016). 2015 participation survey. www.uslacrosse.org/sites/default/files/public/documents/about-us-lacrosse/participation-survey-2015.pdf.

Van Dalen, D.B., & Bennett, B.L. (1971). *A world history of physical education: Cultural, philosophical, comparative.* Englewood Cliffs, NJ: Prentice Hall.

Vertinsky, P. (1990). *The eternally wounded woman: Women, doctors, and exercise in the late nineteenth century.* Manchester, UK: Manchester University Press.

Weber, E. (1970). *Gymnastics and sport in fin-de-siècle France: Opium of the classes?* Paper presented at the 85th annual meeting of the American Historical Association, Boston, MA.

Whorton, J.C. (1982). *Crusaders for fitness: The history of American health reformers.* Princeton, NJ: Princeton University Press.

Wiggins, D.K. (1980). The play of slave children in the plantation communities of the old South, 1820–1860. *Journal of Sport History,* 7(2), 21–39.

Wilson, W. (1977). Social discontent and the growth of wilderness sport in America: 1965–1974. *Quest, 27,* 54–60.

Chapter 6

Acosta, R.V., & Carpenter, L.J. (1994). The status of women in intercollegiate athletics. In S. Birrell & C.L. Cole (Eds.), *Women, sport, and culture* (pp. 111–118). Champaign, IL: Human Kinetics.

Acosta, R.V., & Carpenter, L.J. (2004). Women in intercollegiate sport: A longitudinal, national study—Twenty-seven year update, 1977–2004. www.acostacarpenter.org.

Acosta, R.V., & Carpenter, L.J. (2014). Women in intercollegiate sport: A longitudinal, national study—Thirty-seven year update, 1977–2014. www.acostacarpenter.org.

Andrews, D. (Ed.). (1996). Deconstructing Michael Jordan: Reconstructing postindustrial America [Special issue]. *Sociology of Sport Journal, 13*(4).

Birrell, S., & McDonald, M.G. (2000). Reading sport, articulating power lines. In S. Birrell & M.G. McDonald (Eds.), *Reading sport: Critical essays on power and representation* (pp. 3–13). Boston: Northeastern University Press.

Brady, E., Upton, J., & Berkowitz, S. (2011, November 7). Salaries for college football coaches back on the rise. *USA Today.* www.usatoday.com/sports/college/football/story/2011-11-17/cover-college-football-coaches-salaries-rise/51242232/1.

Carlston, D.E. (1983). An environmental explanation for race differences in basketball performance. *Journal of Sport and Social Issues,* 7(2), 30–51.

Carpenter, L.J., & Acosta, R.V. (2008). *Women in intercollegiate sport: A longitudinal, national study thirty-one year update.* New York: Brooklyn College.

Coakley, J. (2007). *Sports in society: Issues and controversies* (9th ed.). Boston: McGraw-Hill.

Coakley, J. (2009). *Sports in society: Issues and controversies* (10th ed.). Boston: McGraw-Hill.

Douglas, D.D., & Jamieson, K.M. (2006). A farewell to remember: Interrogating the Nancy Lopez farewell tour. *Sociology of Sport Journal, 23*(2), 117–141.

Economic Policy Institute. (2011). The state of working America. Washington, DC: Author. http://stateofworkingamerica.org/inequality.

Eitzen, D.S. (2006). *Fair and foul: Beyond the myths and paradoxes of sport* (3rd ed.). New York: Rowman & Littlefield.

Eitzen, D.S., & Sage, G.H. (2009). *Sociology of North American sport* (8th ed.). Boulder, CO: Paradigm.

Fausto-Sterling, A. (2000). *Sexing the body.* New York: Basic Books.

Gibbs, A. (1997, Winter). Focus groups. *Social Research Update, 19.* http://sru.soc.surrey.ac.uk/SRU19.html.

Gruneau, R. (1983). *Class, sports, and social development.* Amherst: University of Massachusetts Press.

Guttmann, A. (1978). *From ritual to record: The nature of modern sports.* New York: Columbia University Press.

Hargreaves, J. (2001). *Heroines of sport: The politics of difference and identity.* London: Routledge.

Jackson, D.Z. (1989). Calling the plays in Black and White: Will today's Superbowl be Black brawn vs. White brains. *Boston Globe,* A25.

Jackson, D. (1996, March 27). Chasing spirits down the court at NCAA tourney. *Charlotte Observer,* p. 17A.

Jamieson, K., & Smith, M. (2016). *Fundamentals of Sociology of Sport and Physical Activity*. Champaign, IL: Human Kinetics.

King, C.R. (2004). This is not an Indian: Situating claims about Indianness in sporting worlds. *Journal of Sport & Social Issues, 28*, 3–10.

Kinkema, K.M., & Harris, J.C. (1998). MediaSport studies: Key research and emerging issues. In L.A. Wenner (Ed.), *MediaSport* (pp. 27–54). London: Routledge.

Kochman, T. (1981). *Black and white styles in conflict*. Chicago: University of Chicago Press.

Lapchick, R., & Baker, D. (2016, April). The 2015 racial and gender report card: College sport. The Institute for Diversity and Ethics in Sport. http://tidesport.org/.

Lapchick, R., and Bullock, T. (2016, July). The 2016 racial and gender report card: National Basketball Association. The Institute for Diversity and Ethics in Sport. http://tidesport.org/.

Lapchick, R., Davison, E., Grant, C., and Quirarte, R. (2016, August). Gender report card: 2016 international sports report card on women in leadership roles. The Institute for Diversity and Ethics in Sport. http://tidesport.org/.

Lapchick, R., Dominguez, J., Haldane, L., Loomer, E., & Pelts, J. (2014, December). The 2014 racial and gender report card: Major League Soccer. The Institute for Diversity and Ethics in Sport. http://tidesport.org/.

Lapchick, R., Malveaux, C., Davison, E., & Grant, C. (2016, September). The 2016 racial and gender report card: National Football League. The Institute for Diversity and Ethics in Sport. http://tidesport.org/.

Lapchick, R., & Nelson, N. (2015, October). The 2015 Women's National Basketball Association racial and gender report card. The Institute for Diversity and Ethics in Sport. http://tidesport.org/.

Lapchick, R., & Salas, D. (April, 2015). The 2015 racial and gender report card: Major League Baseball. The Institute for Diversity and Ethics in Sport. http://tidesport.org/.

Lorber, J. (1994). *Paradoxes of gender*. New Haven, CT: Yale University Press.

Majors, R. (1990). Cool pose: Black masculinity and sports. In M.A. Messner & D.F. Sabo (Eds.), *Sport, men, and the gender order: Critical feminist perspectives* (pp. 109–126). Champaign, IL: Human Kinetics.

McDonald, M.G. (1996). Michael Jordan's family values: Marketing, meaning, and post-Reagan America. *Sociology of Sport Journal, 13*, 344–365.

McIntosh, P.C. (1963). *Sport in society*. London: Watts.

McKenna, L. (2016, March). The madness of college basketball coaches' salaries. Amid spiraling tuition costs and a growing reliance on part-time faculty, athletic departments pay them millions of dollars per year. *The Atlantic*. www.theatlantic.com/education/archive/2016/03/the-madness-of-college-basketball-coaches-salaries/475146/.

Metheny, E. (1965). *Connotations of movement in sport and dance*. Dubuque, IA: Brown.

National Federation of State High School Associations. (2016). 2015–16 high school athletic participation survey. www.nfhs.org/ParticipationStatistics/ParticipationStatistics.

National Sporting Goods Association (NSGA). (2016). *Sport participation in 2009*. Mt. Prospect, IL: National Sporting Goods Association.

Neal, M.A. (2006). *New black man*. New York: Routledge.

Pandian, A. (2016, July 21). WNBA fines three teams and players for wearing Black Lives Matter t- shirts. *CBS Sports.com*. www.cbssports.com/nba/news/wnba-fines-three-teams-and-players-for-wearing-black-lives-matter-t-shirts/.

Ross, H. (1996, March 10). Waiting for the call. *Greensboro News and Record*, p. C1.

Sage, G.H. (1997). Sport sociology. In J.D. Massengale & R.A. Swanson (Eds.), *History of exercise and sport science* (pp. 109–141). Champaign, IL: Human Kinetics.

Siedentop, D. (1996). Valuing the physically active life: Contemporary and future directions. *Quest, 48*, 266–274.

Sommeiller, E., and Price, M. (2015, January 26). The increasingly unequal states of America: Income inequality by state, 1917 to 2012. Economic Policy Institute. www.epi.org/publication/income-inequality-by-state-1917-to-2012/.

Staurowsky, E.J. (1999, November). American Indian sport imagery and the miseducation of Americans. *Quest*, i, 382–392.

Staurowsky, E.J. (2004). Privilege at play: On the legal and social fictions that sustain American Indian sport imagery. *Journal of Sport & Social Issues, 28*, 11–29.

Stein, P.J., & Hoffman, S. (1978). Sports and male role strain. *Journal of Social Issues, 34*, 136–150.

Stephens, T., & Caspersen, C.J. (1994). The demography of physical activity. In C. Bouchard, R.J. Shephard, & T. Stephens (Eds.), *Physical activity, fitness, and health: International proceedings and consensus statement* (pp. 204–213). Champaign, IL: Human Kinetics.

The Big Lead Staff. (2016, January 5). The top 25 most influential people in sport business. http://thebiglead.com/2016/01/05/top-25-most-influential-people-sports-business/.

The Guardian. (2014, October 8). Indian sprinter, Dutee Chand, appeals against ban for failing gender test. www.theguardian.com/sport/2014/oct/08/sprinter-dutee-chand-appeals-ban-failing-gender-test.

U.S. Census Bureau. (2007). Participation in selected sport activities: 2005. (Table 1222). http://search.census.gov/search?utf8=%E2%9C%93&affiliate=census&query=Participation+in+selected+sport+activities%3A+2005&commit.x=0&commit.y=0.

Wilson, B., & Sparks, R. (1996). "It's gotta be the shoes": Youth, race, and sneaker commercials. *Sociology of Sport Journal, 13*, 398–427.

Wilson, V., and Rodgers, W. M., III. (2016, September 20). Black–white wage gaps expand with rising wage inequality. Economic Policy Institute. www.epi.org/publication/black-white-wage-gaps-expand-with-rising-wage-inequality/.

Women's Sports Foundation. (2009, March). Women's sports and fitness facts and statistics. www.womens-sportsfoundation.org/wp-content/uploads/2016/08/wsf-facts-march-2009.pdf.

Zimbalist, A. (2001). *Unpaid professionals: Commercialism and conflict in big-time college sports.* Princeton, NJ: Princeton University Press.

Zirin, D. (2013, October 21). Why they refused to play: Read the grievance letter of the Grambling State Tigers football team. *The Nation*. www.thenation.com/article/why-they-refused-play-read-grievance-letter-grambling-state-tigers-football-team/.

Chapter 7

Abernethy, B., Baker, J., & Côté, J. (2005). Transfer of pattern recall skills may contribute to the development of sport expertise. *Applied Cognitive Psychology, 19*, 705–718.

Abernethy, B., & Sparrow, W.A. (1992). The rise and fall of dominant paradigms in motor behavior research. In J.J. Summers (Ed.), *Approaches to the study of motor control and learning* (pp. 3–45). Amsterdam: Elsevier.

Abernethy, B., Thomas, K.T., & Thomas, J.R. (1993). Strategies for improving understanding of motor expertise (or mistakes we have made and things we have learned!!). In J.L. Starkes & F. Allard (Eds.), *Cognitive issues in motor learning* (pp. 317–356). Amsterdam: Elsevier.

Adams, J.A. (1987). Historical review and appraisal of research on the learning, retention, and transfer of human motor skills. *Psychological Bulletin, 101*, 41–74.

Albers, A.S., Thomas, J.R., & Thomas, K.T. (2005). Development of rapid aiming movements: Index of difficulty and movement substructure. *Human Movement, 6*, 5–11.

Bayley, N. (1935). The development of motor abilities during the first three years. *Monographs of the Society for Research in Child Development*, Whole No. 1, 1–26.

Blix, M. (1892–1895). Die lange und spannung des muskels. *Skandinavische Archiv Physiologie, 3*, 295–318; *4*, 399–409; *5*, 150–206.

Chamberlin, C., & Lee, T. (1993). Arranging practice conditions and designing instruction. In R.N. Singer, M. Murphey, & L.K. Tennant (Eds.), *Handbook of research on sport psychology* (pp. 213–241). New York: Macmillan.

Chi, M.T.H, & Koeske, R.D. (1983). Network representation of a child's dinosaur knowledge. *Developmental Psychology, 19*, 29–39.

Christina, R.W. (1989). Whatever happened to applied research in motor learning? In J. Skinner et al. (Eds.), *Future directions in exercise and sport science research* (pp. 411–422). Champaign, IL: Human Kinetics.

Christina, R.W. (1992). The 1991 C.H. McCloy Research Lecture: Unraveling the mystery of the response complexity effect in skilled movements. *Research Quarterly for Exercise and Sport, 63*, 218–230.

Clark, J.E., & Phillips, S.J. (1991). The development of intralimb coordination in the first six months of walking. In J. Fagard & P.H. Wolff (Eds.), *The development of timing control and temporal organization in coordinated action* (pp. 245–257). New York: Elsevier Science.

Clark, J.E., & Whitall, J. (1989). What is motor development? The lessons of history. *Quest, 41*, 183–202.

Côté, J., Lidor, R., & Hackfort, D. (2009). ISSP position stand: To sample or to specialize? Seven postulates about youth sport activities that lead to continued participation and elite performance. *International Journal of Sport and Exercise Psychology, 9*, 7–17.

Dennis, W. (1938). Infant development under conditions of restricted practice and a minimum of social stimulation: A preliminary report. *Journal of Genetic Psychology, 53*, 149–158.

Dennis, W., & Dennis, M. (1940). The effect of cradling practices on the age of walking in Hopi children. *Journal of Genetic Psychology, 56*, 77–86.

Espenschade, A. (1960). Motor development. In W.R. Johnson (Ed.), *Science and medicine of exercise and sport* (p. 330). New York: Harper & Row.

Fischman, M.G. (2007). Motor learning and control foundations of kinesiology: Defining the academic core. *Quest, 59*, 67–76.

Fitts, P.M., & Posner, M.I. (1967). *Human performance.* Pacific Grove, CA: Brooks/Cole.

Galton, F. (1876). The history of twins as a criterion of the relative power of nature. *Anthropological Institute Journal, 5*, 391–406.

Gentile, A.M. (1972). A working model of skill acquisition with application to teaching. *Quest*, Monograph XVII, 2–23.

Gesell, A. (1928). Infancy and human growth. New York: Macmillan.

Haken, H., Kelso, J.A.S., & Bunz, H. (1985). A theoretical model of phase transitions in human hand movements. *Biological Cybernetics, 51*, 347–356.

Henry, F.M., & Rogers, D.E. (1960). Increased response latency for complicated movements and a "memory drum" theory of neuromotor reaction. *Research Quarterly, 31*, 448–458.

Hofstein, A., & Lunetta, V.N. (1982). The role of the laboratory in science teaching: Neglected Aspects of research. *Review of Educational Research, 52*, 201–217.

Kelso, J.A.S. (1995). *Dynamic patterns: The self-organization of brain and behavior.* Cambridge, MA: MIT Press.

Locke, E.A., & Latham, G.P. (1985). The application of goal setting to sports. *Sport Psychology Today, 7,* 205–222.

Magill, R.A., & Hall, K.G. (1990). A review of the contextual interference effect in motor skill acquisition. *Human Movement Science, 9,* 241–289.

Malina, R.M. (1984). Physical growth and maturation. In J.R. Thomas (Ed.), *Motor development during childhood and adolescence* (pp. 2–26). Edina, MN: Burgess International.

Malina, R.M., Bouchard, C., & Bar-Or, O. (2004). *Growth, maturation, and physical activity* (2nd ed.). Champaign, IL: Human Kinetics.

McCullagh, P. (1993). Modeling: Learning, developmental, and social psychological considerations. In R.N. Singer, M. Murphey, & L.K. Tennant (Eds.), *Handbook of research on sport psychology* (pp. 106–126). New York: Macmillan.

McGraw, M.B. (1935). *Growth: A study of Johnny and Jimmy.* New York: Appleton-Century-Crofts.

McGraw, M.B. (1939). Later development of children specially trained during infancy: Johnny and Jimmy at school age. *Child Development, 10,* 1–19.

Park, R., Seefeldt, V., Malina, R.M., & Broadhead, G.D. (1996). In Memoriam: G. Lawrance Rarick. *Journal of Physical Education, Recreation &Dance, 67*(1), 16.

Perreault, M.E., & French, K.E. (2015). External-focus feedback benefits free-throw learning in children. *Research Quarterly for Exercise and Sport, 86,* 422–427.

Rikli, R. (2005). Movement and mobility influence on successful aging: Addressing the issue of low physical activity. *Quest, 57,* 46–66.

Rikli, R.E., & Jones, C.J. (2012). Development and validation of criterion-referenced clinically relevant fitness standards for maintaining physical independence in later years. *The Gerontologist, 53,* 255–267.

Rosenbaum, D.A. (1991). *Human motor control.* San Diego: Academic Press.

Salmoni, A.W., Schmidt, R.A., & Walter, C.B. (1984). Knowledge of results and motor learning: A review and critical reappraisal. *Psychological Bulletin, 95,* 355–386.

Schmidt, R.A. (1975). A schema theory of discrete motor skill learning. *Psychological Review, 82,* 225–260.

Schmidt, R.A. & Lee, T.D. (2011). *Motor control and learning: A behavioral emphasis* (5th ed.). Champaign, IL: Human Kinetics.

Schmidt, R.A. & Lee, T.D. (2014). *Motor learning and performance: From principles to application* (5th ed.). Champaign, IL: Human Kinetics.

Seefeldt, V., & Haubenstricker, J. (1982). Patterns, phases, or stages: An analytical model for the study of developmental movement. In J.A.S. Kelso & J.E. Clark (Eds.), *The development of movement control and co-ordination* (pp. 309–318). Chichester, UK: Wiley.

Sherrington, C.S. (1906). *The integrative action of the nervous system.* New Haven, CT: Yale University Press.

Slone, M.R. (1984). R*uth B. Glassow: The cutting edge.* The Academy Papers. www.nationalacademyofkinesiology.org/.../TAP_20_CuttingEdgeinPEandExerciseSci.

Spirduso, W.W., & MacRae, P.G. (1990). Motor performance and aging. In J.E. Birren & K.W. Schaie (Eds.), *The handbook of psychology of aging* (3rd ed., pp. 184–197). San Diego: Academic Press.

Stelmach, G.E., & Nahom, A. (1992). Cognitive-motor abilities of the elderly driver. *Human Factors, 34*(1), 53–65.

Thelen, E., Ulrich, D., & Jensen, J.L. (1990). The developmental origins of locomotion. In M.H. Woollacott & A. Shumway-Cook (Eds.), *Development of posture and gait: Across the lifespan* (pp. 25–47). Columbia: University of South Carolina Press.

Thomas, J.R. (1997). History of motor behavior. In J.D. Massengale & R.A. Swanson (Eds.), *History of exercise and sport sciences* (pp. 203–292). Champaign, IL: Human Kinetics.

Thomas, J.R. (2006). Motor behavior: From telegraph keys and twins to linear slides and stepping. *Quest, 58,* 112–127.

Thomas, J.R., & French, K.E. (1985). Gender differences across age in motor performance: A meta-analysis. *Psychological Bulletin, 98,* 260–282.

Thomas, J.R., & Thomas, K.T. (1989). What is motor development: Where does it belong? *Quest, 41,* 203–212.

Thomas, J.R., & Thomas, K.T. (2008). Principles of motor development for elementary school physical education. *Elementary School Journal, 108,* 181–195.

Thomas, K.T., Gallagher, J.D., & Thomas, J.R. (2001). Motor development and skill acquisition during childhood and adolescence. In R.N. Singer, H.A. Hausenblas, & C.M. Janelle (Eds.), *Handbook of sport psychology* (2nd ed., pp. 20–52). New York: Wiley.

Thomas, K.T., & Thomas, J.R. (1994). Developing expertise in sport: The relation of knowledge to performance. *International Journal of Sport Psychology, 25,* 295–312.

Ulrich, B. (2007). Motor development: Core curricular concepts. *Quest, 59,* 77–91.

Ulrich, B., & Reeve, T.G. (2005). Studies in motor behavior: 75 years of research in motor development, learning, and control. *Research Quarterly for Exercise and Sport, 75,* S62–S70.

Weiss, M.R., & Klint, K.A. (1987). "Show and tell" in the gymnasium: An investigation of developmental differences in modeling and verbal rehearsal of motor skills. *Research Quarterly for Exercise and Sport, 58,* 234–241.

Wyble, B.P., & Rosenbaum, D.A. (2016). Are motor adjustments quick because they don't require detection or because they escape competition? *Motor Control, 20,* 182–186.

Chapter 8

Baltzell, A.L. (Ed.). (2016). *Mindfulness and performance.* New York: Cambridge University Press.

Balyi, I., Way, R., & Higgs, C. (2013). *Long-term athlete development.* Champaign, IL: Human Kinetics.

Beilock, S. (2010). *Choke.* New York: Free Press.

Bell, J.J., & Hardy, J. (2009). Effects of attentional focus on skilled performance in golf. *Journal of Applied Sport Psychology, 21,* 163–177.

Buckworth, J., Dishman, R.K., O'Connor, P.J., & Tomporowski, P.D. (2013). *Exercise psychology* (2nd ed.). Champaign, IL: Human Kinetics.

Castaneda, B., & Gray, R. (2007). Effects of focus of attention on baseball batting performance in players of differing skill levels. *Journal of Sport and Exercise Psychology, 29,* 60–77.

Csikszentmihalyi, M. (1990). *Flow: The psychology of optimal experience.* New York: Harper & Row.

Dweck, C.S. (2006). *Mindset: The new psychology of success.* New York: Ballantine.

Easterbrook, J.A. (1959). The effect of emotion on cue utilization and the organization of behavior. *Psychological Review, 66,* 183–201.

Eys, M.A., Burke, S.M., Carron, A.V., & Dennis, P.W. (2006). The sport team as an effective group. In J.M. Williams (Ed.), *Applied sport psychology: Personal growth to peak performance* (5th ed., pp. 157–173). Boston: McGraw-Hill.

Franz, S.I., & Hamilton, G.V. (1905). The effects of exercise upon the retardation in conditions of depression. *American Journal of Insanity, 62,* 239–256.

Ginsburg, R.D., Smith, S.R., Danforth, N., Ceranoglu, T.A., Durant, S.A., Kamin, H., . . . Masek, B. (2014). Patterns of specialization in professional baseball players. *Journal of Clinical Sport Psychology, 8,* 261–275.

Goffi, C. (1984). *Tournament tough.* London, England: Ebury Press.

Hanton, S., & Jones, G. (1999). The acquisition and development of cognitive skills and strategies: I. Making the butterflies fly in formation. *The Sport Psychologist, 13,* 1–21.

Hellison, D. (2003). *Teaching responsibility through physical activity* (2nd ed.). Champaign, IL: Human Kinetics.

Heuze, J.P., & Brunel, P. (2003). Social loafing in a competitive context. *International Journal of Sport and Exercise Psychology, 1,* 246–263.

Jones, G., Hanton, S., & Connaughton, D. (2007). A framework of mental toughness in the world's best performers. *The Sport Psychologist, 21,* 243–264.

Krane, V., & Williams, J.M. (2015). Psychological characteristics of peak performance. In J.M. Williams & V. Krane (Eds.), *Applied sport psychology: Personal growth to peak performance* (7th ed., pp. 159–175). New York: McGraw-Hill.

Landers, D.M., & Arent, S.A. (2007). Physical activity and mental health. In G. Tenenbaum & R.C. Eklund (Eds.), *Handbook of sport psychology* (3rd ed., pp. 469–491). Hoboken, NJ: Wiley.

Lox, C.L., & Ginis, K.A.M. (2014). *The psychology of exercise: Integrating theory and practice.* New York: Routledge.

Lox, C.L., Ginis, K.A.M., & Petruzzello, S.J. (2014). *The psychology of exercise* (4th ed.). Scottsdale, AZ: Holcomb Hathaway.

Miles, C.A., Wood, G., Vine, S.J., Vickers, J.N., & Wilson, M.R. (2015). Quiet eye training facilitates visuomotor coordination in children with developmental coordination disorder. *Research on Developmental Disabilities, 40,* 31–41.

Mullen, R., & Hardy, L. (2010). Conscious processing and the process goal paradox. *Journal of Sport and Exercise Psychology, 32,* 275–297.

Neil, R., Hanton, S., & Mellalieu, S.D. (2013). Seeing things in a different light: Assessing the effects of a cognitive-behavioral intervention upon the further appraisals and performance of golfers. *Journal of Applied Sport Psychology, 25,* 106–130.

Ostrow, A.C. (Ed.) (2002). *Directory of psychological tests in the sport and exercise sciences* (2nd ed.). Morgantown, WV: Fitness Information Technology.

Rhodes, R.E., & Pfaelli, L.A. (2012). Personality and physical activity. In E.O. Acevedo (Ed.), *The Oxford handbook of exercise psychology* (pp. 195–223). New York: Oxford.

Shields, D.L., & Bredemeier, B.L. (2009). *True competition.* Champaign, IL: Human Kinetics.

Smith, R.E., & Christensen, D.S. (1995). Psychological skills as predictors of performance and survival in professional baseball. *Journal of Sport and Exercise Psychology, 17,* 399–415.

Smith, R.E., Schutz, R.W., Smoll, F.L., & Ptacek, J.T. (1995). Development and validation of a multidimensional measure of sport-specific psychological skills: The Athletic Coping Skills Inventory-28. *Journal of Sport and Exercise Psychology, 17,* 379–398.

Vallerand, R.J., Blanchard, C., Mageau, G.A., Koestner, R., Ratelle, C., Leonard, M., . . . Marsolais, J. (2003). Les passions de l'ame: On obsessive and harmonious passion. *Journal of Personality and Social Psychology, 85,* 756–767.

Vealey, R.S., & Chase, M.A. (2016). *Best practice for youth sport.* Champaign, IL: Human Kinetics.

Vealey, R.S., & Forlenza, S.T. (2015). Understanding and using imagery in sport. In J.M. Williams & V. Krane (Eds.), *Applied sport psychology: Personal growth to peak performance* (7th ed., pp. 240–273). New York: McGraw-Hill.

Vickers, J.N. (2009). Advances in coupling perception and action: The quiet eye as a bidirectional link between gaze, attention, and action. *Progressive Brain Research, 174,* 279–288.

Weiss, M.R., Bolter, N.D., & Kipp, L.E. (2016). Evaluation of The First Tee in promoting positive youth development: Group comparisons and longitudinal trends. *Research Quarterly for Exercise and Sport, 87*, 271–283.

Chapter 9

Atwater, A.E. (1980). Kinesiology/biomechanics: Perspectives and trends. *Research Quarterly for Exercise and Sport, 51*, 193–218.

Behm, D.G., Blazevich, A.J., Kay, A.D., & McHugh, M. (2016). Acute effects of muscle stretching on physical performance, range of motion, and injury incidence in healthy, active individuals: a systematic review. *Applied Physiology, Nutrition, and Metabolism, 41,* 1–11.

Bus, S.A., van Deursen, R.W., Armstrong, D.G., Lewis, J.E.A., Caravaggi, C.F., & Cavanagh, P.R. (2016). Footwear and offloading interventions to prevent and heal foot ulcers and reduce plantar pressure in patients with diabetes: a systematic review. *Diabetes/Metabolism Research and Reviews, 32*(S1), 99–118.

Cash, J. (2015, July 1). Meyerhold's biomechanics for theatre. www.thedramateacher.com/meyerholds-biomechanics-for-theatre/.

Hatze, H. (1974). The meaning of the term "biomechanics." *Journal of Biomechanics, 7*, 189–190.

Hughes, G. (2014). A review of recent perspectives on biomechanical risk factors associated with anterior cruciate ligament injury. *Research in Sports Medicine, 22*:193–212.

Humphrey, J.D. (2003). Review paper: Continuum biomechanics of soft biological tissues. *Proceedings of the Royal Society of London A: Mathematical, Physical and Engineering Sciences, 459*(2029), 3–46.

Klentrou, P. (2016). Influence of exercise and training on critical stages of bone growth and development. *Pediatric Exercise Science, 28*, 178–186.

Knudson, D. (2013). *Qualitative diagnosis of human movement: Improving performance in sport and exercise* (3rd ed.). Champaign, IL: Human Kinetics.

Meister, D. (1999). *The history of human factors and ergonomics.* Mahway, NJ: Lawrence Erlbaum.

National Association for Sport and Physical Education. (2003). Guidelines for undergraduate biomechanics [Guidance document]. Reston, VA: Author.

National Institute of Arthritis and Musculoskeletal and Skin Diseases (NIAMS). (2015). Handout on health: Osteoarthritis. NIH Publication No. 15-4617. www.niams.nih.gov/health_info/osteoarthritis/.

Pitches, J. (2004). *Vsevolod Meyerhold* (2nd ed.). New York, NY: Routledge.

Pope, M.H. (2005). Giovanni Alfonso Borelli the father of biomechanics. *Spine, 15*(30), 2350–2355.

Posse, N. (1890). *The special kinesiology of educational gymnastics.* Boston: Lothrop, Lee and Shepard.

Rooney, B.D., & Derrick, T.R. (2013). Joint contact loading in forefoot and rearfoot strike patterns during running. *Journal of Biomechanics, 46,* 2201–2206.

Roos, E.M. (2005). Joint injury causes knee osteoarthritis in young adults. *Current Opinion in Rheumatology, 17*(2), 195–200.

Schneck, D.J. (2005). Forensic biomechanics. *American Laboratory, 37*(8), Editor's Page. www.castonline.ilstu.edu/mccaw/Forensic%20Biomechanics.pdf.

Silverwood, V., Blagojevic-Bucknall, M., Jinks, C., Jordan, J.L., Protheroe, J., & Jordan, K.P. (2015). Review: Current evidence on risk factors for knee osteoarthritis in older adults: a systematic review and meta-analysis. *Osteoarthritis and Cartilage, 23*(4), 507–515.

Skarstrom, W. (1909). *Gymnastic kinesiology*. Springfield, MA: Bassette.

Sloan, M.R. (1987). Ruth B. Glassow: The cutting edge. Academy Papers No. 20, National Academy of Kinesiology, 120–128. www.nationalacademyofkinesiology.org/1980-1989/1980-1989/academy-papers-no-20-1987.

Taylor, F.W. (1911). *The principles of scientific management.* New York, NY: Harper & Brothers.

Vannini, F., Spalding, T., Andriolo, L., Berruto, M., Denti, M., Espregueira-Mendes, J., Menetrey, J., Peretti, G., Seil, R., & Filardo, G. (2016). Sport and early osteoarthritis: The role of sport in aetiology, progression and treatment of knee osteoarthritis. *Knee Surgery, Sports Traumatology, Arthroscopy, 24*(6), 1786–1796.

Winter, D.A. (1985). *Biomechanics and motor control of human gait.* Waterloo, ON: University of Waterloo Press.

Chapter 10

American College of Sports Medicine. (2004). Nutrition and athletic performance. *Medicine and Science in Sports and Exercise, 41*, 709–731.

American College of Sports Medicine. (2014). *ACSM's guidelines for exercise testing and prescription.* Baltimore: Lippincott Williams & Wilkins.

Armstrong, L.E., Costill, D.L., & Fink, W.J. (1985). Influence of diuretic-induced dehydration on competitive running performance. *Medicine and Science in Sports and Exercise, 17*, 456–461.

Åstrand, P.-O. (1991). Influence of Scandinavian scientists in exercise physiology. *Scandinavian Journal of Medicine and Science in Sports, 1*, 3–9.

Bar-Or, O. (1983). *Pediatric sports medicine for the practitioner.* New York: Springer-Verlag.

Berryman, J.W. (1995). *Out of many, one: A history of the American College of Sports Medicine.* Champaign, IL: Human Kinetics.

Blair, S.N., Kohl, H.W., Paffenbarger, R.S., Clark, D.G., Cooper, K.H., & Gibbons, L.W. (1989). Physical fitness and all-cause mortality: A prospective study of healthy men and women. *Journal of the American Medical Association, 262*(17), 2395–2401.

Booth, F.W. (1989). Application of molecular biology in exercise physiology. In K.B. Pandolf (Ed.), *Exercise and sport sciences reviews* (Vol. 17, pp. 1–27). Baltimore: Williams & Wilkins.

Bouchard, C., Lesage, R., Lortie, G., Simoneau, J.A., Hamel, P., Boulay, M.R., . . . Leblanc, C. (1986). Aerobic performance in brothers, dizygotic and monozygotic twins. *Medicine and Science in Sports and Exercise, 18*, 639–646.

Celsing, F., Blomstrand, E., Werner, B., Pihlstedt, P., & Ekblom, B. (1986). Effects of iron deficiency on endurance and muscle enzyme activity in man. *Medicine and Science in Sports and Exercise, 18*, 156–161.

Centers for Disease Control and Prevention. (2011). Nonfatal traumatic brain injuries related to sports and recreation activities among persons aged ≤19 years—United States, 2001–2009. *Morbidity and Mortality Weekly Report, 60*(39), 1337–1342. www.cdc.gov/mmwr/preview/mmwrhtml/mm6039a1.htm.

Clarkson, P.M., & Haymes, E.M. (1995). Exercise and mineral status of athletes: Calcium, magnesium, phosphorus, and iron. *Medicine and Science in Sports and Exercise, 27*, 831–843.

Coggan, A., & Coyle, E. (1991). Carbohydrate ingestion during prolonged exercise: Effects on metabolism and performance. In J.O. Holloszy (Ed.), *Exercise and sport sciences reviews* (Vol. 19, pp. 1–40). Baltimore: Williams & Wilkins.

Conners, R.T., Morgan, D.W., Fuller, D.K., & Caputo, J.L. (2014). Underwater treadmill training, glycemic control, and health-related fitness in adults with type 2 diabetes. *International Journal of Aquatic Research and Rehabilitation, 8*, 382–396.

Davis, J.M., Burgess, W.A., Sientz, C.A., Bartoli, W.P., & Pate, R.R. (1988). Effects of ingesting 6% and 12% glucose/electrolyte beverages during prolonged intermittent cycling in the heat. *European Journal of Applied Physiology, 57*, 563–569.

Dill, D.B. (1967). The Harvard Fatigue Laboratory: Its development, contributions, and demise. In C.B. Chapman (Ed.), *Physiology of muscular exercise* (pp. 161–170). New York: American Heart Association.

Dill, D.B., Talbott, J.H., & Edwards, H.T. (1930). Studies in muscular activity. VI. Responses of several individuals to a fixed task. *Journal of Physiology, 69*, 267–305.

Donnelly. J.E., Blair, S.N., Jakicic, J.M., Manore, M.M., Rankin, J.W., & Smith, B.K.. (2009). American College of Sports Medicine position stand. Appropriate physical activity intervention strategies for weight loss and prevention of weight regain for adults. *Medicine and Science in Sports and Exercise, 41*(2), 459–471.

Fagard, D. (2001). Exercise characteristics and the blood pressure response to dynamic physical training. *Medicine and Science in Sports and Exercise, 33*, S484–S492.

Gollnick, P.D., Timson, B.F., Moore, R.L., & Riedy, M. (1981). Muscular enlargement and number of fibers in skeletal muscles of rats. *Journal of Applied Physiology, 50*, 936–943.

Gonyea, W.J. (1980). Role of exercise in inducing increases in skeletal muscle fiber number. *Journal of Applied Physiology, 48*, 421–426.

Hagberg, J.M. (1987). Effect of training on the decline of VO2max with aging. *Federation Proceedings, 46*, 1830–1833.

Haymes, E.M., & Wells, C.L. (1986). *Environment and human performance*. Champaign, IL: Human Kinetics.

Hill, A.V., & Lupton, H. (1923). Muscular exercise, lactic acid, and the supply and utilization of oxygen. *QJM, os-16* (62), 135–171.

Hultman, E. (1967). Physiological role of muscle glycogen in man, with special reference to exercise. *Circulation Research, 21*(Suppl. 1), 99–114.

Ivy, J.L., Katz, A.L., & Cutler, C.L. (1989). Muscle glycogen resynthesis after exercise: Effect of time on carbohydrate ingestion. *Journal of Applied Physiology, 64*, 1480–1485.

Kenney, W.L., Wilmore, J.H., & Costill, D.L. (2015). *Physiology of sport and exercise* (6th ed.). Champaign, IL: Human Kinetics.

Kroll, W.P. (1982). *Graduate study and research in physical education*. Champaign, IL: Human Kinetics.

Lawler, P.R., Filion, K.B., Eisenberg, M.J. (2011). Efficacy of exercise-based cardiac rehabilitation post-myocardial infarction: A systematic review and meta-analysis of randomized controlled trials. *American Heart Journal, 162*, 571–584.

Margaria, R., Edwards, H.T., & Dill, D.B. (1933). The possible mechanisms of contracting and paying the oxygen debt and the role of lactic acid in muscular contraction. *American Journal of Physiology, 106*, 689–715.

Morris, J.N., Heady, J.A., Raffle, P.A.B., Roberts, C.G., & Parks, J.W. (1953). Coronary heart disease and physical activity of work. *Lancet, 2*, 1111–1120.

Morrissey, M.C., Harman, E.A., & Johnson, M.J. (1995). Resistance training modes: Specificity and effectiveness. *Medicine and Science in Sports and Exercise, 27*, 648–660.

Moshfegh, A., Goldman, J., & Cleveland, L. (2005). What we eat in America, NHANES 2001–2002. Usual nutrient intake from foods as compared to dietary reference intakes. U.S. Department of Agriculture, Agricultural Research Service. www.ars.usda.gov/ARSUserFiles/80400530/pdf/0102/usualintaketables2001-02.pdf.

NIH Consensus Development Panel on Physical Activity and Cardiovascular Health. (1996). Physical activity and cardiovascular health. *Journal of the American Medical Association, 276*, 241–246.

Ogden, C.L., Carroll, M.D., Kit, B.K., & Flegal, K.M. (2014). Prevalence of childhood and adult obesity in the United States, 2011–2012. *Journal of the American Medical Association, 311*(8), 806–814. doi:10.1001/jama.2014.732

Paffenbarger, R.S. (1994). 40 years of progress: Physical activity, health and fitness. In *40th anniversary lectures* (pp. 93–109). Indianapolis: American College of Sports Medicine.

Pate, R.R., Pratt, M., Blair, S.N., Haskell, W.L., Macera, C.A., Bouchard, C., . . . Wilmore, J.H. (1995). Physical activity and public health: A recommendation from the Centers for Disease Control and the American College of Sports Medicine. *Journal of the American Medical Association, 273*, 402–407.

Physical Activity Guidelines Advisory Committee. (2008). Physical activity guidelines for Americans. Washington, DC: U.S. Department of Health and Human Services.

Rowland, T.W. (1989). Oxygen uptake and endurance fitness in children: A developmental perspective. *Pediatric Exercise Science, 1*, 313–328.

Saltin, B. (1973). Metabolic fundamentals in exercise. *Medicine and Science in Sports, 5*, 137–146.

Sinha, R., Fisch, G., Teague, B., Tamborlane, W.V., Banyas, B., Allen, K., . . . Caprio, S. (2002). Prevalence of impaired glucose tolerance among children and adolescents with marked obesity. *New England Journal of Medicine, 345*, 802–810.

Strong, W.B., Malina, R.M., Blimkie, C.J.R., Daniels, S.R., Dishman, R.K., Gutin, B., . . . Trudeau, F. (2005). Evidence-based physical activity for school-age youth. *Journal of Pediatrics, 146*, 732–737.

Thompson, P.D., Buchner, D., Piña, I.L., Balady, G.J., Williams, M.A., Marcus, B.H., Berra, K., . . . Wenger, N.K. (2003). Exercise and physical activity in the prevention and treatment of atherosclerotic cardiovascular disease. *Arteriosclerosis, Thrombosis, and Vascular Biology, 23*, 42–49.

Timson, B.F., Bowlin, B.K., Dudenhoeffer, G.A., & George, J.B. (1985). Fiber number, area, and composition of mouse soleus muscle following enlargement. *Journal of Applied Physiology, 58*, 619–624.

U.S. Department of Health and Human Services (USDHHS). (1996). *Physical activity and health: A report of the surgeon general.* Atlanta: U.S. Department of Health and Human Services, Centers for Disease Control and Prevention, National Center for Chronic Disease Prevention and Health Promotion.

U.S. Department of Health and Human Services, Centers for Disease Control and Prevention, National Center for Chronic Disease Prevention and Health Promotion, & Division of Nutrition, Physical Activity and Obesity. (2015). Data, trends, and maps. www.cdc.gov/nccdphp/DNPAO/index.html.

Vuori, I.M. (2001). Dose-response of physical activity and low back pain, osteoarthritis, and osteoporosis. *Medicine and Science in Sports and Exercise, 33*, s551–s586.

Willett, W.C., Manson, J.E., Stampler, M.J., Colditz, G.A., Rosner, B., Speizer, F.E., & Hennekena, C.H. (1995). Weight, weight change, and coronary disease in women. *Journal of the American Medical Association, 273*, 461–465.

Zwiren, L.D. (1989). Anaerobic and aerobic capacities of children. *Pediatric Exercise Science, 1*, 31–44.

Chapter 11

Albright, C.E., & Smith, K. (2006, March). Welding—trade or profession? *Techniques, 81*(3), 38.

American Kinesiology Association. (2003). Recommended core areas for coursework in an undergraduate kinesiology curriculum. www.americankinesiology.org/white-papers/white-papers/re-examining-the-undergraduate-core-in-kinesiology-in-a-time-of-change.

Eagan, K., Stolzenberg, E.B., Bates, A.K., Aragon, M.C., Suchard, M.R., & Rios-Aguilar, C. (2015). *The American freshman: National norms fall 2015.* Los Angeles: Higher Education Research Institute, UCLA.

Ericsson, K.A., Prietula, M.J., & Cokely, E.T. (2007). The making of an expert. *Harvard Business Review, 85*(7–8), 114–121.

Etzioni. A. (1969). *The semi-professions and their organization.* New York: Free Press.

Feignbaum, E. (2012). Professionalism and workplace savvy. *The Houston Chronicle.* http://smallbusiness.chron.com/professionalism-workplace-savvy-2909.html.

Friedson, E. (2001). *Professionalism: The third logic.* Chicago: University of Chicago Press.

Greenwood, E. (1957). Attributes of a profession. *Social Work, 2*, 45–55.

Hoffman, S.J. (2011). Examining the big picture. In American Kinesiology Association (Ed.), *Careers in sport, fitness, and exercise* (pp. 12–13). Champaign, IL: Human Kinetics.

Jackson, J.A. (1970). *Professions and professionalization.* London: Cambridge University Press.

Lawson, H.A. (1984). *Invitation to physical education.* Champaign, IL: Human Kinetics.

Lawson, H.A. (1998a, July). *Globalization and the social responsibilities of citizen-professionals.* Address to AIESEP International Conference, Adelphi University.

Lawson, H.A. (1998b). Here today, gone tomorrow: A framework for analyzing the invention, development, transformation, and disappearance of helping fields. *Quest, 50*, 225–237.

Malek, M.H., Nalbone, D.P., Berger, D.E., & Coburn, J.W. (2002). Importance of health science education for personal fitness training. *Journal of Strength and Conditioning Research, 16*(1), 19–24.

McGuigan, P.J. (2007, February). The ongoing quest for professionalism: Consumer decision processes for professional services. *Casualty and Property Insurance Underwriters Journal, 60*(2), 1–8.

National Institute on Alcohol Abuse and Alcoholism. (2014). College drinking. http://pubs.niaaa.nih.gov/publications/CollegeFactSheet/CollegeFactSheet.pdf.

Newell, K.M. (1990). Physical activity, knowledge types, and degree programs. *Quest, 42*, 243–268.

Professional Standards Councils. (n.d.) What is a profession? www.psc.gov.au/what-is-a-profession.

Satell, G. (2010, August 1). What it means to be a professional. Digital Toronto. www.digitaltonto.com/2010/what-it-means-to-be-a-professional/.

Schon, D.A. (1995, November/December). Knowing in action: The new scholarship requires a new epistemology. *Change*, 27–34.

Segar, M.L., Guerin, E., Phillips, E., & Fortier, M. (2016). From vital sign to vitality: Selling exercise so patients want to buy it. *Current Sports Medicine Reports, 15*(4): 276–286.

Chapter 12

Afana, M., Brinkikji, W., Cloft, H., & Salka, S. (2015). Hospitalization costs for acute myocardial infarction patients treated with percutaneous coronary intervention in the United States are substantially higher than Medicare payments. *Clinical Cardiology, 38*(1), 13–19.

Ainsworth, B.E., & Hooker, S.P. (2015). The fusion of public health into kinesiology. *Kinesiology Review, 4*(4), 322–328.

Bradley, K.L. (2013). Workplace wellness programs [Letter to the editor]. *Health Affairs, 32*(8), 1510.

Bulmer, S.M., Howell, J., Ackerman, L., & Fedric, R. (2012). Women's perceived benefits of exercise during and after breast cancer treatment. *Women Health, 52*(8), 771–787.

Burd, S. (2009). How Safeway is cutting health-care costs. *Wall Street Journal*. http://online.wsj.com/article/SD124476808026308603.html.

Centers for Disease Control and Prevention. (2016). Workplace health model. Atlanta, GA: U.S. Department of Health and Human Services. www.cdc.gov/workplacehealthpromotion/model/index.html.

Claxton, G., Rae, M., Panchal, N., Whitmore, H., Damico, A., Kenward, K., & Long, M. (2015). Health benefits in 2015: Stable trends in the employer market. *Health Affairs, 34*(10), 1779–1788.

Commission on Dietetic Registration. (2016). RDN credential—Frequently asked questions. www.cdrnet.org/news/rdncredentialfaq.

DeNavas-Walt, C., Proctor, B.D., & Smith, J.C. (2012). *Income, poverty and health insurance coverage in the United States: 2011: Current population reports*. Washington, DC: U.S. Census Bureau.

Dintiman, G.B., & Greenberg, J.S. (1986). *Health through discovery*. New York: Random House.

Dolbow, D.R., & Figoni, S.F. (2015). Accommodation of wheelchair-reliant individuals by community fitness facilities. *Spinal Cord, 53*(7), 515–519.

Dychtwald, K. (1990). *Age wave*. New York: Bantam Books.

Fabius, R., Thayer, R.D., Konicki, D.L., Yarborough, C.M., Peterson, K.W., Isaac, F., . . . Dreger, M. (2013). The link between workforce health and safety and the health of the bottom line: Tracking market performance of companies that nurture a "culture of health." *Journal of Occupational and Environmental Medicine, 55*(9), 993–1000.

Goetzel, R.Z., Anderson, D.R., Whitmer, R.W., Ozminkowski, R.J., Dunn, R.L., & Wasserman, J. for The Health Enhancement Research Organization (HERO) Research Committee. (1998). The relationship between modifiable health risks and health care expenditures: An analysis of the multi-employer HERO health risk and cost database. *Journal of Occupational and Environmental Medicine, 40*(10), 843–854.

Goetzel, R.Z., Pei, X., Tabrizi, M.J., Henke, R.M., Kowlessar, N., Nelson, C.F., & Metz, R.D. (2012). Ten modifiable health risk factors are linked to more than one-fifth of employer-employee health care spending. *Health Affairs, 31*(11), 2474–2484.

Goetzel, R.Z., Roemer, E.C., Liss-Levinson, R.C., & Samoly, D.K. (2008). *Workplace health promotion: Policy recommendations that encourage employers to support health improvement programs for their workers*. Washington, DC: Partnership for Prevention.

Grossmeier, J., Fabius, R., Flynn, J.P., Noeldner, S.P., Fabius, D., Goetzel, R.Z., & Anderson, D.R. (2016). Linking workplace health promotion best practices and organizational financial performance: Tracking market performance of companies with highest scores on the HERO scorecard. *Journal of Occupational and Environmental Medicine, 58*(1), 16–23.

Hannon, P.A., Garson, G., Harris, J.R., Hammerback, K., Sopher, C.J., & Clegg-Thorp, C. (2012). Workplace health promotion, readiness, and capacity among mid-sized employers in low-wage industries: A national survey. *Journal of Occupational and Environmental Medicine, 54*(11), 1337–1343.

Humes, K., Jones, N., & Ramirez, R. (2011). Overview of race and Hispanic origin: 2010 census briefs. U.S. Census Bureau. www.census.gov/prod/cen2010/briefs/c2010br-02.pdf.

International Health, Racquet & Sportsclub Association. (2015). *IHRSA's profiles of success*. Boston: International Health, Racquet & Sportsclub Association.

Jones, D.S., Podolsky, S.H., & Greene, J.A. (2012). The burden of disease and the changing task of medicine. *New England Journal of Medicine, 36*(25), 2333–2338.

Kaiser Family Foundation. (2012). *Health care costs: A primer*. http://kff.org/report-section/health-care-costs-a-primer-2012-report/.

Linnan, L., Bowling, M., Childress, J., Lindsay, G., Blakey, S.P., Wieker, S., & Royall, P. (2008). Results of the 2004 National Worksite Health Promotion survey. *American Journal of Public Health, 98*(8), 1503–1509.

Merrill, R.M., Aldana, S.G., Pope, J.E., Anderson, D.R., Coberley, C.R., Grossmeier, J.J., & Whitmer, R.W. (2013). Self-rated job performance and absenteeism according to employee engagement, health behaviors, and physical health. *Journal of Occupational and Environmental Medicine, 55*(1), 10–18.

Niebuhr, S., & Grossmeier, J. (2015). Is it time for a broader approach? Recasting the value of "employee health" with a focus on workforce capability. *American Journal of Health Promotion, 29*(6), TAHP6–9.

O'Donnell, M.P., & Harris, J.S. (1994). *Health promotion in the workplace.* Albany, NY: Delmar.

Prochaska, J.O., & DiClemente, C.C. (1986). Toward a comprehensive model of change. In W. Miller & N. Heather (Eds.), *Treating addictive behaviors* (pp. 3–27). New York: Plenum Press.

Smith, J.C., & Medalia, C. (2015). *Current population reports, P60-253: Health insurance coverage in the United States: 2014.* U.S. Census Bureau. Washington, DC: U.S. Government Printing Office.

Tu, H.T., & Mayrell, R.C. (2010). *Research brief: Employer wellness initiatives grow, but effectiveness varies widely.* Washington, DC: National Institute for Health Care Reform.

U.S. Census Bureau. (2015). American fact finder. Washington, DC: U.S. Department of Commerce. https://factfinder.census.gov/faces/nav/jsf/pages/index.xhtml.

U.S. Department of Health and Human Services (USDHHS). (1996). *Physical activity and health: A report of the surgeon general.* Atlanta: U.S. Department of Health and Human Services, Centers for Disease Control and Prevention, National Center for Chronic Disease Prevention and Health Promotion.

U.S. Department of Health and Human Services (USDHHS). (2008). *Physical activity guidelines for Americans.* Atlanta: U.S. Department of Health and Human Services. http://health.gov/paguidelines/.

U.S. Department of Health and Human Services (USDHHS). (2010). *Healthy People 2020.* Atlanta: U.S. Department of Health and Human Services. www.healthypeople.gov.

World Health Organization. (2011). United Nations high-level meeting on noncommunicable disease prevention and control. New York: Author. www.who.int/nmh/events/un_ncd_summit2011/en.

World Health Organization. (2016). Noncommunicable diseases. New York: Author. www.who.int/mediacentre/factsheets/fs355/en/#.

YMCA of the USA. (2016). YMCA's diabetes prevention program. Chicago, IL: YMCA. www.ymca.net/diabetes-prevention/.

Chapter 13

Ambrosi, E., & Barker-Schwartz, K. (1995). The profession's image, 1917–1925, part I: Occupational therapy as represented in the media. *American Journal of Occupational Therapy, 49*(7), 715–719.

Bleakley, C.M., Glasgow, P., & MacAuley, D.C. (2011). PRICE needs updating, should we call the POLICE? *British Journal of Sports Medicine, 46,* 220–221.

Board of Certification (BOC). (2016). Map of state regulatory agencies. www.bocatc.org/index.php/state-regulation.

Erskine, H.E., Whiteford, H.A., & Pike, KM. (2016). The global burden of eating disorders. *Current Opinion in Psychiatry, 29*(6), 346–353.

Greiner, A.C., & Knebel, E. (Eds.). (2003). The core competencies needed for health care professions. In *Health professions education: A bridge to quality.* Washington, DC: National Academies Press. www.ncbi.nlm.nih.gov/books/NBK221519/.

King, H., Campbell, S., Herzog, M., Popoli, D., Reisner, A., & Polikandriotis, J. (2015). Epidemiology of injuries in high school football: Does school size matter? *Journal of Physical Activity and Health, 12*(8), 1162–1167.

Knapen, J., Vancampfort, D., Moriën, Y., & Marchal, Y. (2014). Exercise therapy improves both mental and physical health in patients with major depression. *Disability and Rehabilitation, 37*(16), 1490–1495.

Laitman, B.M., & John, G.R. (2015). Understanding how exercise promotes cognitive integrity in the aging brain. *PLoS Biology, 13*(11), e1002300. doi:10.1371/journal.pbio.1002300

Lee, A.C.W., & Billings, M. (2016). Telehealth implementation in a skilled nursing facility: Case report for physical therapist practice in Washington. *Physical Therapy, 96*(2), 252–259.

Maron, B.J., Haas, T.S., Ahluwalia, A., Murphy, C.J., & Garberich, R.F. (2016). Demographics and epidemiology of sudden deaths in young competitive athletes: From the United States national registry. *The American Journal of Medicine, 129*(11), 1170–1177. http://dx.doi.org/10.1016/j.amjmed.2016.02.031.

National Athletic Trainers' Association. (2015). 2014 salary survey. https://members.nata.org/members1/salarysurvey2014/.

Olivares, P.R., Gusi, N., Prieto, J., & Hernandez-Mocholi, M.A. (2011). Fitness and health-related quality of life dimensions in community-dwelling middle aged and older adults. *Health and Quality of Life Outcomes, 9,* 117.

Payscale. (2017). Strength and conditioning coach salary. www.payscale.com/research/US/Job=Strength_and_Conditioning_Coach/Salary.

Prentice, W.E. (2015). *Principles of athletic training: A competency-based approach* (15th ed.). New York: McGraw-Hill.

Salary.com. (2016). Exercise physiologist salaries. www1.salary.com/Exercise-Physiologist-Salary.html.

Sims, J., Hill, K., Davidson, S., Gunn, J., & Huang, N. (2006). Exploring the feasibility of a community-based strength training program for older people with depressive symptoms and its impact on depressive symptoms. *BMC Geriatrics, 30*(6), 18.

U.S. Bureau of Labor Statistics (BLS). (2016a). Occupational outlook handbook: Athletic trainers. www.bls.gov/ooh/healthcare/athletic-trainers.htm.

U.S. Bureau of Labor Statistics (BLS). (2016b). Occupational outlook handbook: Cardiovascular technologists and technicians. www.bls.gov/ooh/healthcare/diagnostic-medical-sonographers.htm.

U.S. Bureau of Labor Statistics (BLS). (2016c). Occupational outlook handbook: Occupational therapists. www.bls.gov/ooh/healthcare/occupational-therapists.htm.

U.S. Bureau of Labor Statistics (BLS). (2016d). Occupational outlook handbook: Physical therapist assistants and aides. www.bls.gov/ooh/healthcare/physical-therapist-assistants-and-aides.htm.

U.S. Bureau of Labor Statistics (BLS). (2016e). Occupational outlook handbook: Physical therapists. www.bls.gov/ooh/healthcare/physical-therapists.htm.

U.S. Bureau of Labor Statistics (BLS). (2016f). Occupational outlook handbook: Recreational therapists. www.bls.gov/ooh/healthcare/recreational-therapists.htm.

Ussher, M.H., Taylor, A., & Faulkner, G. (2012). Exercise interventions for smoking cessation. *Cochrane Database of Systematic Reviews*, (1), CD02295.

Wardlaw, F.B., McGuire, F.A., & Overby, Z. (2000). Therapeutic recreation: Optimal health treatment for orthopaedic disability. *Orthopedic Nursing, 19*, 56–60.

World Health Organization. (2002). Towards a common language for functioning, disability, and health: The international classification of functioning, disability, and health. Geneva, Switzerland: Author. www.who.int/classifications/icf/training/icfbeginnersguide.pdf.

Chapter 14

Anderson, W.G., & Barrette, G.T. (1978). *What's going on in gym: Descriptive studies of physical education classes* [Monograph #1]. Newtown, CT: Motor Skills.

Bangsoo, J., et al. (2016). The Copenhagen Consensus Conference 2016: Children, youth, and physical activity in schools and during leisure time. *British Journal of Sports Medicine*. Advance online publication. doi:10.1136/bjsports-2016-096325

Berliner, D.C. (1988). *The development of expertise in pedagogy*. Washington, DC: American Association of Colleges for Teacher Education.

Brophy, J.E., & Good, T.L. (1986). Teacher behavior and student achievement. In M.C. Wittrock (Ed.), *Handbook of research on teaching* (3rd ed., pp. 328–375). New York: Macmillan.

Castelli, D.M., & Beighle, A. (2007). Rejuvenating the school environment to include physical activity. *Journal of Physical Education, Recreation and Dance, 78*(5), 25–28.

Castelli, D.M., Centeio E.E., Hwang, J., Barcelona, J.M., Glowacki, E.M., Calvert, H.G., & Nicksic, H.M. (2014). The history of physical activity and academic performance research: Informing the future. *Monographs of the society for research in child development, 79*, 199–148.

Cawley, J., Meyerhoefer, C., & Newhouse, D. (2006, Fall). Not your father's PE: Obesity, exercise, and the role of schools. *Education Next, 4*, 61–66.

Cooper Institute. (2016). *Fitnessgram/activitygram test administration manual* (4th ed.). Champaign, IL: Human Kinetics.

Corbin, C.B., Lindsey, R., Welk, G.J., & Corbin, W.R. (2001). *Fundamental concepts of fitness and wellness.* New York: McGraw-Hill.

Davis, K. (2003). Teaching for gender equity in physical education: A review of the literature. *Women in Sport and Physical Activity Journal, 12*(2), 55–82.

DeMarco, G.M. (1999). Physical education teachers of the year: Who they are, what they think, say, and do. *Teaching Elementary Physical Education, 10*(2), 11–13.

Dodds, P. (1994). Cognitive and behavioral components of expertise in teaching physical education. *Quest, 46*, 153–163.

Donnelly, J.E., Hillman, C.H., Castelli, D., Etnier, J.L., Lee, S.L., Tomporowski, P., Lambourne, K., & Szabo-Reed, A.N. (2016). Physical activity, fitness, cognitive function, and academic achievement in children: A systematic review. *Medicine & Science in Sports & Exercise, 48*, 1197–1222.

Dyson, B.P. (1995). Students' voices in two alternative elementary physical education programs. *Journal of Teaching in Physical Education, 14*, 394–407.

Fink, J., & Siedentop, D. (1989). The development of routines, rules, and expectations at the start of the school year. *Journal of Teaching in Physical Education, 8*, 198–212.

Flory, S.B., & McCaughtry, N. (2011). Culturally relevant physical education in urban schools: Reflecting cultural knowledge. *Research Quarterly for Exercise and Sport, 82*, 49–60.

Gloria, C., Faulk, K., & Steinhardt, M. (2013). Positive affectivity predicts successful and unsuccessful adaptations to stress. *Motivation and Education, 37*, 185–193.

Graber, K.C. (2001). Research on teaching in physical education. In V. Richardson (Ed.), *Handbook of research on teaching* (4th ed., pp. 491–519). Washington, DC: American Educational Research Association.

Graber, K.C., Locke, L.F., Lambdin, D., & Solmon, M.A. (2008). The landscape of elementary school physical education. *The Elementary School Journal, 108*, 151–159.

Graber, K.C., Templin, T.J., Haag, R., & O'Connor, J. (2011). Careers in physical education, sport instruction, coaching, sports officiating, and sport psychology. In American Kinesiology Association (Ed.), *Careers in sport, fitness, and exercise* (pp. 51–74). Champaign, IL: Human Kinetics.

Graber, K.C., & Woods, A.M. (2007). Stepping up to the plate: Physical educators as advocates for wellness policies—Part 2. *Journal of Physical Education, Recreation and Dance, 78*(6), 19–28.

Graham, G. (2001). *Teaching children physical education: Becoming a master teacher.* Champaign, IL: Human Kinetics.

Griffin, P.S. (1984). Girls' participation patterns in a middle school team sports unit. *Journal of Teaching in Physical Education, 4*, 30–38.

Griffin, L., Mitchell, S., & Oslin, J. (1997). *Teaching sports concepts and skills: A tactical games approach*. Champaign, IL: Human Kinetics.

Griffin, P.S. (1985). Teachers' perceptions of and responses to sex equity problems in a middle school physical education program. *Research Quarterly for Exercise and Sport, 56*, 103–110.

Griffin, P.S. (1986). Analysis and discussion: What have we learned? *Journal of Physical Education, Recreation and Dance, 57*(4), 57–59.

Hellison, D. (2011). *Teaching personal and social responsibility through physical activity* (3rd ed.). Champaign, IL: Human Kinetics.

Howie, E.K., & Pate, R.R. (2012). Physical activity and academic achievement in children: A historical perspective. *Journal of Sport and Health Science, 1(3)*, 160–169.

Konukman, F., Agbuga, B., Erdogan, S., Zorba, E, Demirhan, G., & Yilmaz, I. (2010). Teacher-coach role conflict in school-based physical education in USA: A literature review and suggestions for the future. *Biomedical Human Kinetics, 2,* 19–24.

Lanier, J.E., & Little, J.W. (1986). Research on teacher education. In M.C. Wittrock (Ed.), *Handbook of research on teaching* (3rd ed., pp. 527–569). New York: Macmillan.

Lau, E.Y., Wandersman, A.H., & Pate, R.R. (2016). Factors influencing implementation of youth physical activity interventions: An expert perspective. *Translational Journal of the American College of Sports Medicine, 1*(7), 60–70. doi:10.2379-2868/0107/60-70

Locke, L.F. (1975). *The ecology of the gymnasium: What the tourists never see.* Amherst, MA: University of Massachusetts. (ERIC Document Reproduction No. ED 104-823)

Lortie, D. (1975). *Schoolteacher: A sociological study.* Chicago: University of Chicago Press.

Manross, D., & Templeton, C.L. (1997). Expertise in teaching physical education. *Journal of Physical Education, Recreation & Dance, 68*(3), 29–35.

Maslach, C. & Jackson, S. (1986). *Maslach burnout inventory: Manual* (2nd ed.). Palo Alto, CA: Consulting Psychologists Press.

McKenzie, T.L., & Lounsbery, M.A.F. (2009). School physical education: The pill not taken. *American Journal of Lifestyle Medicine, 3,* 219–225.

Metzler, M. (1979). The measurement of academic learning time in physical education (Unpublished doctoral dissertation). Ohio State University, Columbus.

Metzler, M. (1989). A review of research on time in sport pedagogy. *Journal of Teaching in Physical Education, 8*, 87–103.

Ogden, C.L., Carroll, M.D., Fryar, C.D., & Flegal, K.M. (2015). Prevalence of obesity among adults and youth: United States, 2011-2014. *NCHS Data Brief No. 219.* www.cdc.gov/nchs/data/databriefs/db219.pdf.

O'Sullivan, M., & Doutis, P. (1994). Research on expertise: Guideposts for expertise and teacher education in physical education. *Quest, 46,* 176–185.

Rasberry, C.N., Lee, S.M., Robin, L., Laris, B.A., Russell, L.A., Coyle, K.K., & Nihiser, A.J. (2011). The association between school-based physical activity, including physical education, and academic performance: A systematic review of the literature. *Preventive Medicine (52),* S10–20.

Richards, K., Levesque-Bristol, C., Templin, T.J. & Graber, K.C. (2016). The impact of resilience on role stressors and burnout in elementary and secondary teachers. *Social Psychology of Education.* doi:10.1007/s11218-016-9346x

Richards, K., Templin, T., Levesque-Bristol, C., & Blankenship, B. (2014). Understanding differences in role stressors, resilience, and burnout in teacher/coaches and non-coaching teachers. *Journal of Teaching in Physical Education, 33*, 383–402.

Rink, J.E. (2006). *Teaching physical education for learning* (5th ed.). Boston: McGraw-Hill.

Siedentop, D., Hastie, P., & van der Mars, H. (2011). *Complete guide to sport education* (2nd ed.). Champaign, IL: Human Kinetics.

Siedentop, D., & Tannehill, D. (2000). *Developing teaching skills in physical education* (4th ed.). Mountain View, CA: Mayfield.

Society of Health and Physical Educators. (n.d.). Comprehensive school physical activity program. www.shapeamerica.org/cspap/what.cfm.

Society of Health and Physical Educators. (2014). *National standards & grade-level outcomes for K-12 physical education.* Champaign, IL: Human Kinetics.

Society of Health and Physical Educators. (2015a). *Updated standard 1 assessments for PE metrics eBook: Assessing standards 1-6 in elementary school.* Champaign, IL: Human Kinetics.

Society of Health and Physical Educators. (2015b). *Updated standard 1 assessments for PE metrics eBook: Assessing standards 1-6 in secondary school.* Champaign, IL: Human Kinetics.

Society of Health and Physical Educators and American Heart Association. (2016). *Shape of the nation.* www.shapeamerica.org/advocacy/son/2016/upload/Shape-of-the-Nation-2016_web.pdf.

Society of Health and Physical Educators. (2017). *National standards for initial physical education teacher education.* www.shapeamerica.org.

Staurowsky, E., Hogshed-Makar, N., Kane, M., Wughalter, E., Yiamouyiannis, A., & Lerner, P. (2007). Gender equity in physical education and athletics. In S. Klein (Ed.), *Handbook for achieving sex equity through education* (pp. 381–410). Routledge. New York.

Stroot, S. (1996). Organizational socialization: Factors impacting beginning teachers. In S. Silverman & C. Ennis (Eds.), *Studying learning in physical education: Applying research to enhance instruction* (pp. 339–366). Champaign, IL: Human Kinetics.

Stroot, S.A., & Oslin, J.L. (1993). Use of instructional statements by preservice teachers for overhand throwing performance of children. *Journal of Teaching in Physical Education, 13*, 24–45.

Tyson, L. (1996). Context of schools. In S. Silverman & C. Ennis (Eds.), *Studying learning in physical education: Applying research to enhance instruction* (pp. 55–80). Champaign, IL: Human Kinetics.

Woods, A.M., & Graber, K.C. (Eds.). (2007). Stepping up to the plate: Physical educators as advocates for wellness policies—Part I [Special feature]. *Journal of Physical Education, Recreation & Dance, 78*(5), 17–28.

Chapter 15

Barnett, Z. (2015). 2015–2016 coaching changes. FootballScoop. http://footballscoop.com/news/2015-16-head-coaching-changes/.

Ewing, M.E., & Seefeldt, V. (1989). *Participation and attrition patterns in American agency-sponsored and interscholastic sports: An executive summary* [Final report to U.S. Footwear Association]. East Lansing, MI: Institute for Youth Sports.

Hellison, D., Cutforth, N., Kallusky, J., Martinek, T., Parker, M., & Stiehl, J. (2000). *Serving underserved youth through physical activity: Toward a model of university community collaboration*. Champaign, IL: Human Kinetics.

National Collegiate Athletic Association. (2016). Estimated probability of competing in athletics beyond high school interscholastic level. *2016 Guide for the College Bound Student Athlete*. www.NCAA.org.\ compliance.

National Federation of State High School Associations. (2016). High school sports participation increase for 27th consecutive year. www.nfhs.org/articles/high-school-sports-participation-increases-for-27th-consecutive-year/.

Sports coach [Job profile]. (2016). *U.S. News & World Report*. http://money.usnews.com/careers/best-jobs/sports-coach.

USA Today. (2017). NCAA salaries. http://sports.usatoday.com/ncaa/salaries/mens-basketball/coach/.

U.S. Bureau of Labor Statistics. (2015). Occupational outlook handbook: Coaches and scouts. www.bls.gov/ooh/entertainment-and-sports/coaches-and-scouts.htm.

Chapter 16

Andrew Walker, WTA chief marketing officer. (2011, August 29). *Street & Smith's SportsBusiness Journal, 14*(8), 30.

Badenhausen, K. (2011, December 14). The NFL signs TV deal worth $27 billion. *Forbes*. www.forbes.com/sites/kurtbadenhausen/2011/12/14/the-nfl-signs-tv-deals-worth-26-billion/#540e6e162a67.

Bopp, T., Goldsmith, A., & Walker, M. (2014). Commitment to diversity. *Journal of Applied Sport Management, 6*(1), 1–27.

Bradish C.L., & Cronin, J.J. (2009). Corporate social responsibility in sport. *Journal of Sport Management, 23*(6), 691–697.

Brooks, D., Harrison, L., Norris, M., and Norwood, D. (2013). Why should we care about diversity in kinesiology? *Kinesiology Review, 2*(3), 145–156.

Bruening, J., Fuller, R.D., Catrufo, R.J., Madsen, R.M., Evanovich, J., and Wilson-Hill, D.E. (2014). Applying intergroup contact theory to the sport management classroom. *Sport Management Education Journal, 8*(1), 35–46.

Canadian Sport Tourism Alliance (CSTA). (n.d.). About CSTA. http://canadiansporttourism.com/about-csta/about-csta.html.

Charlie Besser, president & CEO, Intersport. (2011, May 2). *Street & Smith's SportsBusiness Journal*. www.sportsbusinessdaily.com/Journal/Issues/2011/05/02/People-and-Pop-Culture/What-I-Like.aspx.

Chelladurai, P. (2006). *Human resource management in sport and recreation* (2nd ed.). Champaign, IL: Human Kinetics.

Collins, L., & Collins, D. (2013). Decision making and risk management in Adventure Sports coaching. *Quest, 65*(1), 72–83.

Commission on Sport Management Accreditation (COSMA). (2016, May). Accreditation principles manual & guidelines for self-study preparation. www.cosmaweb.org/accreditation-manuals.html.

Conrad Smith, COO, Minnesota Lynx. (2011, September 12). *Street & Smith's SportsBusiness Journal, 14*(20), 30.

Dave Butler, CEO, Paciolan. (2011, September 5). *Street & Smith's SportsBusiness Journal, 14*(9), 38.

Deitsch, R. (2011, December 15). NFL, networks win in extended rights deal. www.si.com/nfl/2011/12/15/nfl-rightsdeal.

Fielding, L.W., & Miller, L.K. (1996). Historical eras in sport marketing. In B.G. Pitts & D.K. Stotlar (Eds.), *Fundamentals of sport marketing* (pp. 41–74). Morgantown, WV: Fitness Information Technology.

Fisher, R. (2015, March 2). NBA hires Verizon exec to run point on mobile. *Street & Smith's SportsBusiness Journal, 17*(44), 10.

Ganim, S. (2015, August 12). NCAA punishment is anyone's guess. CNN. www.cnn.com/2015/08/12/us/ncaa-academic-fraud/.

Gibson, H., & Fairley, S. (2014). Sport tourism. In P.M. Pedersen & L. Thibault (Eds.), *Contemporary sport management* (5th ed., pp. 264–288). Champaign, IL: Human Kinetics.

Global esports market report: Revenues to jump to $463M in 2016 as US leads the way. (2016, January 25). Newzoo. https://newzoo.com/insights/articles/global-esports-market-report-revenues-to-jump-to-463-million-in-2016-as-us-leads-the-way/.

Heinze, K.L., Soderstrom, S., & Zdroik, J. (2014). Toward a strategic and authentic corporate social responsibility in professional sport: A case study of the Detroit Lions. *Journal of Sport Management, 28*, 672–686.

Hoang, T.V., Cardinal, B.J., & Newhart, D.W. (2016). An exploratory study of ethnic minority students' constraints to and facilitators of engaging in campus recreation. *Recreational Sports Journal, 40*(1), 69–82.

Host Committee Advisory Group. (n.d.). San Francisco Super Bowl Committee. www.sfbaysuperbowl.com/about/host-committee-advisory-group#707uDC1SM-BrwVIv7.97.

Howard, D.R., & Crompton, J.L. (2014). *Financing sport* (3rd ed.). Morgantown, WV: Fitness Information Technology.

Hurd, A.R., & Anderson, D.M. (2011). *The park and recreation professional's handbook*. Champaign, IL: Human Kinetics.

International Health, Racquet, and Sportsclub Association (IHRSA). (2011). IHRSA's global report on the state of the health club. www.ihrsa.org/industry-research/.

International Health, Racquet and Sportsclub Association. (2016). *About the industry*. www.ihrsa.org/about-the-industry.

Irwin, R.L., Sutton, W.A., & McCarthy, L.M. (2008). *Sport promotion and sales management* (2nd ed.). Champaign, IL: Human Kinetics.

Knowledge@Wharton. (2014, March 25). Do tax incentives for nonprofits provide an unfair advantage? http://knowledge.wharton.upenn.edu/article/tax-incentives-nonprofits-provide-unfair-advantage/.

Li, M., Hofacre, S., & Mahony, D. (2001). *Economics of sport*. Morgantown, WV: Fitness Information Technology.

Mark King, CEO and president, TaylorMade-Adidas Golf. (2011, August 15). *Street & Smith's SportsBusiness Journal, 14*(6), 38.

Mathner, R.P., & Martin, C.L.L. (2012). Sport management graduate and undergraduate students' perceptions of career expectations in sport management. *Sport Management Education Journal, 6*, 21–31.

Meek, A. (1997). An estimate of the size and supported economic activity of the sports industry in the United States. *Sport Marketing Quarterly, 6*(4), 15–22.

Milano, M., & Chelladurai, P. (2011). Gross domestic sport product: The size of the sport industry in the United States. *Journal of Sport Management, 25*, 24–35.

Miller, L.K. (1997). *Sport business management*. Gaithersburg, MD: Aspen.

Mullen, L. (2015, March 2). NBPA launches apps for members. *Street & Smith's SportsBusiness Journal, 17*(44), 10.

Mullin, B.J., Hardy, S., & Sutton, W.A. (2014). *Sport marketing* (4th ed.). Champaign, IL: Human Kinetics.

Muret, D. (2014, September 8). Construction makes a comeback: Wave of new development has the facilities business jumping. *Street & Smith's SportsBusiness Journal, 17*(21), 17–22.

National Association of Sports Commissions (NASC). (n.d.). About NASC. www.sportscommissions.org/About-NASC.

National Collegiate Athletic Association. (2005, February 25). Commitment to diversity programming. *NCAA News, 42*(5), A2.

National Collegiate Athletic Association (NCAA). (2016, July 1). [LSDBi search for major infraction cases from August 25, 2010 to July 1, 2016]. https://web1.ncaa.org/LSDBi/exec/miSearch.

National sponsors. (n.d.). CollegeFootballPlayoff.com. www.collegefootballplayoff.com/sponsors.

North American Society for Sport Management (NASSM). (n.d.). Sport management programs. www.nassm.com/Programs/AcademicPrograms/United_States.

Olmsted, L. (2016, Feb. 12). Fat tire biking is the hot new sport and not just in winter. *Forbes*. www.forbes.com/sites/larryolmsted/2016/02/12/fat-tire-biking-is-the-hot-new-winter-sport-and-not-just-for-winter/#18a4401a7ae2.

Peckham, M. (2016). Why ESPN is so serious about covering esports. *Time*. http://time.com/4241977/espn-esports/.

Pitts, B.G., Fielding, L.W., & Miller, L.K. (1994). Industry segmentation theory and the sport industry: Developing a sport industry segment model. *Sport Marketing, 3*(4), 15–28.

Pitts, B.G., & Stotlar, D.K. (2007). *Fundamentals of sport marketing* (3rd ed.). Morgantown, WV: Fitness Information Technology.

Plunkett Research. (n.d.). Sport industry overview. www.plunkettresearch.com/industries/sports-recreation-leisure-market-research/.

Quinn, G. (2014, April 21). One year after bombing, some are critical of new security measures at Boston Marathon. MassLive. www.masslive.com/news/boston/index.ssf/2014/04/security_at_the_boston_maratho.html.

Rich, K., & Giles, A. (2015). Managing diversity to provide culturally safe sport programming: A case study of the Canadian Red Cross's swim program. *Journal of Sport Management, 29*(3), 305–318.

Sarah Hirshland, senior vice president, Wasserman Media Group. (2010, November 15). *Street & Smith's SportsBusiness Journal.* www.sportsbusinessdaily.com/Journal/Issues/2010/11/20101115/What-I-Like/Sarah-Hirshland-Senior-Vice-President-Wasserman-Media-Group.aspx?hl=trends%20future&sc=0.

Shank, M.D. & Lyberger, M.R. (2015). *Sports marketing: A strategic perspective* (5th ed.). New York: Routledge.

Sharp, L.A., Moorman, A.M., & Claussen, C.L. (2010). *Sport law: A managerial approach* (2nd ed.). Scottsdale, AZ: Holcomb Hathaway.

Sports Business Research Network. (2016). Market research/demographics. www.sbrnet.com/research.aspx?subrid=452.

Statistica. (2016). Consumer purchases of sporting goods in the U.S. from 2002 to 2015. www.statista.com/statistics/200773/sporting-goods-consumer-purchases-in-the-us-since-2004/.

Stoldt, G.C., Dittmore, S.W., & Branvold, S.E. (2012). *Sport public relations* (2nd ed.). Champaign, IL: Human Kinetics.

Stoldt, G.C., Dittmore, S.W., & Pedersen, P.M. (2014). Communication in the sport industry. In P.M. Pedersen & L. Thibault (Eds.), *Contemporary sport management* (5th ed., pp. 338–359). Champaign, IL: Human Kinetics.

Super Bowl 50 sets records across the board. (2016, February 10). San Francisco Super Bowl Committee. www.sfbaysuperbowl.com/super-bowl-50-sets-records-across-the-board#b1Lu3P5ZKvrPyx5j.97.

The Y. (2016). Organizational profile. www.ymca.net/organizational-profile/.

Tony Schiller, EVP and Partner, Paragon Marketing Group. (2011, November 28). *Street & Smith's SportsBusiness Journal, 14*(31), 54.

University of Florida. (2016). Undergraduate catalog 2016–17. https://catalog.ufl.edu/ugrad/current/courses/descriptions/tourism-recreation-and-sport-management.aspx.

University of Minnesota. (2016). University catalogs. www.catalogs.umn.edu/index.html.

U.S. Department of Labor, Bureau of Labor Statistics. (2015, May). Occupational employment statistics: May 2015 national industry-specific occupational employment and wage estimates. NAICS 713940 - Fitness and recreational sports centers. www.bls.gov/oes/current/naics5_713940.htm.

U.S. Department of Labor, Bureau of Labor Statistics. (2015, May). Occupational employment statistics: May 2015 national industry-specific occupational employment and wage estimates. NAICS 711200 - Spectator Sports. www.bls.gov/oes/current/naics4_711200.htm.

Weerakoon, R.K. (2016). Human resource management in sports: A critical review of its importance and pertaining issues. *Physical Culture & Sport. Studies & Research, 69*(1), 15–22.

Wichita State University. (2016). Undergraduate catalog 2016–17. http://webs.wichita.edu/?u=registrar&p=/catalogs/.

INDEX

Note: Page numbers followed by an italicized *f* or *t* refer to the figure or table on that page, respectively.

T

U

V

W

Y

Z

ABOUT THE EDITORS

Shirl J. Hoffman, EdD, is a professor emeritus of kinesiology at the University of North Carolina at Greensboro, where he served as department head for 10 years. He has served at many levels of education: as an elementary physical education teacher, as a college-level coach, and as a professor in both research university and liberal arts college settings.

Hoffman has published extensively on a variety of topics, including sport philosophy and ethics, religion in sport, qualitative analysis and diagnosis of movement, and motor learning and performance. Hoffman is the author of *Good Game: Christianity and the Culture of Sport* (2010) and was the project coordinator for *Careers in Sport, Fitness, and Exercise* (2011), published by the American Kinesiology Association. He also has been a frequent commentator on issues in kinesiology and higher education.

Hoffman is a former editor of *Quest* and was the charter executive director of the American Kinesiology Association (AKA). He is a fellow emeritus of the National Academy of Kinesiology (NAK).

Hoffman and his wife, Claude, reside in Greensboro, North Carolina, and Boone, North Carolina. His photo is © Shirl Hoffman.

Duane V. Knudson, PhD, is a professor of biomechanics in the department of health and human performance at Texas State University. He earned his doctorate at the University of Wisconsin and has held tenured positions at three universities.

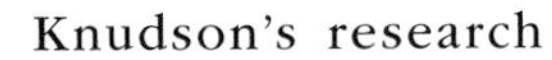

Knudson's research areas are the biomechanics of tennis, stretching, qualitative movement diagnosis, and learning biomechanical concepts. He has authored more than 130 peer-reviewed articles and 14 chapters. Knudson is the author of *Biomechanical Principles of Tennis Technique* (2006), *Fundamentals of Biomechanics, Second Edition* (2007), and *Qualitative Diagnosis of Human Movement, Third Edition* (2013). He has received numerous regional, state, national, and international awards for his scholarship and has been elected a fellow of three scholarly societies. He has served as a department chair, as an associate dean, as president of the American Kinesiology Association, as vice president of publications for the International Society of Biomechanics in Sports, and in numerous other professional leadership roles. Photo courtesy of Texas State University.

ABOUT THE CONTRIBUTORS

Jennifer L. Caputo, PhD, is a professor at Middle Tennessee State University and co-coordinator of the undergraduate and graduate programs in exercise science. She teaches courses in exercise physiology, health and fitness assessment, and exercise prescription; she also directs theses and dissertations in exercise science. Caputo is a Certified Strength and Conditioning Specialist through the National Strength and Conditioning Association and is a licensed medical bone densitometer operator. She received her doctoral degree in exercise physiology from the University of North Carolina at Greensboro; she also holds a master's degree in sport and exercise psychology. Photo courtesy of Jennifer Caputo.

Warren D. Franke, PhD, is a professor and associate chair in the kinesiology department at Iowa State University. He received his doctorate from Virginia Tech. His primary teaching responsibilities center on exercise prescription and exercise programming. His research focuses primarily on identifying the mechanisms that underlie the increased prevalence of heart disease seen in law enforcement officers. He has 30 years of experience in working with a variety of clients, ranging in age from high school students to nonagenarians and ranging in circumstances from cardiac and pulmonary rehabilitation patients to professional athletes. He is also a leader in the SCOUTStrong Healthy Living Initiative of the Boy Scouts of America. Franke is certified as a Program Director (clinical track) by the American College of Sports Medicine. Photo courtesy of Ryan Riley, Iowa State University College of Human Sciences.

Kim C. Graber, EdD, is a professor and associate head in the Department of Kinesiology and Community Health and director of the Campus Honors Program at the University of Illinois at Urbana–Champaign. She received her bachelor's degree from the University of Iowa, her master's from Teachers College Columbia University, and her doctorate from the University of Massachusetts Amherst. Her research focuses on teacher education, children's wellness, and the scholarship of teaching and learning. She is a fellow and the member-at-large on the Research Council of SHAPE America. She has also served as chair of the Research Council, president of the National Association for Sport and Physical Education, and chair of the Curriculum and Instruction Academy. Graber has published dozens of manuscripts in peer-refereed journals and books, has co-authored several books, and regularly presents her scholarship at national and international conferences. She is a University of Illinois Distinguished Teacher-Scholar and received the 2009 Campus Award for Excellence in Undergraduate Teaching. She is Fellow #526 in the National Academy of Kinesiology. Photo courtesy of Kim Graber.

Katherine M. Jamieson, PhD, serves as a professor and chair in the Department of Kinesiology and Health Science at California State University, Sacramento. Jamieson earned her bachelor's and master's degrees in kinesiology from California State University, Fullerton, and completed her doctorate in kinesiology at Michigan State University. Her teaching and research focus on issues related to sport, power, and social stratification; more specifically, her current projects address transnational feminist and postcolonial analyses of the Ladies Professional Golf Association and the global and transnational conditions of elite sport settings and public physical-activity sites. Courses regularly taught by Jamieson include Introduction to Kinesiology, Sport in Society, Qualitative Inquiry in Health and Human Performance, and Critical Analyses for Physically Active Communities. Jamieson is co-author (with Maureen Smith) of *Fundamentals of Sociology of Sport and Physical Activity* (Human Kinetics, 2016). Photo courtesy of Katherine Jamieson.

Scott Kretchmar, PhD, is a professor emeritus of exercise and sport science at Penn State University. He is a founding member of the International Association for the Philosophy of Sport (IAPS) and has served as its president. He has been editor of the *Journal of the Philosophy of Sport,* is a fellow in the National Academy of Kinesiology, and has authored a popular text on the philosophy of sport. He has been named Alliance Scholar by the American Alliance for Health, Physical, Education Recreation and Dance (AAHPERD); Distinguished Scholar by the National Association for Kinesiology and Physical Education in Higher Education; Fraleigh Distinguished Scholar by IAPS (on two occasions); and McCloy Lecturer by the Research Consortium of American Alliance for Health, Physical Education, Recreation and Dance (AAHPERD). Photo courtesy of Scott Kretchmar.

Joseph A. Luxbacher, PhD, has more than 30 years of experience in the fields of health, fitness, and competitive athletics; his PhD is in health, physical, and recreation education. Recently retired from the Department of Athletics at the University of Pittsburgh, Luxbacher has authored more than a dozen books and numerous articles in the areas of sport (soccer), peak athletic performance, fitness, and weight control. He is a former professional soccer player and served as the head men's soccer coach at the University of Pittsburgh. He was twice named Big East Athletic Conference Soccer Coach of the Year. In addition, he was honored as a Letterman of Distinction by the University of Pittsburgh in 2003 and was inducted into the Western Pennsylvania Sports Hall of Fame in 2005. Photo courtesy of Joseph Luxbacher.

Kathy Simpson, PhD, is a professor emerita in the Department of Kinesiology and former director of the Biomechanics Laboratory at the University of Georgia. She received a doctorate in biomechanics from the University of Oregon. Her research focuses on determining how people adapt their movements to varying demands (e.g., impact forces on foot prostheses) and how these adaptations affect the functions and structure of the lower extremity and spine. She has applied this research to areas such as improvement of sport performance in athletes with lower-extremity amputations, movement technique adaptations of individuals who have had spinal fusion surgery, and performance of daily activities by individuals with a hip- or knee-joint replacement or a lower-extremity prosthesis. Simpson has served as the biomechanics section editor for *Research Quarterly for Exercise and Sport,* co-chair of the biomechanics interest group in the American College of Sports Medicine, member of the executive board of the American Society of Biomechanics, and chair of the Biomechanics Academy of the American Alliance for Health, Physical Education, Recreation and Dance. Photo courtesy of Kathy Simpson.

Chad Starkey, PhD, AT, FNATA, is a professor and coordinator of the graduate athletic training program at Ohio University in Athens. A graduate of West Virginia University, he received his master's and doctoral degrees from Ohio University. He is a commissioner for the Commission on Accreditation of Athletic Training Education and has served on the board of directors for the Board of Certification and as chair of the Education Council of the National Athletic Trainers' Association. He has authored several textbooks focused on sports medicine, orthopedic diagnosis, and therapeutic modalities. Photo © College of Health Sciences and Professions/Lauren Dickey.

G. Clayton (Clay) Stoldt, EdD, is associate dean for the College of Education and a professor of sport management at Wichita State University. He teaches classes in sport public relations and sport marketing. Stoldt is co-author of *Sport Public Relations: Managing Stakeholder Communication* (second edition). His research activities have focused on issues in sport public relations, such as crisis communications, the roles of sport public relations professionals, and the application of advanced public relations practices in the field. As associate dean, he provides leadership and support in strategic planning, curricular matters, technology services, and assessment. Stoldt received his doctorate of education from the University of Oklahoma; he also holds a master's degree in sport management and a bachelor's degree in journalism and mass communication. Prior to coming to Wichita State, he worked in the athletic department at Oklahoma City University, where he served as sports information director, radio play-by-play broadcaster, and development officer. He also served as adjunct instructor at both Oklahoma City and the University of Oklahoma, teaching courses in sport management and mass communication. Photo courtesy of Clayton Stoldt.

Richard A. Swanson, PhD, is a professor emeritus of kinesiology at the University of North Carolina at Greensboro, where he was dean of the School of Health and Human Performance for 12 years and a member of the faculty for 27 years. He served previously in faculty and administrative roles at Ohio State University, Wayne State University, and San Francisco State University. His publications include two books, several book chapters, and research articles on the history of physical education, exercise science, and sport in the United States, as well as articles and book chapters on administration in higher education. Swanson is a charter member of the North American Society for Sport History (NASSH). He also served for 30 years as archivist for the National Association for Kinesiology in Higher Education (NAKHE) and served on the editorial board of its journal *Quest* from 2000 through 2015. He was the Dudley Allen Sargent Lecturer for NAKHE in 1997, received its Distinguished Scholar Award in 2008 and its Presidential Award in 2011, and was elected a fellow in 2017. Swanson is also the recipient of the Distinguished Service Award (1992) from the College and University Administrators Council of the American Alliance of Health, Physical Education, Recreation and Dance. He received his BS and MEd degrees from Wayne State University and his PhD from Ohio State University. Photo courtesy of Richard Swanson.

Thomas J. Templin, PhD, is a professor and associate dean for faculty and undergraduate affairs in the School of Kinesiology at the University of Michigan. He is professor emeritus and former head of the Department of Health and Kinesiology at Purdue University, where he worked from 1977 to 2015. He received his education at Indiana University (BS, 1972; MS, 1975) and at the University of Michigan (PhD, 1978). His various honors include the Outstanding Scholar Award in 2012 from the Research on Learning and Instruction in Physical Education Special Interest Group of the American Educational Research Association. He was inducted as a fellow into the National Academy of Kinesiology in 2006. Templin has focused his research on the lives and careers of physical education teachers. He is well known as an author, editor, and reviewer of numerous journal publications and books in physical education, including *A Reflective Approach to Teaching Physical Education* with Don Hellison and *Socialization Into Physical Education: Learning to Teach* with Paul Schempp. He has served in leadership roles and has presented numerous national and international papers for professional organizations such as the American Kinesiology Association; the American Alliance for Health, Physical Education, Recreation and Dance; the International Association for Physical Education in Higher Education; and the American Educational Research Association. He currently serves as president of the American Kinesiology Association, which represents more than 150 member institutions. Photo courtesy of Thomas Templin.

Jerry R. Thomas, EdD, is dean of the College of Education and a professor of kinesiology, health promotion, and recreation at the University of North Texas. Previously, he was a professor of kinesiology and held administrative positions at Iowa State University, Arizona State University, and Louisiana State University. His research in motor development has focused on cognitive factors, expertise, and gender-related differences in children's motor skill. His best-selling graduate textbook, *Research Methods in Physical Activity,* now in its seventh edition, is widely used in the United States and has been translated into seven other languages. He has authored more than 225 research and professional publications, books, and book chapters; presented more than 230 papers to conferences and universities in the United States and other countries; worked as a visiting scholar at institutions in Australia, Korea, China, England, Thailand, and Hong Kong; served as the C.H. McCloy Lecturer and as Alliance Scholar for the American Alliance for Health, Physical Education, Recreation and Dance (AAHPERD); presented numerous named lectures; received the AAHPERD Honor Award (1989–1990); and served as editor in chief of *Research Quarterly for Exercise and Sport* and as a member of that publication's editorial board. Thomas was elected as a fellow of the National Academy of Kinesiology (formerly the American Academy of Kinesiology and Physical Education) in 1984 and served as president in 1991–1992. He has served as president of the North American Society for the Psychology of Sport and Physical Activity and received its Distinguished Scholar Award in 2003 and the President's Award in 2006. He served as president of the Research Consortium of AAHPERD and received its Distinguished Service Award in 2005. He was founding president (2007–2009) of the American Kinesiology Association. Photo courtesy of Jerry Thomas.

Katherine T. Thomas, PhD, is a professor and chair in the School of Health and Kinesiology at Georgia Southern University. She studies factors, particularly skill, that influence physical activity in children. She has been funded to study health and physical activity in women of color and has received more than $1.4 million in external funding for research, service, and teaching. She has published several books, refereed journal articles, and chapters in scholarly books. She served as a founding co-chair of one of the state teams for Action for Healthy Kids, co-chaired the committee that wrote the Iowa Association School Board model local wellness policy, and served as co-chair of the task force that successfully proposed Iowa legislation to examine physical activity and nutrition in schools. She has been involved in two USDA demonstration projects as social scientist and project co-director. She remains focused on informing stakeholders and decision makers in order to influence practice in schools and communities and ensuring that those practices are evidence-based. Photo courtesy of Katherine Thomas.

Cesar R. Torres, PhD, is a professor of kinesiology, sport studies, and physical education at the College at Brockport, State University of New York. He is a past president of the International Association for the Philosophy of Sport and the founding president of the Asociación Latina de Filosofía del Deporte. He is a fellow of the National Academy of Kinesiology and a recipient of the State University of New York Chancellor's Award for Excellence in Scholarship and Creative Activities. He has published several books and numerous articles in academic journals, edited collections, and newspapers, in both English and Spanish. Photo courtesy of Cesar R. Torres.

Robin S. Vealey, PhD, is a professor in the Department of Kinesiology and Health at Miami University in Ohio. Her focal area is sport psychology, and she is particularly interested in self-confidence, mental skills training, and coaching effectiveness. She has authored three books—*Best Practice for Youth Sport, Coaching for the Inner Edge,* and *Competitive Anxiety in Sport*—and is former editor of *The Sport Psychologist.* She has served as a sport psychology consultant for the U.S. Nordic Ski Team, U.S. Field Hockey, elite golfers, and athletes and teams at Miami University. Vealey is a fellow, Certified Consultant, and past president of the Association for the Advancement of Applied Sport Psychology. She is also a fellow of the National Academy of Kinesiology and a member of the United States Olympic Committee Sport Psychology Registry. Photo courtesy of Robin Vealey.

Mark C. Vermillion, PhD, is chair and an associate professor in the Department of Sport Management at Wichita State University, where he teaches courses in sociology of sport, psychology of sport, and ethics in sport. He also teaches in the sociology department, where his courses examine sport, deviance, and violence. His research has addressed topics such as participation factors in college athletics, coaching interactions, disability sport dynamics, and various manifestations of crime and deviance in sport. He received his doctorate in sociology from Oklahoma State University, where he specialized in both crime and deviance and social psychology. He holds a master's degree in sociology from Wichita State University and a bachelor's degree in social science from Kansas State University. In addition to conducting academic research, Vermillion engages in applied research, such as consulting with NCAA- and NJCAA-affiliated athletic departments with regard to survey methods and behavior development plans for student-athletes. He has also worked with local sport organizations, providing consulting and research services for the Air Capital Classic annual golf tournament. Photo courtesy of Mark Vermillion.